Lecture Notes in Computer Science   3787

Commenced Publication in 1973
Founding and Former Series Editors:
Gerhard Goos, Juris Hartmanis, and Jan van Leeuwen

Editorial Board

Dieter Kratsch (Ed.)

Graph-Theoretic Concepts in Computer Science

31st International Workshop, WG 2005
Metz, France, June 23-25, 2005
Revised Selected Papers

 Springer

Volume Editor

Dieter Kratsch
Université Paul Verlaine, Metz
Laboratoire d'Informatique Théorique et Appliquée
UFR MIM Département Informatique
57045 Metz Cedex 01, France
E-mail: kratsch@univ-metz.fr

Library of Congress Control Number: 2005937593

CR Subject Classification (1998): F.2, G.2, G.1.6, G.1.2, E.1, I.3.5

ISSN   0302-9743
ISBN-10   3-540-31000-2 Springer Berlin Heidelberg New York
ISBN-13   978-3-540-31000-6 Springer Berlin Heidelberg New York

Springer is a part of Springer Science+Business Media

springer.com

© Springer-Verlag Berlin Heidelberg 2005
Printed in Germany

Typesetting: Camera-ready by author, data conversion by Scientific Publishing Services, Chennai, India
Printed on acid-free paper   SPIN: 11604686   06/3142   5 4 3 2 1 0

Preface

The 31st International Workshop on Graph-Theoretic Concepts in Computer Science (WG 2005) was held on the campus "Ile du Saulcy" of the University Paul Verlaine-Metz in France. The workshop was organized by the Laboratoire d'Informatique Théorique et Appliquée (LITA) and it took place June 23 – 25 2005. The 94 participants of WG 2005 came from universities and research institutes of 18 different countries.

The WG 2005 workshop continues the series of 30 previous WG workshops. Since 1975, WG has taken place 20 times in Germany, four times in The Netherlands, two times in Austria as well as once in Italy, in Slovakia, in Switzerland and in Czech Republic, and has now been held for the first time in France. The workshop aims at uniting theory and practice by demonstrating how graph-theoretic concepts can be applied to various areas in computer science, or by extracting new problems from applications. The goal is to present recent research results and to identify and explore directions of future research. The talks were given in the "Petit Théatre". They showed how recent research results from algorithmic graph theory can be used in computer science and which graph-theoretic questions arise from new developments in computer science. There were two fascinating invited lectures by Georg Gottlob (Vienna, Austria) and Gregory Kucherov (Nancy, France).

The number of submitted papers was an all-time record of 125. In a careful reviewing process with four reports per submission, the Program Committee selected 38 papers for presentation at the workshop. The Program Committee decided to accept more papers than usual due to the quality of the submissions. Nevertheless, a number of good submissions had to be rejected.

With much pleasure, I thank all those who contributed to the great succes of WG 2005: the authors who submitted their work to the workshop, the speakers, the Program Committee members and the referees. I am indebted to the members of the Local Organization Committee: Michaël Rao, Mathieu Liedloff and Damien Aignel. Without their engagement and the help of various students during the meeting, WG 2005 could not have been such a great success.

Special thanks go to the sponsoring organizations: GDR du CNRS: Algorithmique, Langage et Programmation, GDR du CNRS: Architecture, Réseaux et Systémes, Parallélisme, Laboratoire d'Informatique Théorique et Appliquée de l'Université Paul Verlaine-Metz, UFR MIM de l'Université Paul Verlaine-Metz, Université Paul Verlaine-Metz, Conseil Général de la Moselle, Conseil Régional de Lorraine, Communauté d'Agglomération Metz Métropole (CA2M).

Metz, September 2005   Dieter Kratsch

Organization

The Tradition of WG

1975 U. Pape - Berlin, Germany

1976 H. Noltemeier - Göttingen, Germany

1977 J. Mühlbacher - Linz, Austria

1978 M. Nagl, H. J. Schneider - Castle Feuerstein, Germany

1979 U. Pape - Berlin, Germany

1980 H. Noltemeier - Bad Honnef, Germany

1981 J. Mühlbacher - Linz, Austria

1982 H. J. Schneider, H. Göttler - Neuenkirchen, Germany

1983 M. Nagl, J. Perl - Haus Ohrbeck, Germany

1984 U. Pape - Berlin, Germany

1985 H. Noltemeier - Castle Schwanberg, Germany

1986 G. Tinhofer, G. Schmidt - Bernried, Germany

1987 H. Göttler, H. J. Schneider - Kloster Banz/Staffelstein, Germany

1988 J. van Leeuwen - Amsterdam, The Netherlands

1989 M. Nagl - Castle Rolduc, The Netherlands

1990 R. Möhring - Berlin, Germany

1991 G. Schmidt, R. Berghammer - Fischbachau, Germany

1992 E. W. Mayr - Wiesbaden-Naurod, Germany

1993 J. van Leeuwen - Utrecht, The Netherlands

1994 G. Tinhofer, E. W. Mayr, G. Schmidt - Herrsching, Germany

1995 M. Nagl - Aachen, Germany

1996 G. Ausiello, A. Marchetti-Spaccamela - Como, Italy

1997 R. Möhring - Berlin, Germany

1998 J. Hromkovič - Smolenice, Slovak Republic

1999 P. Widmayer - Ascona, Switzerland

2000 D. Wagner - Konstanz, Germany

2001 A. Brandstädt - Boltenhagen near Rostock, Germany

2002 L. Kučera - Cesky Krumlov, Czech Republic

2003 H. L. Bodlaender - Elspeet, The Netherlands

2004 J. Hromkovič, M. Nagl - Bad Honnef, Germany

2005 D. Kratsch - Metz, France

Program Committee

Hans Bodlaender	Utrecht, The Netherlands
Andreas Brandstädt	Rostock, Germany
Bruno Courcelle	Bordeaux, France
Camil Demetrescu	Rome, Italy
Joseph Diaz	Barcelona, Spain
Fedor Fomin	Bergen, Norway
Pierre Fraigniaud	Paris, France
Martin C. Golumbic	Haifa, Israel
Michel Habib	Montpellier, France
Michael Kaufmann	Tübingen, Germany
Jan Kratochvil	Prague, Czech Republic
Dieter Kratsch	Metz, France (Chair)
Ernst W. Mayr	Munich, Germany
Haiko Müller	Leeds, UK
Takao Nishizeki	Tohoku, Japan
Jeremy Spinrad	Nashville, USA
Ondrej Sykora	Loughborough, UK
Bernhard Westfechtel	Bayreuth, Germany

Additional Reviewers

Luca Allulli, Takao Asano, Yasuhito Asano, Rolf Backofen, Vincenzo Bonifaci, Ulrik Brandes, Hajo Broersma, Luciana Salete Buriol, Frederique Carrere, Dmytro Chibisov, Miroslav Chlebik, Janka Chlebikova, Bogdan Chlebus, Pier Francesco Cortese, Jean-Michel Couvreur, Christophe Crespelle, Paolo Detti, Guido Diepen, Stefan Dobrev, Debora Donato, Frederic Dorn, Feodor Dragan, Zdenek Dvorak, Stefan Eckhardt, Eran E. Edirisinghe, Jack Edmonds, Jens Ernst, Elaine Eschen, Irene Finocchi, Paolo Franciosa, Toshihiro Fujito, Cyril Gavoille, Fanica Gavril, Markus Geyer, Emilio Di Giacomo, Emeric Gioan, Wayne Goddard, Petr Golovach, Martin Golumbic, Jens Gramm, Fabrizio Grandoni, Sylvain Gravier, Jan Griebsch, Alexander Grigoriev, Irith Hartman, Pinar Heggernes, Hongmei He, Stefan Hougardy, Takehiro Ito, Klaus Jansen, Haim Kaplan, Jan Kara, Ton Kloks, Martin Knor, Petr Kolman, Jean-Claude Konig, Ephraim Korach, Arie M. C. A. Koster, Sven Kosub, Vaclav Koubek, Dan Kral, Rastislav Kralovic, Ludek Kucera, Ago Kuusik, Mathieu Latapy, Luigi Laura, Emmanuelle Lebhar, Erik Jan van Leeuwen, Katharina Lehmann, Hoang-Oanh Le, Van Bang Le, Vadim Levit, Moshe Lewenstein, Marina Lipshteyn, Xuan Liu, Vadim Lozin, Moritz G. Maaß, Johann Makowsky, Claudia Malvenuto, Jan Manuch, Martin Mares, Maurice Margenstern, Yuki Matsuo, Ross McConnell, Terry McKee, Daniel Meister, Werner Meixner, Yves Metivier, Kazuyuki Miura, Takaaki Mizuki, Jaroslav Nesetril, Nicolas Nisse, Lhouari Nourine, Johannes Nowak, Gabriel Oksa, Richard Ostertag, Christophe Paul, Uri N. Peled, David Peleg, Jeann-Guy Penaud, Iain Phillips, Ely Porat, Andrzej Proskurowski, Artem

Table of Contents

Hypertree Decompositions:
Structure, Algorithms, and Applications*

Georg Gottlob[1], Martin Grohe[2], Nysret Musliu[1],
Marko Samer[1], and Francesco Scarcello[3]

[1] Institut für Informationssysteme, TU Wien, Vienna, Austria
[2] Institut für Informatik, Humboldt-Universität, Berlin, Germany
[3] D.E.I.S., University of Calabria, Rende (CS), Italy

Abstract. We review the concepts of hypertree decomposition and hypertree width from a graph theoretical perspective and report on a number of recent results related to these concepts. We also show – as a new result – that computing hypertree decompositions is fixed-parameter intractable.

1 Hypertree Decompositions: Definition and Basics

This paper reports about the recently introduced concept of *hypertree decomposition* and the associated notion of *hypertree-width*. The latter is a cyclicity measure for hypergraphs, and constitutes a hypergraph invariant as it is preserved under hypergraph isomorphisms. Many interesting NP-hard problems are polynomially solvable for classes of instances associated with hypergraphs of bounded width. This is also true for other hypergraph invariants such as treewidth, cutset-width, and so on. However, the advantage of hypertree-width with respect to other known hypergraph invariants is that it is more general and covers larger classes of instances of bounded width. The main concepts of hypertree decomposition and hypertree-width are introduced in the present section. A normal form for hypertree decompositions is described in Section 2. Section 3 describes the Robbers and Marshals game which characterizes hypertree-width. In Section 4 we use this game to explain why the problem of checking whether the hypertree-width of a hypergraph is $\leq k$ is feasible in polynomial time for each constant k. However, in Section 5 we show that this problem is fixed-parameter intractable with respect to k. In Section 6 we compare hypertree-width to other relevant hypergraph invariants. In Section 7 we discuss heuristics for computing hypertree decompositions. In Section 8 we show how hypertree decompositions can be beneficially applied for solving constraint satisfaction problems (CSPs). Finally, in Section 9 we list some open problems left for future research. Due to space limitations this paper is rather short, and most proofs are missing. A more thorough treatment can be found in [13,16,2,1,15,17], most of which are available at the Hypertree Decomposition Homepage at http://si.deis.unical.it/~frank/Hypertrees.

* This paper was supported by the Austrian Science Fund (FWF) project: *Nr. P17222-N04, Complementary Approaches to Constraint Satisfaction.* Correspondence to: Georg Gottlob, Institut für Informationssysteme, TU Wien, Favoritenstr. 9-11/184-2, A-1040 Wien, Austria, E-mail: gottlob@acm.org.

D. Kratsch (Ed.): WG 2005, LNCS 3787, pp. 1–15, 2005.

A *hypergraph* is a pair $H = (V(H), E(H))$, consisting of a nonempty set $V(H)$ of *vertices*, and a set $E(H)$ of subsets of $V(H)$, the *hyperedges* of H. We only consider finite hypergraphs. *Graphs* are hypergraphs in which all hyperedges have two elements.

For a hypergraph H and a set $X \subseteq V(H)$, the *subhypergraph induced by* X is the hypergraph $H[X] = (X, \{e \cap X \mid e \in E(H)\})$. We let $H \setminus X := H[V(H) \setminus X]$. The *primal graph* of a hypergraph H is the graph

$$\underline{H} = (V(H), \{\{v, w\} \mid v \neq w, \text{ there exists an } e \in E(H) \text{ such that } \{v, w\} \subseteq e\}).$$

A hypergraph H is *connected* if $\underline{H}$ is connected. A set $C \subseteq V(H)$ is *connected (in H)* if the induced subhypergraph $H[C]$ is connected, and a *connected component* of H is a maximal connected subset of $V(H)$. A sequence of nodes of $V(H)$ is a *path* of H if it is a path of $\underline{H}$.

A *tree decomposition* of a hypergraph H is a tuple (T, χ), where $T = (V(T), E(T))$ is a tree and $\chi : V(T) \longrightarrow 2^{V(H)}$ is a function associating a set of vertices $\chi(t) \subseteq V(H)$ to each vertex t of the decomposition tree T, such that for each $e \in E(H)$ there is a node $t \in V(T)$ such that $e \subseteq \chi(t)$, and for each $v \in V(H)$ the set $\{t \in V(T) \mid v \in \chi(t)\}$ is connected in T.

We assume the tree T in a tree decomposition to be rooted. For every node t, T_t denotes the rooted subtree of T with root t. For each such subtree T_t, let $\chi(T_t) = \bigcup_{v \in V(T_t)} \chi(v)$.

The *width* of a tree decomposition (T, χ) is $\max \{ |\chi(t)| - 1 \mid t \in V(T) \}$, and the *tree-width* of H is the minimum of the widths of all tree decompositions of H.

Observe that (T, χ) is a tree decomposition of H if and only if it is a tree decomposition of $\underline{H}$. Thus a hypergraph has the same tree-width as its primal graph.

Let H be a hypergraph. A *generalized hypertree decomposition of H* is a triple (T, χ, λ), where (T, χ) is a tree decomposition of H and $\lambda : V(T) \longrightarrow 2^{E(H)}$ is a function associating a set of hyperedges $\lambda(t) \subseteq E(H)$ to each vertex t of the decomposition tree T, such that for every $t \in V(T)$ we have $\chi(t) \subseteq \bigcup \lambda(t)$. The *width* of a generalized hypertree decomposition (T, χ, λ) is $\min\{|\lambda(t)| \mid t \in V(T)\}$, and the *generalized hypertree-width* ghw(H) of H is the minimum of the widths of all generalized hypertree decompositions of H.

A *hypertree decomposition* of H is a generalized hypertree decomposition (T, χ, λ) that satisfies the following *special condition*: $(\bigcup \lambda(t)) \cap \chi(T_t) \subseteq \chi(t)$ for all $t \in V(T)$. The *hypertree-width* hw(H) of H is the minimum of the widths of all hypertree decompositions of H.

Example 1. Figure 1 shows a hypergraph H (consisting of 15 hyperedges and 19 vertices) and a tree decomposition of H. A generalized hypertree decomposition and a hypertree decomposition of H are illustrated in Figure 2. The left set within each rectangle represents the λ-labels and the right set represents the χ-labels. The generalized hypertree decomposition violates the special condition, because vertex 13 disappears from node with λ-label $\{h10, h14\}$ and it appears again in a subtree rooted at this node. The generalized hypertree-width of H is 2, whereas its hypertree-width is 3.

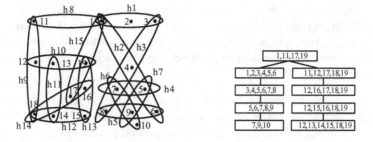

Fig. 1. A hypergraph H (left) and a tree decomposition of H (right)

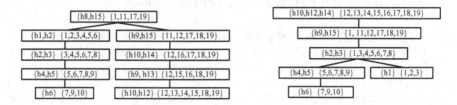

Fig. 2. Generalized hypertree decomposition (left) and hypertree decomposition (right) of H

Example 2. Let H be a the hypergraph with $V(H) = \{1, \dots, n\}$ and

$$E(H) = \big\{\{v, w\} \mid v, w \in V(H) \text{ with } v \neq w\big\} \cup \{V(H)\}.$$

Hence H is the hypergraph obtained from a complete graph with n vertices by adding a hyperedge that contains all vertices. It is easy to see that $\mathrm{hw}(H) = 1$ and $\mathrm{tw}(H) = n - 1$. Moreover, even the treewidth of the bipartite incidence graph of H is $n - 1$.

The structure of many problems can be described by hypergraphs (see also Section 8). Let us informally define a *hypergraph decomposition* as a method of dividing hypergraphs into different parts so that the solution of certain problems whose structure is best described by hypergraphs can be obtained by a polynomial divide-and-conquer algorithm that suitably exploits this division. The *width* of such a decomposition is the size of the largest indecomposable part of this division.

The importance of hypergraph decompositions (be it tree decompositions, hypertree decompositions, or several others) lies in the fact that many problems can be polynomially solved if their associated hypergraph has a low width for the chosen decomposition (see Section 8). The problem is thus to find decompositions that have the following properties:

1. They should be as *general* as possible, i.e., so that the classes of hypergraphs of bounded width are as large as possible. A criterion for comparing the generality of decomposition methods will be given in Section 6.
2. They should be *polynomially computable*. More precisely, for each fixed constant k, we want to be able to check in polynomial time whether a decomposition of width k of an input hypergraph exists.

3. Hypergraph decompositions of bounded width should lead to the polynomial solution of the underlying problem (e.g. of constraint satisfaction problems as described in Section 8). Typically, we expect that for a decomposition of a certain type, the class of problems whose associated hypergraph has width bounded by k can be solved in time $O(n^{O(k)})$.

Several decomposition methods satisfy properties 2 and 3, in particular the method of hypertree decomposition. Hypertree decompositions also satisfy Property 1. By results of [15] and [2], which will be briefly reviewed in Section 6, the method of hypertree decompositions is – so far – the most general method satisfying all three of the above criteria.

2   A Normal Form for Hypertree Decompositions

Let $H = (V(H), E(H))$ be a hypergraph, and let $V \subseteq V(H)$ be a set of vertices and $a, b \in V(H)$. Then a is $[V]$-adjacent to b if there exists an edge $h \in E(H)$ such that $\{a, b\} \subseteq h \setminus V$. A $[V]$-path π from a to b is a sequence $a = a_0, a_1, a_2, \ldots, a_\ell = b$ of vertices such that a_i is $[V]$-adjacent to a_{i+1}, for each $i \in [0, \ell - 1]$. A set $W \subseteq V(H)$ of vertices is $[V]$-connected if, for all $a, b \in W$, there is a $[V]$-path from a to b. A $[V]$-component is a maximal $[V]$-connected non-empty set of vertices $W \subseteq V(H) \setminus V$. For any set C of vertices, let $edges(C) = \{h \in E(H) \mid h \cap C \neq \emptyset\}$.

Let $HD = (T, \chi, \lambda)$ be a generalized hypertree decomposition for H. For any vertex $v \in V(T)$, we will often use v as a synonym of $\chi(v)$. In particular, $[v]$-component denotes $[\chi(v)]$-component; the term $[v]$-path is a synonym of $[\chi(v)]$-path; and so on. We introduce a normal form for generalized hypertree decompositions, and thus also for hypertree decompositions.

Definition 1 ([13]). A generalized hypertree decomposition $HD = (T, \chi, \lambda)$ of a hypergraph H is in *normal form* (NF) if, for each vertex $r \in V(T)$, and for each child s of r, all the following conditions hold:

1. there is (exactly) one $[r]$-component C_r such that $\chi(T_s) = C_r \cup (\chi(s) \cap \chi(r))$;
2. $\chi(s) \cap C_r \neq \emptyset$, where C_r is the $[r]$-component satisfying Condition 1;
3. $(\bigcup \lambda(s)) \cap \chi(r) \subseteq \chi(s)$.

Intuitively, each subtree rooted at a child node s of some node r of a normal form decomposition tree serves to decompose precisely one $[r]$-component.

Theorem 1 ([13]). *For each k-width hypertree decomposition of a hypergraph H there exists a k-width hypertree decomposition of H in normal form.*

3   Robbers and Marshals

In [29], graphs G of treewidth k are characterized by the so called *Robber-and-Cops game* where $k + 1$ cops have a winning strategy for capturing a robber on G. Cops can control vertices of a graph and can fly at each move to arbitrary vertices, say, by using a helicopter. The robber can move (at infinite speed) along paths of G, and will try to escape the approaching helicopter(s), but cannot go over vertices controlled by a cop. It is, moreover, shown that a winning strategy for the cops exists, iff the cops can capture the robber in a *monotone* way, i.e., never returning to a vertex that a cop has previously vacated, which implies that the moving area of the robber is monotonic shrinking. For more detailed descriptions of the game, see [29] or [16].

In order to provide a similarly natural characterization for hypertree-width, a new game, the *Robber and Marshals game (R&Ms game)*, was defined in [16]. A marshal is more powerful than a cop. While a cop can control a single vertex of a hypergraph H only, a marshal controls an entire hyperedge. In the *R&Ms* game, the robber moves on vertices along a path of H (i.e., a path of the primal graph $\underline{H}$) just as in the robber and cops game, but now marshals instead of cops are chasing the robber. During a move of the marshals from the set of hyperedges E to the set of hyperedges E', the robber cannot pass through the vertices in $B = (\bigcup E) \cap (\bigcup E')$, where, for a set of hyperedges $F, \bigcup F$ denotes the union of all hyperedges in F. Intuitively, the vertices in B are those not released by the marshals during the move.

In this game, the set of all marshals is considered to be one player and the robber the other player. The marshals objective is thus to move a marshal (via helicopter) on a hyperedge containing the vertex occupied by the robber. The robber tries to elude capture. As for the robber and cops game, we distinguish between a *general* (not necessarily monotone) and a *monotone* version of the *R&Ms* game. In the monotone version of the game, the marshals have to make sure, that in each step the robber's escape space, i.e., the component in which the robber can freely move around, decreases. The *(monotone) marshal-width* of a hypergraph H, $\mathrm{mw}(H)$ (and mon-$\mathrm{mw}(H)$, respectively), is the least number k of marshals that have a (monotone) winning strategy in the robber and k marshals game played on H (see [1,16] for more precise definitions).

Clearly, for each hypergraph H, $\mathrm{mw}(H) \leq \mathrm{mon\text{-}mw}(H)$. However, unlike for the robber and cops game, the marshal width and the monotone marshal width differ. Adler [1] proved that for each constant k there is a hypergraph H such that $\mathrm{mon\text{-}mw}(H) - \mathrm{mw}(H) = k$.

In [16] it is shown that there is a one-to-one correspondence between the winning strategies for k marshals in the monotone game and the normal-form hypertree decompositions of width at most k.

Theorem 2 ([16]). *A hypergraph H has k-bounded hypertree-width if and only if k marshals have a winning strategy for the monotone R&Ms game played on H.*

4 Computing Hypertree Decompositions

For each constant k it can be decided in polynomial time whether a given hypergraph H has a k-bounded hypertree decomposition. In this section we briefly sketch the algorithm k-decomp which solves this problem in logarithmic space via alternating computations.

The algorithm is best understood via the monotone *R&Ms* game. A typical game situation is depicted in Figure 3, where we assume that the marshals are at some instant in position R, i.e., occupy a set R of k hyperedges, and that the robber is in a component C_R corresponding to this position of the marshals, i.e., in an $[\bigcup R]$-*component* C_R. In

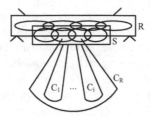

Fig. 3. Marshals moving from position R to position S

the next move, the marshals must chase the robber within C_R. They move to a new position $S \subseteq E(H)$ determined by at most k hyperedges. This move is a correct move in the game iff the following conditions are satisfied: (a) the robber cannot escape from component C_R during or after the move of the marshals, and (b) the escape space of the robber is effectively shrinking.

Condition (a) is mathematically expressed through the statement

$$(a) \quad \forall P \in edges(C_R), \; (P \cap \bigcup R) \subseteq \bigcup S.$$

In fact, since C_R is an $[\bigcup R]$-component, all ways out of it pass through the set of vertices $\bigcup R$ of R. Thus, if there is a way out of C_R, there must be an edge P of $edges(C_R)$ leading from C_R into R. However, by the above condition (a), the robber cannot enter R through this edge P, because the set $P \cap \bigcup R$ of vertices of P that are in R are also in S and remain thus off-limits for the robber both during and after the move of the marshals.

Assuming that condition (a) is satisfied, it is easy to see that to make sure that the escape space has shrunk after the move of the marshals, it suffices to require that the new marshal position S covers at least one vertex of the former escape space C_R, formally:

$$(b) \quad (\bigcup S) \cap C_R \neq \emptyset.$$

In fact, Condition (a) already guarantees that the escape space cannot become larger. Condition (b) requires that some vertex of the former escape space be covered by the marshals after the move, and thus the escape space must shrink. Notice that the original escape space C_R, after the move of the marshals from R to S may be split into several $[\bigcup S]$-components $C_1, C_2, \ldots, C_i, \ldots$

Figure 4 shows (a high-level description of) the algorithm k-decomp. This algorithm tries to construct a winning strategy for k marshals to win the *R&Ms* game on an input hypergraph H. Such a winning strategy is constructed in an alternating fashion by the procedure k-decomposable(C_R, R) which has as parameters a marshals position R (i.e. a set of $\leq k$ hyperedges of H), and an $[\bigcup R]$-*component* C_R which is the current escape space where the robber is to be chased. The procedure guesses (as an existential computation) in Step 1 a marshals position S, and checks in Steps 2.a and 2.b, whether this position is correct according to the above discussed conditions (a) and (b), respectively. The algorithm then determines (in Step 3) the new components determined by the S-position of the marshals and recursively checks if the k marshals have a winning strategy for *each* of

---

ALTERNATING ALGORITHM k-decomp
Input: A non-empty Hypergraph $H = (V(H), E(H))$.
Result: "Accept", if H has k-bounded hypertree-width; "Reject", otherwise.

Procedure k-decomposable(C_R: SetOfVertices, R: SetOfHyperedges)
begin
1) **Guess** a set $S \subseteq E(H)$ of k elements at most;
2) **Check** that all the following conditions hold:
    2.a) $\forall P \in edges(C_R)$, $(P \cap \bigcup R) \subseteq \bigcup S$ and
    2.b) $(\bigcup S) \cap C_R \neq \emptyset$
3) **If** the check above fails **Then Halt** and **Reject; Else**
    Let $\mathcal{C} := \{C \subseteq V(H) \mid C$ is a $[\bigcup S]$-component and $C \subseteq C_R\}$;
4) **If, for each** $C \in \mathcal{C}$, k-decomposable(C, S)
    Then Accept
    Else Reject
end;

begin(* MAIN *)
    Accept if k-decomposable($V(H), \emptyset$)
end.

Fig. 4. A non-deterministic algorithm deciding k-bounded hypertree-width

these components C by calling k-decomposable(C, S). The algorithm accepts if this is the case and rejects otherwise. This part is clearly a universal computation.

The algorithm is initialized (MAIN program) by the call k-decomposable($V(H), \emptyset$) where $V(H)$ is the initial escape space consisting of the entire vertex set of H, and where the initial marshals position is the empty set, i.e., where no hyperedge is occupied by a marshal. The correctness of the algorithm follows easily from the characterization of hypertree-width through the R&Ms game (Theorem 2). A direct proof (not involving the R&Ms game) is given in [16].

A position U of k marshals can be stored as k pointers to (or indices of) hyperedges of H, and, each $[\bigcup U]$-component can be identified through a single vertex. Thus the workspace required at the global level of the initial and each recursive activation of k-decomp is logarithmic in the size of the input hypergraph H. Thus k-decomp can be implemented on an alternating Turing machine using logarithmic workspace, which proves that the associated decision problem is solvable in polynomial time. Actually, a witness of a successful computation corresponds to a hypertree decomposition in NF, thus k-decomp can actually be implemented on a logspace ATM having *polynomially bounded tree-size*, cf. [27], and therefore deciding whether hw(H) $\leq k$ for a hypergraph H is actually in the low complexity class LOGCFL. This is the class of all problems that are logspace-reducible to a context-free language. LOGCFL is a subclass of the class AC^1 of highly parallelizable problems.

Theorem 3 ([13]). *Deciding whether a hypergraph H has k-bounded hypertree-width is in* LOGCFL.

From an accepting computation of the algorithm of Figure 4 we can efficiently extract a NF hypertree decomposition. Since an accepting computation tree of a bounded-treesize logspace ATM can be *computed* in (the functional version of) LOGCFL [12], we obtain the following:

Theorem 4 ([13]). *Computing a k-bounded hypertree decomposition (if any) of a hypergraph H is in* L^{LOGCFL}, *i.e., in functional LOGCFL.*

As for sequential algorithms, a polynomial time algorithm opt-k-decomp which, for a fixed k, decides whether a hypergraph has k-bounded hypertree-width and, in this case, computes an optimal, i.e., smallest width hypertree decomposition in normal form is described in [14]. The opt-k-decomp algorithm is obtained by "uprolling" k-decomp in a sequential bottom-up fashion using polynomial space for storing intermediate results while pruning non-optimal partial decompositions. As for many other decomposition methods, the running time of this algorithm to find the hypergraph decomposition is exponential in the parameter k. More precisely, opt-k-decomp runs in $O(m^{2k}v^2)$ time, where m and v are the number of edges and the number of vertices of H, respectively.

In the next section we will show that the constant k in the exponent of the runtime for computing a hypertree decomposition can most likely not be eliminated.

5   Complexity of Hypertree-Width Computation

In this section we show that determining whether $\text{hw}(H) \leq k$ is NP-complete and actually *fixed-parameter intractable (FP-intractable)* with respect to the parameter k. It follows that, unless some unexpected collapse of FP classes occurs, we cannot eliminate the parameter k from the exponent of the runtime of any algorithm deciding whether a hypergraph H has hypertree-width k, or computing (if possible) a hypertree decomposition of width k of H.

The theory of fixed-parameter tractability or intractability is extensively described in [8]. A problem $\mathcal{P}$ is *fixed-parameter tractable (FP-tractable)* w.r.t. *parameter k* if there exists a function f and a constant c such that $\mathcal{P}$ can be solved in time $O(f(k)n^c)$, where n is the input size and where $f(k)$ depends only on k and c is a fixed constant independent of k. To prove that a problem is *not* fixed-parameter tractable (*FP-intractable*) one usually reduces another problem, known to be FP-intractable, to it via a *parametric reduction* (see [8]). Such a reduction involves a standard polynomial time reduction f between problem instances, and a mapping g between the parameters.

There is a hierarchy W[1], W[2], W[3], ..., the so called *W-hierarchy*, of classes of parameterized problems that are conjectured to be FP-intractable. A well-known FP-intractable problem at the second level W[2] of this hierarchy is the SET COVER problem. An instance of SET COVER consists of a hypergraph $H = (V, E)$ and an integer $k \leq |E|$. The problem is to decide whether there exists a set $K \subseteq E$ of k hyperedges covering $V(H)$, i.e., such that $\bigcup_{e \in K} e = V(H)$. The parameter here is k. By FP-reducing SET COVER to the problem of checking whether $\text{hw}(H) \leq k$, we can prove that the latter is W[2]-hard as well. Given that SET COVER (for non-constant parameter k) is NP-hard, the same transformation gives us as a side result that checking whether $\text{hw}(H) \leq k$ is NP-hard in case k is not constant.

Theorem 5. *The problem of deciding whether for a hypergraph H, $\mathrm{hw}(H) \leq k$ is NP-complete and W[2]-hard wrt. parameter k. The same complexity results hold for determining whether $\mathrm{ghw}(H) \leq k$.*

Proof. We state the proof for hypertree-width (hw). First note that the problem is obviously in NP. To show that it is NP-complete and $W[2]$-hard, it suffices to FP-reduce SET COVER to it. Consider an instance I of SET COVER given by a hypergraph $H = (V, E)$ and an integer $k \leq |E|$. Let us define a new hypergraph $H' = (V', E')$ as follows: $V' = V \times \{1, \ldots, 2k+1\}$,

$$E' = \big\{\{(v, i), (w, j)\} \mid (v, i), (w, j) \in V'\big\} \cup \big\{e \times \{1, 2, \ldots, 2k+1\} \mid e \in E\big\}.$$

We claim that H has a set cover of size $\leq k$ iff H' has hypertree-width $\leq k$.

The "only if" part is almost trivial to see. Indeed, if there exists a set cover K of size k of H, then a hypertree decomposition of width k of H' is constituted by a tree T consisting of a single node t such that $\chi(t) = V'$ and $\lambda(t) = \{e \times \{1, 2, \ldots, 2k+1\} \mid e \in K\}$.

To see the "if" part of the claim, assume there exists a hypertree decomposition (T, χ, λ) of width k of H'. Then, by construction of H', there must exist a decomposition vertex t of T such that $\chi(t) = V'$. In fact, H' contains as subhypergraph the clique obtained by pairwise relating all vertices of V', and thus any tree decomposition of E' must contain a block containing all vertices of V'. Let

$$S = \{e \in E \mid e \times \{1, 2, \ldots, 2k+1\} \in \lambda(t)\}.$$

Then $|S| \leq |\lambda(t)| \leq k$. We will next show that for each $v \in V$ there exists some $e \in S$ such that $v \in e$, thus S is a set cover of size $\leq k$ of H.

Assume thus that there exists a $v \in V$ such that there is no $e \in S$ for which $v \in e$. Then the elements $(v, 1), (v, 2), \ldots, (v, 2k+1)$ of $V' = \chi(t)$ must be covered by edges in $\lambda(t)$ of the form $\{v', w'\}$ where $v', w' \in V'$. But for covering $2k+1$ elements by such pairs, at least $k + 1$ such pairs would be necessary, which contradicts our assumption that $|\lambda(t)| \leq k$.

The reduction from H to H' is computable in time $O(k \cdot |H|)$ and is thus an FP-reduction.

The same arguments apply if we use the notion of generalized hypertree-width (ghw) instead of hypertree-width (hw). In fact, we have nowhere in this proof made use of the special condition which distinguishes hw from ghw.   □

6 Comparing Hypertree-Width to Other Hypergraph Invariants

A hypergraph invariant f is *(at least) as good as* invariant g, if there exists a constant c such that whenever for a hypergraph H, $g(H) = k$, then $f(H) \leq c \cdot k$. We say that f *strongly dominates* g if f is at least as good as g and there is a class $\mathcal{H}$ of hypergraphs for which f is bounded (i.e., $\exists k \, \forall H \in \mathcal{H} : f(H) \leq k$), but g is unbounded. We say that two invariants f and g are *equivalent* if each is as good as the other one.

We start by discussing some hypergraph invariants that are generalizations of sophisticated graph invariants, and then report some results on comparing invariants used in Constraint Satisfaction, Artificial Intelligence and Database Theory to hypertree-width.

Hyperlinkedness. Let H be a hypergraph, $M \subseteq E(H)$ and $C \subseteq V(H)$. C is M-*big*, if it intersects more than half of the edges of M, that is, $|\{e \in M \mid e \cap C \neq \emptyset\}| > \frac{|M|}{2}$. Note that if $S \subseteq E(H)$, then $H \setminus \bigcup S$ has at most one M-big connected component. Let $k \geq 0$ be an integer. A set $M \subseteq E(H)$ is k-*hyperlinked*, if for any set $S \subseteq E(H)$ with $|S| < k$, $H \setminus \bigcup S$ has an M-big component. The largest k for which H contains a k-hyperlinked set is called *hyperlinkedness of* H, hlink(H). Hyperlinkedness is an adaptation of the linkedness of a graph [25] to the setting of hypergraphs.

Brambles. Let H be a hypergraph. Sets $X_1, X_2 \subseteq V(H)$ *touch* if $X_1 \cap X_2 \neq \emptyset$ or there exists an $e \in E(H)$ such that $e \cap X_1 \neq \emptyset$ and $e \cap X_2 \neq \emptyset$. A *bramble of* H is a set B of pairwise touching connected subsets of $V(H)$. This is defined in analogy to brambles of graphs [25]. The *hyper-order of a bramble* B is the least integer k such that there exists a set $R \subseteq E(H)$ with $|R| = k$ and $\bigcup R \cap X \neq \emptyset$ for all $X \in B$. The *hyperbramble number* hbramble-no(H) of H is the maximum of the hyper-orders of all brambles of H.

Theorem 6 ([2]). *For each hypergraph* H, hlink$(H) \leq$ hbramble-no$(H) \leq$ mw(H) $\leq$ ghw$(H) \leq$ mon-mw$(H) =$ hw$(H) \leq 3 \cdot$ hlink$(H) + 1$.

Corollary 1. *The hypergraph invariants* hlink, hbramble-no, mw, ghw, mon-mw, *and* hw *are all equivalent.*

Of particular interest is the result that the generalized hypertree-width ghw(H) of a hypergraph H is at most a factor 3 smaller than the hypertree-width hw(H). This is important, because while it is currently an open problem whether ghw$(H) \leq k$ is decidable in polynomial time for constants k, the notion of generalized hypertree-width is by many considered the best possible measure of cyclicity of a hypergraph. For example, Cohen, Jeavons, and Gyssens [4] recently introduced a general framework for hypergraph decomposition in the context of which they introduced the concept of an *acyclic guarded cover* as their most general considered decomposition guaranteeing tractability of the underlying problems (i.e., satisfying the above Condition 3). It turns out, however, that an acyclic guarded cover can be equivalently defined as the set of nodes of a generalized hypertree decomposition, and that the corresponding notion of width *precisely* coincides with the notion of generalized hypertree-width. This provides further evidence of the naturalness and importance of this notion.

The following hypergraph invariants were considered in AI, and, in particular, in the area of constraint processing.

Biconnected Components (short: BICOMP) [9]. Any graph $G = (V, E)$ can be decomposed into a pair $\langle T, \chi \rangle$, where T is a tree, and the labeling function χ associates to each vertex of T a biconnected component of G. The *biconnected width* of a hypergraph H, denoted by BICOMP-width(H), is the maximum number of vertices over the biconnected components of the primal graph of H.

Cycle Cutset and Hypercutset (short: CUTSET) [5]. A *cycle cutset* of a hypergraph H is a set $S \subseteq V(H)$ such that the subhypergraph of H induced by $V(H) - S$ is acyclic. The CUTSET-width of H is the minimum cardinality over all its possible cycle cutsets. A generalization of this is the method of hypercutsets, short HYPERCUTSET (for a definition, see [15]).

Tree Clustering (short: TCLUSTER) [7]. The *tree clustering* method is based on a triangulation algorithm which transforms the primal graph $G = (V, E)$ of any hypergraph H into a chordal graph G'. The maximal cliques of G' are then used to build the hyperedges of an acyclic hypergraph H'. The *tree-clustering width* (short: TCLUSTER *width*) of H is 1 if H is an acyclic hypergraph; otherwise it is equal to the maximum cardinality over the cliques of the chordal graph G'.

The Hinge Method (HINGE) [19,18]. This is an interesting decomposition method generalizing acyclic hypergraphs. For space reasons, we omit a formal definition. Computing the HINGE-width of a hypergraph is feasible in polynomial time [19,18]. One can also combine the methods HINGE and TCLUSTER, yielding the more general method HINGE^{TCLUSTER}.

Theorem 7 ([15]). *Hypertree-width strongly dominates treewidth,* BICOMP-*width,* CUTSET-*width,* HYPERCUTSET-*width,* TCLUSTER-*width,* HINGE-*width, and* HINGE^{TCLUSTER}-*width.*

7 Heuristics for Hypertree Decomposition

Recall that the algorithm opt-k-decomp decides, for a fixed k, whether a given hypergraph has k-bounded hypertree-width and, if so, computes a hypertree decomposition of minimal width. Although opt-k-decomp runs in polynomial time, it is too slow and needs a huge amount of space when applied to large hypergraphs. Therefore, recent research focuses on heuristic approaches for the construction of hypertree decompositions. Of particular interest is the application of well-known heuristics from other areas to hypertree decomposition.

Recall that a hypertree decomposition is in principle the same as a tree decomposition satisfying two additional conditions. The first one leads from a tree decomposition to a generalized hypertree decomposition and says that for every $t \in V(T)$ it holds that $\chi(t) \subseteq \bigcup \lambda(t)$, and the second one is the *special condition* leading from a generalized hypertree decomposition to a hypertree decomposition. Note that the *special condition* was introduced in order to be able to prove the polynomial runtime of opt-k-decomp. Hence, the *special condition* can be ignored when considering heuristic algorithms, and thus, one actually aims at computing *generalized* hypertree decompositions by using heuristics.

So, when constructing hypertree decompositions via tree decomposition heuristics, there is only one additional condition we have to satisfy. This condition forces the λ-labels to cover the χ-labels. A natural approach to obtain a hypertree decomposition from a tree decomposition is therefore to implement this condition in a straight-forward way by set covering, i.e., to use set covering algorithms in order to compute the λ-labels of the hypertree decomposition based on the χ-labels given by the tree decomposition. In this way, it is possible to use tree decomposition heuristics (together with set covering heuristics) for the heuristic construction of hypertree decompositions.

This approach was firstly applied by McMahan [23] who obtained surprisingly good results within a small amount of time. McMahan used *Bucket Elimination* [6] in combination with several variable ordering heuristics. Obviously, to construct hypertree decompositions in this way, any underlying tree decomposition method can be used. Moreover, also branch decomposition heuristics are applicable [28], since every branch decomposition of width k can be transformed into a tree decomposition of width at most $3k/2$ [26].

Another approach for heuristic hypertree decomposition is dual to the above ones in the sense that we obtain a hypertree where the λ-labels are given and appropriate χ-labels have to be set. This can be easily achieved by building a tree decomposition of the dual graph. The *dual graph* of a hypergraph is obtained by creating a vertex for each hyperedge and connecting two vertices if the corresponding hyperedges have a common vertex. This dual graph, however, has too many edges for our purposes, i.e., the resulting hypertree-width would be higher than necessary. Moreover, a hypertree decomposition resulting from this procedure is always a query decomposition [13] whose width is always larger than or equal to the hypertree-width of a hypergraph. However, by using pre- and post-processing heuristics, it is possible to overcome both problems.

Finally, let us mention a further heuristic approach. It is based on hypergraph clustering resp. hypergraph partitioning. There exist several heuristics in the literature for building clusters of strongly connected hyperedges in a hypergraph such that there are as less hyperedges as possible between the clusters. By using such methods, it is possible to construct a hypertree decomposition in such a way that the clusters are recursively partitioned and in each step a *special hyperedge* is added [21]. During this process, for each cluster a hypertree-node is created whose λ-labels are exactly the hyperedges separating the subclusters of the current cluster. Afterwards, it is possible to connect these hypertree-nodes in such a way that the resulting hypertree is indeed a hypertree decomposition of the hypergraph (cf. [21]).

8   Applications

There are many problems in various domains of Computer Science whose underlying structure is best described as a hypergraph and that are efficiently solvable if this structure is acyclic. We next show that, for most of them, the notion of (generalized) hypertree-width provides a technique for solving efficiently large classes of instances that were believed to be intractable, according to previous known methods.

A very important example of such problems is the NP-hard *Constraint Satisfaction Problem (CSP)*, which is an important goal of AI research. Constraint satisfaction is a central issue of *problem solving* and has an impressive spectrum of applications [24]. A constraint (S_i, R_i) consists of a *constraint scope* S_i, i.e., a list of variables, and an associated *constraint relation* r_i containing the legal combinations of values. A CSP consists of a set $\{(S_1, r_1), (S_2, r_2), \ldots, (S_q, r_q)\}$ of constraints whose variables may overlap. A solution to a CSP consists of an assignment of values to all variables such that all constraints are simultaneously satisfied. By *solving* a CSP we mean determining whether the problem has a solution at all (i.e., checking for *constraint satisfiability*), and, if so, compute one solution.

Example 3. Consider the CSP I^a consisting of constraints $\{C_1^a, \ldots, C_9^a\}$ where, for each constraint C_i^a, the constraint relation r_i^a encodes some required property for the variables occurring together in the corresponding scope S_i^a, and the constraint scopes are the following: $S_1^a(3, 4, 5, 6, 7, 8)$; $S_2^a(12, 16, 17, 18, 19)$; $S_3^a(7, 9, 10)$; $S_4^a(1, 11, 17, 19)$; $S_5^a(1, 2, 3, 4, 5, 6)$; $S_6^a(5, 6, 7, 8, 9)$; $S_7^a(12, 15, 16, 18, 19)$; $S_8^a(12, 13, 14, 15, 18, 19)$; $S_9^a(11, 12, 17, 18, 19)$.

The constraint hypergraph of a CSP I is the hypergraph $H(I)$ whose vertices are the variables of the CSP and whose hyperedges are the sets of all those variables which occur together in a constraint scope. It is well known that CSPs with *acyclic* constraint hypergraphs are polynomially solvable [5]. For instance, our example CSP instance I^a is acyclic, as its hypergraph has a join tree. In fact, it is easy to check that the tree shown in Figure 1 (on the right) is a join tree of hypergraph $H(I^a)$. Intuitively, the efficient behavior of acyclic instances is due to the fact that they can be evaluated by processing any of their join trees bottom-up by performing upward semijoins (in database lingo) [30]. That is, starting from the leaves, for each vertex v of the tree, we may filter out of its parent $p(v)$ the tuples of values from $p(v)$'s constraint relation that do not match with any tuple in the relation of v. At the end, if the relation in the root is not empty, we know that the given instance has a solution. This procedure takes $O(nm \log m)$ time, where m is the size of the largest relation and n is the number of constraints. Note that we do not distinguish here among join tree vertices and constraints, because join tree vertices correspond to hyperedges and hence to constraints (assuming, w.l.o.g., that there is no pair of constraints with exactly the same scopes). Recall that in general even computing small outputs, e.g. just one solution, requires exponential time (unless P $=$ NP) [3], indeed the typical worst case cost for CSP algorithms is $O(m^{n-1} \log m)$.

The idea behind CSP algorithms based on generalized hypertree decompositions is to transform a CSP I into an equivalent acyclic CSP I', by organizing its scopes into a polynomial number of clusters that may suitably be arranged as a tree. Consider a generalized hypertree decomposition of $H(I)$ and some vertex v of this decomposition. We can combine the constraints in $\lambda(v)$ in a unique constraint over the only variables listed in $\chi(v)$. Building this fresh constraint takes $O(m^{|\lambda(v)|-1} \log m)$ time. It is easy to see that, after this phase, we get a new CSP instance I', which is acyclic and solution equivalent to the original instance I. Therefore, we can eventually solve this instance in time $O(n'm^{w-1} \log m)$, where w is the decomposition-width and n' is the number of vertices in the decomposition tree, which is bounded by the number of hypergraph vertices (CSP variables). Note that, for classes of CSPs having small (bounded) width, solving these problems by exploiting hypertree decompositions may lead to a tremendous speed-up. Indeed, hypertrees with the smallest width say to us precisely the best way of combining together constraints of I, in order to obtain a nice acyclic equivalent instance to be solved efficiently.

Example 4. Consider a CSP instance I^c with the following constraint scopes:
$S_1(1, 2, 3)$; $S_2(1, 4, 5, 6)$; $S_3(3, 4, 7, 8)$; $S_4(5, 7)$; $S_5(6, 8, 9)$; $S_6(7, 9, 10)$; $S_7(5, 9)$; $S_8(1, 11)$; $S_9(11, 12, 18)$; $S_{10}(12, 13, 19)$; $S_{11}(13, 14)$; $S_{12}(14, 15, 18)$; $S_{13}(15, 16, 19)$; $S_{14}(16, 17, 18)$; $S_{15}(1, 17, 19)$;

The associated hypergraph $H(I^c)$ is shown in Figure 1. The generalized hypertree-width of this hypergraph is 2 and a decomposition having this (optimal) width is shown in Figure 1, on the left. Following the "instructions" encoded in this decomposition, we build exactly the acyclic instance I^a in Example 3. Then, by exploiting hypertrees, we know that I^c may be solved in $O(9m \log m)$ time, in the worst case, which is clearly quite good, if compared with the traditional worst case $O(m^{14} \log m)$.

Though we focused on constraint satisfiability, all the above considerations immediately apply to a large number of important problems that, as CSP, are efficiently solvable if their hypergraph structure is acyclic. We just mention here a few examples, such as the game theory problem of computing pure Nash equilibria in graphical games [10], and various database problems, e.g., the problem of conjunctive query containment [20], or the problem of evaluating *Boolean conjunctive queries* over a relational database [22] (for a discussion of this and other equivalent problems, see [11]).

9   Open Problems and Future Research

We believe that hypertree decompositions and hypertree-width are interesting concepts deserving further investigations. The following problems are of particular interest: (1) Is it possible to check whether $\mathrm{ghw}(H) \leq k$ in polynomial time for each constant k? (2) Are there other hypergraph invariants (and associated decompositions) that fulfill the three criteria given in Section 1 and that strongly generalize hypertree-width? (3) Can we find a deterministic algorithm for computing a k-width hypertree decomposition whose worst case runtime is significantly better than n^{2k}? (4) Is it possible to find some heuristic method for computing "good" hypertree decompositions for an overwhelmingly large number of realistic examples stemming from various applications?

References

1. I. Adler. Marshals, monotone marshals, and hypertree width. *Journal of Graph Theory* 47, pages 275–296, 2004.
2. I. Adler, G. Gottlob, and M. Grohe. Hypertree-width and related hypergraph invariants. Manuscript, submitted for publication, available from the authors.
3. A. K. Chandra and P. M. Merlin. Optimal implementation of conjunctive queries in relational databases. In *Proc. STOC'77*, pages 77–90, 1977.
4. D. A. Cohen, P. G. Jeavons, and M. Gyssens. A unified theory of structural tractability for constraint satisfaction and spread cut decomposition. In *Proc. IJCAI'05*, pages 72–77, 2005.
5. R. Dechter. Constraint networks. In *Encyclopedia of Artificial Intelligence*, second edition, Wiley & Sons, pages 276–285, 1992.
6. R. Dechter and J. Pearl. Network-based heuristics for constraint satisfaction problems. *Artificial Intelligence* 34(1), pages 1–38, 1987.
7. R. Dechter and J. Pearl. Tree clustering for constraint networks. *Artificial Intelligence* 38(3), pages 353–366, 1989.
8. R. G. Downey and M. R. Fellows. *Parameterized Complexity*. Springer, 1999.
9. E. C. Freuder. A sufficient condition for backtrack–bounded search. *Journal of the ACM* 32(4), pages 755–761, 1985.

10. G. Gottlob, G. Greco, and F. Scarcello. Pure Nash equilibria: Hard and easy games. *Journal of Artificial Intelligence Research (JAIR)*, 2005. To appear. Preliminary version in: *Proc. TARK'03*, 2003.
11. G. Gottlob, N. Leone, and F. Scarcello. The complexity of acyclic conjunctive queries. *Journal of the ACM* 48(3), pages 431–498, 2001. Preliminary version in: *Proc. FOCS'98*, 1998.
12. G. Gottlob, N. Leone, and F. Scarcello. Computing LOGCFL certificates. *Theoretical Computer Science* 270(1-2), pages 761–777, 2002. Preliminary version in: *Proc. ICALP'99*, 1999.
13. G. Gottlob, N. Leone, and F. Scarcello. Hypertree decompositions and tractable queries. *Journal of Computer and System Sciences (JCSS)* 64(3), pages 579–627, 2002. Preliminary version in: *Proc. PODS'99*, 1999.
14. G. Gottlob, N. Leone, and F. Scarcello. On tractable queries and constraints. In *Proc. DEXA'99*, pages 1–15, 1999.
15. G. Gottlob, N. Leone, and F. Scarcello. A comparison of structural CSP decomposition methods. *Artificial Intelligence* 124(2), pages 243–282, 2000. Preliminary version in: *Proc. IJCAI'99*, 1999.
16. G. Gottlob, N. Leone, and F. Scarcello. Robbers, marshals, and guards: Game-theoretic and logical characterizations of hypertree width. In *Proc. PODS'01*, pages 195–206, 2001.
17. G. Gottlob and R. Pichler. Hypergraphs in model checking: Acyclicity and hypertree-width versus clique-width. *Siam Journal of Computing* 33(2), pages 351–378, 2004.
18. M. Gyssens, P. G. Jeavons, and D. A. Cohen. Decomposing constraint satisfaction problems using database techniques. *Artificial Intelligence* 66, pages 57–89, 1994.
19. M. Gyssens, and J. Paredaens. A decomposition methodology for cyclic databases. In *Advances in Database Theory*, vol. 2, pages 85–122, 1984.
20. Ph. G. Kolaitis and M. Y. Vardi. Conjunctive-query containment and constraint satisfaction. *Journal of Computer and System Sciences (JCSS)* 61, pages 302–332, 2000.
21. T. Korimort. Constraint satisfaction problems – Heuristic decomposition. PhD thesis, Vienna University of Technology, April 2003.
22. D. Maier. The theory of relational databases. Computer Science Press, 1986.
23. B. McMahan. Bucket eliminiation and hypertree decompositions. Implementation report, Institute of Information Systems (DBAI), TU Vienna, 2004.
24. J. Pearson and P. G. Jeavons. A survey of tractable constraint satisfaction problems. Technical report CSD-TR-97-15, Royal Halloway University of London, 1997.
25. B. Reed. Tree width and tangles: A new connectivity measure and some applications. In *Surveys in Combinatorics*, volume 241 of LNS, pages 87–162. Cambridge University Press, 1997.
26. N. Robertson and P. D. Seymour. Graph minors. X. Obstructions to tree-decomposition. *Journal of Combinatorial Theory, Series B* 52, pages 153–190, 1991.
27. W. L. Ruzzo. Tree-size bounded alternation. *Journal of Computer and System Sciences (JCSS)* 21(2), pages 218–235, 1980.
28. M. Samer. Hypertree-decomposition via branch-decomposition. In *Proc. IJCAI'05*, pages 1535–1536, 2005.
29. P. D. Seymour and R. Thomas. Graph searching and a min-max theorem for tree-width. *Journal of Combinatorial Theory, Series B* 58, pages 22–33, 1993.
30. M. Yannakakis. Algorithms for acyclic database schemes. In *Proc. VLDB'81*, pages 82–94, 1981.

Combinatorial Search on Graphs Motivated by Bioinformatics Applications: A Brief Survey

Mathilde Bouvel[1], Vladimir Grebinski[2], and Gregory Kucherov[3]

[1] Département d'Informatique, Ecole Normale Supérieure de Cachan, 94235, France
[2] CompuGene Inc., Jamesburg, NJ 08831, USA
[3] INRIA/LORIA, 615, rue du Jardin Botanique, B.P. 101, 54602,
Villers-lès-Nancy, France
Gregory.Kucherov@loria.fr

Abstract. The goal of this paper is to present a brief survey of a collection of methods and results from the area of combinatorial search [1,8] focusing on graph reconstruction using queries of different type. The study is motivated by applications to genome sequencing.

1 Introduction

1.1 Generic Problem and Bioinformatics Application

Assume we have a set of labeled chemicals and some pairs of chemicals can react. Assume we have an experimental tool to detect if a reaction occurs when mixing two or several chemicals together, or a tool that allows us to count how many reacting pairs there are in the mixture. Our goal is to recover all pairs of reacting chemicals with as few experiments as possible.

One important application area for such problems is bioinformatics. For example, obtaining a whole genomic sequence is a crucial first step in the study of an organism. A common practical approach to genome sequencing is to obtain a number of short and possibly overlapping *reads* from the genomic sequence, that are then assembled into *contigs* – contiguous fragments that cover the genome with possible gaps. The problem is then to determine the relative placement of contigs on the genome, i.e. to reconstruct their original order. This step is a accomplished by testing the adjacency of contigs using a so-called *Polymerase Chain Reaction* (PCR). Nowadays, PCR is one of the most ubiquitous tools in molecular biology and can be performed very cheaply, efficiently and almost automatically (see e.g. [2]). It is based on the idea that any region of the genome can be described by a pair of *primers* that can be thought of as short nucleotide sequences bounding this region. If the primers are proximate (within several thousands of nucleotides in practice), the region that they delimit is amplified into a huge number of copies, which can be observed experimentally. Therefore, by picking primer sequences from both ends of each contig, we can reliably test if they are adjacent on the original DNA, under the assumption that the gaps between contigs are of bounded size.

D. Kratsch (Ed.): WG 2005, LNCS 3787, pp. 16–27, 2005.

While the basic PCR allows one to test one pair of primers at a time, the *multiplex PCR* presents an extension that uses several primers simultaneously to determine amplified regions. Since several regions can be amplified simultaneously, this approach can also provide an information of *how many* pairs of primers resulted in an amplification.

In all cases, a very important question in practice is how many reactions are needed in the worst case and how quickly we can perform all of them. Ideally, we want to implement as few reactions as possible and run them in parallel. In this paper we survey some of the results related to such and similar problems.

1.2 Mathematical Formulation and Main Definitions

If chemicals are represented as vertices of a non-oriented graph and a reaction as an edge, we come up with a problem of reconstructing an unknown graph of a given class of graphs. Note that we might also consider that a reaction is triggered by more than two chemicals, which would result in a hypergraph reconstruction problem.

The multiplex PCR problem can lead to two different mathematical formalizations. If the objects ("chemicals") we are dealing with are contigs (i.e. primers coming from both ends of a contig are always tested together), the underlying problem is to reconstruct a Hamiltonian path or a Hamiltonian cycle[1] on K_n (the complete graph with n vertices, where n is the number of *contigs*) [11]. If we are dealing with primers, we face the problem of reconstructing a matching on K_n (where n is the number of *primers*).

Graph Reconstruction Problem. Different kinds of combinatorial search problems on graphs have been considered in the literature (see [1]): identifying an unknown edge or vertex in a given graph, reconstructing a hidden graph of a given class, verifying a property of a hidden graph, and some others. Our interest here will be the following graph reconstruction problem:

Problem 1. *Given a class of graphs $\mathcal{G} = \cup_n \mathcal{G}_n$, where $\mathcal{G}_n$ contains all the graphs of $\mathcal{G}$ on the set of vertices $V = \{1, \ldots, n\}$, we want to reconstruct a hidden graph $G \in \mathcal{G}_n$ for a given n, making as few queries as possible. A query is a subset of V, and the answer we obtain provides us with information about the edges in the subgraph of G induced by the queried subset. This information depends on the model under consideration.*

In the particular case when only two vertices of V can be tested at a time, the query just checks if a specific edge exists in G, and the model is called a *two-vertex model*.

Boolean and Quantitative Models. One type of query is: "For $Q \subseteq V$, is there at least one edge in the subgraph of G induced by Q?". The possible answers being `true` or `false`, this query model is called *boolean*.

---

[1] Depending on whether the genome is linear or circular.

A natural extension of this model admits queries of the following form: "For $Q \subseteq V$, *how many* edges does the subgraph of G induced by Q contain?". This query model is called *quantitative* (or *additive*) since the answer to a query is an integer ranging between 0 and the number of edges of a G.

In both cases, the *complexity* of a problem is defined as the minimum number of queries required to reconstruct a graph of $\mathcal{G}_n$ in the worst case. The complexity depends on n but can also made dependent on other parameters (see [4] for example).

We will be generally interested in finding upper and lower bounds on the complexity of a problem. The information theory provides a simple and powerful method to estimate the lower bound: at least $\log_d |\mathcal{G}_n|$ queries must be made in order to identify a graph from $\mathcal{G}_n$, where d is the maximal number of distinct answers provided by a query.

Adaptive and Nonadaptive Algorithms. Two main kinds of algorithms must be distinguished in the area of combinatorial search: in *adaptive algorithms*, every query potentially depends on the answers obtained to previous queries while in *nonadaptive algorithms*, all queries are independent of each other. A nonadaptive algorithm can be described as a family of subsets of V (queries) or as an vertex-query incidence matrix M ($M_{i,j} = 1$ iff vertex j appears in query i, $M_{i,j} = 0$ otherwise).

Nonadaptive algorithms can be seen as 1-*round algorithms*, i.e. those in which all queries can be made in parallel. From this perspective, adaptive algorithms are multi-round (have an unlimited number of rounds). Intermediate case of *s-round algorithms* composed of s successive nonadaptive stages will also be considered.

In this paper, we present a short survey of different known results on graph reconstruction. From the application perspective, our main motivation is on reconstructing Hamiltonian cycles but we also consider other graph classes such as matchings, stars, cliques, graphs with bounded vertex degree, and others. Two main query models will be considered: the boolean model (Section 2) and the quantitative model (Section 3). For each graph class, we will be interested in the complexity of reconstruction using different types of algorithms.

2   Boolean Model

2.1   Hamiltonian Cycles

Assume we have to reconstruct an unknown Hamiltonian cycle in the complete graph K_n. Under the boolean model, the information theory yields the lower bound $\log_2 \frac{(n-1)!}{2} = \Omega(n \log_2 n)$ as there are $\frac{(n-1)!}{2}$ Hamiltonian cycles on n vertices. The following theorem states that this bound can be reached under particular conditions.

Theorem 1. *The $\Omega(n \log_2 n)$ lower bound on the complexity of Hamiltonian cycle reconstruction can be reached by an adaptive algorithm.*

Note first that if we are restricted to the two-vertex model, any reconstruction algorithm requires $\Omega(n^2)$ queries, as shown in [1].

An adaptive algorithm reconstructing a Hamiltonian cycle H with $2n \log_2 n$ queries has been described in [11]. An interesting fact is that under the boolean model, this complexity cannot be achieved by a nonadaptive algorithm. As showed in [5], $\Omega(n^2)$ queries are necessary for a nonadaptive algorithm to reconstruct a Hamiltonian cycle. The result of [5] is actually more general, and establishes that $\Omega(n^2)$ queries are necessary for a nonadaptive algorithm to reconstruct a graph in one of the following classes: matchings, perfect matchings, graphs isomorphic to a fixed bounded degree graph with $\Omega(n)$ edges, graphs consisting in the disjoint union of a clique of size $n - 3$ and a single edge.

This example illustrates the case when adaptive algorithms are strictly more powerful than nonadaptive algorithms.

2.2   Matchings

A matching is a graph such that each vertex has degree 0 or 1. As mentioned above, any nonadaptive algorithm reconstructing a matching requires a quadratic number of queries. More precisely, at least $\frac{49}{153}\binom{n}{2}$ nonadaptive queries are necessary to reconstruct a matching on K_n [5]. The authors of [5] also prove the upper bound $(\frac{1}{2} + o(1))\binom{n}{2}$ using a construction based on the Wilson theorem [22] on the decomposition of complete graphs into subgraphs isomorphic to a given graph.

As the enumeration of matchings is an open question, it is difficult to compute the exact information-theoretic lower bound. However, we can easily compute the number of perfect matchings[2] of K_n to be $\frac{n!}{2^{\lfloor \frac{n}{2} \rfloor} \cdot \lfloor \frac{n}{2} \rfloor!}$. This provides a lower bound on the number of general matchings, and implies the following information-theoretic lower bound on the reconstruction of matchings: $\log_2\left(\frac{n!}{2^{\lfloor \frac{n}{2} \rfloor} \cdot \lfloor \frac{n}{2} \rfloor!}\right) = (1 + o(1)) \cdot (\frac{n}{2} \log_2 n)$. Even though this bound has been computed for perfect matchings only, it is possible to built an adaptive algorithm reconstructing general matchings and achieving this bound within a constant factor.

The algorithm works in two steps. The first one is adaptive and partitions the set of vertices into $V_1 \uplus V_2$ such that no two vertices in the same V_i are adjacent in the matching. This can be done in n queries by processing vertices one-by-one. The second step can be made nonadaptive. It finds for every $v \in V_1$ the adjacent vertex to v (if it exists) in V_2 using a group testing algorithm to find one "counterfeit coin" among n (see Section 3.1). This group testing problem can be solved within $\lceil \log_2 n \rceil$ nonadaptive queries, yielding a total complexity of $(1 + o(1)) \cdot (n \log_2 n)$ for the entire algorithm. Note that the same algorithm applied to the reconstruction of perfect matchings has an optimal asymptotic complexity $(1 + o(1)) \cdot (\frac{n}{2} \log_2 n)$.

---

[2] A perfect matching is a graph such that the degree of all vertices except possibly one is 1.

2.3   Stars and Cliques

The reconstruction of stars and cliques on n vertices has been studied in [4]. Following that paper, we define S_k to be the set of all stars with a center, k leaves and $n - k - 1$ isolated vertices, and C_k to be the set of all cliques with k vertices and $n - k$ isolated vertices. $S = \cup_{k=0}^{n-1} S_k$ and $C = \cup_{k=1}^{n} C_k$ are respectively the set of all stars and all cliques on n vertices, with an arbitrary number of isolated vertices.

We now examine the information-theoretic lower bound for reconstructing stars and cliques under the boolean model. To estimate the cardinality of S, recall that a star of S_k (for $k \geq 2$) is defined by a center chosen among the n vertices and k leaves chosen among the $n - 1$ remaining vertices. So $|S| = \sum_{k=2}^{n-1} n \cdot \binom{n-1}{k} + \frac{n(n-1)}{2} + 1 = n \cdot (2^{n-1} - 1) - \frac{n(n-1)}{2} + 1$. Consequently, we get the lower bound $\log_2 |S| = (1 + o(1)) \cdot n$ for the complexity of the star reconstruction problem. For cliques, it is clear that $|C| = \sum_{k=0}^{n} \binom{n}{k} = 2^n$, and the information theoretic lower bound is then $\log_2 |C| = (1 + o(1)) \cdot n$.

For both stars and cliques, the $\Omega(n)$ bound can be achieved by the following algorithm composed of two nonadaptive rounds. At the first round, find a starting vertex from which it becomes easy to reconstruct the whole graph: the center of the star or a vertex that belongs to the clique. Finding the center of the star is done through n nonadaptive queries $V \setminus \{i\}$ for $1 \leq i \leq n$. To find a vertex of the clique, we simply ask the queries $Q_i = \{1, \ldots, i\}$ for $2 \leq i \leq n$. At the second round (nonadaptive as well), finish the reconstruction by determining the neighbors of the starting vertex. Each round requires a linear number of queries.

While cliques and stars can be easily reconstructed in two nonadaptive rounds, the situation changes if we are restricted to fully nonadaptive (1-round) algorithms. To reconstruct a star of S with a nonadaptive algorithm, it is necessary, in the worst case, to query each of the $\binom{n}{2}$ pairs of vertices $\{u, v\}$ [4], i.e. the most naive algorithm turns out to be the optimal one in the worst case. In contrast, for cliques, only $\Omega(n \log n)$ nonadaptive queries are needed, and [4] showed the existence of a nonadaptive algorithm reconstructing a clique of C with $\mathcal{O}(n \log^2 n)$ queries.

3   Quantitative Model

We now turn to the quantitative model, much less studied in the literature. We show that under this model, nonadaptive algorithms get all their power and often allow to achieve (or to approach) the lower bound. This is due to powerful combinatorial constructions of $(0, 1)$-matrices verifying certain properties.

3.1   Hamiltonian Cycles

We start again with our initial problem of reconstructing a Hamiltonian cycle on n vertices. As under the quantitative model there are $n + 1$ possible answers to each query $Q \subseteq V$, the information-theoretic lower bound is $\log_{n+1} \frac{(n-1)!}{2} = (1 + o(1)) \cdot n$.

Theorem 2. *Under the quantitative model, there exists an algorithm reconstructing a Hamiltonian cycle in $O(n)$ queries.*

One such algorithm has been presented in [11] and is composed of two steps: an adaptive preparatory step followed by a nonadaptive reconstruction step[3]. We now describe this algorithm.

First Stage. The goal of the first stage is to reduce the problem to the reconstruction of bipartite graphs. By processing all the vertices successively, we transform the Hamiltonian cycle H into a tripartite graph, i.e. we partition the set of vertices V into 3 subsets $V_1 \uplus V_2 \uplus V_3$ such that two vertices in the same subset are not adjacent in H. As each vertex has exactly two neighbors, this transformation can be done in at most $2n$ queries. We are now dealing with the problem of reconstruction of a tripartite graph that we view as three bipartite graphs.

Second Stage. The second stage reconstructs each of the three bipartite graphs in $O(n)$ nonadaptive queries. This crucial step is based on two auxiliary constructions.

First Subproblem. Consider a bipartite graph $(C_1, C_2; E)$ with vertex degree bounded by a constant (2 in our case). Assume that we want to determine the *degrees* of all vertices of C_1 by querying subsets of C_1 together with the whole set C_2. This problem is equivalent to the reconstruction of an unknown vector $v = (v_1, \ldots, v_n)$ with $v_i \in \{0, \ldots, d-1\}$ ($d = 3$ in our case) by querying sums of the form $\sum_{i=1}^{n} \epsilon_i v_i$, $\epsilon_i \in \{0, 1\}$. A nonadaptive algorithm solving this problem corresponds to a $(0, 1)$-matrix M of dimension $k \times n$ (k as small as possible) such that for vectors $v \in \{0, \ldots, d-1\}^n$, all products Mv are distinct. We call such matrix a *d-detecting matrix*.

The information-theoretic lower bound for k is $\log_{(d-1)n+1} d^n = (1 + o(1)) \cdot (\frac{n}{\log_d n})$.

For the particular case $d = 2$, this lower bound can be improved to $(2 + o(1)) \cdot (\frac{n}{\log_2 n})$, as it was shown in [9] (another proof using Kolmogorov complexity can be found in [16]). On the other hand, it has been shown in [17,6] that this bound can be achieved. A decade later, Lindström [21] gave a tricky effective construction of a 2-detecting matrix with $(2 + o(1)) \cdot (\frac{n}{\log_2 n})$ rows using the Möbius function.

In our case, $d = 3$ and a 3-detecting matrix with $(4 + o(1)) \cdot (\frac{n}{\log_2 n})$ rows can be effectively constructed as an extension of the Lindström construction. Furthermore, for an arbitrary constant d, a d-detecting matrix with $(2 + o(1))(\log d \cdot \frac{n}{\log n})$ rows can be effectively constructed, and this is also a lower bound [11].

Second Subproblem. Consider a bipartite graph $(C_1, C_2; E)$ and a vertex $i \in C_1$. We want to determine the vertices of C_2 adjacent to i by querying i together

---

[3] As it will follow from Section 3.4, Hamiltonian cycles can be reconstructed in $O(n)$ fully nonadaptive queries. The two-step construction presented here is for explanatory purposes.

with subsets of C_2. In the case of Hamiltonian cycle, there are exactly two such vertices, but to be more general, we assume that their number is bounded by a constant d. The problem can be viewed as a problem of discovering d counterfeit coins (neighbors of i) among n coins (vertices C_2) and is well-known in the area of group testing [8]. We want to solve it in a nonadaptive way (for reasons that will be clear later) using queries of type "*how many* counterfeit coins does a given subset contain?".

The case of finding one counterfeit among n can be solved by an optimal non-adaptive set of queries $Q_i = \{j |$ the i-th bit of j is $1\}$ for $1 \leq i \leq \lceil \log_2 n \rceil$. However, already for two coins the situation gets more complicated: the information-theoretic lower bound is $\log_3 \binom{n}{2} \approx 1.26 \cdot \log_2 n$ while the best known upper bound for *adaptive algorithms* is $1.44 \cdot \log_2 n$. For *nonadaptive algorithms*, the best known lower and upper bounds are respectively $\frac{5}{3} \cdot \log_2 n$ and $2 \cdot \log_2 n$ [18,20].

For the general problem of finding nonadaptively d counterfeit coins among n, we need to construct a $(0,1)$-matrix A of dimension $k \times n$ (k as small as possible) such that for vectors $v \in \{0,1\}^n$ having at most d 1's, all products Av are distinct. We call such a matrix a d-*separating matrix*. Known upper and lower bounds for the number of rows in a d-separating matrix are respectively $(4+o(1)) \cdot (\frac{d}{\log d} \log n)$ [11] and $(2+o(1)) \cdot (\frac{d}{\log d} \log n)$ [3]. Both are proved using probabilistic arguments, and thus the upper bound is non-constructive. The best known explicit nonadaptive construction uses BCH error-correcting codes and uses $O(d \log_2 n)$ queries. Note also that no better properly adaptive algorithm is known.

Combining the Subproblems. The two techniques presented above (d-detecting and d-separating matrices) allow us to solve the problem of reconstruction of a bipartite graph $(C_1, C_2; E)$ with the degree of each vertex in C_1 bounded by a constant d. Using d-separating matrices, the adjacent vertices of each $i \in C_1$ can be obtained by querying i against $P_1, \ldots, P_m \subseteq C_2$, where $P_1, \ldots, P_m$ do not depend on i. For each P_j, we can determine the degree of each $i \in C_1$ in P_j by querying P_j against $S_1, \ldots, S_\ell \subseteq C_1$ using d-detecting matrices. Again, $S_1, \ldots, S_\ell$ do not depend on P_j. Thus, querying all pairs $S_k \cup P_j$ is sufficient to reconstruct the whole graph. The resulting number of queries is $(2 + o(1))(\log d \frac{n}{\log n})(4 + o(1))(\frac{d}{\log d} \log n) = (8 + o(1))dn$.

This proves the following

Theorem 3. *A (one-sided) d-bounded degree bipartite graph can be reconstructed within $(8+o(1)) \cdot dn$ nonadaptive queries. This matches the lower bound up to a constant factor.*

Turning back to our initial motivation (Theorem 2), a Hamiltonian cycle can be reconstructed within $2n + 3 \cdot 2 \log_2 n \cdot \frac{(4+o(n)) \cdot n}{\log_2 n} = O(n)$ queries asymptotically by a two-stage algorithm. This matches the lower bound up to a constant factor.

3.2   Matchings

As in Section 2.2, consider the lower bound $\frac{n!}{2^{\lfloor \frac{n}{2} \rfloor} \cdot \lfloor \frac{n}{2} \rfloor!}$ on the number of matchings on n vertices. Note that as the number of edges in a matching on n vertices is at most $\lfloor \frac{n}{2} \rfloor$, the maximal number of distinct answers to a query is $\lfloor \frac{n}{2} \rfloor + 1$. Consequently, we can compute an information-theoretic lower bound on the complexity of the matching reconstruction problem under the quantitative model to be $\log_{\lfloor \frac{n}{2} \rfloor + 1} \left(\frac{n!}{2^{\lfloor \frac{n}{2} \rfloor} \cdot \lfloor \frac{n}{2} \rfloor!} \right) = (1 + o(1)) \cdot \frac{n}{2}$.

It is possible to reach this bound, up to a constant factor, by a fully non-adaptive algorithm. This will follow from Section 3.4 where we describe a general nonadaptive algorithm for reconstructing graphs of vertex degree bounded by d within $\mathcal{O}(dn)$ queries.

3.3 Stars and Cliques

Recall from Section 2.3 that the number of stars and cliques on n vertices are respectively $|S| = n \cdot (2^{n-1} - 1) - \frac{n(n-1)}{2} + 1$ and $|C| = 2^n$. The information-theoretic lower bound for reconstructing stars under the quantitative model is then $\log_n \left(n \cdot (2^{n-1} - 1) - \frac{n(n-1)}{2} + 1 \right) = (1 + o(1)) \cdot \left(\frac{n}{\log_2 n} \right)$ and that for cliques is $\log_{\frac{n(n-1)}{2} + 1} (2^n) = \left(\frac{1}{2} + o(1) \right) \cdot \left(\frac{n}{\log_2 n} \right)$.

There exist adaptive algorithms that achieve these bounds within a constant factor. Here we give only a very high-level description of them. Similar to Section 2.3, the algorithms are divided into two main steps, the first one is adaptive and the second one nonadaptive. At the first step, we find, in a logarithmic number of adaptive queries, either the center of the star, or one vertex of the clique. (This can be done using binary search.) $(2 + o(1)) \frac{n}{\log_2 n}$ nonadaptive queries are then sufficient to reconstruct the neighbors of the vertex found in the first stage, using 2-detecting matrices introduced in the first subproblem of Section 3.1 (see [19,13]). For stars, this construction applies immediately and for cliques, we need to transform each query answer from $k + k(k-1)/2$ to k which is done non-ambiguously.

3.4 Bounded Degree Graphs

Theorem 3 states that a (one-sided) d-bounded degree bipartite graph can be reconstructed through $O(dn)$ nonadaptive queries. We now want to use this technique to reconstruct general bounded degree graphs [13]. The idea is to consider a bipartite representation of a graph defined as follows. Given a graph $G = (V, E)$, the bipartite representation of G is $G' = (V_1, V_2; E')$, where V_1 and V_2 are two disjoint copies of V, $E \subseteq V_1 \times V_2$, and $(i, j) \in E$ implies $(i, j) \in E'$ and $(j, i) \in E'$. Note that any edge of G produces two edges in G'. Moreover, if G is d-bounded degree then G' is d-bounded degree too.

We want to query the binary representation through the following queries: "Given $X \subseteq V_1$ and $Y \subseteq V_2$, how many edges are there in G' connecting vertices of X to vertices of Y ? ". We define the corresponding query function $\mu'_{G'}(X, Y) = |E' \cap (X \times Y)|$. A query $\mu'_{G'}(X, Y)$ can be expressed through quantitative queries to the initial graph G, i.e. through the query function $\mu_G(X) = |E \cap (X \times X)|$, for $X \subseteq V$. Using elementary set-theoretic considerations, it can be shown that $\mu'(X, Y) = \mu((X \setminus Y) \cup (Y \setminus X)) - 2\mu(X \setminus Y) - 2\mu(Y \setminus X) + \mu(X) + \mu(Y)$.

By Theorem 3, the binary representation G' can be reconstructed by $O(dn)$ nonadaptive queries $\mu'(X, Y)$. From the observation above, it follows that G' can be reconstructed by $O(dn)$ nonadaptive queries $\mu(X)$.

Theorem 4. *A d-bounded degree graph can be reconstructed within $O(dn)$ non-adaptive queries. This is an asymptotically tight bound.*

3.5   General Graphs

Under the quantitative model, the information-theoretic lower bound for reconstructing general graphs is $\log_{1+\frac{n(n-1)}{2}} 2^{\frac{n(n-1)}{2}} = (\frac{1}{4}+o(1)) \cdot \frac{n^2}{\log_2 n}$. A better lower bound $(\frac{1}{2}+o(1)) \cdot \frac{n^2}{\log_2 n}$ can be obtained using lower bounds for d-detecting matrices (see Section 3.1). As it was shown in [13], this bound can be achieved up to a constant factor using again the bipartite representation of a graph introduced in the previous section.

Consider the bipartite representation $G' = (V_1, V_2; E')$ of a general graph $G = (V, E)$. For each vertex $i \in V_1$, reconstruct its adjacent vertices among $\{1, \ldots, i-1\} \subseteq V_2$ with $(2+o(1)) \cdot \frac{i}{\log_2 i}$ queries of the form $\mu'(\{i\}, W)$, $W \subseteq V_2$, using 2-detecting matrices. Observe that $\mu'(\{i\}, W) = \mu(W \cup \{i\}) - \mu(W \setminus \{i\})$ which allows us to express each query $\mu'(\{v_i^1\}, W)$ through two queries to the original graph G.

The overall complexity of this method for the reconstruction of a general graph is then $\sum_{i=2}^{n}(2 + o(1)) \cdot \frac{i}{\log_2 i} = (2 + o(1))\frac{n^2}{\log_2 n}$. This is within the factor of four from the known lower bound for nonadaptive algorithms.

Theorem 5. *A general graph can be reconstructed within $(2 + o(1))\frac{n^2}{\log_2 n}$ non-adaptive queries. This matches the lower bound up to a constant factor.*

3.6   *k*-Degenerate Graphs and Trees

The general technique used to reconstruct bounded degree graphs (Section 3.4) can be further extended to reconstruct more general *k-degenerate graphs*. An intuitive definition of k-degenerate graphs is as follows: G is k-degenerate if there exists a vertex v of G with vertex degree less than or equal to k such that $G \setminus \{v\}$ has the same property. More formally, a graph G is k-degenerate if vertices V can be ordered $(v_1, v_2, \cdots, v_n)$ such that $deg_{G_i}(v_i) \leq k$, where G_i is the subgraph of G induced by the vertices $\{v_i, v_{i+1}, \cdots, v_n\}$. For example, trees are 1-degenerate as there exists a leaf of vertex degree 1 and after deleting it the graph is still a tree. Another example is provided by planar graphs that are 5-degenerate: there is always a vertex of degree at most 5 and deleting it keeps the graph planar.

Let us first compute the information-theoretic lower bound for the reconstruction of k-degenerate graphs. The number of edges in a k-degenerate graph is clearly less than nk. To obtain a lower bound on the number of k-degenerate graphs, we fix some order on vertices and count number of possibilities to connect v_{k+t} to $v_{k+t+1}, \cdots, v_n$. Since all such choices can be made independently for all $v_1, \cdots, v_{n-k}$, we have $N(n+1,k) \geq \prod_{i=k+1}^{n} \binom{i}{k} \geq \frac{(n!)^k}{k^{nk}}$. The corresponding information-theoretic lower bound is then

$$\log_{nk} N(n+1,k) \geq \frac{nk(\log n - \log k - 1)}{\log n + \log k}.$$

In the case $k \leq n^\alpha$ for some $\alpha < 1$, this bound can be simplified into $\Omega(nk)$. For n sufficiently large, we can prove that this bound is tight, meaning that there exists an algorithm that reconstructs a graph in the class of k-degenerate graphs with $\mathcal{O}(nk)$ queries.

Theorem 6. *k-degenerate graphs on n vertices can be reconstructed by a non-adaptive algorithm using $\mathcal{O}(nk)$ queries, and this bound is tight.*

As in the case of bounded degree graphs (Section 3.4), the algorithm uses the bipartite representation of k-degenerate graphs and the same general technique of reconstructing bipartite graphs. While the bipartite representation here is not of bounded vertex degree, the sum of degrees of all vertices from one side is bounded by nk. Therefore, instead of using d-detecting matrices (first subproblem in Section 3.1), we consider matrices that solve a more general combinatorial search problem, namely the reconstruction of d-*bounded weight vectors* which are vectors with the sum of entries bounded by d. Formally, define the class of d-bounded weight vectors by $\Lambda(n,d) = \{(v_1, \ldots, v_n)|v_i \in \mathbb{N} \text{ and } \sum_{i=1}^{n} v_i \leq d\}$. A nonadaptive algorithm reconstructing d-bounded weight vectors is specified by an object-query incidence matrix M such that $M \cdot v_1 \neq M \cdot v_2$ for all $v_1, v_2 \in \Lambda(n,d)$, $v_1 \neq v_2$. It has been shown in [10] that there exists such a matrix with the number of rows $k(n,d) \leq \frac{4min(n,d)\log\left(C_1 \frac{max(n,d)}{min(n,d)}\right)}{\log min(n,d)+C_2} + C_3 \log d$, for some constants C_1, C_2 and C_3.

Consider now the bipartite representation $G' = (V_1, V_2; E')$ of a k-degenerate graph G. Assume we are given two families $\{Q_j\}_{j=1}^{m}$ and $\{P_i\}_{i=1}^{l}$ that solve the d-bounded weight vector reconstruction problem for $d = k$ and $d = 2nk$ respectively. From the bound on $k(n,d)$ above, it follows that $m = \mathcal{O}(k\frac{\log n}{\log k})$ and $l = \mathcal{O}(n\frac{\log k}{\log n})$ when $n \to \infty$. It can be shown that the set of queries $\{\mu'(P_i, Q_j)\}_{i=1,\ldots,l}^{j=1,\ldots,m}$ reconstructs k-degenerate graphs. The proof, given in [10], combines the ideas of Section 3.1 with an iterative procedure of computing the answers of queries $\mu'(P_i, Q_j)$ that would be obtained after deleting all edges incident to a vertex of degree at most k (by definition of k-degenerate graphs, such a vertex always exists).

The overall complexity of the algorithm is $m \cdot l = \mathcal{O}(nk)$, which proves Theorem 6.

4   Conclusions and Open Problems

Through examples of Hamiltonian cycles, matchings, stars and cliques, the quantitative model has been shown to be more powerful than the boolean model. The following table illustrates this difference and provides lower and upper bounds (for adaptive and nonadaptive algorithms) for the two-vertex, boolean and quantitative models, for the case of Hamiltonian cycle that has been our main applicative motivation.

	lower bound	adaptive	nonadaptive
two-vertex model	$\Omega(n^2)$	$O(n^2)$	$O(n^2)$
boolean model	$\Omega(n\log n)$	$O(n\log n)$	$\Omega(n^2)$
quantitative model	$\Omega(n)$	$O(n)$	$O(n)$

Another important conclusion is that nonadaptive algorithms fully benefit from the quantitative model, and vice versa. Not only the quantitative model allows faster reconstruction algorithms, but also these algorithms can be made nonadaptive, or "almost nonadaptive" (having an important nonadaptive component). Interestingly, under the quantitative model, nonadaptive algorithms often reach the asymptotic lower bound and no properly adaptive algorithm is known to outperform nonadaptive algorithms. This contrasts with the boolean model, where nonadaptive algorithms are usually strictly less powerful than adaptive ones.

The power of nonadaptive algorithms under the quantitative model is due to powerful combinatorial constructions of d-detecting and d-separating matrices (Section 3.4) and their generalizations (Section 3.6).

As far as open questions are concerned, we would like to mention two of them here. One concerns an important technical point: the upper bound for d-separating matrices (Section 3.4). The tight upper bound $O(\frac{d}{\log d}\log n)$ has been proved by a probabilistic nonconstructive argument, and finding an *effective* construction of d-separating matrices with $O(\frac{d}{\log d}\log n)$ rows remains an important open question. Another question is of more general nature: how far can we go with optimal nonadaptive reconstruction under the quantitative model? For example, can we reconstruct in $O(dn)$ queries any graph with $O(dn)$ edges?

To conclude, we get back to the applicative side of our study and mention that many other bioinformatics applications give rise to combinatorial search problems. Such applications include screening clone libraries [15], the FISH (*Fluorescent In Situ Hybridization*) method for chromosome identification [12], determination of exon-intron boundaries in genes [7], probe selection for DNA chips [14], and others. Thus, those applications provide a rich source for new interesting developments of combinatorial search methods in future.

References

1. M. Aigner. *Combinatorial Search*. John Wiley and Sons, 1988.
2. B. Alberts, A. Johnson, J. Lewis, M. Raff, K. Roberts, and P. Walter. *Molecular Biology of the Cell*. Garland Science, 1994.
3. N. Alon. Separating matrices. Private communication, May 1997.

4. N. Alon and V. Asodi. Learning a hidden subgraph. In *Automata, Languages and Programming: 31st International Colloquium, ICALP 2004, Turku, Finland, July 12-16, 2004. Proceedings*, volume 3142 of *Lecture Notes in Computer Science*, pages 110–121. Springer, 2004.
5. N. Alon, R. Beigel, S. Kasif, S. Rudich, and B. Sudakov. Learning a hidden matching. In *Proceedings of the 43rd Annual IEEE Symposium on Foundations of Computer Science, FOCS 2002, Vancouver, BC, Canada, 16-19 November 2002*, pages 197–206. IEEE Computer Society Press, 2002.
6. D.G. Cantor and W.H. Mills. Determination of a subset from certain combinatorial properties. *Can. J. Math*, 18:42–48, 1966.
7. F. Cicalese, P. Damaschke, and U. Vaccaro. Optimal group testing algorithms with interval queries and their application to splice site detection. In *Proc. of the Int. Workshop on Bioinformatics Research and Applications (IWBRA 2005)*, volume 3515 of *Lectures Notes in Computer Science*, pages 1029–1037. Springer, 2005.
8. D. Du and F. Hwang. *Combinatorial group testing and its applications*, volume 3. Series on applied Mathematics, 1993.
9. P. Erdös and A. Rényi. Asymmetric graphs. *Acta Math. Acad. Sci. Hung. Acad. Sci.*, 14:295–315, 1963.
10. V. Grebinski. On the power of additive combinatorial search model. In *Proc. of Computing and Combinatorics, 4th Annual International Conference, CO-COON'98, Taipei, Taiwan, August 12-14, 1998*, volume 1449 of *Lecture Notes in Computer Science*, pages 194–203. Springer, 1998.
11. V. Grebinski and G. Kucherov. Reconstructing a hamiltonian cycle by querying the graph: Application to DNA physical mapping. *Discrete Applied Mathematics*, 88:147–165, 1998.
12. V. Grebinski and G. Kucherov. Reconstructing set partitions. In *Proceedings of the 10th Annual ACM-SIAM Symposium on Discrete Algorithms, SODA '99 (Baltimore, Maryland, January 17-19, 1999)*, pages 915–916. ACM, SIAM, 1999.
13. V. Grebinski and G. Kucherov. Optimal reconstruction of graphs under the additive model. *Algorithmica*, 28:104–124, 2000.
14. G.W Klau, S. Rahmann, A. Schliep, M. Vingron, and K. Reinert. Optimal robust non-unique probe selection using integer linear programming. *Bioinformatics*, 20 (suppl. 1):i186–i193, 2004.
15. E. Knill and S. Muthukrishnan. Group testing problems in experimental molecular biology. Technical Report LAUR-95-1503, Los Alamos National Laboratory, March 1995.
16. M. Li and P. M. B. Vitányi. Kolmogorov complexity arguments in combinatorics. *J. Comb. Theory Series A*, 66(2):226–236, 1994.
17. B. Lindström. On a combinatorial problem in number theory. *Canad. Math. Bull*, 8:477–490, 1965.
18. B. Lindström. Determination of two vectors from the sum. *J. Comb. Theory*, 6:402–407, 1969.
19. B. Lindström. On Möbius functions and a problem in combinatorial number theory. *Canad. Math. Bull.*, 14(4):513–516, 1971.
20. B. Lindström. On b_2 sequences of vectors. *Journal of Number Theory*, 4:261–265, 1972.
21. B. Lindström. Determining subsets by unramified experiments. In editor J.N. Srivastava, editor, *A Survey of Statistical Designs and Linear Models*, pages 407–418. North Holland, Amsterdam, 1975.
22. R. M. Wilson. Decomposition of complete graphs into subgraphs isomorphic to a given graph. In *Congressus Numerantium XV*, pages 647–659, 1975.

Domination Search on Graphs with Low Dominating-Target-Number

Divesh Aggarwal[1], Shashank K. Mehta[1,*], and Jitender S. Deogun[2]

[1] Indian Institute of Technology, Kanpur - 208016, India
{cdubey, skmehta}@cse.iitk.ac.in
[2] University of Nebraska-Lincoln, Lincoln, NE 68588-0115, USA
deogun@cse.unl.edu

Abstract. We settle two conjectures on domination-search, a game proposed by Fomin et.al. [1], one in affirmative and the other in negative. The two results presented here are (1) domination search number can be greater than domination-target number, (2) domination search number for asteroidal-triple-free graphs is at most 2.

1 Introduction

Domination search is a game proposed by Fomin et.al. [1] which is a variant of *node-search-game*, see [2]. It is also a graph variant of polygonal search problem, [3,4,5,6,7]. It is a problem of sweeping out a mobile fugitive out of a graph (think of a house where vertices are rooms) with k guards. A guard at a node can check its node and all nodes adjacent to it. The fugitive can move in zero time from node x to y if there is a path between the nodes which does not pass through the nodes under any guard's watch. In each step one guard can move from its current node to any other vertex. During the move this guard is absent from the graph and fugitive can take the advantage. Search is successful if after a finite number of moves entire graph is cleared of the fugitive.

We present a formal definition of domination-search game differently from the original but it is equivalent to that. Here $N[X]$ denotes the closed neighborhood of the vertex set X. The search algorithm with k *guards* on a graph $G = (V, E)$ places k guards on k vertices initially. $D(0)$ denotes these vertices. In each move one guard is moved from it current position (vertex) to a new position. $D(i)$ denotes the set of vertices where the guards are placed after i moves. Formally, the search is a sequence of k-sets: $D(0), D(1), \ldots, D(M)$, where $D(i-1) \cap D(i)$ is denoted by $S(i)$ and has cardinality $k-1$ for all $i > 0$. A vertex is said to *clear* if it was in the neighborhood of some guard in some previous move and since then no path has been established between this vertex and a contaminated (fugitive may potentially be on it) vertex without passing through the neighborhood of a guard in the current position. We define vertex sets $U_a(i)$ (set of clear vertices after i moves) for $0 \leq i \leq M$ and $U_d(i)$ (clear set during move-i) for $1 \leq i \leq M$. These sets are recursively defined by the following equations. $U_a(0)$

* Partly supported by Ministry of Human Resource Development, Government of India under grant no. MHRD/CS/20030320.

is the closed neighborhood of $D(0)$, i.e., $N[D(0)]$. $U_a(i) = U_d(i) \cup N[D(i) - S(i)]$, and $U_d(i)$ is the set $\{v \in U_a(i-1) : every\ path\ from\ v\ to\ any\ vertex\ in\ V - U_a(i-1)\ passes\ through\ N[S(i)]\}$. Finally, $U_a(M) = V$. The domination search number of a graph G is the smallest k for which such a sequence exists. It is denoted by $ds(G)$.

Domination-search number is found to be strongly related with another graph-parameter, dominating-target number denoted by $dt(G)$. A vertex subset, T, of a graph is said to be a dominating-target [8] if every connected subgraph which contains T, dominates the entire graph. The cardinality of the smallest dominating-target is called the dominating-target number, denoted by $dt(G)$. Fomin et.al. [1] have shown that for arbitrary connected graph $ds(G) \leq 2 \cdot dt(G) + 3$. But they have found that this is perhaps not a tight bound and conjectured that $ds(G) \leq dt(G)$.

In their work Fomin et.al. have also studied $ds(G)$ of graphs of small dominating-target number. These include cocomparability graphs; AT-free graphs [9,10,11,12]; and DP graphs [13,14]. These graph classes are defined as follows. An asteroidal-triple is a set of three vertices such that there is a path between any two vertices without entering the neighborhood of the third. A graph is said to be AT-free if it contains no asteroidal-triple. The family of graphs having dominating target number equal to two is denoted by DP. Their results include (i) $ds(G) \leq 4$ for the DP graphs; (ii) $ds(G)$ of cocomparability graphs is 2; (iii) $ds(G) \leq 2$ for AT-free claw-free graphs; and (iv) $ds(G) \leq 3$ for AT-free graphs. They also conjecture that $ds(G) \leq 2$ for AT-free graphs.

In this work we will show that there exists a DP graph for which domination-search number is greater than 2. This settle the conjecture "$ds(G) \leq dt(G)$" in negative. We also present a domination search algorithm for AT-free graphs with $ds(G) \leq 2$ which settles the second conjecture in affirmative. The paper is organized as follows. Section 2 presents a DP graph and shows that it cannot be searched with two guards. Section 3 describes a partial ordering on graphs which plays an important role in developing the domination-search algorithm for AT-free graphs, presented in Section 4.

2   Lower Bound for DP Graphs

In this section we will establish that domination search cannot be performed on all weak dominating pair graphs (family of graphs with dominating target number being 2) with 2 guards. This will settle the conjecture 23 of [1], "$ds(G) \leq dt(G)$", in negative.

The open neighborhood of a vertex x, $N(x)$, in a graph is the set of vertices adjacent to x. The closed neighborhood, $N[x]$ is $N(x) \cup \{x\}$. If $N[x]$ is not a graph separator then x is called an $extreme$ vertex. The set of all extreme vertices of a graph is denoted by L.

Consider the graph $G_0 = (V, E)$ in figure 1. Observe that $L = \{v_0, v_1, v_2, v_3, v_4, v_8, v_9, v_{10}, v_{11}, v_{12}\}$. Therefore only non-extreme vertices in G_0 are v_5, v_6 and v_7. There are two connected components in the induced subgraph on $V - N[v_5]$,

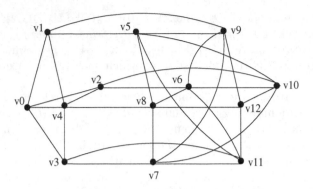

Fig. 1. A dominating pair graph with $ds(G) > 2$

denote them $C_1^{v_5}$ and $C_2^{v_5}$. Due to the symmetry in v_5, v_6 and v_7, $V - N[v_6]$ and $V - N[v_7]$ also have two components: $C_1^{v_6}$, $C_2^{v_6}$ and $C_1^{v_7}$, $C_2^{v_7}$ respectively. $C_1^{v_5} = C_1^{v_6} = C_1^{v_7}$ is the single vertex v_{12}. The second components of each case is given in figure 2 which are indeed isomorphic.

Let us assume that a domination search algorithm for G_0 exists which requires two guards. Let it be expressed by the sequence $A : D(0) = (p_1(0), p_2(0))$, $D(1) = (p_1(1), p_2(1)), \dots, D(M) = (p_1(M), p_2(M))$. Pair $p_1(i), p_2(i)$ denote the vertices where guards are placed after i moves. $\{p_1(i-1), p_2(i-1)\} \cap \{p_1(i), p_2(i)\}$ is a singleton denoted by $S(i)$. For notational convenience we will denote the element in $S(i)$ by $S(i)$ as well without ambiguity. By $U_d(i)$ we denote the set of vertices which are clear (uncontaminated) in the graph during the i-th move when there is only one guard on the graph (at $S(i)$). After the move when there are two guards on the graph the set of clear vertices is denoted by $U_a(i)$. Without loss of generality we assume that this sequence is minimal in the sense that no step of the algorithm is redundant, i.e., no proper subsequence of A is a valid domination search. The graph does not have a dominating set of size two therefore M must be greater than zero.

Proposition 1. $S(i) \notin L$ for all $1 \leq i \leq M$.

Proof. Assume $S(i) \in L$. If $U_d(i)$ was equal to V then there would have been no need for the i-th move. So there is at least one contaminated vertex just before this move. During this move there is only one guard on the graph, at $S(i)$. Since the induced graph on $V - N[S(i)]$ is connected, entire set $V - N[S(i)]$ will get contaminated. So the set of clear vertices after this move will be $N[p_1(i)] \cup N[p_2(i)]$. This state can be achieved at the start of the search by placing the guards at $p_1(i)$ and $p_2(i)$. Therefore we can replace A by $A' : (p_1(i), p_2(i)), (p_1(i+1), p_2(i+1)), \dots, (p_1(M), p_2(M))$ which will also perform the domination search. This violates the minimality condition of A.

Due to symmetry between v_5, v_6, and v_7 we may assume that $S(1) = v_5$ without loss of generality. Suppose $S(i) = v_5$ for $1 \leq i \leq i_0$. During these moves only one guard is moving to clear the parts of $V - N[v_5]$. No single vertex in

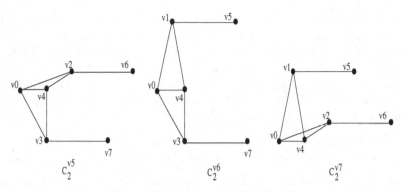

Fig. 2. Components $C_2^{v_5}$, $C_2^{v_6}$, $C_2^{v_7}$

the graph dominates entire $C_2^{v_5}$ so a single guard cannot clear it completely. Therefore during each move upto i_0 entire $C_2^{v_5}$ will be contaminated. In other words $U_d(i) \cap C_2^{v_5} = \emptyset$ for $1 \le i \le i_0$. As a consequence $U_a(i)$ for $1 \le i \le i_0$ cannot be equal to V. This indicates that the search cannot terminate if $S(i)$ remains unchanged at v_5.

Let us suppose $S(i_0 + 1) \ne v_5$. Then $S(i_0 + 1)$ must be either v_6 or v_7. Due to symmetry, we can assume $S(i_0 + 1) = v_6$ with no loss in generality. In the move i_0 one guard stays fixed at v_5 and the the other guard moves to v_6. As discussed above $U_d(i_0)$ does not contain any vertex of $C_2^{v_5}$. $U_a(i_0) = U_d(i_0) \cup N[v_6]$ which does not contain v_0. So $U_a(i_0)$ does not contain v_0. Consequently $C_2^{v_6}$ will be entirely contaminated during move $i_0 + 1$ when guard at v_6 remains fixed. Suppose $S(i) = v_6$ for $i_0 + 1 \le i \le i_1$. As argued above during all these moves $U(i) \cap C_2^{v_6} = \emptyset$ for $i_0 + 1 \le i \le i_1$. So the search cannot terminate with i_1-th move. Once again we may replace v_6 by v_5 or v_7 as the value of $S(i)$ but repeating the argument we conclude that the search will never end. We have following result.

Theorem 1. *Domination search on graph of figure 1 requires at least 3 guards.*

This graph has dominating target number 2 because $\{v_4, v_{12}\}$ is a dominating pair in it. So we establish that $dt(G_0) < ds(G_0)$.

Corollary 1. *The conjecture $ds(G) \le dt(G)$, proposed in [1], for all connected graphs G is incorrect.*

3   A Partial Ordering on Graphs

The domination search algorithm for asteroidal-triple-free graphs proposed in the following section uses two guards. The selection of the successive positions to station the guards is determined based on a partial ordering on the vertices which is described in this section.

Let $G = (V, E)$ be an arbitrary graph and x be any vertex in it. Define relation $\succ_x$ on V as follows. $u \succ_x v$ if (i) u and v are not adjacent and (ii) every path from u to x is intercepted by v, i.e., at least one vertex on each path from u to x belongs to $N[v]$. Observe that condition (ii) can be equivalently stated as: every induced path from u to x is intercepted by v. This relation is reflexive. Define an equivalence relation $\sim_x$ on V as follows. $u \sim_x v$ if there exist $u_1, u_2, \ldots u_k$ such that $u \succ_x u_1 \succ_x u_2 \succ_x \ldots u_k \succ_x v$ and $v \succ_x u_k \succ_x \ldots u_1 \succ_x u$. The equivalence classes, x-classes, induced by $\sim_x$ will be denoted by $[u]_x$ representing the class containing u. The x-class containing x is obviously a singleton.

Observation 2. *Let $u_1, u_2 \in [u]_x$. Then for any induced path from u_1 to x: $u_1 a_1 a_2 \ldots a_m (= x)$, u_2 is adjacent to a_1 and to no other a_i.*

We extend $\succ_x$ to the class, using the same symbol: $[u]_x \succ_x [v]_x$ if there exists $u' \in [u]_x$ and $v' \in [v]_x$ such that $u' \succ_x v'$. The above observation leads to the following result.

Observation 3. *$[u]_x \succ_x [v]_x$ iff for every $u' \in [u]_x$ and $v' \in [v]_x$ either (u', v') is an edge or $u' \succ_x v'$.*

Consider two distinct classes $[u]_x$ and $[v]_x$ such that $[u]_x \succ_x [v]_x$. Let $u' \in [u]_x$ and $v' \in [v]_x$ such that $u' \succ_x v'$. Let $u'' \in [u]_x$ such that $u'' \succ_x u'$. Assume that $u'' \succ_x v'$ is not true. From the previous result u'' must be adjacent to v'. Since $[v]_x$ is distinct from $[u]_x$ there is a path, P, from v' to x which misses u'. Now we have a path $P' = u''v'.P$. We have $u'' \succ_x u'$ and $u' \succ_x v'$ so u' is not adjacent to u'' and v'. Thus entire P' misses u'. This violates $u'' \succ_x u'$. So $u'' \succ_x v'$ must be true. If we have a chain $u^{(k)} \succ_x u^{(k-1)} \ldots u'' \succ_x u'$ then iterative application of the above argument will imply that $u^{(k)} \succ_x v'$. From the definition of the x-classes we have the following observation.

Observation 4. *Let $[u]_x$ and $[v]_x$ be distinct classes such that $[u]_x \succ_x [v]_x$. Each vertex of $[v]_x$ is either adjacent to all vertices of $[u]_x$ or to none.*

Proposition 2. *The relation $\succ_x$ on the equivalence classes is a partial ordering.*

Proof. The reflexivity and anti-symmetry are due to the definitions of $\succ_x$ and $\sim_x$. Next we show the transitivity.

Let $[u]_x \succ_x [v]_x$ and $[v]_x \succ_x [w]_x$. Our goal is to show that $[u]_x \succ_x [w]_x$. If the classes $[u]_x, [v]_x$, and $[w]_x$ are not all distinct, then the claim is true from reflexivity and anti-symmetry. So assume that all three are distinct.

There are u' in $[u]_x$, v' and v'' in $[v]_x$ and $w' \in [w]_x$ such that $u' \succ_x v'$ and $v'' \succ_x w'$. From Observation 4 we also have $v' \succ w'$. Let P be an arbitrary path from u' to x. Since it is intercepted by v', there is a path P' from v' to x in which all vertices, except perhaps v', are from P. w' intercepts P' but it is not adjacent to v' so w' intercepts P. Since P was randomly chosen, w' intercepts all paths from u' to x.

Finally we prove that w' is not adjacent to u'. From our assumption that $[v]_x$ is distinct from $[w]_x$ it follows from anti-symmetry that there is a path P'' from w' to x missing v'. If u' is adjacent to w', then we have a path from u' to w' then follow the path P'' to x. As v' is not adjacent to either u' or w' so this path is not intercepted by v'. This contradicts the fact that $u' \succ_x v'$.

Let x be an arbitrary vertex of G and G' be the induced subgraph on vertex subset V'. Any vertex y of V' will be called x-*minimal* in G' if $[y]_x$ is minimal among all the x-classes which have non-empty intersection with V'.

4 Domination Search on Asteroidal-Triple-Free Graphs

In this section we present a domination-search algorithm for AT-free graphs. We begin with some useful properties of this family.

Lemma 1. *Let $G = (V, E)$ be a connected AT-free graph and x a vertex in it. C is a connected component of the induced graph on $V - N[x]$. Then any x-minimal vertex in C intercepts all paths from any vertex in C to any vertex outside C.*

Proof. Let $y \in C$ is x-minimal. It is sufficient to show that any path from any vertex in C to any vertex in $N[x]$ passes through $N[y]$.

We will first show that any path from any vertex in C to x is intercepted by y. Let z be an arbitrary vertex in C. If $[y]_x$ and $[z]_x$ are related, then $[z]_x \succ_x [y]_x$ because $[y]_x$ is minimal by choice. Then by the definition, y is either adjacent to z or all paths from z to x pass through $N[y]$.

In the second case $[y]_x$ and $[z]_x$ are unrelated. So either (i) y and z are adjacent to each other or (ii) there exists a path from z to x not intercepted by y and a path from y to x not intercepted by z. Thus in case (ii) $\{x, y, z\}$ form an asteroidal triple. This is not possible in AT-free graphs so z must be adjacent to y. This establishes that all paths from z to x pass through $N[y]$.

Finally we consider arbitrary vertex w in $N[x]$. Consider arbitrary path, P, from z to w. If it passes through x, then we already have seen that it must pass through the neighborhood $N[y]$. Assume that P does not contain x. Extend the path to x: $P' = P.x$. It is a path from z to x. Thus P' passes through $N[y]$. y is outside $N[x]$ so x is not in $N[y]$. Therefore some vertex of P must be in $N[y]$.

Corollary 2. *Let $G = (V, E)$ be a connected AT-free graph and x a vertex in it. C is a connected component of the induced graph on $V - N[x]$. Let $y \in C$ is x-minimal in C. Then each connected component of the induced graph on $V - N[y]$ is entirely contained either in C or in $V - C$.*

It has been established that connected AT-free graphs have a pair of vertices, *poles*, such that every path between them dominates the entire graph, [9,10]. We shall use labels p_1 and p_2 for the poles.

Let $G = (V, E)$ be a graph and x be a vertex in it. If $y \in V - N[x]$, then the connected component containing y in the induced graph over $V - N[x]$ will be denoted by $C^x(y)$ and the open neighborhood $N(C^x(y))$ will be denoted by $S^x(y)$. $C^x(y)$ is defined if and only if y is not adjacent to x. A component $C^x(y)$ will be called *deep* if at least one vertex of the component is not adjacent to any vertex of $N[x]$. If a component is not deep, then it will be termed *shallow*. $C^x(p_1)$ and $C^x(p_2)$ will be called *principal* components of x, if defined. All other components of $V - N[x]$ will be termed *secondary*.

Lemma 2. *Let $G = (V, E)$ be a connected AT-free graph and $x \in V$. Then every deep component of the induced graph on $V - N[x]$ must contain exactly one pole.*

Proof. Suppose $C^x(y)$ contains no polar vertex. Assume that $z \in C^x(y)$ such that $N[z]$ is fully contained in $C^x(y)$. Then there exists a path between p_1 and p_2 which does not enter $C^x(y)$. This implies that the path will miss z which is impossible. Thus $C^x(y)$ must be a shallow component.

In case $C^x(y)$ contains both poles, then there exists a path between the poles which does not enter $N[x]$. This path will miss x. Again impossible for an AT-free graph.

Proposition 3. *Let $G = (V, E)$ be a connected AT-free graph with vertex x in it. Let C be a secondary component in the induced subgraph on $V - N[x]$. Then each vertex of C dominates at least one of p_1, p_2, $S^x(p_1)$, and $S^x(p_2)$.*

Proof. In a connected AT-free graph $G = (V, E)$, x is a vertex such that both its principal components are defined, i.e., neither pole is in $N[x]$. Let C be a secondary component of $V - N[x]$ and z be a vertex in C. Suppose there exists $u \in S^x(p_1)$ and $v \in S^x(p_2)$ such that z is adjacent to neither of these vertices. We can build a path from p_1 to p_2: $p_1 \ldots uxv \ldots p_2$ where u, x, and v are the only vertices of the path from $N[x]$. z cannot be adjacent to any vertex of this path other than u, x, and v. But by choice none of the three is adjacent to z so this path misses z. This is impossible.

If every vertex in a secondary component C dominates $S^x(p_2)$ (when p_2 is not adjacent to x) or dominates p_2 (when p_2 is adjacent to x) then C will be called a p_2-sided component.

Proposition 4. *Let $G = (V, E)$ be a connected AT-free graph with non-adjacent poles. Let x be either p_1 or a vertex for which both principal components are defined. Let $y \in C^x(p_2)$ be an x-minimal vertex. If z is a vertex of $C^x(p_2)$ which dominates $S^y(x)$, then z is also x-minimal.*

Proof. $S^y(x)$ is a graph separator which contains at least one vertex of each edge connecting $C^x(p_2)$ with $V - C^x(p_2)$ because all paths between the two pass through $N[y]$. Therefore each component of the induced graph on $V - S^y(x)$ is either completely contained in $C^x(p_2)$ or in $V - C^x(p_2)$. Thus every path from $C^x(p_2)$ to $V - C^x(p_2)$ must touch $S^y(x)$. If $N[z]$ contains $S^y(x)$ then all such path also touch $N[z]$. Therefore z also x-minimal.

Let x and y be vertices in a graph. Then by $|C^x(y)|$ we denote the cardinality of $C^x(y)$ if y is not adjacent to x. If the two vertices are adjacent, then $|C^x(y)|$ is defined to be zero.

Lemma 3. *Let G be a connected AT-free graph and x a vertex which is not adjacent to p_2. Further x is either p_1 or not adjacent to p_1. Let y be x-minimal in $C^x(p_2)$ but different from p_2, $|C^y(p_2)| \geq |C^{y'}(p_2)|$ for all x-minimal y', and p_2 does not dominate $S^y(p_1)$. Then any secondary component of $V - N[y]$ having non-empty intersection with $C^x(p_2)$, is p_2-sided.*

Proof. We consider two cases: $p_2 \notin N[y]$ and $p_2 \in N(y)$ since $y \neq p_2$.

(a) $p_2 \notin N[y]$. Suppose C is a secondary component of induced graph on $V - N[y]$ such that $C \cap C^x(p_2)$ is non-empty. Assume that C is not p_2-sided. Therefore there is a vertex z in C such that it does not dominate $S^y(p_2)$. From Corollary 2 we know that entire C is contained in $C^x(p_2)$ so z belongs to $C^x(p_2)$. From Proposition 3 z dominates $S^y(p_1)$.

Next we show that z does not dominate $S^y(x)$. Assume the contrary. From Proposition 4 it is an x-minimal vertex of $C^x(p_2)$. Since $C^y(p_2) \cap N[z]$ is empty, $C^z(p_2)$ will contain $C^y(p_2)$. In addition, by choice, z does not dominate $S^y(p_2)$ so there is a path from y to p_2 not intercepted by z. Thus y is also contained in $C^z(p_2)$. This implies that $|C^z(p_2)| > |C^y(p_2)|$. But Due to the choice of y, $|C^z(p_2)|$ can never be larger than $|C^y(p_2)|$.

Now we will show that $\{x, z, p_2\}$ is an asteroidal triple. Since p_2 and z belong to $C^x(p_2)$ so there is a path between z and p_2 not passing through $N[x]$.

To show that there is path between z and x which misses p_2 observe that p_2 does not dominate $S^y(p_1)$ so there exists a vertex u in $S^y(p_1)$ which is not adjacent to p_2 but adjacent to z since the latter dominates $S^y(p_1)$. Consider two cases. In the first case $x \in C^y(p_1)$. Consider the path $zu.P$ where P joins u to x and is confined to $C^y(p_1)$. This path misses p_2. In case $x \notin C^y(p_1)$ $N[x]$ contains $S^y(p_1)$ since $N[x]$ separates p_1 from y. Thus x is adjacent to u and xuz is a path that misses p_2.

Finally it needs to be shown that there is a path between x and p_2 not intercepted by z. We have seen that z does not dominate $S^y(x)$ so there is a vertex v in it which is not adjacent to z. Also there is a vertex w in $S^y(p_2)$ not adjacent to z since by choice z does not dominate $S^y(p_2)$. So there is a path $xvyw.P$ where P is a path from w to p_2 contained in $C^y(p_2)$. This path is not intercepted by z. Consequently the entire path, from x to p_2 misses z. Thus $\{x, z, p_2\}$ form an asteroidal set which is not possible.

(b) $p \in N(y)$. Again C is a secondary component of y such that $C \cap C^x(p_2)$ is non-empty. Assume C is not contained in $N[p_2]$. Therefore there is a vertex z in C such that it is not adjacent to p_2. From Corollary 2 we know that entire C is contained in $C^x(p_2)$ so z belongs to $C^x(p_2)$. From Proposition 3 z dominates $S^y(p_1)$.

We will again show that z does not dominate $S^y(x)$. Assume the contrary. From Proposition 4 it is an x-minimal vertex of $C^x(p_2)$. $|C^y(p_2)| = 0$ but $C^z(p_2)$ contains at least p_2 so again $|C^z(p_2)| > |C^y(p_2)|$. But Due to the choice of y, $|C^z(p_2)|$ can never be larger than $|C^y(p_2)|$.

Similar to the proof of part (a) we can show that $\{x, z, p_2\}$ is an asteroidal triple.

Lemma 4. *$G = (V, E)$ is a connected AT-free graph and y is a vertex in it which is not adjacent to pole p_1. Pole p_2 dominates $S^y(p_1)$. Then $\{u, p_2\}$ dominates $V - C^y(p_1)$ where u is any vertex in $C^y(p_1)$.*

Proof. Consider a path $p_2 u.P$ where P is a path to p_1 confined to $C^y(p_1)$. Each vertex of V is dominated by this path. Since vertices of $V - (C^y(p_1) \cup S^y(p_1))$ are not adjacent to any vertex beyond u, $\{u, p_2\}$ dominate them. Further vertices of $S^y(p_1)$ are in the neighborhood of p_2. So $\{u, p_2\}$ dominate $V - C^y(p_1)$.

Algorithm: *Domination search on a connected AT-free graph.*

1. If the poles are adjacent (so $\{p_1, p_2\}$ is a dominating set) then put the two guards at the poles and exit;
2. Place a guard at p_1;
3. Place the second guard at any vertex in $S^{p_1}(p_2)$;
4. Relieve the second guard;
           C1: Vertices of $V - C^{p_1}(p_2)$ are cleared
5. $x = p_1$;
6. While (p_2 is not adjacent to x) Do
7.   { Let vertex u in $C^x(p_2)$ is x-minimal with maximum $|C^u(p_2)|$;
8. $y = u$;
9.   If p_2 dominates $S^y(p_1)$ then
10.   { Place the free guard at p_2 and relieve the guard at x;
            C2: $C^y(p_1)$ being a subset of $V - C^x(p_2)$ remains clear and $N[p_2]$ is also now cleared.
11.   Place the free guard at any vertex in $S^y(p_1)$;
            C3: Entire V is clear.
12.   Exit;
          }
13.   Else
            C4: p_2 does not dominate $S^y(p_1)$ so $y \neq p_2$.
14.   { Place the free guard at y;
15.   Relieve the guard from x;
            C5: All the vertices of $V - C^x(p_2)$ remain clear. In addition $N[y]$ is also cleared.
16.   if p_2 is not adjacent to y
17.   { Place the free guard at any vertex of $S^y(p_2)$ and relieve it;}
18.   else { Place the free guard at p_2 }
            C6: if p_2 is not adjacent to y, then $V - C^y(p_2)$ is clear else entire V is cleared.
19. $x = y$;
        }
            C7: If x is not adjacent to p_2 then vertices of $V - C^x(p_2)$ are cleared else all of V is cleared.
      }
            C8: Entire V is cleared.
20. Exit.

Theorem 5. *The domination-search number of AT-free graphs is at most 2.*

Proof. The algorithm described above performs domination search for any AT-free graph with 2 guards. We prove the correctness of the algorithm by justifying the invariants mentioned in the comments.

C1: In line-2 $N[p_1]$ is cleared. From Proposition 3, line-3 clears all secondary components of p_1. No recontamination of these components occur in line-4 since the first guard is still present at p_1.

C2: From Corollary 2 $C^y(p_1)$ is either entirely contained in $C^x(p_2)$ or in $V - C^x(p_2)$. Due to Lemma 2 p_1 cannot be in $C^x(p_2)$ so $C^y(p_1)$ must be contained in $V - C^x(p_2)$. There is a guard at p_2 and p_2 dominates $S^y(p_1)$ so $C^y(p_1)$ remains clear.

C3: Due to Lemma 4.

C4: Self explanatory.

C5: Due to Corollary 2.

C6: Due to Lemma 3.

C7: Trivial.

C8: Trivial.

The algorithm is monotonic (there is no recontamination and at least one more vertex is cleared in each pass of the loop), due to C5, as long as the condition of line-9 is not true. When the condition is true the algorithm terminates after executing lines 10, 11, and 12. Therefore the algorithm always terminates.

References

1. Fomin, F.V., Kratsch, D., Muller, H.: On the domination search number. Discrete Applied Mathematics **127** (2003) 565–580
2. Bienstock, D.: Graph searching, path-width, tree-width and related problems (a survey). Discrete Mathematics and Theoretical Computer Science **5** (1991) 33–49
3. Suzuki, I., Yamashita, M.: Searching for a mobile intruder in a polygonal region. SIAM Journal of Computing **21** (1992) 863–888
4. Guibas, L., Latombe, J., Lavalle, S., Lin, D., Motwani, R.: A visibility based pursuit evasion problem. International Journal of Computational Geometry and Applications **9** (1999) 471–493
5. Lavalle, S., B.H.Simov, Slutzki, G.: An algorithm for searching a polygonal region with flash light. In: Proceedings of the 16th annual symposium on computational geometry. (2000) 260–269
6. Crass, D., Suzuki, I., Yamashita, M.: Searching for a mobile intruder in a corridor-the open edge variant of the polygon search problem. International Journal of Computational Geometry and Applications **5** (1995) 397–412
7. J.H.Lee, Park, S., Chwa, K.: Searching a polygonal room with one door by a 1-searcher. International Journal of Computational Geometry and Applications **10** (2000) 201–220
8. Kloks, T., Kratsch, D., Muller, H.: On the structure of graphs with bounded asteroidal number. Graphs and Combinatorics **17** (2001) 295–306
9. D.G.Corneil, Olariu, S., Stewart, L.: Asteroidal triple-free graphs. SIAM Journal of Discrete Mathematics **10** (1997) 399–430
10. D.G.Corneil, Olariu, S., Stewart, L.: Linear time algorithms for dominating pairs in asteroidal triple-free graphs. SIAM Journal of Computing **28** (1999) 1284–1297
11. Kloks, T., Kratsch, D., Muller, H.: Approximating the bandwidth of asteroidal triple-free graphs. Journal of Algorithms **32** (1999) 41–57
12. Brandstadt, A., Le, V., Spinrad, J. SIAM Monograph on Discrete Mathematics and Applications. Society for Industrial and Applied Mathematics (1999)
13. Deogun, J.S., Kratsch, D.: Dominating pair graphs. SIAM Journal of Discrete Mathematics **15** (2002) 353–366
14. Przulj, N., Corneil, D.G., Kohler, E.: Hereditary dominating pair graphs. Discrete Applied Mathematics **134** (2004) 239–261

Fully Dynamic Algorithm for Recognition and Modular Decomposition of Permutation Graphs

Christophe Crespelle and Christophe Paul

CNRS - *Département Informatique*, LIRMM, Montpellier
{crespell, paul}@lirmm.fr

Abstract. This paper considers the problem of maintaining a compact representation ($O(n)$ space) of permutation graphs under vertex and edge modifications (insertion or deletion). That representation allows us to answer adjacency queries in $O(1)$ time. The approach is based on a fully dynamic modular decomposition algorithm for permutation graphs that works in $O(n)$ time per edge and vertex modification. We thereby obtain a fully dynamic algorithm for the recognition of permutation graphs.

1 Introduction

The *dynamic recognition and representation problem* (see e.g. [10]) for a family $\mathcal{F}$ of graphs aims to maintain a characteristic representation of dynamically changing graphs as long as the modified graph belongs to $\mathcal{F}$. The input of the problem is a graph $G \in \mathcal{F}$ with its representation and a series of modifications. Any modification is of the following: inserting or deleting a vertex (along with the edges incident to it), inserting or deleting an edge. Several authors have considered the dynamic recognition and representation problem for various graph families. [8] devised a fully dynamic recognition algorithm for chordal graphs which handles edge operations in $O(n)$ time. For proper interval graphs [7], each update can be supported in $O(d + \log n)$ time where d is the number of edges involved in the operation. Cographs, a subfamily of permutation graphs, have been considered in [10] where any modification (edge or vertex) is supported in $O(d)$ time, where d is the number of edges involved in the modification. This latter result has recently been generalised to directed cographs in [3].

This paper deals with the family of permutation graphs. Our algorithm maintains an $O(n)$ space canonical representation based on modular decomposition which enables us to answer adjacency queries in $O(1)$ time. It should be noted that in [9] a purely incremental algorithm is presented for computing the modular decomposition tree of any graph. It runs in $O(n)$ time per vertex insertion. Unfortunately, it is based on a partial representation of the graph compromising the possibility of any vertex deletion. Therefore such an algorithm cannot be applied for efficient fully dynamic recognition of permutation graphs. Our algorithm also performs in $O(n)$ time per operation, but supports insertion as well as deletion of vertices and edges. Let us note that a modification of the input

D. Kratsch (Ed.): WG 2005, LNCS 3787, pp. 38–48, 2005.

graph may lead to $O(n)$ changes in the modular decomposition tree. Therefore our algorithm does not present any complexity extra cost in the maintain of the modular decomposition tree.

2 Preliminaries

2.1 Modular Decomposition

Theory of modular decomposition of graphs has been widely developed since Gallai first introduced it in [5]. Here, we give some known definitions and results that we use in the following. Let $G = (V, E)$ be a graph. The neighbourhood of a vertex $x \in V$ is denoted $N(x)$ and its non-neighbourhood $\overline{N}(x)$. A subset $S \subsetneq V$ of vertices is *uniform* w.r.t. to vertex $x \in V \setminus S$ if $S \subseteq N(x)$ or $S \subseteq \overline{N}(x)$ (otherwise S is *mixed*). A *module* of a graph $G = (V, E)$ is a subset of vertices $M \subseteq V$ which is uniform w.r.t. any vertex $x \in V \setminus M$. It also follows from definition that V and $\{x\}, x \in V$ are modules of G, namely the *trivial modules*. A graph is *prime* if all its modules are trivial. A module M is *strong* if it does not overlap any module M', that is $M \cap M' = \emptyset$ or $M \subseteq M'$ or $M' \subseteq M$. Therefore, the inclusion order of the strong modules of a graph defines a tree, called the *modular decomposition tree* T. The leaves of T correspond to the singleton vertex sets of G (L_x stands for $\{x\}$) and its root is the whole vertex set of G. In the following, a node p of the modular decomposition tree could be identified with the strong module $P = V(p)$ it represents. Denoting T_p the subtree of T rooted at p, P is the set of leaves of T_p. $\mathcal{C}(p)$ is the set of children in T of p.

Thanks to the well-known modular decomposition theorem (see [1] for references), any non-leaf node p of the modular decomposition tree is labelled as follows: *parallel* if $G[P]$ is not connected; *series* if $\overline{G}[P]$ is not connected; and *prime* otherwise (the three cases are disjoint). The label of node p is denoted $label(p)$. The series and parallel nodes are also called *degenerate* nodes. We call *maximal strong modules* of a graph $G = (V, E)$ the strong modules of G maximal wrt. inclusion and distinct from V. It is well known that the children $p_1 \ldots p_k$ of p (i.e. the maximal strong modules of $G[P]$) are respectively in the parallel case, the connected components of $G[P]$, in the series case the co-connected components of G (i.e. the connected components of $\overline{G}[P]$) and in the prime case, the maximal modules of $G[P]$ distinct from P. Given a graph G, we denote $\mathcal{MSM}(G)$ the set of maximal strong modules of G.

Given a set $\mathcal{F}$ of disjoint modules, let $F \subseteq V$ be a set of vertices such that for any $M \in \mathcal{F}, |F \cap M| = 1$. The *quotient graph* $G/\mathcal{F}$ is the subgraph induced by the vertices of $(V \setminus \cup_{M \in \mathcal{F}} M) \cup F$. From the modular decomposition theorem, the quotient $G/\mathcal{MSM}(G)$ of G by the set of its maximal strong modules is either a stable (parallel case) or a clique (series case) or a prime graph. If with each prime node p of the modular decomposition tree T, we associate a representation of the quotient $G[P]/\mathcal{MSM}(G[P])$, then adjacency queries between any pair of vertices x, y can be answered by a search in T and in the quotient graphs.

2.2 Permutation Graphs

If π is a linear order on the vertices, $\pi(x)$ denotes the rank of vertex x in π while $\pi^{-1}(i)$ is the vertex at rank i. Permutation graphs are those graphs for which there exists a pair (π_1, π_2) of linear order on the vertex set such that x and y are adjacent iff $\pi_1(x) < \pi_1(y)$ and $\pi_2(y) < \pi_2(x)$. For a graph G, such a pair $\mathcal{R} = (\pi_1, \pi_2)$ is a *realiser* of G. If $\overline{\pi}_2$ denotes the reverse order of π_1, then $\overline{\mathcal{R}} = (\pi_1, \overline{\pi}_2)$ is a realiser of $\overline{G}$. For a complete introduction to permutation graphs, one can see [6]. It is known that, if G is a prime graph, then its realiser is unique up to reversal and exchange[1] (the reader should refer to [1] for more details on permutation graphs). Moreover, a graph G is a permutation graph iff the quotient graphs associated with the prime nodes of its modular decomposition tree are permutation graphs. It follows that associating the modular decomposition tree T with the realiser of each of its prime nodes provides an $O(n)$ space canonical representation of a permutation graph G, called hereafter the *full modular representation* of G.

Since the full modular representation contains a realiser for each prime node of T, it is well known that a realiser of the whole graph G can be retrieved in $O(n)$ time by a simple search of T. As our dynamic algorithm works in $O(n)$ time per operation, a realiser of G can be maintained without any extra cost. That guarantees the possibility of answering at any time adjacency queries in $O(1)$ time.

An *interval* of a linear order π on V is a set of consecutive elements of V in π. Given a pair (π_1, π_2) of linear orders, a *common interval* is a set I that is an interval of π_1 and of π_2. Recently, [11] proposed an $O(n + K)$ algorithm computing all common intervals of a pair of linear orders, K being the number of common intervals. A common interval is *strong* if it does not overlap any other common interval. Clearly common intervals of a realiser $\mathcal{R} = (\pi_1, \pi_2)$ of a permutation graph G are modules of G. The converse is false, but:

Proposition 1. *[4] The strong modules of a permutation graph $G = (V, E)$ are exactly the strong common intervals of any of its realiser $\mathcal{R}$.*

2.3 Dynamic Arc Operations

Unfortunately an edge modification may imply $O(n)$ changes in the modular decomposition tree. As we propose an $O(n)$ time algorithm for the vertex insertion and for the vertex deletion operations, inserting or deleting an edge e incident to vertex x will be handled by first removing x and then re-inserting x with the updated neighbourhood.

3 Vertex Deletion

Let $G' = G - x$ be the graph resulting from the deletion of a vertex x in the graph G. Since the family of permutation graphs is hereditary, removing x reduces to

---

[1] That is (π_2, π_1), $(\overline{\pi}_1, \overline{\pi}_2)$ and $(\overline{\pi}_2, \overline{\pi}_1)$ are considered as the same realiser as (π_1, π_2).

compute the full modular representation of G' from the one of G. We shall distinguish the case where p, the parent node of x in T, is a prime node from the case where p is a degenerate node.

Degenerate Case. This is the easy case to handle. If x has at least two siblings, then the leaf L_x is removed from T. Assume x has only one sibling say q_2. If q_2 is a leaf, L_x and p are removed from T and q_2 becomes a child of q_1 replacing p (i.e. if q_1 is a prime node, then q_2 takes the place of p in the associated realiser). Assume q_2 is a non-leaf node. If q_1 and q_2 are both series nodes or both parallel nodes, then L_x, p and q_2 are removed from T and the children of q_2 are made children of q_1. Otherwise L_x and p are removed from T and q_2 becomes a child of q_1 replacing p. Such an update of the full modular representation can be done in $O(|\mathcal{C}(q_2)|) = O(n)$ since it leaves unchanged the quotient graphs of the prime nodes.

Prime Case. The removal of x may create some modules in $G'[P']$ (where $P' = P \setminus \{x\}$). We show it can be tested in $O(n)$ time. Moreover if $G'[P']$ is not a prime graph, the updated full modular representation can be computed within the same complexity.

Lemma 1. *Let $G = (V, E)$ be a prime permutation graph and x be a vertex. The non trivial strong modules of $G' = G - x$ can be partitioned in two families (possibly empty) totally ordered by inclusion.*

This is a consequence of Proposition 1 which implies that if $\mathcal{R} = (\pi_1, \pi_2)$ is the realiser of G, then for any strong module M' of G', $I = M' \cup \{x\}$ is an interval of π_1 or π_2. Therefore G' contains $O(n)$ strong modules. Moreover, as there is at most two non-trivial maximal strong modules, the root of the modular decomposition tree T' of G' has at most two non-leaf children, and each internal nodes of T' have at most one non-leaf child. The next lemma complete the information about degenerate internal nodes of T'.

Lemma 2. *Let $G = (V, E)$ be a prime permutation graph and x be a vertex. Every degenerate node of the modular decomposition tree of $G' = G - x$ has at most two children which are leaves.*

It follows that the number of modules (not necessarily strong) of G' is $O(n)$ so there is also $O(n)$ common intervals of the realiser $\mathcal{R}'$ of G'. Therefore applying [11]'s algorithm will cost $O(n)$ time to find the common intervals of $\mathcal{R}'$. From that algorithm the two families of strong common intervals (or equivalently modules) can be retrieved in $O(n)$ time. Moreover from Lemma 2 given a common interval it is possible to find its label (series, parallel or prime) in $O(1)$ time. As the realiser of each prime node of T' can be easily extracted from $\mathcal{R}$, the full modular representation of G' can be computed in $O(n)$.

Theorem 1. *Updating the full modular representation of a permutation graph under vertex deletion costs $O(n)$ time.*

4   Vertex Insertion

Given a graph $G = (V, E)$, a vertex $x \notin V$ and a subset $N(x) \subseteq V$, let us define $G' = G + x$ as the graph on vertex set $V \cup \{x\}$ with edge set $E \cup \{\{x, y\} \mid y \in N(x)\}$). Each node p of the modular decomposition tree T of G is assigned a type w.r.t. x : *linked* (resp. *notlinked*) if $P = V(p)$ is uniform w.r.t. x and $P \subseteq N(x)$ (resp. $P \subseteq \overline{N}(x)$), and *mixed* otherwise. $\mathcal{C}_l(p)$ (resp. $\mathcal{C}_{nl}(p)$) stands for the set of children of p which are typed *linked* (resp. *notlinked*) and $\mathcal{C}_m(p)$ for the set of children of p which are typed *mixed*. For $t \in \{m, l, nl\}$, we denote $F_t(p) = \bigcup_{f \in \mathcal{C}_t(p)} V(f)$.

4.1   Modular Decomposition Tree of $G + x$

Insertion Node. To compute the modular decomposition tree T' of $G' = G + x$, we can restrict our attention to a subtree T_q of T rooted at a certain node q, called the *insertion node*. q is such that T_q contains all the modifications implied by the insertion of x. Moreover, in T', x will be inserted as a child or a grand-child of node q' representing set $Q' = Q \cup \{x\}$. The discussion bellow gives the definition of q and shows that inserting x in G reduces to insert x in $G[Q]$.

Definition 1. *A node p of T is a* proper *node iff either p is uniform wrt. x, or p is a mixed node with a unique mixed child f such that $F \cup \{x\}$ is a module of $G'[P \cup \{x\}]$. Otherwise p is a* non-proper *node.*

From Definition 1, any mixed node p has at least one non-proper descendant. Indeed p always enjoys a mixed descendant having only uniform children. It follows that if any node of T is proper, then the vertex set is uniform w.r.t. x. That is x is either a universal vertex or an isolated vertex. Therefore inserting x preserves the property of being a permutation graph and the full modular representation is easy to update. That case will not be considered anymore in the following.

Definition 2. *The insertion node q is the lca of non-proper nodes of T.*

Lemma 3. *The insertion node q is such that $Q' = Q \cup \{x\}$ is a strong module of $G' = G + x$.*

Since Q is a strong module of G and $Q' = Q \cup \{x\}$ is a strong module of $G' = G + x$, then $G'/\{Q'\} = G/\{Q\}$. That is the changes implied by the insertion of x are located in T_q. Moreover, the permutation graphs family is hereditary and closed under substitution, it follows that:

Lemma 4. $G' = G + x$ *is a permutation graph iff $G'[Q'] = G[Q] + x$ is a permutation graph.*

From Lemma 4 and the discussion above, we conclude that inserting x in G reduces to insert x in $G[Q]$.

Modular Decomposition Tree of $G'[Q']$. As the insertion node q is non-proper, it can either be: 1) a degenerate node with no mixed child but with uniform children of both types (i.e. *linked* and *notlinked*); or 2) a degenerate node with at least one mixed child; or 3) a prime node with no mixed child but a child being a twin of x in the quotient of q; or 4) a prime node with no child being a twin of x in the quotient of q. In cases 1) and 3), q is said to be *cut* (and *uncut* in cases 2) and 4)).

The case where the insertion node is a cut degenerate node (case 1) above) is similar to the case, considered by [2], of maintaining the modular decomposition tree of a cograph under vertex insertion. If q is a series (resp. parallel) node, the root q' of $T'_{q'}$ is a series (resp. parallel) node. The children of q' are those children of q typed *linked* (resp. *notlinked*) and a new parallel node q'_1. The children of q'_1 are $\{x\}$ and the remaining children of q, i.e. those typed *notlinked* (resp. *linked*).

The case where the insertion node is a cut prime node (case 3) above) is quite easy to deal with. In the children of q, the twin f of x is replaced by a new degenerate node q_1 (i.e. q_1 takes the place of f in the realiser of q). The label of q_1 is series if f is typed linked, and parallel if f is typed *notlinked*. x and f are made children of q_1.

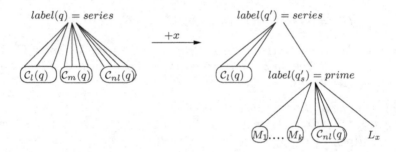

Fig. 1. Updating the modular decomposition tree when the insertion node is a series node. The modules $M_1 \ldots M_k$ are the maximal uniform modules of $G[Q_s]$.

Let us now consider the case where the insertion node q is uncut. Let us define the vertex set Q_s as the set Q if q is a prime node and as the set $F_m(q) \cup F_{nl}(q)$ (resp. $F_m(q) \cup F_l(q)$) if q a series node (resp. parallel node). The modular decomposition tree $T'_{q'}$ of $G'[Q']$ is organised as follows. If q is a prime node, then q' represents the nodes of $Q'_s = Q_s \cup \{x\}$. If q is degenerate, then q' is degenerate and has the same label than q. If q is a series (resp. parallel) node, then the set of children of q' is $\{q'_s\} \cup C_l(q)$ (resp. $\{q'_s\} \cup C_{nl}(q)$) where q'_s is a new node representing vertices of Q'_s. Theorem 2 states on the modular decomposition of $G'[Q'_s]$.

Theorem 2. *Let x be a vertex to be inserted in a graph G. If the insertion node q of the modular decomposition tree T of G is uncut, then $G'[Q'_s]$ is connected and co-connected. And the maximal strong modules of $G'[Q'_s]$ are $\{x\}$ and the maximal uniform (w.r.t. x) modules of $G[Q_s]$.*

Notice that the modular decomposition tree of $G'[M]$, where M is a maximal uniform module of $G[Q_s]$, is the part of T restricted to M. Therefore the whole modular decomposition tree T' of G' follows from discussion above.

4.2   Dynamic Characterisation of Permutation Graphs

As we ask G' to be a permutation graph, the mixed nodes of T_q cannot be spread anywhere in the tree. Lemma 5 claims that there are at most two branches of mixed nodes in T_q beginning at q. These two branches correspond to the two families of Lemma 1.

Lemma 5. *If G' is a permutation graph then the insertion node q has at most two mixed children and any node $p \neq q$ of T_q has at most one mixed child.*

Unfortunately, Lemma 5 is not a sufficient condition for G' being a permutation graph. Theorem 3 gives necessary and sufficient conditions. Given a graph $G = (V, E)$, $S \subsetneq V$ and $y \in V \setminus S$, we denote $G - yS = (V,\ E \setminus \{\{y, z\} \mid z \in S\})$. If p is a node of T_q, then set $P' = P \cup \{x\}$. Since the maximal strong modules of $G[P]$ are uniform wrt. x in $G'[P'] - xF_m(p)$, they are modules of $G'[P'] - xF_m(p)$. We denote

$$\widetilde{G'}_p = (G'[P'] - xF_m(p))/(\mathcal{MSM}(G[P]) \cup \{\{x\}\}).$$

Theorem 3. *Let x be a vertex to be inserted in a permutation graph G. Then $G' = G + x$ is a permutation graph iff either the insertion node q of the modular decomposition tree T of G is cut; or if q is uncut then:*

1. *q satisfies one of the following conditions :*
   (a) q has two mixed children f_1 and f_2, and $\widetilde{G'}_q$ is a permutation graph admitting a realiser $\mathcal{R} = (\pi_1, \pi_2)$ such that x and f_1 are consecutive in π_1, and x and f_2 are consecutive in π_2.
   (b) q has a unique mixed child f_1, and $\widetilde{G'}_q$ is a permutation graph admitting a realiser $\mathcal{R} = (\pi_1, \pi_2)$ such that x and f_1 are consecutive in π_1.
   (c) q has no mixed child and $\widetilde{G'}_q = G'[P']/(\mathcal{MSM}(G[P]) \cup \{\{x\}\})$ is a permutation graph.
2. *and any node $p \neq q$ of T_q satisfies one of the two following conditions :*
   (a) p has a unique mixed child f_1, and $\widetilde{G'}_p$ is a permutation graph admitting a realiser $\mathcal{R} = (\pi_1, \pi_2)$ such that x and f_1 are consecutive in π_1, and x is the first element of π_2.
   (b) p has no mixed child, and $\widetilde{G'}_p$ is a permutation graph admitting a realiser $\mathcal{R} = (\pi_1, \pi_2)$ such that x is the first element of π_2.

Due to space limitation, we only prove that the above conditions are sufficient.

Proof. $\Leftarrow$: We first show by induction that any node p of T_q different from q is such that $G'[P']$ is a permutation graph admitting a realiser $\mathcal{R}$ such that x is the first element of π_2. If p is a leaf of T_q, it trivially satisfies the inductive hypothesis. Let $p \neq q$ be a node of T_q such that its children satisfy the inductive hypothesis. If p has a unique mixed child f_1, it satisfies condition 2a of Theorem 3. According to the inductive hypothesis, $G'[F_1']$ is a permutation graph and admits a realiser $\mathcal{R}_1 = (\tau_1, \tau_2)$ such that x is the first element of τ_2. To obtain a realiser of $G'[P']/(\mathcal{MSM}(G[P]) \setminus \{F_1\})$ such that x is the first element of π_2, the realiser $\mathcal{R} = (\pi_1, \pi_2)$ of $\widetilde{G'}_p$ is modified as follows: in π_1, substitute τ_1 for the interval $\{x, f_1\}$; and in π_2 substitute, τ_2 restricted to F_1 for f_1. Composing the resulting realiser with the realisers of the $(G[F])_{f \in \mathcal{C}(p) \setminus \{f_1\}}$, we obtain a realiser of $G'[P']$ which satisfies the inductive hypothesis. The case where p has no mixed child follows as a particular case of the previous one. This ends the induction.

If q has two mixed children f_1 and f_2, it satisfies condition 1a of Theorem 3. By the previous induction $G'[F_1']$ and $G'[F_2']$ are permutation graphs. They respectively admit a realiser $\mathcal{R}_1 = (\tau_1, \tau_2)$ and $\mathcal{R}_2 = (\sigma_1, \sigma_2)$ such that x is the first element of τ_2 and σ_2. In the realiser $\mathcal{R} = (\pi_1, \pi_2)$ of $\widetilde{G'}_q$, if f_2 occurs after f_1 in π_2, we reverse both orders of $\mathcal{R}_1$, and if f_2 occurs before f_1 in π_1, we reverse both orders of $\mathcal{R}_2$. To obtain a realiser of $G'[Q']/(\mathcal{MSM}(G[Q]) \setminus \{F_1, F_2\})$, $\mathcal{R}$ is modified as follows: in π_1, substitute τ_1 for the interval $\{x, f_1\}$, and σ_2 restricted to F_2 for f_2; and in π_2, substitute σ_1 for the interval $\{x, f_2\}$, and τ_2 restricted to F_1 for f_1. Composing the resulting realiser with the realisers of the $(G[V(f)])_{f \in \mathcal{C}(p) \setminus f_1, f_2}$, we obtain a realiser of $G'[Q']$. We therefore prove that $G'[Q']$ is a permutation graph. By Lemma 4 we can conclude that G is a permutation graph. The cases where p has a single or no mixed child follow as a particular cases of the above discussion.   □

4.3 Algorithm and Complexity

Data-Structure. The realiser $\mathcal{R} = (\pi_1, \pi_2)$ associated with a prime node p of the modular decomposition tree will be stored in two doubly linked lists representing the two linear orders π_1 and π_2. Each cell of a list represents a child c of p. There are two symmetric pointers between c and the cell. Moreover each cell contains its rank in the list (namely $\pi_1(c)$ or $\pi_2(c)$).

Routine *InsPrime*. As a prime permutation graph G has a unique realiser $\mathcal{R} = (\pi_1, \pi_2)$, $G + x$ is a permutation graph iff x can be inserted in $\mathcal{R}$. Routine *InsPrime* performs, if possible, that insertion.

Lemma 6. *Let $\mathcal{R} = (\pi_1, \pi_2)$ be the realiser of a prime permutation graph G and $x \notin V$ a vertex to be inserted. $G + x$ is a permutation graph iff $N(x)$ and $\overline{N}(x)$ can be respectively partitioned into $N_1(x), N_2(x)$ and $\overline{N}_1(x), \overline{N}_2(x)$ such that:*

$$\forall u_1 \in N_1(x) \cup \overline{N}_1(x), v_1 \in N_2(x) \cup \overline{N}_2(x), u_1 <_{\pi_1} v_1$$

$$\forall u_2 \in N_2(x) \cup \overline{N}_1(x), v_2 \in N_1(x) \cup \overline{N}_2(x), u_2 <_{\pi_2} v_2$$

An *initial common interval* of a realiser $\mathcal{R} = (\pi_1, \pi_2)$ is a common interval of $\mathcal{R}$ containing both $\pi_1^{-1}(1)$ and $\pi_2^{-1}(1)$. In order to find the partitions of $N(x)$ and $\overline{N}(x)$ satisfying Lemma 6, Routine *InsPrime* makes use of the next corollary.

Corollary 1. *If $\overline{N}_1(x) \neq \varnothing$ (resp. $N_1(x) \neq \varnothing$) then $\overline{N}_1(x)$ is an initial common interval of $\mathcal{R}[\overline{N}(x)]$ (resp. $\mathcal{R}[N(x)]$), the restriction of $\mathcal{R}$ to $\overline{N}(x)$ (resp. $N(x)$).*

Notice that an initial common interval of $\mathcal{R}[\overline{N}(x)]$ defines a partition $\overline{N}_1(x)$, $\overline{N}_2(x)$ of $\overline{N}(x)$ (and similarly for $N(x)$). The number of initial common intervals of a realiser is $O(n)$.

Routine *InsPrime* computes in $O(n)$ time the sets of initial common intervals of $\mathcal{R}[\overline{N}(x)]$ and of $\overline{\mathcal{R}}[N(x)]$. Then, it checks if there exists a pair of partitions $N_1(x), N_2(x)$ and $\overline{N}_1(x), \overline{N}_2(x)$ satisfying Lemma 6. Testing a given pair of partitions can be done in $O(1)$ time by comparing the ranks of the last elements of N_1 (resp. N_2) and $\overline{N}_1$ in π_1 (resp. π_2) with ranks of the first elements of N_2 (resp. N_1) and $\overline{N}_2$. Scanning π_1, a pair of partitions satisfying the condition of Lemma 6 can be found in $O(n)$ time.

Notice that $G' = G + x$ may not be prime. If it is the case, then x has a twin vertex in G' (i.e. a vertex y s.t. $N(y) \setminus \{x\} = N(x) \setminus \{y\}$). As $\{x, y\}$ is therefore a strong module of G', by Proposition 1, x and y are consecutive in both linear orders of the realiser of G'. It follows that testing the existence of a twin can be done in $O(1)$ time if x has been inserted.

To summarise, if $G + x$ is a permutation graph, then in $O(n)$ time, Routine *InsPrime* returns a pair of doubly linked lists, the realiser of $G + x$, and outputs the twin of x if it exists. Notice that the ranks of the cells are not maintained in these lists.

The Typing Routine. In a bottom-up process, each node p of T receives a type (*linked*, *notlinked* or *mixed*). A leaf L_y of T is typed *linked* if $y \in N(x)$ and *notlinked* otherwise. The type of an inner node p of T depends on the types of its children. If the children of p all have the same type, p inherits that type, otherwise p is typed *mixed*. Since the number of nodes in T is $O(n)$, the typing routine runs in $O(n)$ time.

Finding the Insertion Node q. The purpose of this step is to find the insertion node q, in the case where the root r of T is typed *mixed*. By Lemma 2, q is the *lca* of the non-proper nodes of T. Any node p of the unique path between r and q is *mixed* and proper if $p \neq q$. Since, by Definition 1, any proper mixed node has a unique mixed child, finding the insertion node can be done by a top-down search of the modular decomposition tree T. The search stops when the current node p is non-proper, which can be tested as follows. If p is a series node (resp. parallel node), then p is proper iff all its children but one are typed *linked* (resp. *notlinked*) and the remaining child is *mixed*. If p is a prime node, p is proper iff x has a twin in the quotient of p, which can be checked by Routine *InsPrime*. In both cases, testing whether p is a proper node can be done in $O(|\mathcal{C}(p)|)$. As T contains $O(n)$ nodes, the search finds the insertion node q in $O(n)$ time.

Maintaining the Full Modular Representation. We now determine if $G'[Q']$ is a permutation graph or not, and in the positive, update its full modular representation (i.e. its modular decomposition tree and the realisers of the prime nodes).

If the insertion node q has more than two mixed children, from Lemma 5, $G'[Q']$ is not a permutation graph: the algorithm stops. If q is cut, from earlier discussion $G'[Q']$ is always a permutation graph (see Section 4.1). In that case, the realisers of the prime nodes are not modified. Therefore $T'_{q'}$ can be computed in $O(|\mathcal{C}(q)|)$ as described in Section 4.1. When q is uncut, the nodes of T_q have to fulfil the conditions stated in Theorem 3. To simplify the presentation, let us present our algorithm as three-step process. But notice in practice these three steps can be merged into a single one.

- For each node p of T_q, we check whether p fulfils the condition of Theorem 3. If p is a degenerate node having the right number of mixed children (0, 1 or 2 depending on $p = q$), then $\widetilde{G}'_p$ always enjoys a realiser satisfying Theorem 3 (see Figure 2). If p is prime, using Routine $InsPrime$, we insert x in the realiser associated to p by making x adjacent to $\mathcal{C}_l(p)$ and non-adjacent to $\mathcal{C}_m(p) \cup \mathcal{C}_{nl}(p)$. There may be two different positions to insert x (only if has a twin vertex). We then test if at least one of the possible positions fulfils the conditions of Theorem 3 which simply consists in testing the position of x in the realiser returned by $InsPrime$ (extremity in an order and/or consecutiveness with the mixed children). That can be done in $O(1)$. Since we handle only the quotients of the nodes p of T_q, each of which being processed in $O(|\mathcal{C}(p)|)$ time, this first steps runs in $O(n)$ time.
- Theorem 2 states that the maximal strong modules of $G'[Q'_s]$ are $\{x\}$ and the maximal uniform modules of $G[Q_s]$. These maximal uniform modules can be found in $O(n)$ time by a search in T_q since M is a maximal uniform module iff there exists a mixed node p descendant of the insertion node q such that either p is degenerate and $M = F_l(p)$ or $M = F_{nl}(p)$; or p is prime and M is the vertex set of some uniform child of p. By Theorem 2, these modules will be represented by the children nodes of the new prime node q'_s. Recall that the modular decomposition tree of $G'[M]$ is inherited from the modular decomposition tree T of G.
- The last step computes the realiser $\mathcal{R}_s$ of the quotient of $G'[Q'_s]$ by its maximal strong modules. Notice that in the intermediate realisers computed along the process, the ranks of the cells in the lists are not maintained.

  To that aim, we applied the bottom-up process, described in the proof of Theorem 3, on the modular decomposition tree T where each maximal uniform module has first been contracted into a single vertex (i.e. replaced

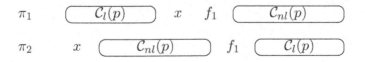

π_1 ($\mathcal{C}_l(p)$) x f_1 ($\mathcal{C}_{nl}(p)$)

π_2 x ($\mathcal{C}_{nl}(p)$) f_1 ($\mathcal{C}_l(p)$)

Fig. 2. The unique realiser of $\widetilde{G}'_p$ (if p is a series node) that fulfils condition 2a of Theorem 3. For a parallel node p, $\mathcal{C}_{nl}(p)$ and $\mathcal{C}_l(p)$ has to be exchanged in π_2.

by a leaf in the tree T). For a prime mixed node p, the realiser of $\widetilde{G}'_p$ is given
by Routine $InsPrime$. For a degenerate node p, the realiser of $\widetilde{G}'_p$ is the one
depicted in Figure 2. As the realisers are encoded by pairs of doubly linked
lists, the substitution operation used in the proof of Theorem 3 can be done
in $O(1)$ time. It follows that the realiser $\mathcal{R}_s$ can be computed in $O(n)$ time.
Finally to maintain the data-structure, a scan of the lists of $\mathcal{R}_s$ allows to get
the ranks of the cells.

Theorem 4. *Updating the full modular representation of a permutation graph
under vertex insertion costs $O(n)$ time.*

References

1. A. Brandstädt, V.B. Le, and J.P. Spinrad. *Graph Classes: a Survey*. SIAM Monographs on Discrete Mathematics and Applications. Society for Industrial and Applied Mathematics, 1999.
2. D.G. Corneil, Y. Perl, and L.K. Stewart. A linear time recognition algorithm for cographs. *SIAM Journal on Computing*, 14(4):926–934, 1985.
3. C. Crespelle and C. Paul. Fully-dynamic recognition algorithm and certificate for directed cographs. In *30th Int. Workshop on Graph Theoretical Concepts in Computer Science (WG04)*, number 3353 in Lecture Notes in Computer Science, pages 93–104, 2004.
4. F. de Montgolfier. *Décomposition modulaire des graphes - Théorie, extensions et algorithmes*. PhD thesis, Université de Montpellier 2, France, 2003.
5. Tibor Gallai. Transitiv orientierbare graphen. *Acta Math. Acad. Sci. Hungar.*, 18:25–66, 1967.
6. Martin Charles Golumbic. *Algorithmic Graph Theory and Perfect Graphs*. Academic Press, New York, 1980.
7. P. Hell, R. Shamir, and R. Sharan. A fully dynamic algorithm for recognizing and representing proper interval graphs. *SIAM Journal on Computing*, 31(1):289–305, 2002.
8. L. Ibarra. Fully dynamic algorithms for chordal graphs. In *10th ACM-SIAM Annual Symposium on Discrete Algorithm (SODA'03)*, pages 923–924, 1999.
9. J.H. Muller and J.P. Spinrad. Incremental modular decomposition algorithm. *Journal of the Association for Computing Machinery*, 36(1):1–19, 1989.
10. R. Shamir and R. Sharan. A fully dynamic algorithm for modular decomposition and recognition of cographs. *Discrete Applied Mathematics*, 136(2-3):329–340, 2004.
11. Takeaki Uno and Mutsunori Yagiura. Fast algorithms to enumerate all common intervals of two permutations. *Algorithmica*, 26(2):290–309, 2000.

Approximating Rank-Width
and Clique-Width Quickly

Sang-il Oum

Princeton University, Princeton NJ 08544, USA*

Abstract. Rank-width is defined by Seymour and the author to investigate clique-width; they show that graphs have bounded rank-width if and only if they have bounded clique-width. It is known that many hard graph problems have polynomial-time algorithms for graphs of bounded clique-width, however, requiring a given decomposition corresponding to clique-width (k-expression); they remove this requirement by constructing an algorithm that either outputs a rank-decomposition of width at most $f(k)$ for some function f or confirms rank-width is larger than k in $O(|V|^9 \log |V|)$ time for an input graph $G = (V, E)$ and a fixed k. This can be reformulated in terms of clique-width as an algorithm that either outputs a $(2^{1+f(k)} - 1)$-expression or confirms clique-width is larger than k in $O(|V|^9 \log |V|)$ time for fixed k.

In this paper, we develop two separate algorithms of this kind with faster running time. We construct a $O(|V|^4)$-time algorithm with $f(k) = 3k + 1$ by constructing a subroutine for the previous algorithm; we may now avoid using general submodular function minimization algorithms used by Seymour and the author. Another one is a $O(|V|^3)$-time algorithm with $f(k) = 24k$ by giving a reduction from graphs to binary matroids; then we use an approximation algorithm for matroid branch-width by Hliněný.

1   Preliminaries

In this paper, all graphs are simple, undirected, and finite.

Cut-Rank Functions. For a matrix $M = (m_{ij} : i \in R, j \in C)$ over a field F, if $X \subseteq R$ and $Y \subseteq C$, let $M[X, Y]$ denote the submatrix $(m_{ij} : i \in X, j \in Y)$. For a graph G, let $A(G)$ be its adjacency matrix over GF(2).

Definition 1. *Let G be a graph. For two disjoint subsets $X, Y \subseteq V(G)$, we define $\rho_G^*(X, Y) = \mathrm{rk}(A(G)[X, Y])$ where rk is the matrix rank function; and we define the* cut-rank *function ρ_G of G by letting $\rho_G(X) = \rho_G^*(X, V(G) \setminus X)$ for $X \subseteq V(G)$.*

Both ρ and ρ^* satisfy submodular inequalities.

---

* Current address. Georgia Institute of Technology, Atlanta, GA 30332, USA.

D. Kratsch (Ed.): WG 2005, LNCS 3787, pp. 49–58, 2005.

Proposition 2 (Oum and Seymour [1]). *Let G be a graph. Let X_1, Y_1, X_2, Y_2 be subsets of $V(G)$ such that $X_1 \cap Y_1 = X_2 \cap Y_2 = \emptyset$. Then,*

$$\rho_G^*(X_1, Y_1) + \rho_G^*(X_2, Y_2) \geq \rho_G^*(X_1 \cap X_2, Y_1 \cup Y_2) + \rho_G^*(X_1 \cup X_2, Y_1 \cap Y_2).$$

Moreover, if $X_1, X_2 \subseteq V(G)$, then

$$\rho_G(X_1) + \rho_G(X_2) \geq \rho_G(X_1 \cap X_2) + \rho_G(X_1 \cup X_2).$$

Rank-Width. A *subcubic tree* is a tree with at least two vertices such that every vertex is incident with at most three edges. A *leaf* of a tree is a vertex incident with exactly one edge. A *rank-decomposition* of a graph $G = (V, E)$ is a pair $(T, \mathcal{L})$ of a subcubic tree T and a bijective function $\mathcal{L} : V \to \{t : t$ is a leaf of $T\}$. (If $|V| \leq 1$, then G admits no rank-decomposition.)

For an edge e of T, the connected components of $T \setminus e$ induce a partition (X, Y) of the set of leaves of T. The *width* of an edge e of a rank-decomposition $(T, \mathcal{L})$ is $\rho_G(\mathcal{L}^{-1}(X))$. The *width* of $(T, \mathcal{L})$ is the maximum width of all edges of T. The *rank-width* $\mathrm{rwd}(G)$ of G is the minimum width of a rank-decomposition of G. (If $|V| \leq 1$, we define $\mathrm{rwd}(G) = 0$.)

Let $\mathrm{cwd}(G)$ be the *clique-width* of a graph G. Clique-width is defined by Courcelle and Olariu [2]. In this paper, we do not need its definition if we just remember the following proposition.

Proposition 3 (Oum and Seymour [1]). *For a graph G, $\mathrm{rwd}(G) \leq \mathrm{cwd}(G) \leq 2^{\mathrm{rwd}(G)+1} - 1$.*

Local Complementation. For two sets A and B, let $A \Delta B = (A \setminus B) \cup (B \setminus A)$.

Definition 4. *Let $G = (V, E)$ be a graph and $v \in V$. The graph obtained by applying local complementation at v to G is*

$$G * v = (V, E \Delta \{xy : xv, yv \in E, x \neq y\}).$$

For an edge $uv \in E$, the graph obtained by pivoting uv is defined by $G \wedge uv = G * u * v * u$. We say that H is locally equivalent *to G if G can be obtained by applying a sequence of local complementations to G.*

A pivoting is well-defined because $G * u * v * u = G * v * u * v$ if u and v are adjacent [3]. The following observation is fundamental.

Proposition 5 (Oum [3]). *Let $G' = G * v$. Then for every $X \subseteq V(G)$,*

$$\rho_G(X) = \rho_{G'}(X).$$

The following lemma will be used in Sect. 2.

Lemma 6 (Oum [3]). *Let G be a graph and $v \in V(G)$. Suppose that (X_1, X_2) and (Y_1, Y_2) are partitions of $V(G) \setminus \{v\}$. If w is a neighbor of v, then*

$$\rho_{G \setminus v}(X_1) + \rho_{G \wedge vw \setminus v}(Y_1) \geq \rho_G(X_1 \cap Y_1) + \rho_G(X_2 \cap Y_2) - 1.$$

Matroids. Since we will use matroids in Sect. 4, let us review matroid theory. For general matroid theory, we refer to Oxley's book [4]. We call $\mathcal{M} = (E, \mathcal{I})$ a *matroid* if E is a finite set and $\mathcal{I}$ is a collection of subsets of E, satisfying

(i) $\emptyset \in \mathcal{I}$
(ii) If $A \in \mathcal{I}$ and $B \subseteq A$, then $B \in \mathcal{I}$.
(iii) For every $Z \subseteq E$, maximal subsets of Z in $\mathcal{I}$ all have the same size $r(Z)$.
    We call $r(Z)$ the *rank* of Z.

An element of $\mathcal{I}$ is called *independent* in $\mathcal{M}$. We let $E(\mathcal{M}) = E$. A matroid $\mathcal{M} = (E, \mathcal{I})$ is *binary* if there exists a matrix N over GF(2) such that E is a set of column vectors of N and $\mathcal{I} = \{X \subseteq E : X \text{ is linearly independent}\}$. The *connectivity* function $\lambda_{\mathcal{M}}$ of $\mathcal{M}$ is $\lambda_{\mathcal{M}}(X) = r(X) + r(E \setminus X) - r(E) + 1$.

    Let $G = (V, E)$ be a bipartite graph with a bipartition $V = A \cup B$. Let $Bin(G, A, B)$ be the binary matroid on V, represented by the $A \times V$ matrix

$$\left(I_A \; A(G)[A, B] \right),$$

where I_A is the $A \times A$ identity matrix. If $\mathcal{M} = Bin(G, A, B)$, then G is called a *fundamental graph* of $\mathcal{M}$.

Branch-Width. A *branch-decomposition* of a matroid $\mathcal{M}$ is a pair $(T, \mathcal{L})$ of a subcubic tree T and a bijective function $\mathcal{L} : E(\mathcal{M}) \to \{t : t \text{ is a leaf of } T\}$. (If $|E(\mathcal{M})| \leq 1$, then $\mathcal{M}$ admits no rank-decomposition.)

    For an edge e of T, the connected components of $T \setminus e$ induce a partition (X, Y) of the set of leaves of T. The *width* of an edge e of a branch-decomposition $(T, \mathcal{L})$ is $\lambda_{\mathcal{M}}(\mathcal{L}^{-1}(X))$. The *width* of $(T, \mathcal{L})$ is the maximum width of all edges of T. The *branch-width* bw$(\mathcal{M})$ of $\mathcal{M}$ is the minimum width of a branch-decomposition of $\mathcal{M}$. (If $|V| \leq 1$, we define bw$(\mathcal{M}) = 1$.) Branch-width has been defined by Robertson and Seymour [5].

    The following proposition links branch-width of binary matroids with rank-width of bipartite graphs.

Proposition 7 (Oum [3]). *Let $G = (V, E)$ be a bipartite graph with a bipartition $V = A \cup B$ and let $\mathcal{M} = Bin(G, A, B)$. Then for every $X \subseteq V$, $\lambda_{\mathcal{M}}(X) = \rho_G(X) + 1$.*

Corollary 8 (Oum [3]). *Let $G = (V, E)$ be a bipartite graph with a bipartition $V = A \cup B$ and let $\mathcal{M} = Bin(G, A, B)$. Then the branch-width of $\mathcal{M}$ is one more than the rank-width of G.*

2   Approximating Rank-Width Quickly

In this section, we show that, for fixed k, there is a $O(n^4)$-time algorithm that, with a n-vertex graph, outputs a rank-decomposition of width at most $3k + 1$ or confirms that the input graph has rank-width larger than k. Oum and Seymour [1] use general submodular function minimization algorithms [6] to find Z minimizing the cut-rank function $\rho_G(Z)$ with $X \subseteq Z \subseteq V(G) \setminus Y$ for given disjoint subsets X, Y of $V(G)$ such that $|X|, |Y| \leq 3k$. If this can be done in time γ, then we obtain an $O(n(n^2 + \gamma))$-time algorithm to outputs a rank-decomposition of width at most $3k + 1$ or confirms that the input graph has rank-width larger than k. In [1], γ is $O(n^8 \log n)$, and therefore the $O(n^9 \log n)$-time algorithm is obtained.

To obtain a $O(n^4)$-time algorithm, we construct a direct combinatorial algorithm that minimizes the cut-rank function. Jim Geelen suggested the use of blocking sequences for this problem (private communication, 2005).

We first define *blocking sequences*, introduced by J. Geelen [7]. Let G be a graph and A, B be two disjoint subsets of $V(G)$. A sequence $v_1, v_2, \ldots, v_m$ of vertices in $V(G) \setminus (A \cup B)$ is called a *blocking sequence* for (A, B) in G if it satisfies the following:

(i) $\rho_G^*(A, B \cup \{v_1\}) > \rho_G^*(A, B)$.
(ii) $\rho_G^*(A \cup \{v_i\}, B \cup \{v_{i+1}\}) > \rho_G^*(A, B)$ for all $i \in \{1, 2, \ldots, m-1\}$.
(iii) $\rho_G^*(A \cup \{v_m\}, B) > \rho_G^*(A, B)$.
(iv) No proper subsequence satisfies (i)—(iii).

The following proposition is used in most applications of blocking sequences.

Proposition 9. *Let G be a graph and A, B be two disjoint subsets of $V(G)$. The following are equivalent:*

(i) *There is no blocking sequence for (A, B) in G.*
(ii) *There exists Z such that $A \subseteq Z \subseteq V(G) \setminus B$ and $\rho_G(Z) = \rho_G^*(A, B)$.*

Proof. (i)→(ii): We assume that $a, b \notin V(G) \setminus (A \cup B)$ by relabeling. Let $k = \rho_G^*(A, B)$. We construct the *auxiliary digraph* $D = (\{a, b\} \cup (V(G) \setminus (A \cup B)), E)$ from G such that for $x, y \in V(G) \setminus (A \cup B)$,

i) $(a, x) \in E$ if $\rho_G^*(A, B \cup \{x\}) > k$,
ii) $(x, b) \in E$ if $\rho_G^*(A \cup \{x\}, B) > k$,
iii) $(x, y) \in E$ if $\rho_G^*(A \cup \{x\}, B \cup \{y\}) > k$.

Since there is no blocking sequence for (A, B) in G, there is no directed path from a to b in D. Let J be a set of vertices in $V(G) \setminus (A \cup B)$ having a directed path from a in D. We show that $Z = J \cup A$ satisfies $\rho_G(Z) = k$.

To prove this, we claim that $\rho_G^*(A \cup X, B \cup Y) = k$ for all $X \subseteq J$, $Y \subseteq V(G) \setminus (Z \cup B)$. We proceed by induction on $|X| + |Y|$. If $|X| \leq 1$ and $|Y| \leq 1$, then we have $\rho_G^*(A \cup X, B \cup Y) = k$ by the construction of J.

If $|X| > 1$, then for all $x \in X$ we have

$$\rho_G^*(A \cup X, B \cup Y) + \rho_G^*(A, B \cup Y) \leq$$
$$\rho_G^*(A \cup (X \setminus \{x\}), B \cup Y) + \rho_G(A \cup \{x\}, B \cup Y) = 2k,$$

because $\rho_G^*(A \cup \{x\}, B \cup Y) = k$ by induction. So, $\rho_G^*(A \cup X, B \cup Y) = k$.

Similarly if $|Y| > 1$, then for all $y \in Y$ we have $\rho_G^*(A \cup X, B \cup Y) + \rho_G^*(A \cup X, B) \le \rho_G^*(A \cup X, B \cup (Y \setminus \{y\})) + \rho_G(A \cup X, B \cup \{y\}) = 2k$, and therefore $\rho_G^*(A \cup X, B \cup Y) = k$.

(ii)→(i): Suppose that there is a blocking sequence $v_1, v_2, \ldots, v_m$. Then, $v_m \notin Z$ because $\rho_G^*(A \cup \{v_m\}, B) > \rho_G(Z)$. Similarly $v_1 \in Z$ because $\rho_G^*(A, B \cup \{v_1\}) > \rho_G(Z)$. Therefore there exists $i \in \{1, 2, \ldots, m-1\}$ such that $v_i \in Z$ but $v_{i+1} \notin Z$. But this is a contradiction, because $\rho_G(Z) < \rho_G^*(A \cup \{v_i\}, B \cup \{v_{i+1}\}) \le \rho_G^*(Z, V(G) \setminus Z) = \rho_G(Z)$. □

Lemma 10. *Let G be a graph (V, E) and A, B be two disjoint subsets of V such that $\rho_G^*(A, B) = k$ and $|A|, |B| \le l$. Let $n = |V|$. There is a polynomial-time algorithm to either*

- *obtain a graph G' locally equivalent to G with $\rho_{G'}^*(A, B) > k$, or*
- *obtain a set Z such that $A \subseteq Z \subseteq V \setminus B$ and $\rho_G(Z) = k$.*

The running time of this algorithm is $O(n^3)$ if l is fixed or $O(n^4)$ if l is not fixed.

Proof. If there is no blocking sequence for (A, B) in G, then $\min_{A \subseteq Z \subseteq V \setminus B} \rho(Z) = k$ by Proposition 9. In this case, we obtain Z by finding a set of vertices reachable from A in the auxiliary graph.

Therefore, we may assume that there is a blocking sequence $v_1, v_2, \ldots, v_m$. We will find another graph G' locally equivalent to G such that $\mathrm{rk}_{G'}(A, B) > k$. Since $\mathrm{rk}_G(A \cup \{v_m\}, B) = k+1$, there is a vertex $w \in B$ adjacent to v_m.

(1) We claim that $v_1, v_2, \ldots, v_{m-1}$ is a blocking sequence of (A, B) in $G \wedge v_m w$ if $m > 1$.

By applying Lemma 6 for $G[A \cup B \cup \{v_1, v_m\}]$, a subgraph of G induced on $A \cup B \cup \{v_1, v_m\}$, we have

$$\rho_{G \wedge v_m w}^*(A, B \cup \{v_1\}) + \rho_G^*(A \cup \{v_1\}, B)$$
$$\ge \rho_G^*(A, B \cup \{v_1, v_m\}) + \rho_G^*(A \cup \{v_1, v_m\}, B) - 1.$$

Since $\rho_G^*(A, B \cup \{v_1, v_m\} \ge \rho_G^*(A, B \cup \{v_1\}) \ge k+1$, $\rho_G^*(A \cup \{v_1, v_m\}, B) \ge \rho_G^*(A \cup \{v_m\}, B) \ge k+1$, and $\rho_G^*(A \cup \{v_1\}, B) = k$, we obtain that $\rho_{G \wedge v_m w}^*(A, B \cup \{v_1\}) \ge k+1$.

By applying the same inequality we obtain that

$$\rho_{G \wedge v_m w}^*(A \cup \{v_i\}, B \cup \{v_{i+1}\}) + \rho_G^*(A \cup \{v_i, v_{i+1}\}, B)$$
$$\ge \rho_G^*(A \cup \{v_i\}, B \cup \{v_{i+1}, v_m\}) + \rho_G^*(A \cup \{v_i, v_{i+1}, v_m\}, B) - 1 \ge 2k+1$$

for each $i \in \{1, 2, 3, \ldots, m-2\}$ and therefore $\rho_{G \wedge v_m w}^*(A \cup \{v_i\}, B \cup \{v_{i+1}\}) \ge k+1$.

Moreover, $\rho_{G \wedge v_m w}^*(A \cup \{v_{m-1}\}, B) + \rho_G^*(A \cup \{v_{m-1}\}, B) \ge \rho_G^*(A \cup \{v_{m-1}\}, B \cup \{v_m\}) + \rho_G^*(A \cup \{v_{m-1}, v_m\}, B) - 1 \ge 2k+1$ and therefore $\rho_{G \wedge v_m w}^*(A \cup \{v_{m-1}\}, B) \ge k+1$.

We prove one lemma to be used later. If X and Y are disjoint subsets of V such that $A \subseteq X$, $B \subseteq Y$, $v_m \notin X \cup Y$ and $\rho_G^*(X, Y) = k$, then $\rho_{G \wedge v_m w}^*(X, Y) = \rho_G^*(X, Y \cup \{v_m\})$ because

$$\rho^*_{G \wedge v_m w}(X, Y) + \rho^*_G(X, Y) \geq \rho^*_G(X, Y \cup \{v_m\}) + \rho^*_G(X \cup \{v_m\}, Y) - 1$$
$$\geq \rho^*_G(X, Y \cup \{v_m\}) + k = \rho^*_{G \wedge v_m w}(X, Y \cup \{v_m\}) + \rho^*_G(X, Y).$$

By letting $X = A \cup \{v_{m-1}\}$ and $Y = B$, we obtain that $\rho^*_{G \wedge v_m w}(A \cup \{v_{m-1}\}, B) = \rho^*_G(A \cup \{v_{m-1}\}, B \cup \{v_m\}) \geq k+1$. We also obtain $\rho^*_{G \wedge v_m w}(A, B \cup \{v_i\}) = k$ for each $i > 1$ by letting $X = A$, $Y = B \cup \{v_i\}$. Similarly we obtain $\rho^*_{G \wedge v_m w}(A \cup \{v_i\}, B \cup \{v_j\}) = k$ for i, j such that $1 \leq i < i+1 < j \leq m-1$.

Therefore, $v_1, v_2, \ldots, v_{m-1}$ is a blocking sequence for (A, B) in $G \wedge v_m w$.

(2) If $m = 1$ then we obtain $\rho^*_{G \wedge v_1 w}(A, B) \geq k+1$, by applying the previous lemma with letting $X = A$ and $Y = B$.

(3) For each k, we claim that we can obtain another graph G' locally equivalent to G with $\rho^*_{G'}(A, B) > k$ or find Z satisfying $A \subset Z \subseteq V \setminus B$ and $\rho_G(Z) = k$.

If l is fixed, then we can test an adjacency in the auxiliary graph (defined in the proof of Proposition 9) in constant time by calculating rank of matrices of size no bigger than $(l+1) \times (l+1)$, and therefore it takes $O(n^2)$ time to construct the auxiliary digraph. If l is not fixed, then it takes $O(n^4)$ time to construct the auxiliary digraph for finding a blocking sequence. We first obtain the diagonalized matrix R obtained by applying elementary row operations to the matrix $M[A, B]$ in $O(n^3)$ time. For each vertex v not in $A \cup B$, we calculate the rank of $M[A \cup \{v\}, B]$ by using the stored matrix in $O(n^2)$ time. Similarly we calculate the rank of $M[A, B \cup \{v\}]$ by storing the matrix obtained by applying elementary column operations to $M[A, B]$. To check whether $\rho^*_G(A \cup \{x\}, B \cup \{y\}) > k$, it is enough to see when $\rho^*_G(A \cup \{x\}, B) = \rho^*_G(A, B \cup \{y\}) = k$. We first store the rows of the original matrices to each column of R and then we obtain the linear combination of rows of $M[A, B]$ giving $M[\{x\}, B]$. By the same linear combination, we check whether rows of $M[A, \{y\}]$ gives $M[\{x\}, \{y\}]$. It takes $O(n^2)$ time for each $x, y \in V \setminus (A \cup B)$ and therefore we construct the auxiliary digraph in $O(n^4)$ time (if l is not fixed).

To find a blocking sequence, it is enough to find a shortest path in this digraph and it takes $O(n^2)$ time. If there is no blocking sequence, then we find Z in $O(n^2)$ time by choosing all vertices reachable from A by a directed path.

We pick a neighbor of v_m in B and obtain $G \wedge v_m w$ in $O(n^2)$ time. By (1), $G \wedge v_m w$ has a blocking sequence $v_1, v_2, \ldots, v_{m-1}$ for (A, B). We apply this kind of pivoting m times so that in the new graph G' we have $\rho^*_{G'}(A, B) > k$. Since $m \leq n$, we obtain G' in $O(n^3)$ time. $\square$

Theorem 11. *Let l be a fixed constant. Let G be a graph (V, E) and A, B be two disjoint subsets of V such that $|A|, |B| \leq l$. Then, there is a $O(|V|^3)$-time algorithm to find Z with $A \subseteq Z \subseteq V \setminus B$ having the minimum cut-rank.*

Proof. We apply the algorithm given by Lemma 10 until it finds a cut. We use the algorithm at most l times, and so the running time is at most $O(|V|^3)$. $\square$

We state the following theorem for the sake of its own interest. We will not use this for the purpose of approximating rank-width since we have the previous theorem.

Theorem 12. *Let G be a graph (V, E) and A, B be two disjoint subsets of V. Then, there is a $O(|V|^5)$-time algorithm to find Z with $A \subseteq Z \subseteq V \setminus B$ having the minimum cut-rank.*

Proof. We apply the algorithm given by Lemma 10 until it finds a cut. We use the algorithm at most $|V|$ times, and so the running time is at most $O(|V|^5)$. □

Combining with Oum and Seymour [1], we obtain the following.

Theorem 13. *For given k, there is an algorithm, for the input graph $G = (V, E)$, that either concludes that $\mathrm{rwd}(G) > k$ or outputs a rank-decomposition of G of width at most $3k + 1$; and its running time is $O(|V|^4)$.*

Since we can convert the rank-decomposition of width k to a $(2^{k+1}-1)$-expression (a decomposition related to clique-width) in $O(|V|^2)$ time [1], we obtain the following corollary.

Corollary 14. *For given k, there is an algorithm, for the input graph $G = (V, E)$, that either concludes that $\mathrm{cwd}(G) > k$ or outputs a $(2^{3k+2}-1)$-expression of G; and its running time is $O(|V|^4)$.*

3   Graphs to Bipartite Graphs

Courcelle [8] shows that Seese's conjecture [9] is true if and only if it is true for bipartite graphs by using a certain graph transformation B from graphs to bipartite graphs which we describe in the following lemma. He proves that there exist two functions f_1 and f_2 such that $f_1(\mathrm{rwd}(G)) \leq \mathrm{rwd}(B(G)) \leq f_2(\mathrm{rwd}(G))$, but does not have explicit constructions of f_1 and f_2. We give a concrete bound on rank-width. We will use this lemma in Sect. 4.

Fig. 1. K_3 and $B(K_3)$

Lemma 15. *Let $G = (V, E)$ be a graph. Let $B(G) = (V \times \{1, 2, 3, 4\}, E')$ be a bipartite graph obtained from G as follows:*

(i) *if $v \in V$ and $i \in \{1, 2, 3\}$, then (v, i) is adjacent to $(v, i + 1)$ in $B(G)$,*
(ii) *if $vw \in E$, then $(v, 1)$ is adjacent to $(w, 4)$ in $B(G)$.*

Then we have $\frac{1}{4} \mathrm{rwd}(G) \leq \mathrm{rwd}(B(G)) \leq \max(2 \mathrm{rwd}(G), 1)$.

Proof. (1) Let us show that $\mathrm{rwd}(B(G)) \leq \max(2 \mathrm{rwd}(G), 1)$. If $\mathrm{rwd}(G) = 0$, then $\mathrm{rwd}(B(G)) = 1$. Now we may assume that $\mathrm{rwd}(G) > 0$ and we claim that $\mathrm{rwd}(B(G)) \leq 2 \mathrm{rwd}(G)$. Let $(T, \mathcal{L})$ be a rank-decomposition of G of width k. Let N be the set of leaves of T. Let T' be a tree such that $V(T') = (V(T) \times \{0\}) \cup (N \times \{1, 2, 3, 4, 12, 34\})$ and

  (i) if $vw \in E(T)$, then $(v, 0)$ is adjacent to $(w, 0)$ in T',
  (ii) for all $v \in N$, $(v, 12)$ is adjacent to both $(v, 1)$ and $(v, 2)$ in T',
 (iii) for all $v \in N$, $(v, 34)$ is adjacent to both $(v, 3)$ and $(v, 4)$ in T',
 (iv) for all $v \in N$, $(v, 0)$ is adjacent to both $(v, 12)$ and $(v, 34)$ in T'.

Informally speaking, we obtain T' from T by replacing each leaf with a rooted binary tree having four leaves. For each vertex (v, i) of $B(G)$, we define $\mathcal{L}'((v, i)) = (\mathcal{L}(v), i) \in V(T')$. Then $(T', \mathcal{L}')$ is a rank-decomposition of $B(G)$.

We claim that the width of $(T', \mathcal{L}')$ is at most $2k$.

For each edge $e = vw \in E(T)$, let (X, Y) be a partition of N induced by the connected components of $T \setminus e$. Then, the edge $(v, 0)(w, 0)$ of $E(T')$ induces a partition $(X \times \{1, 2, 3, 4\}, Y \times \{1, 2, 3, 4\})$ of $N \times \{1, 2, 3, 4\}$. We observe that $\mathcal{L}'^{-1}(X \times \{1, 2, 3, 4\}) = \mathcal{L}^{-1}(X) \times \{1, 2, 3, 4\}$. It is easy to see that

$$\rho_{B(G)}(\mathcal{L}'^{-1}(X \times \{1, 2, 3, 4\})) = 2\rho_G(\mathcal{L}^{-1}(X)) \le 2k.$$

We now consider remaining edges of T'. Each of them induces a partition (X, Y) of leaves of T' such that $|X| \le 2$ or $|Y| \le 2$. So, $\rho_{B(G)}(\mathcal{L}'^{-1}(X)) \le 2$. Therefore we obtain that the width of $(T', \mathcal{L}')$ is at most $2k$.

(2) We claim that $\mathrm{rwd}(G) \le 4\,\mathrm{rwd}(B(G))$. Let $(T, \mathcal{L})$ be a rank-decomposition of $B(G)$ of width k. Let e be an edge of T, and (X, Y) be a partition of leaves of T induced by connected components of $T \setminus e$.

For four subsets A_1, A_2, A_3, A_4 of V, we denote $A_1 | A_2 | A_3 | A_4 = (A_1 \times \{1\}) \cup (A_2 \times \{2\}) \cup (A_3 \times \{3\}) \cup (A_4 \times \{4\})$ to simplify our notation. Let $\mathcal{L}^{-1}(X) = A_1 | A_2 | A_3 | A_4$. Let $B_i = V \setminus A_i$ for $i \in \{1, 2, 3, 4\}$.

It is easy to observe, for each $i \in \{1, 2, 3\}$, that $\rho^*_{B(G)}((A_i \times \{i\}) \cup (A_{i+1} \times \{i+1\}), (B_i \times \{i\}) \cup (B_{i+1} \times \{i+1\})) = |A_i \cap B_{i+1}| + |B_i \cap A_{i+1}| = |A_i \Delta A_{i+1}|$. Since $\rho_{B(G)}(A_1 | A_2 | A_3 | A_4) = \rho^*_{B(G)}(A_1 | A_2 | A_3 | A_4, B_1 | B_2 | B_3 | B_4) \le k$, we have, for each $i \in \{1, 2, 3\}$,

$$|A_i \Delta A_{i+1}| \le \rho_{B(G)}(A_1 | A_2 | A_3 | A_4) \le k.$$

By adding these inequalities for all i, we obtain that $|A_1 \Delta A_4| \le 3k$.

We also observe that $\mathrm{rk}(M[A_4, B_1]) = \rho_{B(G)}(A_4 \times \{4\}, B_1 \times \{1\}) \le k$. Let M be an adjacency matrix of G. Then we have the following bound of $\rho_G(A_1)$:

$$\rho_G(A_1) = \mathrm{rk}(M[A_1, B_1]) \le \mathrm{rk}(M[A_4 \cup (A_4 \Delta A_1), B_1])$$
$$\le \mathrm{rk}(M[A_4, B_1]) + \mathrm{rk}(M[A_4 \Delta A_1, B_1]) \le 4k.$$

Let T' be the minimal subtree of T containing all leaves in $\mathcal{L}(V \times \{1\})$. Let $\mathcal{L}'(v) = \mathcal{L}((v, 1))$ for all vertices v of G. Then $(T', \mathcal{L}')$ is a rank-decomposition of G and its width is at most $4k$.   □

4   Approximating Rank-Width More Quickly

In this section, we show another algorithm that approximate rank-width as in Sect. 2, but in $O(n^3)$ time with a worse approximation ratio. We take a different approach based on a simple observation in Sect. 3. We use the following algorithm for binary matroids developed by Hliněný [10].

Theorem 16 (Hliněný [10–Theorem 4.12]). *For fixed k, there is a $O(n^3)$-time algorithm that, for a given binary matroid with n elements, obtains a branch-decomposition of width at most $3k + 1$ or confirms that the given matroid has branch-width larger than $k + 1$. We assume that binary matroids are given by their matrix representations.*

This algorithm can be used to approximate rank-width of a bipartite graph G because we can run this algorithm for binary matroids having G as a fundamental graph. By Lemma 15, we obtain a bipartite graph $B(G)$ for each graph G such that $\frac{1}{4}\mathrm{rwd}(G) \leq \mathrm{rwd}(B(G)) \leq \max(2\,\mathrm{rwd}(G), 1)$. Moreover we can construct $B(G)$ in $O(n^2)$ time when $n = |V(G)|$ and transform the rank-decomposition of $B(G)$ of width m into rank-decomposition of G of width at most $4m$ in linear time by the proof of Lemma 15. Therefore, we obtain the following algorithm.

Corollary 17. *For fixed k, there is a $O(n^3)$-time algorithm that, for a given graph with n vertices, obtains a rank-decomposition of width at most $24k$ or confirms that the rank-width of the input graph is larger than k.*

Proof. Let $G = (V, E)$ be the input graph. We may assume that $E(G) \neq \emptyset$. First we construct $B(G)$ in $O(n^2)$ time. We run the algorithm of Theorem 16 with an input $\mathcal{M} = Bin(B(G), V \times \{1, 3\}, V \times \{2, 4\})$ and a constant $2k$.

If it confirms that branch-width of $\mathcal{M}$ is larger than $2k + 1$, then rank-width of $B(G)$ is larger than $2k$, and therefore the rank-width of G is larger than k.

If it outputs the branch-decomposition of $\mathcal{M}$ of width at most $6k + 1$, then the output is a rank-decomposition of $B(G)$ of width at most $6k$. This can be transformed into a rank-decomposition of G of width at most $24k$ in linear time by using an argument of Lemma 15.   □

5   Discussions

Many applications of clique-width are polynomial-time algorithms to solve graph problems when inputs are restricted to graphs of bounded clique-width. Most of them ([11,12,13,14,15]) require k-expression of the input graph as an input to take an advantage of tree-structures (except Johnson [16]). But by using [1], we do not need k-expressions as an explicit input, because we can generate a $(2^{1+f(k)} - 1)$-expression in polynomial time and provide it as an input. The result of this paper will make this preprocessing much faster.

In [17], Courcelle and the author show that there is a $O(|V|^9 \log |V|)$-time algorithm that recognizes graphs of rank-width at most k for an input graph $G = (V, E)$ and a fixed k; they use an approximation algorithm by Seymour and the author [1] as a first step, and it is the slowest part of their algorithm. By the result of this paper, we obtain the following.

Theorem 18. *For fixed k, there is a $O(n^3)$-time algorithm to check that the input graph with n vertices has rank-width at most k.*

But it is still open whether, for fixed k, we can construct a rank-decomposition of width at most k if there are any in polynomial time.

Acknowledgment. The author would like to thank Jim Geelen for our valuable discussions.

References

1. Oum, S., Seymour, P.: Approximating clique-width and branch-width. submitted (2004)
2. Courcelle, B., Olariu, S.: Upper bounds to the clique width of graphs. Discrete Appl. Math. **101** (2000) 77–114
3. Oum, S.: Rank-width and vertex-minors. J. Combin. Theory Ser. B (2005) to appear.
4. Oxley, J.G.: Matroid theory. Oxford University Press, New York (1992)
5. Robertson, N., Seymour, P.: Graph minors. X. Obstructions to tree-decomposition. J. Combin. Theory Ser. B **52** (1991) 153–190
6. Iwata, S., Fleischer, L., Fujishige, S.: A combinatorial strongly polynomial algorithm for minimizing submodular functions. Journal of the ACM (JACM) **48** (2001) 761–777
7. Geelen, J.F.: Matchings, matroids and unimodular matrices. PhD thesis, University of Waterloo (1995)
8. Courcelle, B.: The monadic second-order logic of graphs XV: On a conjecture by D. Seese. submitted (2004)
9. Seese, D.: The structure of the models of decidable monadic theories of graphs. Ann. Pure Appl. Logic **53** (1991) 169–195
10. Hliněný, P.: A parametrized algorithm for matroid branch-width. submitted (2002)
11. Wanke, E.: k-NLC graphs and polynomial algorithms. Discrete Appl. Math. **54** (1994) 251–266
12. Courcelle, B., Makowsky, J.A., Rotics, U.: Linear time solvable optimization problems on graphs of bounded clique-width. Theory Comput. Syst. **33** (2000) 125–150
13. Espelage, W., Gurski, F., Wanke, E.: How to solve NP-hard graph problems on clique-width bounded graphs in polynomial time. In: Graph-theoretic concepts in computer science (Boltenhagen, 2001). Volume 2204 of Lecture Notes in Comput. Sci. Springer, Berlin (2001) 117–128
14. Gerber, M.U., Kobler, D.: Algorithms for vertex-partitioning problems on graphs with fixed clique-width. Theoret. Comput. Sci. **299** (2003) 719–734
15. Kobler, D., Rotics, U.: Edge dominating set and colorings on graphs with fixed clique-width. Discrete Appl. Math. **126** (2003) 197–221
16. Johnson, J.L.: Polynomial time recognition and optimization algorithms on special classes of graphs. PhD thesis, Vanderbuilt University (2003)
17. Courcelle, B., Oum, S.: Vertex-minors, monadic second-order logic, and a conjecture by Seese. submitted (2004)

Computing the Tutte Polynomial on Graphs of Bounded Clique-Width

Omer Giménez[1,*], Petr Hliněný[2,**], and Marc Noy[1,***]

[1] Department of Applied Mathematics, Technical University of Catalonia,
Jordi Girona 1-3, 08034 Barcelona, Spain
{omer.gimenez, marc.noy}@upc.edu
[2] Department of Computer Science, FEI, Technical University of Ostrava,
17. listopadu 15, 708 33 Ostrava, Czech Republic
petr.hlineny@vsb.cz

Abstract. The Tutte polynomial is a notoriously hard graph invariant, and efficient algorithms for it are known only for a few special graph classes, like for those of bounded tree-width. The notion of clique-width extends the definition of cograhs (graphs without induced P_4), and it is a more general notion than that of tree-width. We show a subexponential algorithm (running in time $\exp O(n^{2/3})$) for computing the Tutte polynomial on cographs. The algorithm can be extended to a subexponential algorithm computing the Tutte polynomial on on all graphs of bounded clique-width. In fact, our algorithm computes the more general U-polynomial.

Keywords: Tutte polynomial, cographs, clique-width, subexponential algorithm, U polynomial.

2000 Math Subjects Classification: 05C85, 68R10.

1 Introduction

The Tutte polynomial $T(G; x, y)$ of a graph G is a powerful invariant with many applications, not only in graph theory but also in other fields such as knot theory and statistical physics. One important feature of the Tutte polynomial is that by evaluating $T(G; x, y)$ at special points in the plane one obtains several parameters of G. For example, $T(G; 1, 1)$ is the number of spanning trees of G and $T(G; 2, 1)$ is the number of forests (that is, spanning acyclic subgraphs) of G.

A question that has received much attention is whether the evaluation of $T(G; x, y)$ at a particular point of the (x, y) plane can be done in polynomial

* Supported by Beca Fundació Crèdit Andorrà and Project BFM2001-2340.
** Supported by Czech research grant GAČR 201/05/050 and partly by the program "Information Society" of the Czech Academy of Sciences, project No. 1ET101940420.
*** Supported by Project BFM2001-2340.

D. Kratsch (Ed.): WG 2005, LNCS 3787, pp. 59–68, 2005.

time. Jaeger, Vertigan and Welsh [8] showed that evaluating the Tutte polynomial of a graph is #P-hard at every point except those lying on the hyperbola $(x-1)(y-1) = 1$ and eight special points, including at $(1,1)$ which gives the number of spanning trees. In each of the exceptional cases the evaluation can be done in polynomial time. On the other hand, the Tutte polynomial can be computed in polynomial time for graphs of bounded tree-width. This was obtained independently by Andrzejak [2] and Noble [11]. Recently Hliněný [7] has obtained the same result for matroids of bounded branch-width representable over a fixed finite field, which is a substantial generalization of the previous results. See [5] for additional references on this subject.

In this paper we study the problem of computing the Tutte polynomial for cographs and, more generally, for graphs of bounded clique-width. A graph has clique-width $\leq k$ if it can be constructed using k labels and the following four operations: 1) create a new vertex with label i; 2) take the disjoint union of several labeled graphs; 3) add all edges between vertices of label i and label j; and 4) relabel all vertices with label i to have label j. An expression defining a graph G built from the above four operations using k labels is a *k-expression* for G. A *cograph* is a graph of clique-width at most two; equivalently, it is a graph containing no induced path P_4 on four vertices.

Although a class of graphs with bounded tree-width has also bounded clique-width, the converse is not true. For instance, complete graphs have clique-width two. It is well-known that all problems expressible in monadic second order logic of incidence graphs become polynomial time solvable when restricted to graphs of bounded tree-width. For bounded clique-width less is true: all problems become polynomial time solvable if they are expressible in monadic second-order logic using quantifiers on vertices but not on edges (adjacency graphs) [3].

Our main results are as follows:

Theorem 1.1. *The Tutte polynomial of a cograph with n vertices can be computed in time* $\exp\left(O(n^{2/3})\right)$.

Theorem 1.2. *Let G be a graph with n vertices of clique-width k along with a k-expression for G as an input. Then the Tutte polynomial of G can be computed in time* $\exp\left(O(n^{1-1/(k+2)})\right)$.

Theorem 1.2 is not likely to hold for the class of all graphs, since it would imply the existence of a subexponential algorithm for 3-coloring, hence also for 3-SAT; that is considered highly unlike in the Computer Science community. Of course, the main open question is whether there exists a *polynomial time* algorithm for computing the Tutte polynomial of graphs of bounded clique-width. We discuss this issue in the last section.

In fact, our algorithms compute not only the Tutte polynomial, but the so-called U polynomial (see [12]), which is a stronger polynomial invariant. Moreover, we may skip the requirement of having a k-expression for G as an input in Theorem 1.2, if we do not care about an asymptotic behaviour in the exponent: Just to prove a subexponential upper bound we may use the approximation algorithm for clique-width by Oum and Seymour [13,14].

Since our algorithms are quite complicated, for an illustration, we first present in Section 2 a simplified algorithm computing the number of forests in a cograph, that is, evaluating $T(G; 2, 1)$ for graphs of clique-width ≤ 2. (This is #P-hard on all graphs [8].) In Section 3 we extend the algorithm to the computation of the full Tutte polynomial on cographs. Finally, our main result, Theorem 1.2 is proved in details in the long version [6].

2   Forests in Cographs

The class of *cographs* is defined recursively as follows:

1. A single vertex is a cograph.
2. A disjoint union of two cographs is a cograph.
3. A complete union of two cographs is a cograph.

Here a *complete union* of two graphs $G \oplus H$ means the operation of taking a disjoint union $G \dot\cup H$, and adding all edges between $V(G)$ and $V(H)$. A cograph G can be represented by a tree, whose internal nodes correspond to operations 2) and 3) above, and whose leaves correspond to single vertices. We call such a tree an *expression* for G.

For example, all cliques are cographs, and the complement of a cograph is a cograph again. Cographs have long history of theoretical and algorithmic research. In particular, they are known to be exactly the graphs without induced paths on four vertices (P_4-free).

Let us call a *signature* a multiset of positive integers. The *size* $\|\boldsymbol{\alpha}\|$ of a signature $\boldsymbol{\alpha}$ is the sum of all elements in $\boldsymbol{\alpha}$, respecting repetition in the multiset. A signature $\boldsymbol{\alpha}$ of size n is represented by the *characteristic vector* $\boldsymbol{\alpha} = (a_1, a_2, \ldots, a_n)$, where there are $a_i \geq 0$ elements i in $\boldsymbol{\alpha}$, and $\sum_{i=1}^{n} i \cdot a_i = n$. (On the other hand, the *cardinality* of $\boldsymbol{\alpha}$ is $|\boldsymbol{\alpha}| = \sum_{i=1}^{n} a_i$, as usual.) An important fact we need is:

Recall that $\Theta(f)$ is a usual shortcut for all functions having the same asymptotic growth rate as f.

Lemma 2.1. *There are $2^{\Theta(\sqrt{n})}$ distinct signatures of size n.*

Proof. Each signature actually corresponds to a partition of n into an unordered sum of positive integers. It is well-known [10–Chapter 15] that there are $2^{\Theta(\sqrt{n})}$ of those. ∎

We call a *double-signature* a multiset of ordered pairs of non-negative integers, excluding the pair $(0, 0)$. The *size* $\|\boldsymbol{\beta}\|$ of a double-signature $\boldsymbol{\beta}$ is the sum of all $(x + y)$ for $(x, y) \in \boldsymbol{\beta}$, respecting repetition in the multiset. We, moreover, need to prove:

Lemma 2.2. *There are $\exp\left(\Theta(n^{2/3})\right)$ distinct double-signatures of size n.*

Lemma 2.2 is a particular case of Lemma 5.1, which is proved in [6].

Lemma 2.3. *A double-signature $\boldsymbol{\beta}$ of size n has at most $\exp\left(O(n^{2/3})\right)$ different submultisets (i.e. of different characteristic vectors).*

Proof. Just count all double-signatures of size $\leq n$. ∎

2.1 Forest Signature Table

Let us now consider a graph G and a forest $U \subset G$. The signature $\boldsymbol{\alpha}$ of U is the multiset of sizes of the connected components of U. (Obviously, $\boldsymbol{\alpha}$ has size $|V(G)|$ if U spans all the vertices.) We call a *(spanning) forest signature table* of the graph G a vector $\boldsymbol{T}$ (realized as an array $\boldsymbol{T}[\ldots]$); such that $\boldsymbol{T}$ records, for each signature $\boldsymbol{\alpha}$ of size $|V(G)|$, the number of spanning forests $U \subset G$ having signature $\boldsymbol{\alpha}$ (as $\boldsymbol{T}[\boldsymbol{\alpha}]$). For simplicity we usually skip the word "spanning" if it is clear from the context. We are going to compute the forest signature table of a cograph G recursively along the way G has been constructed. For that we describe two algorithms.

Let us denote by Σ_G the set of all signatures of size $|V(G)|$. It is important to keep in mind that signatures are considered as multisets, which concerns also set operations. For instance, a *multiset union* $\boldsymbol{\gamma} \uplus \boldsymbol{\delta}$ is obtained as the sum of the characteristic vectors of $\boldsymbol{\gamma}$ and $\boldsymbol{\delta}$, and a *multiset difference* $\boldsymbol{\gamma} \setminus \boldsymbol{\delta}$ is defined by the non-negative difference of those.

Algorithm 2.4. *Combining the spanning forest signature tables of graphs F and G into the one of the disjoint union $H = F \dot{\cup} G$.*

Input: Graphs F, G, and their forest signature tables $\boldsymbol{T}_F, \boldsymbol{T}_G$.
Output: The forest signature table $\boldsymbol{T}_H$ of $H = F \dot{\cup} G$.

create *empty table $\boldsymbol{T}_H$ of forest signatures of size $|V(H)|$*;
for *all signatures* $\boldsymbol{\alpha}_F \in \Sigma_F$, $\boldsymbol{\alpha}_G \in \Sigma_G$ **do**
   set $\boldsymbol{\alpha} = \boldsymbol{\alpha}_F \uplus \boldsymbol{\alpha}_G$ *(a multiset union)*;
   add $\boldsymbol{T}_H[\boldsymbol{\alpha}] \mathrel{+}= \boldsymbol{T}_F[\boldsymbol{\alpha}_F] \cdot \boldsymbol{T}_G[\boldsymbol{\alpha}_G]$;
done.

The running time of this algorithm is proportional to the number of pairs of signatures (α_F, α_G), which is $\exp\left(O(n^{2/3})\right)$, where $n = |V(H)|$; this is due to Lemma 2.2 and the fact that we have the $O(\ )$ expression in the exponent.

The second algorithm is, on the other hand, much more complicated. It involves double-signatures in the following meaning: Consider a graph H with vertices partitioned into two parts $V(H) = V_1 \cup V_2$, and a forest $U \subset H$. The double-signature of U (wrt. V_1, V_2) is the multiset of pairs $\left(|V(C) \cap V_1|, |V(C) \cap V_2|\right)$ over all connected components C of U.

The idea behind the algorithm is to obtain the double-signatures (for $V_1 = V(F)$ and $V_2 = V(G)$) of the spanning forests in $H = F \oplus G$ from the signatures of the spanning forests in F and G. For every pair of forests $U_F \subset F$ and $U_G \subset G$, the algorithm iteratively counts the different ways in which each component of U_G can be joined to components of U_F. During the process, double signatures are needed to distinguish between former vertices of F and of G in already joined components. In fact, the algorithm works with pairs of signatures $\boldsymbol{\alpha}_F$ and $\boldsymbol{\alpha}_G$, that is, with whole classes of forests instead of particular forests. We also remark that a submultiset is considered among all possible selections of repeated elements, like if they were pairwise distinct.

Algorithm 2.5. *Combining the spanning forest signature tables of graphs F and G into the one of the complete union $H = F \oplus G$.*

Input: Graphs F, G, and their forest signature tables $\boldsymbol{T}_F, \boldsymbol{T}_G$.
Output: The forest signature table $\boldsymbol{T}_H$ of $H = F \oplus G$.

```
create empty table T_H of forest signatures of size |V(H)|;
for all signatures α_F ∈ Σ_F, α_G ∈ Σ_G do
    set z = |V(F)|;
    create empty table X of forest double-signatures of size z;
    // Imagine particular forests U_F ⊂ F, U_G ⊂ G of signature α_F, α_G,
    // and a selected component C ⊂ U_G of size c.
    set X[double-signature {(a,0) : a ∈ α_F}] = 1;
    for each c ∈ α_G (with repetition) do
        create empty table X' of forest double-signatures of size z + c;
        for all double signatures β of size z s.t. X[β] > 0 do
(†)         for all submultisets γ ⊆ β (with repetition) do
                set d_1 = Σ_{(x,y)∈γ} x,  d_2 = Σ_{(x,y)∈γ} y;
                set double-signature β' = (β \ γ) ⊎ {(d_1, d_2 + c)};
(*)             add X'[β'] += X[β] · Π_{(x,y)∈γ} cx;
            done
        done
        set X = X', z = z + c;   dispose X';
    done
    for all double-signatures β of size |V(H)| do
        set signature α_0 = {x + y : (x, y) ∈ β};
        add T_H[α_0] += X[β] · T_F[α_F] · T_G[α_G];
    done
done.
```

Proof of Algorithm 2.5. We now explain the algorithm, and show its correctness. It is better understandable if one imagines particular forests (representatives) $U_F \subset F$ and $U_G \subset G$ in the place of the signatures $\boldsymbol{\alpha}_F$ and $\boldsymbol{\alpha}_G$ chosen in the first `for` cycle. Then one may routinely verify that all subsequent computations depend only on the forest signatures $\boldsymbol{\alpha}_F$, $\boldsymbol{\alpha}_G$ (not on the particular forests), and hence it is correct to finally multiply the computed values in $\boldsymbol{X}$ by the numbers $\boldsymbol{T}_F[\boldsymbol{\alpha}_F] \cdot \boldsymbol{T}_G[\boldsymbol{\alpha}_G]$.

In the tables $\boldsymbol{X}, \boldsymbol{X}'$ we iteratively compute the numbers of all spanning forests in H that result by adding some edges between the forests U_F and U_G (stored by their double signatures). We consider an arbitrary order $C_1, C_2, \ldots, C_k$ on the connected components of U_G. For $i = 1, 2, \ldots, k$, we take the component C_i, and count all possible ways how to connect C_i by selected edges to a subset (†) of components of each of the previously constructed forests on $V(F \cup C_1 \cup \ldots \cup C_{i-1})$ which are recorded in the table $\boldsymbol{X}$. The other ends of those selected edges are considered only among vertices in $V(F)$. (Recall that the complete union $H = F \oplus G$ has added *all* edges between $V(F)$ and $V(C_i)$.) We then record (*) numbers of all the new forests on $V(F \cup C_1 \cup \ldots \cup C_i)$ in a new table $\boldsymbol{X}'$ that will play the role of $\boldsymbol{X}$ in the next iteration.

Saying precisely, after finishing iteration $i = 1, 2, \ldots, k$ described in the previous paragraph, each entry $\boldsymbol{X}'[\beta]$ equals the number of all forests U' of signature β spanning $V(F \cup C_1 \cup \ldots \cup C_i)$ such that $U' \upharpoonright V(F) = U_F$ and $U' \upharpoonright V(G) = U_G \upharpoonright C_1 \cup \ldots \cup C_i$. That follows easily by an induction from the previous arguments. At the end we count each spanning forest $U \subseteq H$ such that $U \upharpoonright V(F) = U_F$ and $U \upharpoonright V(G) = U_G$ exactly once. Finally, the double-signatures in the table $\boldsymbol{X}$ partition the vertices into $V(F)$ and $V(G)$, but that is no longer needed. So we "simplify" them – we record the resulting numbers only by the (single) forest signatures in the resulting table $\boldsymbol{T}_H$.   ∎

2.2   Time Analysis

Lemma 2.6. *A modified implementation of Algorithm 2.5 runs in time* $\exp\big(O(n^{2/3})\big)$ *where* $n = |V(H)|$.

Proof. Since we have $O(\ )$ in the exponent, it is enough to verify that each of the `for` cycles in Algorithm 2.5 is iterated at most $\exp\big(O(n^{2/3})\big)$ times. That follows from Lemma 2.1 for the first cycle, and it is clear for the second cycle. For the third nested cycle it follows from Lemma 2.2.

A problem may occur in the fourth nested cycle 'for all submultisets $\gamma \subseteq \beta$' if β consists, say, of $n/2$ copies of the element 2. Then there are up to $\exp\big(\Theta(n)\big)$ submultisets γ to consider. Fortunately, the results of the subsequent computation depend only on the characteristic vector of γ. Hence it is enough to consider (much less of) pairwise different submultisets $\gamma \subseteq \beta$ (cf. Lemma 2.3), and then multiply the resulting number by all possible choices (combinations) of repeated elements of γ from β. Formally, the program line (†) now reads

$$\text{for all different submultisets } \boldsymbol{\gamma} \subseteq \boldsymbol{\beta} \text{ do},$$

and the line (*) reads

$$\text{add } \boldsymbol{X}'[\beta'] \mathrel{+}= \boldsymbol{X}[\beta] \cdot \prod_{(x,y)\in\gamma} cx \cdot \prod_{(x,y)\in\langle\beta\rangle} \binom{\mu_\beta(x,y)}{\mu_\gamma(x,y)},$$

where $\langle\boldsymbol{\alpha}\rangle$ denotes the ordinary set formed by elements of a multiset $\boldsymbol{\alpha}$, and $\mu_\alpha z$ is the repetition of an element z in $\boldsymbol{\alpha}$. The statement is proved.   ∎

We remark that the improvement discussed in the proof of previous Lemma 2.6 have been fully incorporated in the subsequent algorithms.

Theorem 2.7. *The number of spanning forests in an n-vertex cograph can be computed in time* $\exp\big(O(n^{2/3})\big)$.

Proof. Consider a cograph G and a tree expression defining it. The forest signature table of a single vertex is trivial, and by Algorithms 2.4 and 2.5 (Lemma 2.6), the forest signature tables of a union or a complete union of two cographs can be computed in time claimed. Finally, knowing the forest signature table T of G, the number of all spanning forests of G is computed by adding up the entries of T. ∎

3   The Tutte Polynomial of a Cograph

The Tutte polynomial can be defined in a number of equivalent ways. For our purposes, given a graph $G = (V, E)$ we define the Tutte polynomial as

$$T(G; x, y) = \sum_{F \subseteq E} (x - 1)^{r(E) - r(F)} (y - 1)^{|F| - r(F)},$$

where $r(F) = |V| - k(F)$ and $k(F)$ is the number of connected components of the spanning subgraph induced by the edge-subset F. It is clear that knowing $T(G; x, y)$ is the same as knowing, for every i and j, how many spanning subgraphs with the edge set F in G are there with $|F| = i$ and $k(F) = j$.

Consider a spanning subgraph $W \subset G$ determined on $V(W) = V(G)$ by an arbitrary subset $F \subset E(G)$, $F = E(W)$. The sizes of the connected components of W define a signature of size $|V(G)|$. In the *(spanning) subgraph signature table* S of G, for each signature α of size $|V(G)|$ and each number of edges $f \in \{0, 1, 2, \ldots, |E(G)|\}$, we record the number $S[\alpha, f]$ of all spanning subgraphs of G having f edges and having component sizes according to the signature α. We shortly denote by $\gamma \lceil_i$ the multiset formed by all the i-th coordinates (repetitions accounted for) of the elements of a double-signature γ.

In order to prove Theorem 1.1 we need analogues of Algorithms 2.4 and 2.5 for computing subgraph signature tables. The algorithm for disjoint unions is again straightforward and we omit it; the one for complete unions comes next.

Algorithm 3.1. *A modification of Algorithm 2.5 for computing the (spanning) subgraph signature table of the complete union $H = F \oplus G$.*

Besides adding edge number as the second index to the signature tables, the only other major difference of this algorithm from Algorithm 2.5 is that the single line (*) is replaced with another **for** cycle calling a procedure **CellSel** of further Algorithm 3.2.

Input: Graphs F, G, and their subgraph signature tables S_F, S_G.
Output: The subgraph signature table S_H of $H = F \oplus G$.

create *empty table S_H of subgraph signatures of size $|V(H)|$*;
for all $\alpha_F \in \Sigma_F$, and $e_F = 0, 1, \ldots, |E(F)|$ s.t. $S_F[\alpha_F, e_F] > 0$ **do**
    for all $\alpha_G \in \Sigma_G$, and $e_G = 0, \ldots, |E(G)|$ s.t. $S_G[\alpha_G, e_G] > 0$ **do**
        set $z = |V(F)|$;
        create *empty table Y of subgraph double-signatures of size z*;

```
set Y[double-signature {(a,0) : a ∈ αF}, eF] = 1;
for each c ∈ αG (with repetition) do
    create empty table Y′ of subgraph double-sign. of size z + c;
    for all β of size z, and e s.t. Y[β,e] > 0 do
        for all different submultisets γ ⊆ β do
            set r =      ∏        (μβ(x,y));
                     (x,y)∈⟨β⟩     (μγ(x,y))
            set d₁ = ‖γ⌈₁‖ = Σ(x,y)∈γ x,  d₂ = ‖γ⌈₂‖ = Σ(x,y)∈γ y;
            set double-signature β′ = (β \ γ) ⊎ {(d₁, d₂ + c)};
            for f = |γ|, |γ| + 1,..., c · d₁ do
                set multiset D = c · (γ⌈₁) = {cx : (x,y) ∈ γ};
                call Algorithm 3.2: p = CellSel(D, f);
                add Y′[β′, e + f] += Y[β,e] · r · p;
            done
        done
    done
    set Y = Y′, z = z + c; dispose Y′;
done
for all double-sign. β of size |V(H)|, and f, s.t. Y[β,f] > 0 do
    set signature α₀ = {x + y : (x,y) ∈ β};
    add SH[α₀, f + eG] += Y[β,f] · SF[αF,eF] · SG[αG,eG];
done
    done
done.
```

Proof of Algorithm 3.1. This algorithm is similar to the improved version of Algorithm 2.5 (cf. Lemma 2.6), and so we only sketch the proof here. The main new difficulty lies in counting the different ways in which a connected component of c vertices in α_G can be connected with f edges to the selected components of signatures $(x,y) \in \gamma$. Recall that when counting forests we had no such difficulty, since we joined the component of α_G to each component of γ with exactly one edge; thus we used exactly $f = |\gamma|$ edges chosen in $\prod_{(x,y)\in\gamma} cx$ different ways. The procedure 'CellSel(D, f)' counts this for spanning subgraphs, and we defer the explanation to Algorithm 3.2.

Finally, notice that the edge numbers in tables Y, Y' do not account for the edges from $E(G)$, since we do not know how many edges has each one of the components of α_G. Those edges are summed up at the end, when obtaining the signatures for H from the double-signatures stored in Y. ∎

Algorithm 3.2. *Computing the number of cellular selections: We are selecting ℓ elements from the union $C_1 \cup C_2 \cup \ldots \cup C_k$, where C_i for $i = 1, 2, \ldots, k$ are pairwise disjoint cells of sizes $d_i = |C_i|$, and we require that some element is selected from every cell.*

Input: A multiset $D = \{d_1, d_2, \ldots, d_k\}$ of cell sizes, and a number ℓ.
Output: The number $\texttt{CellSel}(D, \ell)$ of all such possible selections.

```
create table u[1..k][1..ℓ], filled with 0;
for j = 1, 2, ..., d₁ do   set u[1][j] = (d₁ j);
set z = d₁;
for i = 2, 3, ..., k do
   add z += dᵢ;
   for j = i, i + 1, ..., min(ℓ, z) do
      for s = 1, 2, ..., min(j − (i − 1), dᵢ) do
         add u[i][j] += u[i − 1][j − s]· (dᵢ s);
      done
   done
done
return u[k][ℓ].
```

Proof of Algorithm 3.2. Let $u_{i,j} = u[i][j]$ be the number of cellular selections of j elements chosen among the first i cells. These numbers satisfy the recurrence relation

$$u_{i,j} = \sum_{s=1}^{r} u_{i-1,j-s} \cdot \binom{d_i}{s}$$

where r is the maximum number of elements than can be selected from the i-th cell to obtain a total of j elements. Since the i-th cell has d_i elements available, and the $i-1$ previous cells contributed at least one element each to the resulting j elements, it follows that $r = \min\{j - (i - 1), d_i\}$.

Algorithm 3.2 just applies the previous recurrence in a correct order, and avoids useless computations like with values of j too small or too large. It runs in $O(k\ell^2)$ steps.   ∎

Proof of Theorem 1.1. As in Theorem 2.7, the subgraph signature table S of a cograph can be computed in time proportional to the number of all possible double-signatures of size n, i.e. in $\exp\left(O(n^{2/3})\right)$. Then, summing the entries of S, we compute the numbers of spanning subgraphs with a given number of edges and a number of components. As we have remarked previously, these numbers give (efficiently) the Tutte polynomial.   ∎

The U polynomial of an n-vertex graph G is defined in [12] as

$$U(G; \mathbf{x}, y) = \sum_{F \subseteq E} x_{n_1} \cdots x_{n_k} (y - 1)^{|F| - r(F)},$$

where $n_1, \ldots, n_k$ are the vertex sizes of the components of the spanning subgraph (V, F). If we let $x_1 = \cdots = x_n = x - 1$ in the expression above, we recover the Tutte polynomial $T(G; x, y)$ up to a power of $x - 1$. It is clear that the subgraph signature table of a graph is precisely equivalent to the U polynomial, hence in the statement of Theorem 1.1 we can replace "U polynomial" for "Tutte polynomial".

4   Concluding Remarks

We have shown that the Tutte and U polynomials can be computed in subexponential time for cographs, and more generally for graphs with bounded clique-width [6]. Such a result is very unlikely to hold for all graphs. Of course, the important question of whether the Tutte polynomial can be computed in polynomial time, or the problem is #P-hard even for graphs of bounded clique-width, remains open. (The U polynomial is obviously not computable in polynomial time due to its size.)

On the other hand, the *chromatic* polynomial for graphs of bounded clique-width can be computed in polynomial time (although not FPT). This follows by adapting the algorithm in [9] for computing the chromatic number, keeping track also of the number of r-colorings for $r = 1, \ldots, n$, where n is the number of vertices. To our knowledge, that is possibly the only currently known natural example of graph classes other than chordal graphs, where the chromatic polynomial can be computed in polynomial time, but the complexity of computing the Tutte polynomial is undecided.

References

1. G.E. Andrews, *The theory of partitions*, Cambridge U. Press, Cambridge, 1984.
2. A. Andrzejak, *An Algorithm for the Tutte Polynomials of Graphs of Bounded Treewidth*, Discrete Math. 190 (1998), 39–54.
3. B. Courcelle, J.A. Makowsky, U. Rotics, *Linear Time Solvable Optimization Problems on Graphs of Bounded Clique-Width*, Theory Comput. Systems 33 (2000), 125–150.
4. B. Courcelle, S. Olariu, *Upper bounds to the clique width of graphs*, Discrete Appl. Math. 101 (2000), 77–114.
5. O. Giménez, M. Noy, *On the complexity of computing the Tutte polynomial of bicircular matroids*, Combin. Probab. Computing, to appear.
6. O. Giménez, P. Hliněný, M. Noy, *Computing the Tutte Polynomial on graphs of Bounded Clique-Width*, manuscript, 2005.
7. P. Hliněný, *The Tutte Polynomial for Matroids of Bounded Branch-Width*, Combin. Probab. Computing, to appear (2005).
8. F. Jaeger, D.L. Vertigan, D.J.A. Welsh, *On the Computational Complexity of the Jones and Tutte Polynomials*, Math. Proc. Camb. Phil. Soc. 108 (1990), 35–53.
9. D. Kobler, U. Rotics, *Edge dominating set and colorings on graphs with fixed clique-width*, Discrete Applied Math. 126 (2003), 197–221.
10. J.H. van Lint, R.M. Wilson, *A Course in Combinatorics*, Cambridge University Press, Cambridge, 1992.
11. S.D. Noble, *Evaluating the Tutte Polynomial for Graphs of Bounded Tree-Width*, Combin. Probab. Computing 7 (1998), 307–321.
12. S.D. Noble, D.J.A. Welsh, *A weighted graph polynomial from chromatic invariants of knots*, Ann. Inst. Fourier (Grenoble) 49 (1999), 1057–1087.
13. Sang-Il Oum, P.D. Seymour, *Approximating Clique-width and Branch-width*, submitted, 2004.
14. Sang-Il Oum, *Approximating Rank-width and Clique-width Quickly*, In: WG 2005, Proccedings, Lecture Notes in Computer Science, to appear (2005).

Minimizing NLC-Width is NP-Complete

(Extended Abstract)

Frank Gurski and Egon Wanke

Heinrich-Heine-University Düsseldorf,
Institute of Computer Science, D-40225 Düsseldorf, Germany
{gurski-wg, wanke-wg}@acs.uni-duesseldorf.de

Abstract. We show that a graph has tree-width at most $4k-1$ if its line graph has NLC-width or clique-width at most k, and that an incidence graph has tree-width at most k if its line graph has NLC-width or clique-width at most k. In [9] it is shown that a line graph has NLC-width at most $k+2$ and clique-width at most $2k+2$ if the root graph has tree-width k. Using these bounds we show by a reduction from tree-width minimization that NLC-width minimization is NP-complete.

1 Introduction

The clique-width of a graph is defined by a composition mechanism for vertex-labeled graphs [7]. The operations are the vertex disjoint union, the addition of edges between vertices controlled by a label pair, and the relabeling of vertices. The clique-width of a graph G is the minimum number of labels needed to define it. The NLC-width of a graph is defined by a composition mechanism similar to that for clique-width [19]. Every graph of clique-width at most k has NLC-width at most k and every graph of NLC-width at most k has clique-width at most $2k$ [12]. The only essential difference between the composition mechanisms of clique-width bounded graphs and NLC-width bounded graphs is the addition of edges. In an NLC-width composition the addition of edges is combined with the union operation. This union operation applied to two graphs G and J is controlled by a set S of label pairs such that for every pair $(a, b) \in S$ all vertices of G labeled by a will be connected with all vertices of J labeled by b. Both concepts are useful, because it is sometimes much more comfortable to use NLC-width expressions instead of clique-width expressions and vice versa, respectively.

Clique-width and NLC-width bounded graphs are particularly interesting from an algorithmic point of view. A lot of NP-complete graph problems can be solved in polynomial time for graphs of bounded clique-width. For example, all graph properties expressible in monadic second order logic with quantifications over vertices and vertex sets (MSO_1-logic) are decidable in linear time on clique-width bounded graphs [6] if a corresponding decomposition for the graph is given as an input. The MSO_1-logic has been extended by counting mechanisms which allow the expressibility of optimization problems concerning maximal or minimal vertex sets [6]. All graph problems expressible in extended MSO_1-logic can be

D. Kratsch (Ed.): WG 2005, LNCS 3787, pp. 69–80, 2005.
© Springer-Verlag Berlin Heidelberg 2005

solved in polynomial time on clique-width bounded graphs. Furthermore, there are a lot of NP-complete graph problems which are not expressible in extended MSO_1-logic (like Hamiltonicity, various partition problems, and bounded degree subgraph problems) but which can also be solved in polynomial time on clique-width bounded graphs [19,8,14,9].

The recognition problem for graphs of clique-width or NLC-width at most k for fixed integers k is still open for $k \geq 4$ and $k \geq 3$, respectively. Clique-width of at most 3 is decidable in polynomial time [4]. NLC-width of at most 2 is decidable in polynomial time [13]. Clique-width of at most 2 and NLC-width 1 is decidable in linear time [5]. In this paper we show that NLC-width minimization is NP-complete, which was open up to now.

The paper is organized as follows. In Section 2, we recall the definitions of clique-width, NLC-width, and tree-width. In Section 3, we show that a graph has tree-width at most $4k-1$ if its line graph[1] has NLC-width or clique-width at most k. In Section 4, we show that an incidence graph[2] has tree-width at most k if its line graph has NLC-width or clique-width at most k. In [9] it is shown that a line graph has NLC-width at most $k+2$ and clique-width at most $2k+2$ if the root graph has tree-width k. This in connection with the result of Section 4 is used to show by a reduction from tree-width minimization that minimizing NLC-width is NP-complete.

2   Preliminaries

Let $[k] := \{1, \ldots, k\}$ be the set of all integers between 1 and k. We work with finite undirected vertex labeled *graphs* $G = (V_G, E_G, \mathrm{lab}_G)$, where V_G is a finite set of *vertices* labeled by some mapping $\mathrm{lab}_G : V_G \to [k]$ and $E_G \subseteq \{\{u, v\} \mid u, v \in V_G,\ u \neq v\}$ is a finite set of *edges*. The labeled graph consisting of a single vertex labeled by $a \in [k]$ is denoted by $\bullet_a$.

The notion of clique-width is defined by Courcelle and Olariu in [7].

Definition 1 (Clique-width, [7]). *Let k be some positive integer. The class CW_k of labeled graphs is recursively defined as follows.*

1. *The single vertex graph $\bullet_a$ for some $a \in [k]$ is in CW_k.*
2. *Let $G = (V_G, E_G, \mathrm{lab}_G) \in CW_k$ and $J = (V_J, E_J, \mathrm{lab}_J) \in CW_k$ be two vertex disjoint labeled graphs. Then $G \oplus J := (V', E', \mathrm{lab}')$ defined by $V' := V_G \cup V_J$, $E' := E_G \cup E_J$, and*

$$\mathrm{lab}'(u) := \begin{cases} \mathrm{lab}_G(u) & \text{if } u \in V_G \\ \mathrm{lab}_J(u) & \text{if } u \in V_J \end{cases}$$

is in CW_k.

---

[1] The line graph $L(G)$ of a graph G has a vertex for every edge of G and an edge between two vertices if the corresponding edges of G are adjacent [20].

[2] The incidence graph $I(G)$ of a graph G is the graph we get if we replace every edge $\{u, v\}$ of G by a new vertex w and two edges $\{u, w\}, \{w, v\}$.

3. *Let $a, b \in [k]$ be two distinct integers and $G = (V_G, E_G, lab_G) \in CW_k$ be a labeled graph then*
   (a) $\rho_{a \to b}(G) := (V_G, E_G, lab')$ defined by

   $$lab'(u) := \begin{cases} lab_G(u) & \text{if } lab_G(u) \neq a \\ b & \text{if } lab_G(u) = a \end{cases}$$

   is in CW_k and
   (b) $\eta_{a,b}(G) := (V_G, E', lab_G)$ defined by

   $$E' := E_G \cup \{\{u, v\} \mid u, v \in V_G, \ u \neq v, \ lab(u) = a, \ lab(v) = b\}$$

   is in CW_k.

The notion of NLC-width[3] is defined by Wanke in [19].

Definition 2 (NLC-width, [19]). *Let k be some positive integer. The class NLC_k of labeled graphs is recursively defined as follows.*

1. *The single vertex graph $\bullet_a$ for some $a \in [k]$ is in NLC_k.*
2. *Let $G = (V_G, E_G, lab_G) \in NLC_k$ and $R : [k] \to [k]$ be a function, then $\circ_R(G) := (V_G, E_G, lab')$ defined by $lab'(u) := R(lab_G(u))$ is in NLC_k.*
3. *Let $G = (V_G, E_G, lab_G) \in NLC_k$ and $J = (V_J, E_J, lab_J) \in NLC_k$ be two vertex disjoint labeled graphs and $S \subseteq [k]^2$ be a set of label pairs, then graph $G \times_S J := (V', E', lab')$ defined by $V' := V_G \cup V_J$,*

   $$E' := E_G \cup E_J \cup \{\{u, v\} \mid u \in V_G, \ v \in V_J, \ (lab_G(u), lab_J(v)) \in S\},$$

   and

   $$lab'(u) := \begin{cases} lab_G(u) & \text{if } u \in V_G \\ lab_J(u) & \text{if } u \in V_J \end{cases}$$

   is in NLC_k.

The *clique-width (NLC-width)* of a labeled graph G is the least integer k such that $G \in CW_k$ ($G \in NLC_k$, respectively). An expression built with the operations $\bullet_a, \oplus, \rho_{a \to b}, \eta_{a,b}$ for integers $a, b \in [k]$ is called a *clique-width k-expression*. An expression built with the operations $\bullet_a, \times_S, \circ_R$ for $a \in [k]$, $S \subseteq [k]^2$, and $R : [k] \to [k]$ is called an *NLC-width k-expression*. Every clique-width expression (NLC-width expression) has by its recursive definition a tree structure which we call the *clique-width expression tree (NLC-width expression tree*, respectively). A vertex labeled graph G has *linear clique-width (linear NLC-width)* at most k if it can be defined by a clique-width k-expression (NLC-width k-expression, respectively) in that at least one argument of every operation $\oplus$ (every operation $\times_S$, respectively) is a single labeled vertex $\bullet_a$ [11].

The notion of tree-width and path-width is defined by Robertson and Seymour in [18] and [17], respectively.

---

[3] The abbreviation NLC results from the *node label controlled* embedding mechanism originally defined for graph grammars.

Definition 3 (Tree-width and path-width, [18,17]). *A tree decomposition of a graph $G = (V_G, E_G)$ is a pair $(\mathcal{X}, T)$ where $T = (V_T, E_T)$ is a tree and $\mathcal{X} = \{X_u \mid u \in V_T\}$ is a family of subsets $X_u \subseteq V_G$ one for each node u of T such that*

1. *$\bigcup_{u \in V_T} X_u = V_G$,*
2. *for every edge $\{v_1, v_2\} \in E_G$, there is some node $u \in V_T$ such that $v_1 \in X_u$ and $v_2 \in X_u$, and*
3. *for every vertex $v \in V_G$ the subgraph of T induced by the nodes $u \in V_T$ with $v \in X_u$ is connected.*

The width *of a tree decomposition $(\mathcal{X} = \{X_u \mid u \in V_T\}, T = (V_T, E_T))$ is $\max_{u \in V_T} |X_u| - 1$. A tree decomposition $(\mathcal{X}, T)$ is called a* path decomposition *if T is a path. The* tree-width *(path-width) of a graph G is the smallest integer k such that there is a tree decomposition (a path decomposition, respectively) $(\mathcal{X}, T)$ for G of width k.*

The *line graph* $L(G)$ of a graph G has a vertex for every edge of G and an edge between two vertices if the corresponding edges in G have a common vertex [20]. Graph G is called the *root graph* of $L(G)$. For any line graph with at least 4 edges the root graph is unique and can be found in linear time [15].

The *incidence graph* $I(G) = (V_{I(G)}, E_{I(G)})$ of a graph $G = (V_G, E_G)$ is the graph with vertex set $V_{I(G)} = V_G \cup E_G$ and edge set $E_{I(G)} = \{\{u, e\} \mid u \in V_G, e \in E_G, u \in e\}$. The incidence graph of G is the graph we get, if we replace every edge $\{u, v\}$ of G by a new vertex w and two edges $\{u, w\}, \{w, v\}$.

3   Line Graphs of Bounded NLC-Width

Tree-width bounded graphs can also be defined by a merging procedure of so-called *terminal graphs*, which are also called *sourced graphs*. This is a well-known property of tree-width bounded graphs, see also [2]. We will define terminal graphs with edge labels, because this will allow us to define in an easy way the edge labeled root graphs of vertex labeled line graphs.

Let k, l be two positive integers. An *l-labeled k-terminal graph* is a system

$$G = (V_G, E_G, P_G, \mathrm{lab}_G),$$

where (V_G, E_G) is a graph, $P_G = (u_1, \ldots, u_k)$ is a sequence of $k \geq 0$ distinct vertices of V_G, and $\mathrm{lab}_G : E_G \to [l]$ is an edge labeling. The vertices in sequence P_G are called *terminal vertices* or *terminals* for short. The vertex u_i, $1 \leq i \leq k$, is the *i-th terminal* of G. The other vertices in $V_G - P_G$ are called *inner vertices*. The class $\mathrm{TM}_{k,l}$ of l-labeled k-terminal graphs is recursively defined as follows.

1. The terminal graph $\overbrace{\bullet \cdots \bullet}^{r}$, $1 \leq r \leq k$, consisting of r terminals is in $\mathrm{TM}_{k,l}$.
2. The terminal graph $\bullet \overset{a}{\text{—}} \bullet$, $a \in [l]$, consisting of two terminals u, v and an edge $\{u, v\}$ labeled by a is in $\mathrm{TM}_{k,l}$ for $k \geq 2$.

3. Let $G = (V_G, E_G, P_G, \text{lab}_G) \in \text{TM}_{k,l}$, $P = (u_1, \ldots, u_r)$, and $f : [r] \to [r]$, be a bijection. Then the l-labeled r-terminal graph $G|^f = (V_G, E_G, P', \text{lab}_G)$ with $P' = (u_{f(1)}, \ldots, u_{f(r)})$ is in $\text{TM}_{k,l}$.

4. Let $G = (V_G, E_G, P_G, \text{lab}_G) \in \text{TM}_{k,l}$, $P = (u_1, \ldots, u_r)$, and $s \in [r]$. Integer s is also called a *decrement*. Then the l-labeled $(r - s)$-terminal graph $G|_s = (V_G, E_G, P', \text{lab}_G)$ with $P' = (u_1, \ldots, u_{r-s})$ is in $\text{TM}_{k,l}$.

5. Let $G = (V_G, E_G, P_G, \text{lab}_G) \in \text{TM}_{k,l}$ and $R : [l] \to [l]$ be a *relabeling mapping*. Then the terminal graph $\circ_R(G) = (V_G, E_G, P_G, \text{lab}')$ with $\text{lab}'(e) = R(\text{lab}_G(e))$ for all $e \in E_G$ is in $\text{TM}_{k,l}$.

6. Let $H = (V_H, E_H, P_H, \text{lab}_H) \in \text{TM}_{k,l}$, $J = (V_J, E_J, P_J, \text{lab}_J) \in \text{TM}_{k,l}$, and $|P_H| \leq |P_J|$. Then terminal graph $H \times J$ defined as follows is in $\text{TM}_{k,l}$.

   (a) Take the disjoint union of (V_H, E_H, lab_H) and (V_J, E_J, lab_J), and identify the i-th terminal from H with the i-th terminal from J.

   (b) An edge e from $H \times J$ is labeled by $\text{lab}_{H \times J}(e) = \text{lab}_H(e)$ if it is from H and by $\text{lab}_{H \times J}(e) = \text{lab}_J(e)$ if it is from J.

   (c) The i-th terminal of $H \times J$ is the i-th terminal of J.

   (d) Multiple edges are eliminated by removing the corresponding edges from H.

An expression built with the operations $\overbrace{\bullet \cdots \bullet}^{r}$, $\bullet \overset{a}{-} \bullet$, $|^f$, $|_s$, $\circ_R$, and $\times$ is called a *terminal k, l-expression*. The terminal graph defined by a terminal k, l-expression X is denoted by $\text{val}(X)$. It is easy to see that $\text{TM}_{k+1,1}$ defines exactly the set of graphs of tree-width at most k, see [10].

Let $G = (V_G, E_G, P_G, \text{lab}_G)$ be an edge labeled terminal graph, $\mathcal{G} = (V_\mathcal{G}, E_\mathcal{G}, \text{lab}_\mathcal{G})$ be a vertex labeled graph, and $\pi : E_G \to V_\mathcal{G}$ be a bijection such that 1.) for every $e_1, e_2 \in E_G$, e_1 and e_2 have a common vertex if and only if $\pi(e_1)$ and $\pi(e_2)$ are adjacent in $\mathcal{G}$, and 2.) for every $e \in E_G$, $\text{lab}_G(e) = \text{lab}_\mathcal{G}(\pi(e))$. Then $\mathcal{G}$ is called the *labeled line graph* of G, and G is called a *labeled terminal root graph* for $\mathcal{G}$.

The next theorem shows a very tight connection between the tree-width of a graph and the NLC-width of its line graph.

Theorem 1. *If a line graph has NLC-width at most k, then its root graph has tree-width at most $4k - 1$.*

Proof Sketch. Let us first observe what happens if we insert edges between two vertex labeled line graphs by an NLC-width operation. Let $G = (V_G, E_G, \text{lab}_G)$ be an edge labeled graph with at least two edges. Let $\mathcal{G} = (V_\mathcal{G}, E_\mathcal{G}, \text{lab}_\mathcal{G}) \in \text{NLC}_k$ be the vertex labeled line graph of G defined by some bijection $\pi : E_G \to V_\mathcal{G}$.

Every induced subgraph of $\mathcal{G}$ defines by bijection π a unique subgraph of G in that every vertex is incident with at least one edge. Assume $\mathcal{G} = \mathcal{H} \times_S \mathcal{J}$ for some $S \subseteq [k]^2$ and two non-empty vertex labeled graphs $\mathcal{H}$ and $\mathcal{J}$. Since $\mathcal{H}$ and $\mathcal{J}$ are induced subgraphs of $\mathcal{G}$, we know that they are line graphs of two subgraphs H and J of G. Since $\mathcal{H}$ and $\mathcal{J}$ are vertex disjoint, we know that H and J are edge disjoint. Since $\mathcal{H}$ and $\mathcal{J}$ have at least one vertex, we know that H and J have at least one edge. Assume further that every pair $(a, b) \in S$ defines

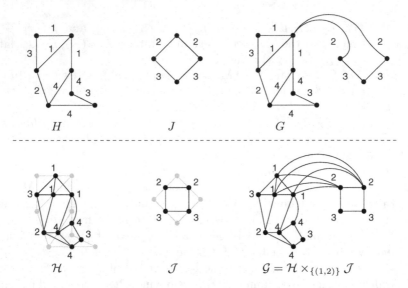

Fig. 1. An NLC-width composition $\mathcal{H} \times_{\{(1,2)\}} \mathcal{J}$ of two vertex labeled line graphs $\mathcal{H}$ and $\mathcal{J}$. The labels at the edges of H, J, and G represent the labels of the corresponding vertices of $\mathcal{H}$, $\mathcal{J}$, and $\mathcal{G}$ specified by bijection π.

at least one edge between a vertex of $\mathcal{H}$ and a vertex of $\mathcal{J}$. If S is nonempty, then in G at least one edge of H has a common vertex with at least one edge of J.

We now show that G can be defined by a vertex disjoint union of H and J and then identifying at most $4k$ vertices from H with at most $4k$ vertices from J. A simple example of such a composition $\mathcal{H} \times_S \mathcal{J}$ is shown in Figure 1.

For a label $a \in [k]$ let G_a, H_a, and J_a be the subgraphs of G, H, and J, respectively, defined by the edges e (and their end vertices) labeled by a. Let $(a, b) \in S$ be a pair of S. Then the operation $\times_S$ connects every vertex of $\mathcal{H}$ labeled by a with every vertex of $\mathcal{J}$ labeled by b. Thus, in root graph G every edge of H_a has a common vertex with every edge of J_b. Let $e = \{u, v\}$ be any edge of H_a. Then every edge of J_b either contains vertex u or vertex v. If J_b has three or more edges, then at least two of them must have a common vertex. By the same argumentation, if H_a has three or more edges then at least two of them must have a common vertex. Thus, H_a and J_b have at most two connected components. If H_a has two connected components, then all edges of every connected component have exactly one common vertex, because an edge of J_b can only contain one vertex from every of the two connected components of H_a. If H_a is connected then it contains no simple path with 6 vertices and no simple cycle with 3 or 5 vertices.

This observation leeds to a case distinction which divides all subgraphs H_a, $a \in [k]$, of H into 8 distinct types as illustrated in Figure 2. Type 8 of Figure 2 represents all graphs that have neither a vertex u such that all edges are incident with u nor two non-adjacent vertices u, v such that every edge is incident with u or v.

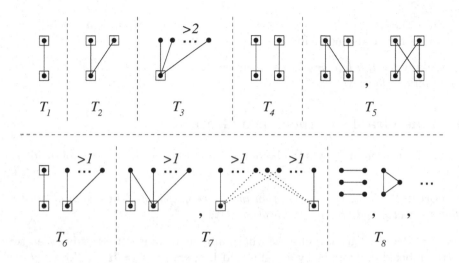

Fig. 2. Eight types for the subgraphs H_a and J_b of H and J, respectively. The specific vertices are framed by squares.

Graphs of Type 1, 2, 3, and 5 have one connected component. Graphs of Type 4 and 6 have two connected components. Graphs of Type 7 have one or two connected components. Every graph of Type 1 to 7 has at most 4 *specific* vertices of which some can be in both graphs, in H_a and in J_b. In Figure 2, these specific vertices are framed by squares.

Since the edges of G are labeled by at most k labels, it follows that at most $4k$ vertices of H are contained in J. That is, at most $4k$ vertices of H and at most $4k$ vertices of J have to be identified to define G from a vertex disjoint union of H and J. Graph G itself has also at most $4k$ vertices which can be identified with other vertices during further composition steps.

This allows us to define for an arbitrary NLC-width k-expression X that defines a line graph a mapping σ that associates for every subexpression X' of X a terminal $4k, k$-expression $\sigma(X')$ such that $\mathrm{val}(\sigma(X'))$ is the edge labeled terminal root graph of $\mathrm{val}(X')$.

1. If $X = \bullet_a$ for some $a \in [k]$ then let $\sigma(X) = \bullet \stackrel{a}{\longrightarrow} \bullet$.
2. If $X = \circ_R(X')$ for some relabeling $R : [k] \to [k]$ then let $\sigma(X) = \circ_R(\sigma(X'))$.
3. If $X = X_1 \times_S X_2$ for some $S \subseteq [k]^2$ then $\sigma(X)$ can be defined by

$$\sigma(X) = ((\sigma(X_1) \times (\sigma(X_2) \times \overbrace{\bullet \cdots \bullet}^{r})|^{f_1})|^{f_2})|_s$$

with two bijections f_1, f_2, a decrement s, and some $r \leq 4k$. $\sigma(X)$ can be defined as above with some $r \leq 4k$, although not all terminals of $\mathrm{val}(\sigma(X_1))$ need to be identified with terminals of $\mathrm{val}(\sigma(X_2))$ via $\mathrm{val}(\overbrace{\bullet \cdots \bullet}^{r})$, or vice versa, for the complete proof of this non trivial fact see [10]. □

Since the NLC-width of a graph is always less than or equal to its clique-width [12], Theorem 1 also holds for line graphs of clique-width at most k.

Corollary 1. *If a line graph has clique-width at most k, then its root graph has tree-width at most $4k - 1$.*

4   Line Graphs of Incidence Graphs

The next theorem improves the bound of Theorem 1 for line graphs of incidence graphs.

Theorem 2. *If the line graph of an incidence graph has NLC-width at most k, then its root graph has tree-width at most k.*

Proof Sketch. Let us now observe what happens if we insert edges between two vertex labeled line graphs by an NLC-width operation $\mathcal{G} = \mathcal{H} \times_S \mathcal{J}$, $S \subseteq [k]^2$ if the root graphs G, H, and J of $\mathcal{G}$, $\mathcal{H}$, and $\mathcal{J}$, respectively, are incidence graphs. Let again G_a, $a \in [k]$, be the terminal subgraph of a terminal graph G defined by the edges (and their end vertices) labeled by a.

Since any incidence graph (and also any subgraph of an incidence graph) has no cycle of length < 6 and that every edge of an incidence graph (and also any edge of a subgraph of an incidence graph) has one end vertex of degree at most 2, every subgraph G_a, $a \in [k]$, of G can be divided into four types as illustrated in Figure 3, see [10]. Type 4 of Figure 3 represents all incidence graphs with two non-adjacent vertices u, v and an edge not incident with u or v. If G_a is of Type 4, then no vertex of G_a needs to be a terminal of G.

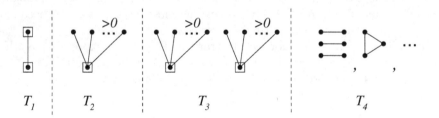

Fig. 3. Four types for the subgraphs G_a of a terminal incidence graph G. The specific vertices are framed by squares.

The same argumentation as in the proof of Theorem 1 now shows that for an arbitrary NLC-width k-expression X that defines a line graph of an incidence graph there is a mapping σ that associates for every subexpression X' of X a terminal $2k$, k-expression $\sigma(X')$ such that $\mathrm{val}(\sigma(X'))$ is the edge labeled terminal root graph of $\mathrm{val}(X')$.

We next transform $\sigma(X)$ into a terminal $2k$, k-expression Y such that every subexpression defines a connected terminal graph. This is possible, because the final root graph $\sigma(X)$ is connected, see [10]. Now every subexpression Y' of Y is of the form

1. $Y' = \bullet \xrightarrow{a} \bullet$ for some $a \in [k]$,
2. $Y' = Y'_1|^f$ for some bijection f,
3. $Y' = Y'_1|_s$ for some decrement s,
4. $Y' = \circ_R(Y'_1)$ for some relabeling R, or
5. $Y' = ((Y'_1 \times (Y'_2 \times \overbrace{\bullet \cdots \bullet}^{r})|^{f_1})|^{f_2})|_s$ for bijections f_1, f_2, some $r \le 2k$, and a decrement s.

These subexpressions define connected terminal graphs. For every of these subexpressions Y' there is an NLC-width k-expression X' such that val(Y') is the edge labeled root graph of the vertex labeled line graph val(X').

Now we will show that Y can be transformed into an equivalent terminal $k+1, k$-expression. Let Y' be a subexpressions of Y of the form stated above and let $G = $ val(Y'). Let again G_a for some $a \in [k]$ be the terminal subgraph of G defined by the edges (and their end vertices) labeled by a.

1. If all subgraphs G_a, $a \in [k]$, of G are of Type 1 of Figure 3, then G has at most k edges. Since G is connected, it has at most $k+1$ terminals.
2. If all subgraphs G_a, $a \in [k]$, of G are of Type 1, 2, or 4 of Figure 3, and at least one of these subgraphs is of Type 2 or 4, then G has at least one inner vertex. In this case G has at most k terminals, see [10].
3. If some subgraph G_a, $a \in [k]$, of G is of Type 3, then two vertices u_a, v_a of G_a are terminals of G. If u_a, v_a are not adjacent in the root graph val(Y) we can remove them from the terminal vertex list. Otherwise we know that during any further composition these two vertices will get incident only with the missing edge $\{u_a, v_a\}$. We now modify the expression as follows.
A subgraph of Type 3 can only be created in the following two cases.
(a) Let
$$G = \circ_R(H)$$
be a graph such that G has a subgraph G_a, $a \in [k]$ of Type 3, but H has no subgraph of Type 3. Then H is connected and at least one inner vertex, and thus H has at most k terminals. We insert the edge between u_a and v_a now by
$$G = (((\bullet \xrightarrow{a} \bullet \times \circ_R (H)|^{f_1})|^{f_2})|_s)|^{f_3}$$
with three bijections f_1, f_2, f_3 and a decrement $s = 2$. The decrement $s = 2$ removes the two vertices u_a, v_a from the terminal vertex list. (This can be done for all subgraphs G_a, $a \in [k]$, of G of Type 3 step by step.)
(b) Let
$$G = (H \times (J \times \overbrace{\bullet \cdots \bullet}^{r})|_{f_1})|_{f_2}$$
be a graph such that G has a subgraph G_a of Type 3, but H and J have no subgraphs of Type 3. Then H and J are connected and have at least

one inner vertex, thus H and J have at most k terminals. Let u_a from H and v_a from J. We insert the edge between u_a, v_a of G_a by

$$G = ((H|^{f_3} \times ((J|^{f_2} \times (\bullet \overset{a}{\text{---}} \bullet \times \overbrace{\bullet \cdots \bullet}^{r'})|^{f_1})|_{s_1} \times \overbrace{\bullet \cdots \bullet}^{r})|^{f_4})|_{s_2})|^{f_5}$$

with bijections f_1, f_2, f_3, f_4, f_5 and decrements $s_1 = 1, s_2 = 1$. If J has k' terminals then $r' = k' + 1$. Let u_a be from H and v_a be from J. One end vertex of edge $\bullet \overset{a}{\text{---}} \bullet$ will be identified with the terminal v_a of J. Decrement $s_1 = 1$ will remove this vertex from the terminal vertex list. The other end vertex of edge $\bullet \overset{a}{\text{---}} \bullet$ will then be identified with u_a from H. The final restriction $s_2 = 1$ will remove this vertex from the terminal vertex list. (This can be done for all subgraphs G_a, $a \in [k]$, of G of Type 3 step by step in the same way.)
In both cases, the composition step which originally inserts the edge between u_a and v_a will be omitted.

Now the resulting composition is set up with terminal graphs that have at most $k + 1$ terminals.   □

Since the NLC-width of a graph is always less than or equal to its clique-width [12], Theorem 2 also holds for line graphs of incidence graphs of clique-width at most k.

Corollary 2. *If the line graph of an incidence graph has clique-width at most k, then its root graph has tree-width at most k.*

5   The NP-Completeness of NLC-Width Minimization

Since a graph G has tree-width k if and only if its incidence graph $I(G)$ has tree-width k, see for example [16], Theorem 1, 2, Corollary 1, 2 and the results of [10] together now imply the following bounds.

(1.) $\dfrac{\text{tree-width}(G)+1}{4} \leq$ NLC-width$(L(G))$ $\leq$ tree-width$(G) + 2$

(2.) $\dfrac{\text{tree-width}(G)+1}{4} \leq$ clique-width$(L(G))$ $\leq 2 \cdot$ tree-width$(G) + 2$

(3.) $\dfrac{\text{path-width}(G)+1}{4} \leq$ linear-NLC-width$(L(G))$ $\leq$ $2 \cdot$ path-width(G)

(4.) $\dfrac{\text{path-width}(G)+1}{4} \leq$ linear-clique-width$(L(G))$ $\leq 2 \cdot$ path-width$(G) + 1$

(5.) tree-width$(G) \leq$ NLC-width$(L(I(G)))$ $\leq$ tree-width$(G) + 2$

(6.) tree-width$(G) \leq$ clique-width$(L(I(G)))$ $\leq 2 \cdot$ tree-width$(G) + 2$

(7.) $\dfrac{\text{path-width}(G)+1}{2} \leq$ linear-NLC-width$(L(I(G))) \leq 2 \cdot$ path-width$(G) + 2$

(8.) $\dfrac{\text{path-width}(G)+1}{2} \leq$ linear-clique-width$(L(I(G))) \leq 2 \cdot$ path-width$(G) + 3$

Inequality (5.) can be used to show that NLC-width minimization is NP-complete.

Theorem 3. *Given a graph G and an integer k, the problem to decide whether G has NLC-width at most k is NP-complete.*

Proof. The problem to decide whether a given graph has NLC-width at most k is obviously in NP.

For a graph $G = (V, E)$ and some integer $r > 1$ let G^r be the graph G in that every vertex u is replaced by a clique C_u with r vertices and every edge $\{u, v\}$ is replaced by all edges between the vertices of C_u and C_v. That is, $G^r = (V_r, E_r)$ has vertex set $V_r = \{u_{i,j} \mid u_i \in V, j \in \{1, \ldots, r\}\}$ and edge set

$$E_r = \{\{u_{i,j}, u_{i',j'}\} \mid j, j' = 1, \ldots, r \text{ and } i = i' \vee \{u_i, u_{i'}\} \in E)\}.$$

Bodlaender et al. have shown in [3], that G has tree-width k if and only if G^r has tree-width $r(k + 1) - 1$.

Arnborg et al. have shown in [1] that tree-width minimization is NP-complete. That is, given a graph G and an integer k, the problem to decide whether G has tree-width at most k, is NP-complete.

For a given graph G, we first construct the graph G^3, then the incidence graph $I(G^3)$, and then the line graph $L(I(G^3))$. This can be done in polynomial time. If G has tree-width k, then G^3 has tree-width $3k + 2$, and $I(G^3)$ has tree-width $3k + 2$. By Theorem 2 graph $L(I(G^3))$ has NLC-width at least $3k + 2$ and by Theorem 3 of [9] NLC-width at most $3k + 4$. That is, tree-width$(G) = \left\lfloor \frac{\text{NLC-width}(L(I(G^3))) - 2}{3} \right\rfloor$. Thus, a graph G has tree-width at most k if and only if $L(I(G))$ has NLC-width at most $3k + 4$ which completes our proof.   □

In [3] it is also shown that there is no polynomial time approximation algorithm for tree-width with constant difference guarantee, unless P = NP, and that for every ϵ, $0 < \epsilon < 1$, there is no polynomial time algorithm that computes for a given graph G a tree decomposition of width k such that $k - \text{tree-width}(G) \leq |V_G|^\epsilon$, unless P = NP. Inequality (5.) can be used again to show similar results for NLC-width approximation, see [10].

Corollary 3.

1. *For every positive integer c there is no polynomial time approximation algorithm that computes for a given graph G an NLC-width k-expression such that $k - NLC\text{-}width(G) \leq c$, unless $P = NP$.*
2. *For every ϵ, $0 < \epsilon < \frac{1}{2}$, there is no polynomial time approximation algorithm that computes for a given graph G an NLC-width k-expression such that $k - NLC\text{-}width(G) \leq |V_G|^\epsilon$, unless $P = NP$.*

References

1. S. Arnborg, D.G. Corneil, and A. Proskurowski. Complexity of finding embeddings in a k-tree. *SIAM Journal of Algebraic and Discrete Methods*, 8(2):227–284, 1987.
2. S. Arnborg, B. Courcelle, A. Proskurowski, and D. Seese. An algebraic theory of graph reduction. *Journal of the ACM*, 40(5):1134–1164, 1993.

3. H.L. Bodlaender, J.R. Gilbert, H. Hafsteinsson, and T. Kloks. Approximating treewidth, pathwidth, frontsize, and shortest elimination tree. *Journal of Algorithms*, 18(2):238–255, 1995.
4. D.G. Corneil, M. Habib, J.M. Lanlignel, B. Reed, and U. Rotics. Polynomial time recognition of clique-width at most three graphs. In *Proceedings of Latin American Symposium on Theoretical Informatics*, volume 1776 of *LNCS*, pages 126–134. Springer-Verlag, 2000.
5. D.G. Corneil, Y. Perl, and L.K. Stewart. A linear recognition algorithm for cographs. *SIAM Journal on Computing*, 14(4):926–934, 1985.
6. B. Courcelle, J.A. Makowsky, and U. Rotics. Linear time solvable optimization problems on graphs of bounded clique-width. *Theory of Computing Systems*, 33(2):125–150, 2000.
7. B. Courcelle and S. Olariu. Upper bounds to the clique width of graphs. *Discrete Applied Mathematics*, 101:77–114, 2000.
8. W. Espelage, F. Gurski, and E. Wanke. How to solve NP-hard graph problems on clique-width bounded graphs in polynomial time. In *Proceedings of Graph-Theoretical Concepts in Computer Science*, volume 2204 of *LNCS*, pages 117–128. Springer-Verlag, 2001.
9. F. Gurski and E. Wanke. Vertex disjoint paths on clique-width bounded graphs (Extended abstract). In *Proceedings of Latin American Symposium on Theoretical Informatics*, volume 2976 of *LNCS*, pages 119–128. Springer-Verlag, 2004.
10. F. Gurski and E. Wanke. Line graphs of bounded clique-width. Manuscript, available at "http://www.acs.uni-duesseldorf.de/~gurski", submitted, 2005.
11. F. Gurski and E. Wanke. On the relationship between NLC-width and linear NLC-width. Manuscript, accepted for Theoretical Computer Science, 2005.
12. Ö. Johansson. Clique-decomposition, NLC-decomposition, and modular decomposition - relationships and results for random graphs. *Congressus Numerantium*, 132:39–60, 1998.
13. Ö. Johansson. NLC$_2$-decomposition in polynomial time. *International Journal of Foundations of Computer Science*, 11(3):373–395, 2000.
14. D. Kobler and U. Rotics. Edge dominating set and colorings on graphs with fixed clique-width. *Discrete Applied Mathematics*, 126(2-3):197–221, 2003.
15. P.G.H. Lehot. An optimal algorithm to detect a line graph and output its root graph. *Journal of the ACM*, 21(4):569–575, 1974.
16. V. Lozin and D. Rautenbach. The tree- and clique-width of bipartite graphs in special classes. Technical Report RRR 33-2004, Rutgers University, 2004.
17. N. Robertson and P.D. Seymour. Graph minors I. Excluding a forest. *Journal of Combinatorial Theory, Series B*, 35:39–61, 1983.
18. N. Robertson and P.D. Seymour. Graph minors II. Algorithmic aspects of tree width. *Journal of Algorithms*, 7:309–322, 1986.
19. E. Wanke. k-NLC graphs and polynomial algorithms. *Discrete Applied Mathematics*, 54:251–266, 1994.
20. H. Whitney. Congruent graphs and the connectivity of graphs. *American Journal of Mathematics*, 54:150–168, 1932.

Channel Assignment and Improper Choosability of Graphs

Frédéric Havet and Jean-Sébastien Sereni

MASCOTTE, I3S-CNRS/INRIA/UNSA, 2004 Route des Lucioles, BP93
F-06902, Sophia-Antipolis Cedex, France
{fhavet, sereni}@sophia.inria.fr

Abstract. We model a problem proposed by Alcatel, a satellite building company, using improper colourings of graphs. The relation between improper colourings and maximum average degree is underlined, which contributes to generalise and improve previous known results about improper colourings of planar graphs.

1 Introduction

In this paper, we investigate the following problem proposed by Alcatel, a satellite building company. A satellite sends information to receivers on earth, each of which is listening on a frequency. Technically it is impossible to focus the signal sent by the satellite exactly on receiver. So part of the signal is spread in an area around it creating noise for the other receivers displayed in this area and listening on the same frequency. A receiver is able to distinguish the signal directed to it from the extraneous noises it picks up if the sum of the noises does not become too big, i.e. does not exceed a certain threshold T. The problem is to assign frequency to the receivers in such a way that each receiver gets its dedicated signal properly. We investigate this problem in the fundamental case where the noise area at a receiver does not depend on the frequency and where the "noise relation" is symmetric that is if a receiver u is in the noise area of a receiver v then v is in the noise area of u. Moreover the intensity I of the noise created by a signal is independent of the frequency and the receiver. Hence to distinguish its signal from noises, a receiver must be in the noise area of at most $k = \lfloor \frac{T}{I} \rfloor$ receivers listening signals on the same frequency.

We model this problem in a graph colouring problem. We define a *noise graph*: the vertices are the receivers and we put an edge between u and v if u is in the noise area of v (and v in the noise area of u). The frequencies are represented by colours. So assigning frequencies to receivers is equivalent to k-improper colouring the noise graph. Indeed the *impropriety* of a vertex v of a graph G under the colouring c, denoted by $\mathrm{im}_G^c(v)$, is the number of neighbours of v coloured $c(v)$. A colouring is *k-improper* if all the vertices have impropriety at most k under it. Note that 0-improper colouring is the usual notion of proper colouring.

Due to some practical reasons (as, for instance, the specific environment of a receiver), the colour of each vertex v must be chosen among a list of colours

D. Kratsch (Ed.): WG 2005, LNCS 3787, pp. 81–90, 2005.

$L(v)$ (that represents the frequencies allowed for that receiver). Formally, given a graph G, an l-*list-assignment* L of G is an function which assigns to each vertex of G a list of at least l colours. An L-*colouring* of G is a vertex colouring in which each vertex v is assigned a colour of the list $L(v)$. G is k-*improper L-colourable* if there exists a k-improper L-colouring of G. G is said to be k-*improper l-choosable* if such a colouring exists for any l-list-assignment.

Improper choosability of planar graphs has been widely studied. In particular, any planar graph is known to be 0-improper 5-choosable [8] and 2-improper 3-choosable [3,6]. It is conjectured that any planar graph is 1-improper 4-choosable. Škrekovski [7] studied k-improper 2-choosability of planar graphs in relation with their girth (the *girth* of a graph G is the size of a smallest cycle of G). Denoting by g_k the smallest integer such that every planar graph of girth at least g_k is k-improper 2-choosable, he proved $6 \leq g_1 \leq 9$, $5 \leq g_2 \leq 7$, $5 \leq g_3 \leq 6$ and $\forall k \geq 4, g_k = 5$. In this paper, we study improper colourings of (not necessarily planar) graphs in relation with their density. Not only does this approach generalise and improve the results of [7] concerning planar graphs, but it also has practical interest since the noise graphs modelling Alcatel's networks have bounded density.

The *average degree* of a graph G, denoted by $\mathrm{Ad}(G)$, is the sum of the degree of each vertex divided by the number of vertices. The *maximum average degree* of G, denoted by $\mathrm{Mad}(G)$, is the maximum of the average degree of each of its subgraphs (including G). If G is not a forest, the *heart* of G, denoted by $h(G)$, is the biggest subgraph of G in which every vertex has degree at least 2. It can be obtained by consecutive removing of vertices of degree 1.

Proposition 1. *If G is not a forest, then* $\mathrm{Mad}(G) = \mathrm{Mad}(h(G))$.

Proof. As $h(G)$ is a subgraph of G, $\mathrm{Mad}(G) \geq \mathrm{Mad}(h(G))$. Let H be a subgraph of G such that $\mathrm{Mad}(G) = \mathrm{Ad}(H)$. Then H is not a forest since otherwise we would have $\mathrm{Mad}(G) < 2$ and G would be a forest. So $h(H)$ is defined and it is a subgraph of $h(G)$. Moreover, $h(H)$ has minimum degree at least 2, so adding to it vertices of degree 1 cannot increase its average degree: let H' be a supergraph obtained from $h(H)$ by adding $k \geq 1$ vertices of degree 1. We assume that $h(H)$ has n vertices. Then

$$\mathrm{Ad}(H') = \frac{n \times \mathrm{Ad}(h(H)) + 2k}{n+k} = \mathrm{Ad}(h(H)) + \frac{2k - k \times \mathrm{Ad}(h(H))}{n+k} \leq \mathrm{Ad}(h(H))$$

since $\mathrm{Ad}(h(H)) \geq 2$. So $\mathrm{Mad}(h(G)) \geq \mathrm{Ad}(h(H)) \geq \mathrm{Ad}(H) = \mathrm{Mad}(G)$.

Let $M(k,l)$ be the greatest real such that every graph of maximum average degree less than $M(k,l)$ is k-improper l-choosable. Obviously, $M(k_1,l) \leq M(k_2,l)$ if $k_1 \leq k_2$. We have that $M(k,1) = \frac{2k+2}{k+2}$ since a graph is k-improper 1-choosable if and only if it has maximum degree at most k (and a graph of maximum degree at least $k+1$ contains the star S_{k+1} as a subgraph, so it has maximum average degree at least $\frac{2k+2}{k+2}$). If $l \geq 2$, first note that any tree is 0-improper 2-choosable. Furthermore, for any $k \geq 0$, a graph G which is not a forest is k-improper 2-choosable if and only if its heart is. Hence, we shall restrict the study to graphs with minimum degree at least 2.

In the following section, we show:

Theorem 1. *For all $k \geq 0$, all graphs of maximum average degree less than $\frac{4k+4}{k+2}$ are k-improper 2-choosable.*

Theorem 2. *For all $k \geq 1$, $M(k,2) \leq \dfrac{4k^2 + 6k + 4}{k^2 + 2k + 2} = 4 - \dfrac{2k+4}{k^2 + 2k + 2}$.*

We then generalise Theorem 1:

Theorem 3. *For all $l \geq 2$ and all $k \geq 0$, all graphs of maximum average degree less than $\frac{l(l+2k)}{l+k}$ are k-improper l-choosable.*

Corollary 1. *For any fixed l, $\lim\limits_{k \to +\infty} M(k,l) = 2l$.*

Using Euler's formula, one can show that if G is a planar graph with minimum degree at least 2 and girth at least g, then $\mathrm{Mad}(G) < \frac{2g}{g-2}$. So Theorem 1 immediately implies:

Corollary 2. *Let G be a planar graph of girth g.*

1. *If $g \geq 8$ then G is 1-improper 2-choosable, so $g_1 \leq 8$.*
2. *If $g \geq 6$ then G is 2-improper 2-choosable, so $g_3 \leq g_2 \leq 6$.*
3. *If $g \geq 5$ then G is 4-improper 2-choosable, so $g_k \leq 5$ for $k \geq 4$.*

Some proofs are omitted or just sketched. The detailed proofs are presented in [4].

2   Improper 2-Choosability

2.1   Lower Bound for $M(k,2)$

In this subsection, we shall prove Theorem 1. Note that if $k = 0$ then Theorem 1 holds trivially. Indeed a graph with maximum average degree less than 2 contains no cycle and so it is a forest. Hence it is 2-choosable. Furthermore $M(0,2) \leq 2$ since an odd cycle is not 2-colourable, so $M(0,2) = 2$. For bigger values of k, we will need the following preliminary definitions and results:

Definition 1. *If $v \in V(G)$ then $d_G(v)$ denotes the degree of v in the graph G. For all positive integer p, a vertex of degree equal to (resp. at most, resp. at least) p is called a p-vertex (resp. ($\leq p$)-vertex, resp. ($\geq p$)-vertex). For $S \subseteq V(G)$ (resp. $E \subseteq E(G)$) we denote by $G-S$ (resp. $G-E$) the induced subgraph of G obtained by removing the vertices (resp. edges) of S (resp. E) from $V(G)$ (resp. $E(G)$). If $S = \{v\}$ and $E = \{uv\}$, we shall note $G-v = G-S$ and $G-uv = G-E$. The union (resp. intersection) of the graphs G_1 and G_2 is the graph $G = G_1 \cup G_2$ (resp. $G = G_1 \cap G_2$) such that $V(G) = V(G_1) \cup V(G_2)$ (resp. $V(G) = V(G_1) \cap V(G_2)$) and $E(G) = E(G_1) \cup E(G_2)$ (resp. $E(G) = E(G_1) \cap E(G_2)$). Let D be a digraph* and u one of its vertices. An *outneighbour* (resp. *inneighbour*) of u in D is a vertex v of D such that there exists an arc from u to v (resp. from v to u) in D. The *outdegree* (resp. *indegree*) of u in D, denoted by $d_D^+(u)$ (resp. $d_D^-(u)$), is the number of outneighbours (resp. inneighbours) of u in D. The *degree* of u is $d_D(u) = d_D^-(u) + d_D^+(u)$; it is the degree of u in the underlying undirected graph.

A graph is said to be $(k, 2)$-*minimal* if it is not k-improper 2-choosable but each of its proper subgraphs is.

The idea of the proof of Theorem 1 is to consider a $(k, 2)$-minimal graph and apply a discharging procedure, the rule of which is to discharge $\frac{k}{k+2}$ along the arcs of a discharging digraph which is obtained using the following process:

1. Orient each edge uv where v is a 2-vertex from u to v.
2. If $k \geq 3$, orient each edge uv where v is a 3-vertex from u to v.
3. While there is an unoriented edge uv where v an i-vertex with outdegree $i - 1$ for some $k + 2 \leq i < \frac{3k}{2} + 2$, we orient it from u to v.

The digraph D induced by the oriented edges is called a *discharging digraph* of G.

The aim of the next lemmata is to establish some properties of such a discharging digraph.

Lemma 1 (Škrekovski [7]). *Let $k \geq 1$ and let G be a $(k, 2)$-minimal graph. Then G has minimum degree at least 2 and two $(\leq k + 1)$-vertices are not adjacent.*

Definition 2. If u and v are two vertices of a digraph D, a (u, v)-*dipath* is a directed path from u to v. The *outsection* of u in D, denoted $A_D^+(u)$, is the set of vertices v such that there is a (u, v)-dipath in D.

An *arborescence* is an oriented tree in which every path is directed from a vertex called the *root*. Note that in an arborescence every vertex except the root has indegree 1. The *leaves* of the arborescence are the vertices of outdegree 0. A vertex which is neither a leaf nor the root is an *internal vertex*. A *quasi-arborescence* is a directed graph obtained from an arborescence by identifying some leaves.

Lemma 2. *Let D be a discharging digraph of a $(k, 2)$-minimal graph, and $k \geq 1$.*

- *D has no 2-circuit since by Lemma 1 two $(\leq k + 1)$-vertices cannot be adjacent. So it has no circuit at all.*
- *If $k \leq 2$, only vertices of degree 2 or $k + 2$ have indegree more than zero.*
- *Every 2-vertex has indegree 2 in D and if $k \geq 3$, every 3-vertex has indegree 3.*
- *For every vertex u, $A_D^+(u)$ is a quasi-arborescence whose leaves have degree 2 (resp. 2 or 3) in G if $k \leq 2$ (resp. $k \geq 3$). In particular, the indegree of the leaves in $A_D^+(u)$ is at most 2 (resp. 3).*

Definition 3. A quasi-arborescence is a $(k, 2)$-*quasi-arborescence* if and only if every vertex has outdegree at most $\max\{2, 2k - 1\}$ and every leaf has indegree at most $\min\{k, 3\}$.

Lemma 3. *Let $k \geq 2$. Let Q be a $(k, 2)$-quasi-arborescence rooted at u and L a 2-list-assignment of Q. Then any L-colouring of the leaves can be extended to a k-improper L-colouring of D such that u has impropriety at most $k - 1$.*

Proof. By induction on the number of vertices of Q, the result being trivially true if $|V(Q)| = 1$.

Suppose now that $|V(Q)| > 1$ and the result holds for smaller k-quasi-arborescences. Let $v_1, \ldots, v_s$ be the outneighbours of u in Q. Note that $Q - u$ is the union of s $(k, 2)$-quasi-arborescences Q_i, $1 \leq i \leq s$ rooted at v_i that are disjoint except possibly on their leaves.

Let c be an L-colouring of the leaves of Q. Then by induction it can be extended to a k-improper L-colouring of each of the Q_i so that $im(v_i) \leq k - 1$. Since a leaf of Q has indegree at most $\min\{k, 3\}$ and $im_Q(x) = im_{Q_i}(x)$ for every vertex of Q_i which is not a leaf, then the union of these colourings is a k-improper L-colouring of Q such that $im(v_i) \leq k - 1, 1 \leq i \leq s$.

Now, one of the two colours of $L(u)$, say α, is assigned to at most $k - 1$ neighbours of u since $s \leq 2k - 1$. Thus setting $c(u) = \alpha$, we obtain the desired colouring. $\square$

Obviously, the above result cannot be extended for $k = 1$ because it is hopeless to extend every L-colouring of the leaves in a colouring such that the root has impropriety 0. However, the following weaker result holds:

Lemma 4. *Let Q be a $(1, 2)$-quasi-arborescence rooted at u, L a 2-list-assignment of Q with $L(u) = \{\alpha, \beta\}$ and c an L-colouring of S, the set of leaves of Q with indegree 1. One of the following holds:*

(i) c can be extended to a 1-improper L-colouring of Q such that $im(u) = 0$;
(ii) c can be extended to two different 1-improper L-colourings of Q c_1 and c_2 such that $c_1(v) = c_2(v)$ if $v \neq u$.

Lemma 5. *Let $k \geq 3$. Let D be a discharging digraph of a $(k, 2)$-minimal graph G.*

(i) Every i-vertex with $4 \leq i \leq k + 1$ has outdegree zero.
(ii) Every i-vertex with $k + 2 \leq i \leq 2k + 1$ has outdegree less than i.

Proof. (i) Suppose, for a contradiction, that v is a vertex contradicting the assertion and let u be an outneighbour of v. Note that u is a $(< \frac{3k}{2} + 2)$-vertex by definition of a discharging digraph.

Let L be a 2-list-assignment of G. Let S be the set of leaves of $A_D^+(u)$. By minimality, let c be a k-improper L-colouring of $G - A_D^+(u)$.

$A_D^+(u)$ is a $(k, 2)$-quasi-arborescence: since u is dominated by v in D, u has outdegree less than $\frac{3k}{2} + 1$, and so at most $2k - 1$. Thus, by Lemma 3, we can extend c to $G - vu$ so that $im(u) \leq k - 1$. Since the leaves have degree at most $3 \leq k$, the impropriety of the leaves is at most $3 \leq k$. So we obtain a k-improper L-colouring of $G - uv$.

If $c(u) \neq c(v)$ or $im_{G-uv}(v) \leq k - 1$ then c is a k-improper L-colouring of G. Otherwise all the $k + 1$ neighbours of v are coloured by the same colour so recolouring v with its other allowed colour yields a k-improper L-colouring of G.

Hence G is k-improper 2-choosable which is a contradiction.

(ii) Suppose, for a contradiction, that v is an i-vertex contradicting the assertion. Let L be 2-list-assignment of G and c a k-improper L-colouring of $G - v$. There is a colour of $L(v)$, say α, that is assigned to at most k neighbours of v. Let $v_1, \ldots, v_s$ be these neighbours.

Let $G' = G - \bigcup_{j=1}^{s} A_D^+(v_j)$. And set $c' = c$ for every vertex of G' and every leaf of the $A_D^+(v_j)$. By Lemma 3 applied to each $A_D^+(v_j)$ (which are disjoint except possibly on their leaves), we can extend c' into a k-improper L-colouring of $G - v$ so that $im(v_j) \leq k - 1$ for $1 \leq j \leq s$. Now by definition of c', the only neighbours of v that may be assigned α by c' are those of $\{v_1, \ldots, v_s\}$. Hence setting $c'(v) = \alpha$, the L-colouring c' is k-improper.

Hence G is k-improper 2-choosable which is a contradiction.

Analogously, one can prove the following two lemmata:

Lemma 6. *Let D be a discharging digraph of a $(2,2)$-minimal graph G.*

(i) The outdegree of a 3-vertex is zero.
(ii) If v is an i-vertex with $i \in \{4,5\}$ then its outdegree is less than i.

Lemma 7. *Let D be a discharging digraph of a $(1,2)$-minimal graph G. There is no 3-vertex with outdegree 3 in D.*

Proof (of Theorem 1). Let G be a $(k,2)$-minimal graph and D a discharging digraph of G. We start with a charge $w(v) = d(v)$ on each vertex and we apply the following discharging rule: every vertex gives $\frac{k}{k+2}$ to each of its outneighbours.

Let us examine the new charge $w'(v)$ of a vertex v, regarding its degree:

- If v is a 2-vertex then it has indegree 2 so its new charge is $w'(v) = 2 + \frac{2k}{k+2} = \frac{4k+4}{k+2}$.
- If v is a 3-vertex and $k \geq 3$, then it has indegree 3 so its new charge is $w'(v) = 3 + 3 \times \frac{k}{k+2} = \frac{6k+6}{k+2} > \frac{4k+4}{k+2}$. If v is a 3-vertex and $k = 2$ then it has outdegree 0 by Lemma 6 and indegree 0 by the construction and hence $w'(v) = 3$.
- If $4 \leq d(v) \leq k + 1$, $(k \geq 3)$, then by Lemma 5 (i), v has outdegree 0 so its charge is $d(v) \geq 4 > \frac{4k+4}{k+2}$.
- If $k+2 \leq d(v) < \frac{3k}{2}+2$ then either v has outdegree at most $d(v)-2$ and so its new charge is at least $d(v) - (d(v)-2) \times \frac{k}{k+2} = \frac{2d(v)}{k+2} + \frac{2k}{k+2} \geq 2 + \frac{2k}{k+2} = \frac{4k+4}{k+2}$, or by Lemmata 5-7, it has outdegree $d(v) - 1$. In this case, by definition of a discharging digraph, v has indegree 1 so its new charge is: $d(v) - (d(v) - 1) \times \frac{k}{k+2} + \frac{k}{k+2} = d(v) - (d(v) - 2) \times \frac{k}{k+2} \geq \frac{4k+4}{k+2}$.
- If $\frac{3k}{2} + 2 \leq d(v) \leq 2k+1$, $(k \geq 2)$, then by Lemmata 5 and 6, v has outdegree at most $d(v) - 1$. So $w'(v) \geq d(v) - (d(v) - 1) \times \frac{k}{k+2} = \frac{2d(v)}{k+2} + \frac{k}{k+2} \geq \frac{3k+4+k}{k+2} = \frac{4k+4}{k+2}$.

- If $d(v) \geq 2k + 2$, then $w'(v) \geq d(v)(1 - \frac{k}{k+2}) = \frac{2d(v)}{k+2} \geq \frac{4k+4}{k+2}$.

Hence $\mathrm{Mad}(G) \geq \frac{1}{|V|} \sum_{v \in V} d(v) = \frac{1}{|V|} \sum_{v \in V} w'(v) \geq \frac{4k+4}{k+2}$.

2.2   Upper Bound for $M(k, 2)$

Let us fix $k \geq 1$. In this subsection, we shall construct a family of graphs $(G_n^k)_{n \geq 1}$ such that for all $n \geq 1$:

- G_n^k is not k-improper 2-colourable.
- $\mathrm{Mad}(G_n^k) = \dfrac{2n(4k^2 + 6k + 4) + 4k^2 + 6k + 2}{2n(k^2 + 2k + 2) + (k+1)^2}$.

Hence we will deduce Theorem 2. We denote by H_k the graph composed of two adjacent vertices u and v also connected by $k + 1$ internally disjoint paths of length 2. Take k copies of H_k and create the graph F_k by identifying the vertices v of each copy. Note that F_k has one vertex of degree $k(k + 2)$, k vertices of degree $k + 2$ and $k(k + 1)$ vertices of degree 2. Now we take $2n + 1$ copies of F_k and we join the vertices v of each copy creating a cycle of size $2n + 1$. At last we make a subdivision of all the edges of the cycle but one so as to obtain the graph G_n^k.

Lemma 8. G_n^k is not k-improper 2-colourable.

As it is easily seen, the maximum average degree of G is its average degree, which is equal to M_n^k.

3   Improper l-Choosability, $l \geq 2$

3.1   Lower Bound for $M(k, l)$

In this subsection, we shall prove Theorem 3. The result of the theorem is trivial if $k = 0$ since a graph of maximum average degree less than l is $(l-1)$-degenerate (i.e. each of its subgraphs has a vertex of degree at most $l - 1$). Hence it is l-choosable. For bigger values of k, we will need some preliminary results.

Definition 4. A graph is said to be (k, l)-minimal if it is not k-improper l-choosable but each of its proper subgraphs is.

Lemma 9. Let G be a graph, L a list-assignment and c an L-colouring. If a vertex v has impropriety at least $d(v) - |L(v)| + 2$ under c, then there exists an L-colouring c' of G such that $c'(u) = c(u)$ if $u \neq v$ and $\mathrm{im}_{c'}(v) = 0$.

We now generalise Lemmata 1, 3 and 4.

Lemma 10. Let $k \geq 1$ and let G be a (k, l)-minimal graph. Then G has minimum degree at least l and two $(\leq l + k - 1)$-vertices are not adjacent.

Definition 5. Let G be a (k, l)-minimal graph. We partially orient G using the following process:

1. Orient each edge uv where v is a $(\leq l+k-1)$-vertex from u to v.
2. While there is an i-vertex v with outdegree exactly $i - l + 1$ and indegree 0 for some $l + k \leq i < l + k + \frac{k}{l}$, we orient one of its unoriented incident edges uv from u to v.

The digraph D induced by the oriented edges is called a *discharging digraph* of G. Note that only vertices of degree less than $l + k + \frac{k}{l}$ can have indegree more than zero, and for $i \leq l + k - 1$, every i-vertex has indegree exactly i in D.

A quasi-arborescence rooted at u is a (k, l)-*quasi-arborescence* if and only if every vertex has outdegree at most $\max\{2, 2k - 1\}$ and every leaf has indegree at most $l + k - 1$.

Lemma 11. *Let $k \geq 2$ and let Q be a (k, l)-quasi-arborescence rooted at u. Let L be a list-assignment of Q such that $|L(v)| \geq \max\{1, d_Q(v) - k + 1\}$ if v is a leaf and $|L(v)| \geq 2$ otherwise. We denote by S the set of leaves that have indegree at least $k + 1$ in Q (and hence a colour-list of size at least 2). Any L-colouring of the leaves extends in an L-colouring of Q such that:*

— $im(u) \leq k - 1$.
— $\forall v \notin S, im(v) \leq k$.

Furthermore, possibly by recolouring some vertices of S, this L-colouring of G can be made k-improper.

The above result cannot be extended for $k = 1$. However the following result holds:

Lemma 12. *Let Q be a $(1, l)$-quasi-arborescence rooted at u and L any list-assignment of Q such that $|L(v)| \geq 2$ if v is not a leaf, and $|L(v)| \geq d_Q(v)$ otherwise. We denote by S the set of leaves with indegree at least 2. Let c be an L-colouring of the leaves. One of the followings holds:*

(i) *c can be extended to an L-colouring of Q such that $im(u) = 0$ and $im(v) \leq 1$ if $v \notin S$;*
(ii) *c can be extended to two different L-colourings of Q c_1 and c_2 such that $c_1(v) = c_2(v)$ if $v \neq u$ and $im^{c_i}(v) \leq 1$ if $v \notin S$.*

Furthermore, possibly by recolouring vertices of S, all these L-colourings can be made 1-improper.
Moreover, if $|L(u)| \geq 3$ then (i) holds.

Using these results, we can say more about the structure of a discharging digraph. The following lemma generalises Lemma 2.

Lemma 13. *Let D be a discharging digraph of a (k, l)-minimal graph G.*

(i) *Every vertex u with $l + k \leq d(u) \leq l + 2k - 1$ has outdegree at most $d(u) - l + 1$. In particular, D is acyclic.*

(ii) For every vertex u with indegree 1, $A_D^+(u)$ is a (k,l)-quasi-arborescence. In particular, the indegree of the leaves in $A_D^+(u)$ is at most $l + k - 1$.

Proof (of Theorem 3). Let G be a (k,l)-minimal graph and D a discharging digraph of G. We start with a charge $w(v) = d(v)$ on each vertex and we apply the following discharging rule: every vertex gives $\frac{k}{l+k}$ to each of its outneighbours. One can check that, by using Lemma 13, the new charge of every vertex is at least $l + \frac{lk}{l+k}$.

3.2 Upper Bound for $M(k,l)$

In this subsection we shall construct for all $l \geq 2$ and all $k \geq 1$, a graph G_l^k which is not k-improper l-colourable. So its maximum average degree will give an upper bound for $M(k,l)$. To construct G_2^k, take $k + 1$ copies of H_k (defined in Subsection 2.2) and identify their vertex v. We define G_l^k, $l \geq 3$, inductively. First we create the graph M_l^k by taking k copies of G_{l-1}^k and adding a vertex w which we join to every other vertices. Then we take $l - 1$ copies $M^1, \ldots, M^{l-1}$ of M_l^k and we join all the vertices $w_1, \ldots, w_{l-1}$ (so that they form a complete graph of size $l - 1$). Now, we add $k + 2$ vertices $z_0, z_1, \ldots, z_{k+1}$ each joined to each of the w_i, $1 \leq i \leq l - 1$. Last we add the edges $z_0 z_i$ for $1 \leq i \leq k + 1$.

Lemma 14. *For all $l \geq 2$ and all $k \geq 1$, the graph G_l^k is not k-improper l-colourable.*

Proposition 2. $\mathrm{Mad}(G_l^k)$ *tends to $2l$ as k tends to infinity.*

Proof. It is clear that the maximum average degree of G_l^k is its average degree.

The number of vertices of G_l^k is $n_l^k = 2l + (l+1)k + \sum_{i=2}^{l} \frac{(l-1)!}{(l-i)!} k^i$. Indeed n_l^k satisfies: $n_2^k = k^2 + 3k + 3$ and $\forall l \geq 3, n_l^k = (k \times n_{l-1}^k + 1) \times (l - 1) + k + 2$. In particular, as a polynomial in k, $n_l^k \sim (l-1)!k^l$.

Let s_l^k denotes the sum of the degrees of the vertices in G_l^k. s_l^k satisfies: $s_2^k = 4k^2 + 10k + 6$ and $s_l^k = (l-1)(k \times s_{l-1}^k + 2k \times n_{l-1}^k + l + k) + (l+1)k + 2l$ if $l \geq 3$. Hence it is a polynomial in k of degree l. Furthermore, denoting by c_l^k its dominant coefficient, we have: $c_2^k = 4$ and $\forall l \geq 3, c_l^k = (l-1) \times c_{l-1}^k + 2k \times (l-1)!$. Thus $c_l^k = 2l!$. So $s_l^k \sim 2l!k^l$.

Hence the limit of $\mathrm{Mad}(G_l^k)$ as k tends to infinity is $2\frac{l!}{(l-1)!} = 2l$.

Corollary 1 immediately follows from Theorem 3 and Proposition 2.

References

1. Appel, K. and Haken, W.: Every planar map is four colourable, Part I: Discharging. Illinois J. Math. **21** (1977) 429–490.
2. Appel, K. and Haken, W. and Koch, J.: Every planar map is four colourable, Part II: Reducibility. Illinois J. Math. **21** (1977) 491–567.

3. Eaton, N. and Hull, T.: Defective list colorings of planar graphs. Bull. Inst. Combin. Appl. **25** (1999) 79–87.
4. Havet, F. and Sereni, J.-S.: Improper choosability and maximum average degree. Internal Report INRIA RR-5164, ftp://ftp.inria.fr/INRIA/publication/publi-ps-gz/RR/RR-5164.ps.gz, (2004).
5. Lih, K.-W. and Song, Z. and Wang, W. and Zhang, K.: A Note on List Improper Coloring Planar Graphs. Appl. Math. Let. **14** (2001) 269–273.
6. Škrekovski, R.: List improper colouring of planar graphs. Comb. Prob. Comp. **8** (1999) 293–299.
7. Škrekovski, R.: List improper colorings of planar graphs with prescribed girth. Discrete Math. **214** (2000) 221–233.
8. Thomassen, C.: Every planar graph is 5-choosable. J. Comb. Theory B **62** (1994) 180–181.

Computing Treewidth and Minimum Fill-in for Permutation Graphs in Linear Time

Daniel Meister

Bayerische Julius-Maximilians-Universitaet Wuerzburg,
97074 Wuerzburg, Germany
`meister@informatik.uni-wuerzburg.de`

Abstract. A chordal graph H is a triangulation of a graph G, if H is obtained by adding edges to G. If no proper subgraph of H is a triangulation of G, then H is a minimal triangulation of G. A potential maximal clique of G is a set of vertices that induces a maximal clique in a minimal triangulation of G. We will characterise the potential maximal cliques of permutation graphs and give a characterisation of minimal triangulations of permutation graphs in terms of sets of potential maximal cliques. This results in linear-time algorithms for computing treewidth and minimum fill-in for permutation graphs.

1 Introduction

Treewidth and minimum fill-in are among the most interesting graph parameters. The treewidth of a graph is a measure for the treelikeness of a graph. By the minimum fill-in, the degree of chordality can be measured in a certain sense. Both parameters can be formulated as embedding problems into chordal graphs where, in the case of treewidth, the clique number, in the case of minimum fill-in, the number of edges is to be minimized. Treewidth plays a big role in algorithm design, since many hard problems can be solved efficiently for graphs of bounded treewidth. Minimum fill-in has applications in matrix elimination [13]. The decision problems TREEWIDTH and MINIMUM FILL-IN are NP-complete even for co-bipartite graphs [1], [15].

A triangulation of a graph G is a chordal graph H where G and H have the same vertex set and G is a subgraph of H. H is called a minimal triangulation of G, if no proper subgraph of H is a triangulation of G. Minimal triangulations were first studied by Rose, Tarjan, Lueker [14]. Since then, the study of minimal triangulations has attracted a considerable community of researchers. An explanation seem to be the following observations: The treewidth of a graph equals the smallest clique size minus 1 among its minimal triangulations, and the minimum fill-in of a graph is the smallest number of additional edges among its minimal triangulations. So, treewidth and minimum fill-in become special problems on minimal triangulations.

An early interesting result about minimal triangulations of special graph classes was the result by Bodlaender and Möhring that every minimal triangulation of a cograph is a cograph [4]. Since chordal cographs are interval graphs,

D. Kratsch (Ed.): WG 2005, LNCS 3787, pp. 91–102, 2005.

minimal triangulations of cographs are interval graphs. This result was improved by Bodlaender, Kloks, Kratsch and extended to permutation graphs in the sense that every minimal triangulation of a permutation graph is an interval graph [2]. A first endpoint in this series was set by Möhring's theorem for AT-free graphs: every minimal triangulation of an AT-free graph is an interval graph [10], which was extended by Parra and Scheffler to a complete characterisation of AT-free graphs [12].

The works of Bodlaender, Kloks, Kratsch and Bodlaender, Kloks, Kratsch, Müller resulted in $\mathcal{O}(\text{tw}(G) \cdot n_G)$- and $\mathcal{O}(n^2)$-time algorithms for treewidth and minimum fill-in for permutation graphs, respectively, [2], [3]. For both problems on permutation graphs we will give linear-time algorithms, which improve the stated time bounds. The main part of this paper is dedicated to characterising the minimal triangulations of a permutation graph in an appropriate manner. The central idea is to consider the potential maximal cliques. The notion of a potential maximal clique of a graph was introduced by Bouchitté and Todinca [5]. A *potential maximal clique* of a graph is a set of vertices that is a maximal clique in a minimal triangulation of the graph. We will show that the set of potential maximal cliques of a permutation graph can be generated in linear time. We will define the *potential maximal cliques graph* of a permutation graph, which is a directed graph containing an arc for every potential maximal clique, and show that a maximal path of the potential maximal cliques graph corresponds exactly to a minimal triangulation. Adding weights to vertices and arcs, treewidth and minimum fill-in can be solved in linear time by simply exploiting the obtained and so-called *weighted* potential maximal cliques graph. Our approach to solve treewidth and minimum fill-in can be understood as an improvement of the approach used in [2] and [3], which is based on the fact that the weighted potential maximal cliques graph has less vertices than the auxiliary graph in [2] and [3] and can be generated in linear time.

The paper is organized as follows. Section 2 contains basic definitions and results. In Section 3, we will consider minimal separators and identify special scanlines as appropriate representations. In Section 4, we will characterise the potential maximal cliques of a permutation graph and define the potential maximal cliques graph. We will show the correspondence between the minimal triangulations of a permutation graph and the maximal paths in its potential maximal cliques graph. Finally, in Section 5, we will conclude by obtaining linear-time algorithms for computing treewidth and minimum fill-in for permutation graphs. Proofs may be omitted or sketched due to space restrictions.

2   Preliminaries

We will consider only simple and finite graphs that may be directed or undirected. For a directed graph $G' = (V, A)$, V denotes the set of vertices, and A denotes the set of arcs. Arcs are denoted as (u, v), which means that u is start vertex and v is end vertex. For an undirected graph $G = (V, E)$, V and E denote the sets of vertices and edges, respectively. Since we mostly deal with undirected graphs, most of the following definitions are for undirected graphs. An edge is denoted as uv, which means that vertices u and v are *adjacent*. For $S \subseteq V$, the subgraph of G *induced* by S is denoted as $G[S]$; we write $G \setminus S$ instead of $G[V \setminus S]$. $G \cup F$ for a set F of edges is short for $(V, E \cup F)$; $G - e =_{\text{def}} (V, E \setminus \{e\})$, and $G \setminus F =_{\text{def}} (V, E \setminus F)$. The *neighbourhood* of a vertex $u \in V$, denoted by $N_G(u)$, is the set of vertices adjacent to u in G; $N_G[u] =_{\text{def}} N_G(u) \cup \{u\}$. For $A \subseteq V$, the neighbourhood of A is $N_G(A) =_{\text{def}} \bigcup_{v \in A} N_G(v) \setminus A$. For further definitions, we also refer to [7]. A *path* of length k in G is a sequence $(x_0, \ldots, x_k)$ of $k+1$ different vertices of G where x_{i-1} and x_i are adjacent for every $i \in \{1, \ldots, k\}$; a similar definition holds for G'. A path $(x_1, \ldots, x_k)$ of G is a *cycle* of length k if $x_1 x_k \in E$. A *chord* in a cycle C of G is an edge of G between two non-consecutive vertices in C; a chord is *unique* in C if it is the only edge in G that is a chord in C. C is *chordless* if there is no edge in G that is a chord in C. G is *chordal* if there is no chordless cycle of length at least 4 in G. An *interval graph* on n vertices is the intersection graph of a family of n closed intervals of the real line. Interval graphs are chordal. A triple u, v, w of pairwise non-adjacent vertices of G is an *asteroidal triple*, *AT* for short, if there is a path between any two of them not containing a neighbour of the third. G is *AT-free* if G does not contain three vertices that form an AT. Interval graphs are exactly the chordal graphs that are AT-free [9].

Permutation Graphs. Let $[n]$, $n \geq 1$, denote the set of the numbers $1, 2, \ldots, n$. Let $\pi : [n] \to [n]$ be a bijection. We also say that π is a *permutation sequence over* $[n]$. The graph $G(\pi)$ has vertex set $[n]$, and two vertices $u, v \in [n]$, $u \neq v$, are adjacent if and only if $(u - v)(\pi^{-1}(u) - \pi^{-1}(v)) < 0$. A *permutation graph over* $[n]$ is a graph $G(\pi)$ for some permutation sequence π over $[n]$. A graph G is a *permutation graph* if there is $n \geq 1$ such that G is isomorphic to a permutation graph over $[n]$. Permutation graphs can be represented by *permutation diagrams*; we refer the reader to [7] for more details about permutation diagrams. Since we often use the permutation diagram representation, we identify line segments and the corresponding vertices of the graph. This will never cause confusion.

Separators and Triangulations. Let $G = (V, E)$ be a graph. A graph $H = (W, F)$ is a *triangulation* of G if H is chordal, $V = W$ and $E \subseteq F$. H is a *minimal triangulation* of G if there is no triangulation of G that is a proper subgraph of H.

Theorem 1 ([14]). *Let $G = (V, E)$ and $H = G \cup F$ be graphs where H is chordal and $E \cap F = \emptyset$. Then, H is a minimal triangulation of G if and only if, for every $e \in F$, e is unique chord in a cycle of length 4 in H.*

Let $S \subseteq V$. We call S a c, d-*separator* of G for two non-adjacent vertices $c, d \in V$ if c and d are in different connected components of $G \setminus S$. S is a *minimal* c, d-*separator* of G if there is no c, d-separator S' such that $S' \subset S$. S is a *minimal separator* of G if S is a minimal c, d-separator for some non-adjacent vertices $c, d \in V$. A connected component C of $G \setminus S$ is S-*full* if every vertex of S has a neighbour in C; S-*full components* of G are S-full connected components of $G \setminus S$. Then, S is a minimal separator of G if and only if G has two S-full components [7]. S is a minimal c, d-separator of G if and only if c and d are in different S-full components of G. Two minimal separators S_1 and S_2 of G *cross* if there are two connected components in $G \setminus S_1$ that contain vertices from S_2. The crossing relation is symmetric [12].

Theorem 2 ([12]). *Let $G = (V, E)$ be a graph.*

1. *Let S be a maximal set of pairwise non-crossing minimal separators of G. Graph H is obtained from G by making all separators in S into cliques. Then, H is a minimal triangulation of G.*
2. *Let H be a minimal triangulation of G and let S be the set of minimal separators of H. Then, S is a maximal set of pairwise non-crossing minimal separators of G and H originates from G by completing into cliques the separators in S.*

Theorem 3 ([8]). *Let $G = (V, E)$ be a graph and let H be a minimal triangulation of G.*

1. *For every pair c, d of non-adjacent vertices of H, every minimal c, d-separator in H is a minimal c, d-separator in G.*
2. *For every minimal separator S of H and every connected component C in $H \setminus S$, $V(C)$ induces a connected component in $G \setminus S$.*

Minimal triangulations can characterise graph classes.

Theorem 4 ([10], [12]). *A graph G is AT-free if and only if every minimal triangulation of G is AT-free, i.e., an interval graph.*

Let $A_1, \ldots, A_k$ be the maximal cliques of an interval graph G. A *consecutive clique arrangement* for G is a sequence $A_{\pi(1)}, \ldots, A_{\pi(k)}$ for π a permutation over $[k]$ such that, for every vertex u in G, the maximal cliques containing u appear consecutively in the sequence. A graph is an interval graph if and only if it has a consecutive clique arrangement [6].

Theorem 5 ([8]). *Let $G = (V, E)$ be an interval graph with consecutive clique arrangement $A_1, \ldots, A_k$. The sets $A_i \cap A_{i+1}$, $i \in [k-1]$, are exactly the minimal separators of G.*

3   Special Scanlines

Minimal separators of permutation graphs can be represented efficiently in the permutation diagram. For this purpose, Bodlaender, Kloks, Kratsch introduced *scanlines* [2], which play a central role in our subsequent studies.

Definition 1. *Let $G = G(\pi)$ be a permutation graph over $[n]$. A **scanline** of G is a pair (a, e) where $a, e \in \{0.5, 1.5, \ldots, n+0.5\}$.*

Let $G = G(\pi)$ be a permutation graph over $[n]$, and let $\mathfrak{D}$ be its permutation diagram. Let $s = (a, e)$ and $s' = (a', e')$ be scanlines of G. We say that $s \leq s'$ if and only if $a \leq a'$ and $e \leq e'$; $s < s'$ if and only if $s \leq s'$ and $s \neq s'$. By $\mathrm{int}(s)$ we mean the set of vertices $x \in [n]$ such that $(a - x)(e - \pi^{-1}(x)) < 0$. In the world of permutation diagrams, s can be thought of as a line segment and $\mathrm{int}(s)$ is the set of vertices intersecting with s. We say that a vertex x is *to the left of* s in $\mathfrak{D}$ if it does not intersect with s and is smaller than a. Similarly, we define what it means to be *to the right of* s. It is easy to see that $\mathrm{int}(s)$ is a separator of G if there are vertices to the left and right of s. For $C_1, C_2 \subseteq [n]$, s is *between* C_1 and C_2, if every vertex of C_1 is to the left of s and every vertex of C_2 is to the right of s, or if every vertex of C_2 is to the left of s and every vertex of C_1 is to the right of s. A scanline s is *special* if $\mathrm{int}(s)$ is a minimal separator in G and s is between C_1 and C_2 where C_1 and C_2 induce S-full components of G. Such scanlines have been used by Parra and Scheffler to represent minimal separators of d-trapezoid graphs [11]. The following lemma is an extension of a result by Bodlaender, Kloks, Kratsch and can be proved similarly.

Lemma 1 ([2]). *Let $G = G(\pi)$ be a permutation graph over $[n]$. Let $C_1, C_2 \subseteq V$ induce connected subgraphs of G. If $N_G[C_1] \cap C_2 = \emptyset$, then there is a special scanline s between C_1 and C_2 such that $S =_{\mathrm{def}} \mathrm{int}(s)$ is a minimal u, v-separator in G for some vertices $u \in C_1$ and $v \in C_2$. In particular, if S is a minimal u, v-separator for $u, v \in V$, then there is a special scanline s of G between $\{u\}$ and $\{v\}$ such that $\mathrm{int}(s) = S$.*

The *enclosure* of scanline $s = (a, e)$ where $a, e \in \{1.5, \ldots, n{-}0.5\}$ is the set $\mathrm{en}(s) =_{\mathrm{def}} \{a{-}0.5, a{+}0.5, \pi(e{-}0.5), \pi(e{+}0.5)\}$. So, the enclosure may contain two, three or four vertices.

Lemma 2. *Let $G = G(\pi)$ be a permutation graph over $[n]$. A scanline s of G is special if and only if $\mathrm{en}(s) \cap \mathrm{int}(s) = \emptyset$.*

Proof. Let $s = (a{+}0.5, e{+}0.5)$ be a scanline, $a, e \in [n{-}1]$, and $S =_{\mathrm{def}} \mathrm{int}(s)$. If $\mathrm{en}(s) \cap S = \emptyset$, then a and $\pi(e)$ belong to one connected component C_1 of $G \setminus S$ and $a{+}1$ and $\pi(e{+}1)$ belong to another connected component C_2 of $G \setminus S$. Since every vertex in S has neighbours in C_1 and C_2, C_1 and C_2 are S-full components of G and s is a special scanline. Now, let $u \in \mathrm{en}(s) \cap S$. Then, u cannot have a neighbour in the connected components of $G \setminus S$ to the left or the right of s, hence s is not a special scanline.

Corollary 1. *Let $G = G(\pi)$ be a permutation graph over $[n]$, and let m be the number of edges of G. Then, G has at most $\min\{n{+}m, \binom{n}{2}{-}m\}$ special scanlines, and they can be listed in linear time.*

Proof. Every special scanline $s = (a{+}0.5, e{+}0.5)$ can be defined by $\{a, \pi(e)\}$ and $\{a, \pi(e{+}1)\}$. In the former case, these sets can be of cardinality 1 or 2, and if it contains two vertices these vertices are adjacent. Thus, G contains at most $n + m$ special scanlines. In the latter case, the set always contains two non-adjacent vertices. So, G contains at most $\binom{n}{2} - m$ special scanlines. For listing the special scanlines, check for every vertex $x \in [n]$ and every edge uv, $u < v$, whether $(x+0.5, \pi^{-1}(x)+0.5)$ and $(v+0.5, \pi^{-1}(u)+0.5)$ are special scanlines by applying Lemma 2.

Motivated by Theorem 2, the crossing relation of minimal separators is of great importance when studying minimal triangulations.

Definition 2. *Two scanlines s_1 and s_2 of a permutation graph $G = G(\pi)$ intersect if and only if neither $s_1 \leq s_2$ nor $s_2 < s_1$.*

Lemma 3. *Let $G = G(\pi)$ be a permutation graph. Let s_1 and s_2 be special scanlines of G, and let $S_1 =_{\text{def}} \text{int}(s_1)$ and $S_2 =_{\text{def}} \text{int}(s_2)$.*

1. *If s_1 and s_2 do not intersect then S_1 and S_2 are non-crossing.*
2. *If s_1 and s_2 intersect then S_1 and S_2 cross.*

Proof. Let s_1 and s_2 do not intersect. We assume $s_1 < s_2$. Let $C_1, \ldots, C_\ell$ be the connected components of $G \setminus S_1$, $\ell \geq 2$, where the vertices in C_i are smaller than the vertices in C_{i+1} for all $i \in [\ell-1]$. Suppose S_1 and S_2 cross. There is $r \geq 1$ such that s_2 intersects with vertices from C_r and C_{r+1}. Note that the vertices in $\text{en}(s_2)$ that are to the right of s_2 are not contained in $\text{int}(s_1) = S_1$, so that there is a path between a vertex from C_r and a vertex from C_{r+1}, and C_r and C_{r+1} are not different connected components of $G \setminus S_1$. Hence, S_1 and S_2 do not cross. For the converse, let s_1 and s_2 intersect. Observe that the vertices of $\text{en}(s_1)$ belong to two connected components of $G \setminus S_1$, and one vertex from each of these components belongs to S_2. Hence, S_2 crosses S_1.

4   Potential Maximal Cliques and Minimal Triangulations

Bouchitté and Todinca introduced the notion of potential maximal cliques [5].

Definition 3. *Let $G = (V, E)$ be a graph. A set $C \subseteq V$ of vertices is a **potential maximal clique** of G if and only if there is a minimal triangulation H of G such that C is a maximal clique in H.*

Let $G = G(\pi)$ be a permutation graph over $[n]$. For the following considerations, $s_0 =_{\text{def}} (0.5, 0.5)$ and $s_e^n =_{\text{def}} (n+0.5, n+0.5)$ are also special scanlines of G. Let s_1 and s_2 be two special scanlines of G such that $s_1 < s_2$. By $G[s_1, s_2]$ we denote the subgraph of G induced by the vertices that intersect with s_1 or s_2 or that lie between s_1 and s_2, i.e., to the right of s_1 and to the left of s_2. With $G[s_1, s_2]$ we associate the permutation diagram of G reduced to only the vertices of $G[s_1, s_2]$. The scanlines s_1 and s_2 are *neighbours* if there is no special scanline s' of G such that $s_1 < s' < s_2$. We say that s_1 is a *left neighbour* of s_2 and s_2 is a *right neighbour* of s_1. By $N_\pi^<(s)$ we denote the set of left neighbours of special scanline s of G.

Lemma 4. *Let $G = (V, E) = G(\pi)$ be a permutation graph over $[n]$. Let s_1, s_2 and s_1', s_2' be special scanlines of G where $s_1 \in N_\pi^<(s_2)$ and $s_1' \in N_\pi^<(s_2')$. Let $S_1 =_{\text{def}} \text{int}(s_1)$ and $S_2 =_{\text{def}} \text{int}(s_2)$, and let $C =_{\text{def}} V(G[s_1, s_2])$. Then:*

1. $u \in C \setminus (S_1 \cup S_2) \implies N_G[u] = C$
2. $u \in S_1 \setminus S_2$ and $v \in S_2 \setminus S_1 \implies uv \in E$
3. $C = V(G[s_1', s_2']) \implies s_1' = s_1$ and $s_2' = s_2$.

Proof. Let $u, v \in C$ such that $uv \notin E$. Let $u \in C \setminus (S_1 \cup S_2)$. If $v \in C \setminus (S_1 \cup S_2)$, there is a special scanline between u and v by Lemma 1 that does not intersect with neither s_1 nor s_2. If $v \in S_1$, there is a vertex in the enclosure of s_1 that is to the left of s_1 and adjacent to v such that there is a special scanline between $\{v, w\}$ and $\{u\}$ by Lemma 1, and this scanline does not intersect with neither s_1 nor s_2. The case $v \in S_2$ is similar to $v \in S_1$. If $u \in S_1 \setminus S_2$ and $v \in S_2 \setminus S_1$, let x and z be the vertices of the enclosures of s_1 and s_2, respectively, that do not belong to C and that are adjacent to u and v, respectively, such that there is a special scanline between $\{u, x\}$ and $\{v, z\}$ by Lemma 1, that is between s_1 and s_2. Since s_1 and s_2 are neighbours there cannot be a special scanline between s_1 and s_2. Finally, for claim 3, observe that s_1' does not intersect with neither s_1 nor s_2, since otherwise $C' =_{\text{def}} V(G[s_1', s_2'])$ would contain vertices from the enclosure of s_1 or s_2. Analogously, s_2' does not intersect with neither s_1 nor s_2. So, $\{s_1, s_1', s_2, s_2'\}$ is a set of pairwise non-intersecting special scanlines. Since C' must contain a vertex to the right of s_1 that belongs to the enclosure of s_1, $s_1 < s_2'$; similarly, $s_1' < s_2$, and this is only possible if $s_1' = s_1$ and $s_2' = s_2$.

Let s_1 and s_2 be special scanlines of G where $s_1 \in N_\pi^<(s_2)$. A vertex $x \in [n]$ is an *inner vertex* of $G[s_1, s_2]$ if $x \in V(G[s_1, s_2])$ and $x \notin \text{int}(s_1) \cup \text{int}(s_2)$. Inner vertices play a special role in our study, since $N_G[x] = V(G[s_1, s_2])$ by Lemma 4. For a potential maximal clique C of G, we say that $x \in C$ is an *inner vertex* of C if $N_G[x] = C$.

Lemma 5. *Let $G = G(\pi)$ be a permutation graph over $[n]$. Let s_1, s_2, s be special scanlines of G where $s_1 \in N_\pi^<(s_2)$ and $G[s_1, s_2]$ contains an inner vertex x. If s intersects with s_1 or s_2, then $x \in \text{int}(s)$. In particular, if x is inner vertex in $G[s_1', s_2']$ for s_1' and s_2' special scanlines of G, $s_1' \in N_\pi^<(s_2')$, then $s_1' = s_1$ and $s_2' = s_2$.*

Proof. Observe that x is to the right of s_1 and to the left of s_2. Let s intersect with s_1. Two non-adjacent vertices of the enclosure of s belong to $\text{int}(s_1)$, and by Lemma 4, they are neighbours of x. Then, x belongs to $\text{int}(s)$. Similarly, x belongs to $\text{int}(s)$, if s intersects with s_2. If x is inner vertex also in $G[s_1', s_2']$, then $V(G[s_1', s_2']) = N_G[x] = V(G[s_1, s_2])$ and $s_1 = s_1'$ and $s_2 = s_2'$ by Lemma 4.

Lemma 6. *Let $G = (V, E)$ be a graph, and let H be a minimal triangulation of G. Let $u \in V$ such that there is only one maximal clique C in H containing u. Then, u is an inner vertex of C.*

Proof. Note that all neighbours of u are contained in C and that u cannot be endpoint of an additional edge.

Theorem 6. *Let $G = G(\pi)$ be a permutation graph over $[n]$. $C \subseteq [n]$ is a potential maximal clique of G if and only if there are special scanlines s_1 and s_2 of G such that $s_1 \in N_\pi^<(s_2)$ and $C = V(G[s_1, s_2])$.*

Proof. First, we prove the "only if" part. Let H be a minimal triangulation of G, and let $C \subseteq V$ induce a maximal clique in H. By Theorem 4, H is an interval graph. Let $A_1, \ldots, A_k$, $k \geq 1$, be a consecutive clique arrangement for H. If $k = 1$, then G is complete, and $C = V(G[s_0, s_e^n])$. Let $k \geq 2$. Let $C = A_i$ for $1 \leq i \leq k$, and let $S_1 =_{\text{def}} A_{i-1} \cap A_i$ and $S_2 =_{\text{def}} A_i \cap A_{i+1}$; we assume $A_0 =_{\text{def}} A_{k+1} =_{\text{def}} \emptyset$. By Theorems 5 and 3, S_1 and S_2 are minimal separators of G. Let C have an inner vertex. Let $S =_{\text{def}} S_1 \cup S_2$, and let $C' =_{\text{def}} C \setminus S$. Remember that $N_G[u] = C$ for each $u \in C'$ due to Lemma 6. Let

$$a_1 =_{\text{def}} \max\{x < \min C' : x \in [n] \setminus C\} \cup \{0\}$$
$$a_2 =_{\text{def}} \min\{x > \max C' : x \in [n] \setminus C\} \cup \{n+1\}$$
$$e_1 =_{\text{def}} \max\{i < \min \pi^{-1}(C') : i \in [n] \setminus \pi^{-1}(C)\} \cup \{0\}$$
$$e_2 =_{\text{def}} \min\{i > \max \pi^{-1}(C') : i \in [n] \setminus \pi^{-1}(C)\} \cup \{n+1\},$$

and let $s_1 =_{\text{def}} (a_1+0.5, e_1+0.5)$ and $s_2 =_{\text{def}} (a_2-0.5, e_2-0.5)$. Every vertex in C' is to the right of s_1 and to the left of s_2 by definition of a_1, a_2, e_1, e_2, hence between s_1 and s_2. Observe that $a_1+1, \ldots, a_2-1$ and $\pi(e_1+1), \ldots, \pi(e_2-1)$ are vertices of C. Then, s_1 and s_2 are special scanlines of G. Furthermore, every vertex in S has a neighbour to the left of s_1 or to the right of s_2, since between s_1 and s_2 there are only vertices of C. Hence, $S \subseteq \text{int}(s_1) \cup \text{int}(s_2)$. Suppose there is $x \in (\text{int}(s_1) \cup \text{int}(s_2)) \setminus S$. By the aforesaid, $x \in \text{int}(s_1) \cap \text{int}(s_2)$ and must be a neighbour of $u \in C'$, hence $x \in C$ and $x \in S$, which is a contradiction. So, $\text{int}(s_1) \cup \text{int}(s_2) = S$, and $V(G[s_1, s_2]) = C$. Finally, suppose there is a special scanline s of G such that $s_1 < s < s_2$. Then, s must have a common endpoint with s_1 and s_2, and C contains two vertices u, v that are non-adjacent in G. By Theorems 2 and 5, there is a minimal separator $S' = A_j \cap A_{j+1}$ in G that contains u and v. Let s' be a special scanline of G such that $S' = \text{int}(s')$. It holds that s' cannot intersect with s_1 or s_2. If s' intersects with both scanlines, a vertex to the left of s_1 would be adjacent to the vertices in C' in H, which is not possible. If s' intersects with either s_1 or s_2, it must have a common endpoint with the other scanline, and again, the vertices from C' would be adjacent in H with a vertex to the left of s_1 or to the right of s_2. Hence, s' has a common endpoint with s_1 and s_2 and intersects with s. But then, $C' \subseteq S'$, which contradicts the assumption about C. So, $s_1 \in N_\pi^<(s_2)$.

Now, let $C = S_1 \cup S_2$. Then, $1 < i < k$. Let $u \in C \setminus S_1$, $v \in A_{i-1} \setminus S_1$ and $w \in C \setminus S_2$, $x \in A_{i+1} \setminus S_2$. We assume $v < u$; if $u < v$ we use $A_k, \ldots, A_1$ as consecutive clique arrangement for H and the same vertices with their new meanings. It holds that v, w, u, x induce a P_4 in H and $uw \in E$. First, we show that $w < x$. If $x < v$ then S_1 is an x, u-separator of G, which is not possible by Theorem 3. If $v < x$ then S_2 is a w, v-separator of G, which is also not possible. Hence $w < x$. Due to Theorem 3, $C \setminus S_1$ is contained in a connected component of $G \setminus S_1$ induced by D_1. Let $a_1 =_{\text{def}} \min(D_1)$ and $e_1 =_{\text{def}} \min \pi^{-1}(D_1)$ and $s_1 =_{\text{def}} (a_1-0.5, e_1-0.5)$. Observe that a_1 and $\pi(e_1)$ are to the right of s_1. Since $a_1-1 \notin D_1$ and $\pi(e_1-1) \notin D_1$ and $a_1-1, \pi(e_1-1) \notin S_1 \subseteq C$, a_1-1 and $\pi(e_1-1)$ are to the left of s_1, hence s_1 is a special scanline. Let D_1' induce the connected component of $G \setminus S_1$ containing v. Since every vertex in S_1 has a neighbour in

D_1 and D'_1, $S_1 = \text{int}(s_1)$. By a similar construction using S_2, define D_2 and D'_2 and $s_2 =_{\text{def}} (a_2+0.5, e_2+0.5)$, where $a_2 =_{\text{def}} \max(D_2)$ and $e_2 =_{\text{def}} \max \pi^{-1}(D_2)$. It holds that $S_2 = \text{int}(s_2)$. Since $u < w$ or $\pi^{-1}(u) < \pi^{-1}(w)$, $a_1 < a_2$ or $e_1 < e_2$, and since S_1 and S_2 are non-crossing minimal separators, $s_1 < s_2$ by Lemma 3. Hence, $C \subseteq V(G[s_1, s_2])$. Suppose there is $z \in V(G[s_1, s_2]) \setminus C$. Since $z \notin C$, $z \in A_1 \cup \cdots \cup A_{i-1}$ or $z \in A_{i+1} \cup \cdots \cup A_k$. In the former case, S_1 is a z, u-separator, in the latter case, S_2 is a w, z-separator, hence, z is not contained in D_1 or D_2, so that $z < a_1$ or $a_2 < z$. Finally, suppose there is a special scanline s of G such that $s_1 < s < s_2$. For every pair of vertices $a, b \in C$, if $a \in S_1 \setminus S_2$ and $b \in S_2 \setminus S_1$, then $ab \in E$. So, s_1 and s_2 have a common endpoint. If $a_1 = a_2 + 1$, then there is $j \in \{e_1, \ldots, e_2-1\}$ such that $\pi(e_1), \ldots, \pi(j) \in S_2$ and $\pi(j+1), \ldots, \pi(e_2) \in S_1$. If $e_1 = e_2 + 1$, then there is $j \in \{a_1, \ldots, a_2-1\}$ such that $a_1, \ldots, j \in S_2$ and $j+1, \ldots, a_2 \in S_1$. But then, there cannot be a special scanline s such that $s_1 < s < s_2$, and $s_1 \in N_\pi^<(s_2)$.

For the "if" part, let s_1 and s_2 be special scanlines such that $s_1 \in N_\pi^<(s_2)$, and let $C =_{\text{def}} V(G[s_1, s_2])$. There are $u, v \in C$ such that $u \notin \text{int}(s_1)$ and $v \notin \text{int}(s_2)$; u and v may be identical. Let G' emerge from G by completing $\text{int}(s_1)$ and $\text{int}(s_2)$ into cliques. Due to Lemma 4, C induces a clique in G'. Since no vertex to the left of s_1 is adjacent to u and no vertex to the right of s_2 is adjacent to v, C is a maximal clique in G'. Obtain H' from G' by making every connected component of $G' \setminus C$ complete. C induces a maximal clique in H'. H' is a triangulation of G'. There is a subgraph H of H' that is a minimal triangulation of G'. C induces a maximal clique in H. Since every minimal triangulation of G' is a minimal triangulation of G, C is a potential maximal clique of G.

Corollary 2. *A permutation graph on $n \geq 1$ vertices and with m edges has $\mathcal{O}(n + m)$ potential maximal cliques.*

Proof. Let $G = G(\pi)$ be a permutation graph over $[n]$. A potential maximal clique $G[s_1, s_2]$ of G has an inner vertex or s_1 and s_2 have a common endpoint. Due to Lemma 5, G has at most n potential maximal cliques with inner vertices. Let s be a special scanline of G. Then, s has at most two right neighbours that share an endpoint with s.

The *potential maximal cliques graph* of a permutation graph $G(\pi)$ is a directed graph that is denoted by $\mathcal{PC}(\pi)$ and defined as follows: $\mathcal{PC}(\pi)$ has a vertex for every special scanline of $G(\pi)$, and there is an arc from vertex u to vertex v, if $s_u \in N_\pi^<(s_v)$, where s_u and s_v denote the special scanlines the vertices u and v are labelled with, respectively. Hence, there is a 1-1 correspondence between the potential maximal cliques of $G(\pi)$ and the arcs in $\mathcal{PC}(\pi)$.

Lemma 7. *Let $G = G(\pi)$ be a permutation graph over $[n]$.*

1. *The potential maximal cliques graph $\mathcal{PC}(\pi)$ of G can be computed in linear time.*
2. *$\mathcal{PC}(\pi)$ is acyclic, and a topological ordering can be computed in linear time.*
3. *Vertex sequence $(x_0, \ldots, x_k)$, $k \geq 0$, is a maximal path in $\mathcal{PC}(\pi)$ if and only if*

- $s_0, s_1, \ldots, s_{k-1}, s_e^n$ *are the labels of* $x_0, \ldots, x_k$, *respectively*
- $\{s_0, s_1, \ldots, s_{k-1}, s_e^n\}$ *is a maximal set of pairwise non-intersecting special scanlines of* G.

Proof. For claims 1 and 2, remember that the special scanlines of G can be listed in linear time. We add the scanlines to the permutation diagram of G, delete the line segments corresponding to vertices of G and obtain a representation from which we can derive $\mathcal{PC}(\pi)$ in linear time. Ordering the scanlines according to their lower endpoints (respecting the order by the upper endpoints) gives a topological ordering. For claim 3, let $(x_0, \ldots, x_k)$ be a maximal path in $\mathcal{PC}(\pi)$. Obviously, the labels of x_0 and x_k are s_0 and s_e^n, respectively, since $s_0 < s < s_e^n$ for every special scanline s of G. Let $s_1, \ldots, s_{k-1}$ be the labels of $x_1, \ldots, x_{k-1}$, respectively. By definition of $\mathcal{PC}(\pi)$, $\{s_0, s_1, \ldots, s_{k-1}, s_e^n\}$ is a maximal set of pairwise non-intersecting special scanlines of G. Conversely, let $S =_{\mathrm{def}} \{s_0, \ldots, s_k\}$ be a maximal set of pairwise non-intersecting special scanlines of G where $s_0 < \cdots < s_k$. Then, $s_{i-1} \in N_\pi^{\leq}(s_i)$ for $i \in [k]$, and $s_0 = s_0$ and $s_k = s_e^n$. This, however, uniquely defines a maximal path in $\mathcal{PC}(\pi)$.

Theorem 7. *Let* $G = G(\pi)$ *be a permutation graph over* $[n]$. *An interval graph* H *over* $[n]$ *is a minimal triangulation of* G *if and only if there is a maximal path* $(x_0, \ldots, x_k)$ *in* $\mathcal{PC}(\pi)$ *such that* $A_1, \ldots, A_k$ *is a consecutive clique arrangement for* H *where* $A_i =_{\mathrm{def}} V(G[s_{i-1}, s_i])$ *and the special scanlines* s_{i-1} *and* s_i *are the labels of* x_{i-1} *and* x_i *in* $\mathcal{PC}(\pi)$ *for* $i \in [k]$.

Proof. Let $(x_0, \ldots, x_k)$ be a maximal path in $\mathcal{PC}(\pi)$, and let $s_0, \ldots, s_k$ be the labels of $x_0, \ldots, x_k$, respectively. By Lemma 7, $s_0 = s_0$ and $s_k = s_e^n$. Let $A_i =_{\mathrm{def}} V(G[s_{i-1}, s_i])$ for $i \in [k]$, and let H be the interval graph defined by the consecutive clique arrangement $A_1, \ldots, A_k$. Note that $A_i \cap A_{i+1} = \mathrm{int}(s_i)$ for $i \in [k-1]$. Let $u, v \in [n]$ be vertices in H, $u < v$. Let $uv \in E$. There is a largest $j \in [k]$ such that u or v is to the right of s_{j-1}. Since neither u nor v is to the right of s_j, $u, v \in A_j$. So, H is a triangulation of G. If $uv \in E(H) \setminus E$, there is $j \in [k-1]$ such that $u, v \in A_j \cap A_{j+1}$, since otherwise u and v are adjacent in G by Lemma 4. Then, uv is the unique chord in a cycle of length 4 in H. Hence, H is a minimal triangulation of G by Theorem 1.

Now, let H be a minimal triangulation of G, and let $A_1, \ldots, A_k$ be a consecutive clique arrangement for H. Let $S_0 =_{\mathrm{def}} S_k =_{\mathrm{def}} \emptyset$ and $S_i =_{\mathrm{def}} A_i \cap A_{i+1}$, $i \in [k-1]$. By Theorem 5, $S_1, \ldots, S_{k-1}$ are the minimal separators of H, and by Theorem 2, $\{S_1, \ldots, S_{k-1}\}$ is a maximal set of pairwise non-crossing minimal separators of G. Since $A_1, \ldots, A_k$ are potential maximal cliques of G, there are special scanlines $s_1, s_1', s_2, \ldots, s_k'$ of G such that $A_i = V(G[s_i, s_i'])$ and $s_i \in N_\pi^{\leq}(s_i')$ for $i \in [k]$. Suppose there are $i, j \in [k]$, $i < j$, such that s_j or s_j' intersects with s_i or s_i'. Due to Lemma 5, A_i and A_j do not contain inner vertices. So, $1 < i < j < k$ and $A_i = S_{i-1} \cup S_i$ and $A_j = S_{j-1} \cup S_j$. By the proof of Theorem 6, $\{\mathrm{int}(s_i), \mathrm{int}(s_i')\} = \{S_{i-1}, S_i\}$ and $\{\mathrm{int}(s_j), \mathrm{int}(s_j')\} = \{S_{j-1}, S_j\}$. So, by Theorem 2 and Lemma 3, s_i, s_i', s_j, s_j' are pairwise non-intersecting, and $S =_{\mathrm{def}} \{s_1, \ldots, s_k'\}$ is a set of pairwise non-intersecting special scanlines of G. There is a maximal set S' of pairwise non-intersecting special scanlines of G

such that $S \subseteq S'$ and $\mathcal{C}_S =_{\text{def}} \{A_1, \ldots, A_k\} \subseteq \mathcal{C}_{S'}$ where $\mathcal{C}_{S'}$ denotes the set of potential maximal cliques of G defined on the path of $\mathcal{PC}(\pi)$ determined by S'. By the first part of this proof, $\mathcal{C}_{S'}$ defines a minimal triangulation H' of G, and since $\mathcal{C}_S \subseteq \mathcal{C}_{S'}$, H is a subgraph of H'. Then, H and H' are equal, and $\mathcal{C}_S$ and $\mathcal{C}_{S'}$ are equal, and the theorem holds.

5   Treewidth and Minimum Fill-In

Let $G = (V, E)$ be a graph. Treewidth and minimum fill-in can be defined as follows:

$$\text{tw}(G) =_{\text{def}} \min\{\omega(H) : H \text{ is a minimal triangulation of } G\} - 1$$
$$\text{mfi}(G) =_{\text{def}} \min\{|E(H)| - |E(G)| : H \text{ is a minimal triangulation of } G\}.$$

In the same manner the *pathwidth* and *interval completion* problems can be defined by replacing "minimal triangulation" by "triangulation that is an interval graph". Since minimal triangulations of AT-free graphs are interval graphs, treewidth and pathwidth as well as chordal completion (minimum fill-in) and interval completion describe the same problems on AT-free graphs.

Our algorithms work on the *weighted* potential maximal cliques graph. Let $G = G(\pi)$ be a permutation graph. The weighted potential maximal cliques graph of G is the potential maximal cliques graph of G where the vertices are assigned the numbers of vertices of the corresponding minimal separators and the arcs are assigned the numbers of vertices of the corresponding potential maximal cliques.

Theorem 8. *Let $G = G(\pi)$ be a permutation graph. The weighted potential maximal cliques graph of G can be computed in linear time.*

Treewidth and minimum fill-in can be solved on the weighted potential maximal cliques graph by finding shortest paths.

Theorem 9. *Treewidth and minimum fill-in for permutation graphs can be computed in linear time.*

Proof. Let $G = G(\pi)$ be a permutation graph over $[n]$. In linear time, the weighted potential maximal cliques graph of G can be computed. By Theorem 7, the treewidth of G is realised on a maximal path of the weighted potential maximal cliques graph of G with the least maximal arc weight. For computing the minimum fill-in, consider the following observation. Let $A_1, \ldots, A_k$ be a consecutive clique arrangement for an interval graph H. For every $i \in [k]$:

$$|E(H[A_0 \cup \cdots \cup A_i])| + |E(H[A_{i-1} \cap A_i])| = |E(H[A_0 \cup \cdots \cup A_{i-1}])| + |E(H[A_i])|,$$

where $A_0 =_{\text{def}} \emptyset$. Since $H[A_{i-1} \cap A_i]$ and $H[A_i]$ are complete graphs and the numbers of vertices of these graphs are known as weights, it is easy to determine the numbers of edges. In linear time, a path on which the smallest number of edges among the minimal triangulations of G is realised can be found.

Acknowledgements. I thank anonymous referees for suggestions concerning the presentation of the material.

References

1. ST. ARNBORG, D.G. CORNEIL, A. PROSKUROWSKI, *Complexity of finding embeddings in a k-tree*, SIAM Journal on Algebraic and Discrete Methods 8, pp. 277–284, 1987.
2. H.L. BODLAENDER, T. KLOKS, D. KRATSCH, *Treewidth and Pathwidth of Permutation Graphs*, SIAM Journal on Discrete Mathematics 8, No. 4, pp. 606–616, 1995.
3. H.L. BODLAENDER, T. KLOKS, D. KRATSCH, H. MÜLLER, *Treewidth and minimum fill-in on d-trapezoid graphs*, Journal of Graph Algorithms and Applications 2, No. 3, pp. 1–23, 1998.
4. H.L. BODLAENDER, R.H. MÖHRING, *The pathwidth and treewidth of cographs*, SIAM Journal on Discrete Mathematics 6, pp. 181–188, 1993.
5. V. BOUCHITTÉ, I. TODINCA, *Treewidth and Minimum Fill-in: Grouping the Minimal Separators*, SIAM Journal on Computing 31, pp. 212–232, 2001.
6. D.R. FULKERSON, O.A. GROSS, *Incidence matrices and interval graphs*, Pacific Journal of Mathematics 15, pp. 835–855, 1965.
7. M.CH. GOLUMBIC, *Algorithmic Graph Theory and Perfect Graphs*, Academic Press, New York, 1980.
8. T. KLOKS, D. KRATSCH, J. SPINRAD, *On treewidth and minimum fill-in of asteroidal triple-free graphs*, Theoretical Computer Science 175, pp. 309–335, 1997.
9. C.G. LEKKERKERKER, J.CH. BOLAND, *Representation of finite graphs by a set of intervals on the real line*, Fundamenta Mathematicae 51, pp. 45–64, 1962.
10. R.H. MÖHRING, *Triangulating graphs without asteroidal triples*, Discrete Applied Mathematics 64, pp. 281–287, 1996.
11. A. PARRA, P. SCHEFFLER, *How to Use the Minimal Separators of a Graph for its Chordal Triangulation*, Proceedings of the 22nd International Colloquium on Automata, Languages and Programming, ICALP95, Lecture Notes in Computer Science 944, pp. 123–134, Springer-Verlag, 1995.
12. A. PARRA, P. SCHEFFLER, *Characterizations and algorithmic applications of chordal graph embeddings*, Discrete Applied Mathematics 79, pp. 171–188, 1997.
13. D.J. ROSE, *Triangulated Graphs and the Elimination Process*, Journal of Mathematical Analysis and Applications 32, pp. 597–609, 1970.
14. D.J. ROSE, R.E. TARJAN, G.S. LUEKER, *Algorithmic aspects of vertex elimination on graphs*, SIAM Jounal on Computing 5, pp. 266–283, 1976.
15. M. YANNAKAKIS, *Computing the minimum fill-in is NP-complete*, SIAM Journal on Algebraic and Discrete Methods 2, pp. 77–79, 1981.

Roman Domination over Some Graph Classes[*]

Mathieu Liedloff[1], Ton Kloks, Jiping Liu[2], and Sheng-Lung Peng[3]

[1] Université Paul Verlaine - Metz,
Laboratoire d'Informatique Théorique et Appliquée,
57045 Metz Cedex 01, France
liedloff@sciences.univ-metz.fr
[2] Department of Mathematics and Computer Science,
The university of Lethbridge,
Alberta, T1K 3M4, Canada
liu@cs.uleth.ca
[3] Department of Computer Science and Information Engineering,
National Dong Hwa University,
Hualien, Taiwan, R.O.C
lung@csie.ndhu.edu.tw

Abstract. A Roman dominating function of a graph $G = (V, E)$ is a function $f : V \rightarrow \{0, 1, 2\}$ such that every vertex x with $f(x) = 0$ is adjacent to at least one vertex y with $f(y) = 2$. The weight of a Roman dominating function is defined to be $f(V) = \sum_{x \in V} f(x)$, and the minimum weight of a Roman dominating function on a graph G is called the Roman domination number of G.

In this paper we answer an open problem mentioned in [2] by showing that the Roman domination number of an interval graph can be computed in linear time. We also show that the Roman domination number of a cograph can be computed in linear time. Besides, we show that there are polynomial time algorithms for computing the Roman domination numbers of AT-free graphs and graphs with a d-octopus.

1 Introduction

Let $G = (V, E)$ be an undirected and simple graph. A *Roman dominating function* is a function $f : V \rightarrow \{0, 1, 2\}$ such that every vertex x with $f(x) = 0$ is adjacent to at least one vertex y with $f(y) = 2$. The *weight* of a Roman dominating function is $f(V) = \sum_{x \in V} f(x)$. The minimum weight of a Roman dominating function on a graph G is called the *Roman domination number* of G and is denoted by $\gamma_R(G)$.

Roman domination has been introduced in [2] as a new variety of the classical domination problem having both historical and mathematical interest, particularly in the field of server placements [15]. We refer to [2,6,10,11,12,16,17] for

[*] The second and the third authors were partially supported by NSERC of Canada. The second author was supported also by the National Science Council of Taiwan under grant NSC 93-2811-M-002-004. The first author is the corresponding author.

D. Kratsch (Ed.): WG 2005, LNCS 3787, pp. 103–114, 2005.

more background on the historical importance of the Roman domination problem and various mainly graph-theoretic results not mentioned here.

The complexity of the Roman domination problem when restricted to interval graphs was mentioned as an open problem in [2]. In this paper we show that there are linear time algorithms to compute the Roman domination number for interval graphs and cographs. We also show that there are polynomial time algorithms for computing the Roman domination numbers of AT-free graphs and graphs with a d-octopus. The paper is organized as follows. Section 2 gives some preliminaries about our problem. The results for interval graphs and cographs are presented in Sections 3 and 4, respectively. In Section 5, we present polynomial time algorithms for computing the Roman domination numbers of AT-free graphs and graphs with a d-octopus.

2　Preliminaries

Let $G = (V, E)$ be an undirected and simple graph. For a vertex x of G we denote by $N(x)$ the neighborhood of x in G and by $N[x] = N(x) \cup \{x\}$ the closed neighborhood of x. The distance $d_G(x, y)$ between two vertices x and y is the length of a shortest path joining these two vertices.

A dominating set D of a graph $G = (V, E)$ is a subset of vertices such that every vertex of $V - D$ has at least one neighbor in D. The minimum cardinality of a dominating set of G is said to be the domination number of G, and it is denoted by $\gamma(G)$. An independent set in a graph G is a subset of pairwise non-adjacent vertices.

Now let us summarize some useful facts on Roman domination [2].

Theorem 1 ([2]). $\gamma(G) \leq \gamma_R(G) \leq 2\gamma(G)$.

Lemma 1 ([2].). If G is a graph of order n, then $\gamma_R(G) = \gamma(G)$ if and only if $G = \overline{K_n}$, i.e., G is an independent set with n vertices.

Definition 1. A 2-packing is a set $S \subseteq V$ such that for every pair $x, y \in S$ $N[x] \cap N[y] = \varnothing$. The maximum cardinality of a 2-packing in G is called the 2-packing number of G.

Theorem 2 ([2]). Let f be a minimum weighted Roman dominating function of a graph G without isolated vertices. Let V_i, $i = 0, 1, 2$, be the set of vertices x with $f(x) = i$. Let f be such that $|V_1|$ is the minimum. Then V_1 is a 2-packing and there is no edge between V_1 and V_2.

Theorem 3 ([2]). For any non-trivial connected graph G,

$$\gamma_R(G) = \min\{|S| + 2\gamma(G - S) \mid S \text{ is a 2-packing}\}.$$

Remark 1. A 2-packing S can serve as V_1 and a dominating set in $G - S$ as V_2. Notice that the weight of a Roman dominating function is $|V_1| + 2|V_2|$.

Definition 2. *We call* (V_1, V_2) *a Roman pair of a graph* G *if* (V_1, V_2) *is a solution induced by a minimum weighted Roman dominating function of the graph* G.

We refer the reader to [1,8] for definitions and properties of graph classes not given in this paper.

3   Roman Domination on Interval Graphs

Throughout this section we assume that $G = (V, E)$ is connected. Clearly, if G is disconnected then, obviously, $\gamma_R(G)$ is the *sum* of the Roman domination numbers of its components.

Definition 3. *A graph* $G = (V, E)$ *is an* interval graph *if there exists a set* $\{I_v \mid v \in V\}$ *of intervals of the real line such that* $I_u \cap I_v \neq \varnothing$ *iff* $uv \in E$.

Both I_v and v can be used to represent the vertex v in an interval graph. Let $l(v)$ and $r(v)$ denote the values of the left and right end points of the interval $I_v = [l_v, r_v]$, respectively. A model of an interval graph is normalized if $\cup_{v \in V} \{l(v), r(v)\} = \{1, 2, \ldots, 2n\}$. In the following we assume that a normalized model of the graph is part of the input.

Our linear time algorithm to compute the Roman domination number of an interval graph uses dynamic programming and passes through the interval collection from left to right to enumerate all the potential optimum solutions (V_1, V_2).

3.1   Structure of an Optimum Solution

In this section, we examine the structure of an optimum solution.

Lemma 2. *For every interval graph there exists a Roman pair* (V_1, V_2) *such that no interval in* V_2 *is properly contained in another interval.*

Lemma 3. *If* (V_1, V_2) *is a Roman pair, then* V_2 *contains no clique of size* 3 *or more.*

Proof. Let $\{i_1, i_2, i_3\} \subseteq V_2$ be a clique of size three. By Lemma 2, there is no interval which is properly contained in another interval. Without loss of generality, we assume $l(i_1) < l(i_2) < l(i_3) < r(i_1) < r(i_2) < r(i_3)$. Then we obtain that $N[i_2] \subseteq N[i_1] \cup N[i_3]$. That is, $(V_1, V_2 \setminus \{i_2\})$ is a Roman pair of G which is a contradiction.   □

Lemma 4. *If* (V_1, V_2) *is a Roman pair, then the connected components induced by* V_2 *are paths.*

Proof. By Lemma 2, each connected component induced by V_2 is a proper interval graph. Hence, it is chordal and it does not contain a claw, i.e., $K_{1,3}$. Together with Lemma 3, our lemma holds.   □

We can use this last result in the following way: in order to find a set V_2 of an optimum solution, we only have to consider certain shortest paths between some pairs of vertices. Now, we characterize the set V_1 of an optimum solution.

Definition 4. *Let (V_1, V_2) be a Roman pair of an interval graph G. Intervals $\mathcal{J} \subseteq V_1$ are consecutive iff between the leftmost and rightmost end points of $\mathcal{J}$ there is no end point of an interval $I \in V_2$.*

Lemma 5. *There exists a Roman pair (V_1, V_2) with the property that V_1 is an independent set, and there is no subset $\mathcal{J} \subseteq V_1$ containing more than two consecutive intervals.*

Proof. By Theorem 2, we have a Roman pair (V_1, V_2) with V_1 being an independent set. Let $\{a, b, c\} \subseteq V_1$ be a set of three consecutive intervals in V_1. Suppose that $l(a) < r(a) < l(b) < r(b) < l(c) < r(c)$. Since (V_1, V_2) is of minimum weight, we have $\forall v \in N(b)$, $v \notin V_1$ and $v \notin V_2$. Consequently, if $v \in N(b)$ there must exist a $w \in N(v)$ such that $w \in V_2$. However $\{a, b, c\}$ are consecutive, therefore, we have $r(w) < l(a)$ (or resp. $r(c) < l(w)$). As a result of $v \in N(a)$ (resp. $v \in N(c)$), there exists a solution with $f(v) = 2$ and $f(a) = f(b) = 0$ (resp. $f(b) = f(c) = 0$). Consequently if we have a solution with three consecutive intervals, there exists a solution (V_1, V_2) of same weight such that V_1 contains no more than two consecutive intervals.   □

3.2   Description of the Algorithm

Previous results show us how to build a potential solution (V_1, V_2). Indeed, we have seen that connected components induced by V_2 are paths and each of these paths can be preceded or followed by at most two consecutive intervals of V_1. So, our algorithm goes through the interval collection in a left-right fashion. An optimum solution, i.e, a solution whose weight is the minimum over all possible solutions, will be one of the solutions found by the algorithm with minimum value of $|V_1| + 2|V_2|$. The algorithm uses dynamic programming in order to intelligently test every possible solution with respect to the structure established by previous lemmas.

For any given normalized interval graph $G = (V, E)$ of order n, the algorithm treats intervals increasingly according to their right end points. Corresponding to a right end point d ($0 \le d \le 2n$) of an interval, we define a sub-solution (V_1', V_2') by

1. $V_1', V_2' \subseteq \{i \in V : r(i) \le d\}$,
2. (V_1', V_2') is a solution of minimum weight over all the solutions for the graph $G[S]$, where $S = \{v \in V : l(v) \le d\}$, such that the interval i with $r(i) = d$ belongs to V_2'.

Clearly, at the beginning of the algorithm no intervals are yet considered and we define for $d = 0$ the sub-solution $(V_1', V_2') = (\varnothing, \varnothing)$. Then, for each step, we start with a current integer d and its corresponding sub-solution (V_1', V_2'), and we construct an *extension* (V_1'', V_2'') of (V_1', V_2') corresponding to a new d', where $d' > d$. According to previous lemmas, there are three possible cases:

1. add two intervals i_1 and i'_1 to V'_1 and one interval i_2 to V'_2 such that $(V''_1, V''_2) = (V'_1 \cup \{i_1, i'_1\}, V'_2 \cup \{i_2\})$ is a sub-solution corresponding to $d' = r(i_2)$ (see procedure Add-intervals-first-choice),
2. add one interval i_1 to V'_1 and one interval i_2 to V'_2 such that $(V''_1, V''_2) = (V'_1 \cup \{i_1\}, V'_2 \cup \{i_2\})$ is a sub-solution corresponding to $d' = r(i_2)$ (see procedure Add-intervals-second-choice),
3. add one interval i_2 to V'_2 such that $(V''_1, V''_2) = (V'_1, V'_2 \cup \{i_2\})$ is a sub-solution corresponding to $d' = r(i_2)$ (see procedure Add-intervals-third-choice).

The first choice corresponds to adding two consecutive intervals to V'_1 and then starting a new path in V''_2. In the second case, we add one interval to V'_1 and begin a new path in V''_2. In the last case, we add only one interval to V'_2 which extends an existing path in V'_2 or begins a new path in V''_2.

Now, we provide another result which will be used in the construction of some sub-solutions.

Lemma 6. *Let d be an integer such that $1 \leq d \leq 2n$. Suppose we have a sub-solution (V'_1, V'_2) for the set of all intervals i with $l(i) < d$. Let i_1 and i'_1 be such that $r(i_1) = \min\{r(i) : l(i) > d\}$ and $r(i'_1) = \min\{r(i) : l(i) > r(i_1)\}$. Let w be such that $r(w) = \min\{r(i) : l(i) > d \wedge i \neq i_1 \wedge i \neq i'_1\}$. If $w \in N(i_1)$, then there exists an optimum solution (V''_1, V''_2) where i_1 and i'_1 are not two consecutive intervals in V''_1.*

Proof. By the construction of i_1, i'_1 and w, we have that $d < l(i_1)$, $d < l(i'_1)$, $d < l(w)$ and $r(i_1) < r(w)$. Since $w \in N(i_1)$, then $l(w) < r(i_1) < r(w)$. There are two cases.

1. $w \in N(i'_1)$. Then there exists an alternative solution with $w \in V''_2$ and $i_1, i'_1 \notin V''_1$.
2. $w \notin N(i'_1)$. Then we have $r(w) < l(i'_1)$ and there are three sub-cases:
   (a) $w \in V''_2$. Then i_1 and i'_1 are not consecutive.
   (b) There exists a $v \in N(w)$ such that $v \in V''_2$ ($w \in V''_0$). Then $l(v) < r(w) < r(v)$ and $v \in V''_2$. If both i_1 and i'_1 are in V''_1, then i_1 and i'_1 cannot be consecutive since at least one end of v is between them.
   (c) $w \in V''_1$. In this case i_1 cannot be in V''_1, thus i_1 and i'_1 cannot be consecutive. $\square$

3.3   Preprocessing Data

In order to achieve a linear-time algorithm, we do some pre-processing so that when we run the program, the necessary data is available in constant time. In particular, the following operations must be done in constant time in order to obtain the claimed time bound.

- find i, j, k such that $r(i) = \min\{r(v) : l(v) > d\}$, $r(j) = \min\{r(v) : l(v) > d \wedge v \neq i\}$ and $r(k) = \min\{r(v) : l(v) > d \wedge v \neq i \wedge v \neq j\}$ for a fixed d,
- find i such that $r(i) = \max\{r(v) : v \in N[x]\}$ for a fixed x,
- check whether $N[x] \cap N[y] \neq \varnothing$ for two intervals x and y such that $r(x) < r(y)$ (for this operation we only have to find i such that $r(i) = \max\{r(v) : v \in N[x]\}$ and then check whether $i \in N[y]$).

Sort Intervals According to Their Right End Points (SIRE). The collection of n intervals is given in a normalized interval model. We sort the intervals in an array D of size $2n$ such that $D[i] = \begin{cases} j & \text{if } \exists j \text{ s.t. } r(j) = i, \\ \text{NIL otherwise.} \end{cases}$ in time $O(n)$ using bucket sort.

Find Three Intervals with Lowest Right End Points (ILRE). Now, we use the array D to build another 2-dimensional array $MinR$ which contains for each value $d \in \{0, 1, \ldots, 2n\}$ the first, second, and third interval whose right end points are the first, second, and third lowest, respectively, and such that their left end points are greater than d.

Find Intervals with Greatest Right End Points (IGRE). Finally, we calculate for each interval $i \in \{1, \ldots, n\}$ its neighbor which has the greatest right end point, or the interval i if there is no such a neighbor, in an array $MaxR$.

The three procedures SIRE, ILRE and IGRE have been shown in detail in [13], and each takes $O(n)$ time.

3.4   A Linear-Time Algorithm

Using the structure of an optimum solution described by previous lemmas of this section and some results stated in section 2 (in particular Theorem 2), we are ready to present a linear-time algorithm for solving the Roman domination problem on interval graphs. An optimum solution can be easily constructed by standard techniques.

Procedure Add-intervals-first-choice(d)
Data: An integer d such that a corresponding sub-solution (V_1', V_2') has already been computed.
Result: An extension of the sub-solution (V_1', V_2') constructed using the first case (add two intervals to V_1' and one interval to V_2').

$i_1 \leftarrow MinR[d][1]$
if $i_1 \neq$ *NIL* **then**
> $i_1' \leftarrow MinR[r(i_1)][1]$
> **if** $i_1' \neq$ *NIL* **then**
> > **if** $MaxR[i_1]$ *does not intersect* i_1' **then**
> > > $w \leftarrow MinR[d][2]$
> > > **if** $w = i_1'$ **then** $w \leftarrow MinR[d][3]$
> > > **if** $w \neq$ *NIL* **then**
> > > > **if** i_1 *does not intersect* w **then**
> > > > > $i_2 \leftarrow MaxR[w]$
> > > > > **if** i_1 *does not intersect* i_2 ***and*** i_1' *does not intersect* i_2 **then**
> > > > > > $Weight[r(i_2)] \leftarrow \min\{Weight[r(i_2)], Weight[d] + 4\}$
> >
> > **else** $Weight[2n] \leftarrow \min\{Weight[2n], Weight[d] + 2\}$

Procedure Add-intervals-second-choice(d)
Data: An integer d such that a corresponding sub-solution (V_1', V_2') has already
been computed.
Result: An extension of the sub-solution (V_1', V_2') constructed using the second
case (add one interval to V_1' and one interval to V_2').

$i_1 \leftarrow MinR[d][1]$
if $i_1 \neq NIL$ then
   | $w \leftarrow MinR[d][2]$
   | if $w \neq NIL$ then
   | | $i_2 \leftarrow MaxR[w]$
   | | if i_1 *does not intersect* i_2 then
   | | $\lfloor Weight[r(i_2)] \leftarrow \min\{Weight[r(i_2)], Weight[d] + 3\}$
   | else $Weight[2n] \leftarrow \min\{Weight[2n], Weight[d] + 1\}$

Procedure Add-intervals-third-choice(d)
Data: An integer d such that a corresponding sub-solution (V_1', V_2') has already
been computed.
Result: An extension of the sub-solution (V_1', V_2') constructed using the third case
(add one interval to V_2').

$w \leftarrow MinR[d][1]$
if $w \neq NIL$ then
   | $i_2 \leftarrow MaxR[w]$
   | $Weight[r(i_2)] \leftarrow \min\{Weight[r(i_2)], Weight[d] + 2\}$
else $Weight[2n] \leftarrow \min\{Weight[2n], Weight[d]\}$

Algorithm Roman-Dom(normalized interval model of a graph G ; $\gamma_R(G)$)
Data: An interval graph represented by a normalized model.
Result: The Roman domination number γ_R of the input interval graph.
Construct the data structures D, $MinR$ and $MaxR$

for $i = 1$ to $2n$ do $Weight[i] \leftarrow 2n$
$Weight[0] \leftarrow 0$
Add-intervals-first-choice(0)
Add-intervals-second-choice(0)
Add-intervals-third-choice(0)
for $i = 1$ to $2n$ do
   | if $D[i] \neq NIL$ *and* $Weight[r(D[i])] \neq 2n$ then
   | Add-intervals-first-choice($r(D[i])$)
   | Add-intervals-second-choice($r(D[i])$)
   | $\lfloor$Add-intervals-third-choice($r(D[i])$)

return $\gamma_R(G) = Weight[2n]$

Theorem 4. *The Roman domination problem can be solved in $O(n)$ time on
any interval graph with a normalized interval model.*

Proof. The correctness of the algorithm follows from the lemmas stated in Subsections 3.1 and 3.2
    We note that it takes linear time to construct D, $MinR$ and $MaxR$, and
it takes constant time to process each of the procedures **Add-intervals-first-**

choice, **Add-intervals-second-choice**, and **Add-intervals-third-choice**. The complexity of the algorithm **Roman-Dom** is dominated by the second **for** loop. Therefore, the complexity of the algorithm is $O(n)$.   □

4   Roman Domination on Cographs

In this section we describe an algorithm to compute the Roman domination number of a cograph G. We may assume that G is connected, since otherwise $\gamma_{\mathsf{R}}(G)$ equals the sum of the Roman domination numbers of its components.

If G is connected then G is the *join* of two graphs G_1 and G_2. Clearly, *any* 2-packing of G consists of at most one vertex since G is P_4-free. By Theorem 3 the Roman domination number of G can be computed by taking the minimum over all vertices x of $2\gamma(G-x)+1$ and $2\gamma(G)$. It is well-known that the domination number of a cograph can be computed in linear time. Thus, we can compute the Roman domination number of G in $O(n(m+n))$ time, where n and m are the numbers of the vertices and edges of G respectively. However, we can obtain a linear time algorithm by using the structure of cotree.

It is well-known that any cograph G can be represented by a cotree $\mathcal{T}$ [9]. In $\mathcal{T}$, each leaf represents a vertex of G and each internal node represents either a **join** or a **union**. For any two vertices u and v, if (u,v) is an edge of G, then the lowest common ancestor of u and v in $\mathcal{T}$ is a **join** node. Since G is connected, the root of $\mathcal{T}$ is a **join** node. We may assume that $\mathcal{T}$ is a binary tree. For a node v, let $\mathcal{T}_v$ denote the subtree of $\mathcal{T}$ rooted at v. Let G_v denote the subgraph defined by $\mathcal{T}_v$. Now, our algorithm is as follows.

For a cograph G, we traverse its corresponding cotree $\mathcal{T}$ from leaves to the root. Let $(V_1(G_v), V_2(G_v))$ be a Roman pair of G_v. Initially, every leaf w is in $V_1(G_w)$ and $V_2(G_w)$ is empty, i.e., $\gamma_{\mathsf{R}}(G_w) = 1$. Now let us consider an internal node u in $\mathcal{T}$, let l (respectively, r) be its left (respectively, right) child. That is, G_u is the resulting cograph by applying union or join operation on G_l and G_r. If u is a **union** node, then $(V_1(G_u), V_2(G_u)) = (V_1(G_l) \cup V_1(G_r), V_2(G_l) \cup V_2(G_r))$ is a Roman pair of G_u. If u is a **join** node, we do the following. Without loss of generality, let $\gamma_{\mathsf{R}}(G_l) \le \gamma_{\mathsf{R}}(G_r)$.

1. $\gamma_{\mathsf{R}}(G_l) = \gamma_{\mathsf{R}}(G_r)$. If at least one of $V_2(G_l)$ and $V_2(G_r)$ is not empty, say $V_2(G_l) \neq \emptyset$, then set $V_1(G_r) = V_2(G_r) = \emptyset$. We do this because every vertex in G_r is dominated by a vertex $v \in V_2(G_l)$.

   If both $V_2(G_l)$ and $V_2(G_r)$ are empty, then we move any vertex $v \in V_1(G_l)$ to $V_2(G_l)$. We then set $V_1(G_r) = V_2(G_r) = \emptyset$ for the same reason.

2. $\gamma_{\mathsf{R}}(G_l) < \gamma_{\mathsf{R}}(G_r)$. If $V_2(G_l) = \emptyset$, again we move a vertex $v \in V_1(G_l)$ to $V_2(G_l)$. Since every vertex in G_r is dominated by v, we set $V_1(G_r) = V_2(G_r) = \emptyset$.

   If $V_2(G_l) \neq \emptyset$, then we set $V_1(G_r) = V_2(G_r) = \emptyset$ for the same reason.

In any one of above cases, if $2|V_2(G_l)| + |V_1(G_l)| > 4$, then (i) keep only one vertex in $V_2(G_l)$, (ii) set $V_1(G_l) = \emptyset$, and (iii) arbitrarily select a vertex in G_r and add it to $V_2(G_r)$. Finally, let $V_i(G_u) = V_i(G_l) \cup V_i(G_r)$ for $i = 1, 2$. It is not hard to see that $\gamma_R(G) \leq 4$ for any connected cograph G. We have the following theorem.

Theorem 5. *The Roman domination number of a cograph can be computed in linear time.*

Proof. For the correctness, we show it by induction on the height of $\mathcal{T}$. In the base case that we consider the height equal to 0. Since every vertex w is an isolated vertex, $\gamma_R(G_w) = 1$. Thus, $(\{w\}, \emptyset)$ is the Roman pair of G_w. Assume that for any node v in $\mathcal{T}$ with height equal to h, we can compute a Roman pair $(V_1(G_v), V_2(G_v))$ for G_v. Now, consider a node u with height $h + 1$. Let l and r be its left and right children in $\mathcal{T}$, respectively. If u is a **union** node, it is easy to check that $(V_1(G_l) \cup V_1(G_r), V_2(G_l) \cup V_2(G_r))$ is a Roman pair of G_u. We now consider the case that u is a **join** node. Without loss of generality, we assume that $\gamma_R(G_l) \leq \gamma_R(G_r)$. By the definition, every vertex in G_r is adjacent to any vertex of G_l. If $V_2(G_l)$ is not empty, then every vertex is dominated by a vertex in $V_2(G_l)$. Thus $(V_1(G_l), V_2(G_l))$ can Roman dominate G_u. If $V_2(G_l)$ is empty, we can promote a vertex in $V_1(G_l)$ to $V_2(G_l)$ such that it can dominate G_r. Since $\gamma_R(G_l) \leq \gamma_R(G_r)$, we can obtain a better solution by doing so. However, it will increase the weight of the Roman dominating function. If $|V_1(G_l)| + 2|V_2(G_l)| \leq 4$, then $(V_1(G_l), V_2(G_l))$ is a Roman pair of G_u. If $|V_1(G_l)| + 2|V_2(G_l)| > 4$, we select a vertex v_l from $V_2(G_l)$ and arbitrarily select a vertex v_r from G_r. Since v_l dominates G_r and v_r dominates G_l, $(\emptyset, \{v_l, v_r\})$ is a Roman pair of G_u. This show the correctness of our algorithm.

For the time complexity, we implement each dominating set using a linked list with *front* and *tail* pointers. Thus the Roman pair of a **union** node can be computed in constant time. For a **join** node, it costs constant time to empty a set. For the other operations, at most constant number of vertices are updated. Thus, the overall time complexity is linear. $\qquad\square$

Remark 2. In [2] a graph G is called *Roman* if $\gamma_R(G) = 2\gamma(G)$. It is proved that a graph G is Roman if and only if $\gamma(G) \leq \gamma(G - S) + \frac{|S|}{2}$ for *every* 2-packing S in G. It follows that a connected cograph G is Roman if and only if $\gamma(G) = \gamma(G - x)$ for *every* vertex x. Since, in [2] it is posed as an open problem to determine Roman graphs other than trees[4], it would be of interest to know which cographs satisfy this equality. Notice that a large subclass of Roman cographs can be constructed as follows: Take *any* cograph G and construct a graph H by replacing every vertex of G by a *true twin*. It is easy to check that H is a cograph[5], and furthermore for every vertex x in H, $\gamma(H) = \gamma(H - x)$.

5   Roman Domination on **AT**-Free Graphs and Graphs with a d-Octopus

In this section we study the Roman domination problem on AT-free graphs and graphs with d-octopus. Our approaches are based on algorithms for the

---

[4] A constructive characterization of Roman trees is given in [10].
[5] Any induced P_4 would lead to an induced P_4 in G.

domination problem in [7,14]. First we provide some preliminaries on AT-free graphs and d-octopus.

Definition 5. *Three vertices x, y and z of a graph $G = (V, E)$ form an as-teroidal triple, AT for short, if for any two of the three vertices there is a path between them that avoids the neighborhood of the third. A graph is said to be AT-free if it does not contain an AT.*

Definition 6. *A pair of vertices x and y is a* dominating pair *of a graph G, if the vertex set of any path between x and y in G is a dominating set in G.*

Theorem 6 ([4]). *Any connected AT-free graph has a dominating pair.*

Definition 7. *A path $P = (x = x_0, x_1, \ldots, x_d = y)$ is a dominating shortest path, DSP for short, of a graph $G = (V, E)$ if*

1. *P is a shortest path between x and y in G,*
2. *$\{x_0, x_1, \ldots, x_d\}$ is a dominating set of G.*

Corollary 1 ([14]). *Every connected AT-free graph has a DSP.*

Definition 8. *A d-octopus O of a graph $G = (V, E)$ is a subgraph of G such that*

1. *the vertices of O is a dominating set of G,*
2. *there are vertices $r, v_1, v_2, \ldots, v_d$ of G, and for each $i \in \{1, \ldots, d\}$ there is a shortest path P_i from r to v_i in G such that O is the union of the paths $P_1, P_2, \ldots, P_d$.*

We call the common end point r of the d shortest paths the root *of the d-octopus O. Note that the paths need not to be disjoint.*

Remark 3. A graph with a DSP is a 1-octopus graph.

The following results are Roman domination versions of Lemma 33 in [7] and Theorem 4 in [14] "replacing D by V_2".

Theorem 7. *Let $G = (V, E)$ be a graph with a d-octopus with root x. Let H_0, H_1, ..., H_l be the levels of the BFS-tree with the root x. Then G has a Roman pair (V_1, V_2) such that:*

$$\bigwedge_{i \in \{0,1,\ldots,l\}} \bigwedge_{j \in \{0,1,\ldots,l-i\}} \left| V_2 \cap \bigcup_{s=i}^{i+j} H_s \right| \leq (j+5)d - 1. \tag{1}$$

Theorem 8. *Let $G = (V, E)$ be a connected AT-free graph. There is a vertex x which can be determined in linear time such that if H_0, H_1, ..., H_l are the levels of the BFS-tree with the root x, then G has a Roman pair (V_1, V_2) such that:*

$$\bigwedge_{i \in \{0,1,\ldots,l\}} \bigwedge_{j \in \{0,1,\ldots,l-i\}} \left| V_2 \cap \bigcup_{s=i}^{i+j} H_s \right| \leq j+3. \tag{2}$$

A Polynomial Time Algorithm:

Our algorithm uses dynamic programming to compute a Roman pair through the levels of a BFS-tree. A subsolution computed during the execution of the algorithm is a set $S \subseteq \bigcup_{j=0}^{i-1} H_j$ chosen up to a fixed level $i - 1 \in \{1, 2, \ldots, l-1\}$. Information of any subsolution S that we must store during the execution are the vertices that belong to the last two current levels (i.e, $S \cap (H_{i-2} \cup H_{i-1})$). Consequently, the number of vertices from V_2 that a Roman pair (V_1, V_2) might have in any three consecutive BFS-levels is important for the complexity of the algorithm. The previous theorems guarantee that this number is at most 5 for connected AT-free graphs and at most $7d - 1$ for graphs with a d-octopus.

The algorithm $rp_k(G)$, where k is a fixed positive integer, computes a Roman pair of the given connected graph G. If G has a vertex x and a Roman pair (V_1, V_2) such that at most k vertices of V_2 belong to any three consecutive levels of the BFS-tree which has x as a root, then $rp_k(G)$ outputs a Roman pair for G.

Algorithm $rp_k(G)$
  $D \leftarrow V$
  $val(D) \leftarrow |V|$ /* initialization: every vertex of V is in V_1, this is a
    trivial Roman dominating set   */
  forall $x \in V$ **do**
    Compute the BFS-level of vertex x
    $H_0 = \{x\}$, $H_1 = N(x)$, $\ldots$, $H_l = \{u \in V : d_G(x, u) = l\}$
    $i \leftarrow 1$
    Initialize the queue A_1 to contain an ordered triple $(S, S, val(S))$ for all
    nonempty subsets S of $N[x]$ satisfying $|S| \leq k$ with $val(S) \leftarrow 2|S|$
    Add to the queue A_1 the ordered triple $(\varnothing, \varnothing, 1)$
    while $A_i \neq \varnothing$ *and* $i < l$ **do**
      $i \leftarrow i + 1$
      forall *triples* $(S, S', val(S'))$ *in the queue* A_{i-1} **do**
        forall $U \subseteq H_i$ *with* $|S \cup U| \leq k$ **do**
          $R \leftarrow (S \cup U) \backslash H_{i-2}$
          $R' \leftarrow S' \cup U$
          $val(R') \leftarrow val(S') + 2|U| + |H_{i-1} \backslash N[S \cup U]|$
          if *there is no triple in* A_i *with first entry* R **then**
            Insert $(R, R', val(R'))$ in the queue A_i
          if *there is a triple* $(P, P', val(P'))$ *in* A_i *such that* $P = R$ *and*
          $val(R') < val(P')$ **then**
            Replace $(P, P', val(P'))$ in A_i by $(R, R', val(R'))$
    Among all triples $(S, S', val(S'))$ in the queue A_l, determine one with
    minimum value $v = val(S') + |H_l \backslash N[S]|$, say $(B, B', val(B'))$
    if $v < val(D)$ **then** $D \leftarrow B'$; $val(D) \leftarrow v$
  return $(V_1, V_2) = (V \backslash N[D], D)$

Theorem 9. *Algorithm* $rp_k(G)$ *computes a Roman pair of the given connected graph G in time $O(n^{k+2})$ if G has a Roman pair (V_1, V_2) and a vertex $x \in V$ such that at most k vertices of V_2 belong to any three consecutive BFS-levels of x.*

Proof. The correctness can be seen easily and the analysis of the running time is the same as the Theorem 5 in [14]. □

Theorem 10. *There is an $O(n^{7d+1})$-time algorithm to compute Roman pairs for graphs with a d-octopus. In particular, there is an $O(n^7)$-time algorithm to calculate Roman pairs for graphs having a DSP and there is an $O(n^6)$-time algorithm to compute Roman pairs for AT-free graphs.*

Proof. By combining Theorems 7 and 9 and using the results in [3,5,14] we obtain the theorem (see [13] for more details). □

Acknowledgement. We would like to thank Dieter Kratsch for his helpful comments and advices.

References

1. Brändstadt, A., V. Le, and J. P. Spinrad, *Graph classes: A survey*, SIAM Monogr. Discrete Math. Appl., Philadelphia, 1999.
2. Cockayne, E. J., P. A. Jr. Dreyer, S. M. Hedetniemi, and S. T. Hedetniemi, Roman domination in graphs, *Discrete Math.* **278**, (2004), pp. 11–22.
3. Corneil, D. G., S. Olariu, and L. Stewart, Linear time algorithms for dominating pairs in asteroidal triple-free graphs, *SIAM J. Comput.* **28**, (1999), pp. 1284–1297.
4. Corneil, D. G., S. Olariu, and L. Stewart, Asteroidal triple-free graphs, *SIAM J. Discrete Math.* **10**, (1997), pp. 399–430.
5. Deogun, J. S. and D. Kratsch, Diametral path graphs, *Proceedings of WG'95*, *LNCS* **1017**, (1995), pp. 344–357.
6. Fernau, H., Roman domination: a parameterized perspective, Manuscript.
7. Fomin, F. V., D. Kratsch, and H. Müller, Algorithms for graphs with small octopus, *Discrete Appl. Math.* **134**, (2004), pp. 105–128.
8. Golumbic, M. C., *Algorithmic graph theory and perfect graphs*, Academic Press, New York, 1980.
9. Habib, M. and C. Paul, A simple linear time algorithm for cograph recognition, *Discrete Appl. Math.* **145**, (2005), pp. 183–197.
10. Henning, M. A., A characterization of Roman trees, *Discuss. Math. Graph Theory* **22**, (2002), pp. 225–234.
11. Henning, M, A., Defending the Roman empire from multiple attacks, *Discrete Math.* **271**, (2003), pp. 101–115.
12. Henning, M. A. and S. T. Hedetniemi, Defending the Roman empire–A new strategy, *Discrete Math.* **266**, (2003), pp. 239–251.
13. Kloks, T., M. Liedloff, J. Liu and S. L. Peng, Roman domination in some special classes of graphs, Technical Report TR-MA-04-01, Nov. 2004, University of Lethbridge, Alberta, Canada.
14. Kratsch, D., Domination and total domination on asteroidal triple-free graphs, *Discrete Appl. Math.* **99**, (2000), pp. 111–123.
15. Pagourtzis, A., P. Penna, K. Schlude, K. Steinhfel, D. S. Taylor and P. Widmayer, Server placements, Roman domination and other dominating set variants, *IFIP TCS Conference Proceedings* **271**, (2002), pp. 280–291.
16. ReVelle, C. S. and K. E. Rosing, Defenders imperium Romanum: A classical problem in military strategy, *Amer. Math. Monthly* **107**, (2000), pp. 585–594.
17. Stewart, I., Defend the Roman empire! *Sci. Amer.* **281**, (1999), pp. 136–139.

Algorithms for Comparability of Matrices in Partial Orders Imposed by Graph Homomorphisms

Jiří Fiala[1], Daniël Paulusma[2], and Jan Arne Telle[3]

[1] Charles University, Faculty of Mathematics and Physics,
DIMATIA and Institute for Theoretical Computer Science (ITI)[*],
Malostranské nám. 2/25, 118 00, Prague, Czech Republic
`fiala@kam.mff.cuni.cz`
[2] Department of Computer Science, University of Durham,
Science Laboratories, South Road,
Durham DH1 3LE, England
`daniel.paulusma@durham.ac.uk`
[3] Department of Informatics, University of Bergen,
N-5020 Bergen, Norway
`telle@ii.uib.no`

Abstract. Degree refinement matrices have tight connections to graph homomorphisms that locally, on the neighborhoods of a vertex and its image, are constrained to three types: bijective, injective or surjective. If graph G has a homomorphism of given type to graph H, then we say that the degree refinement matrix of G is smaller than that of H. This way we obtain three partial orders. We present algorithms that will determine whether two matrices are comparable in these orders. For the bijective constraint no two distinct matrices are comparable. For the injective constraint we give a PSPACE algorithm, which we also apply to disprove a conjecture on the equivalence between the matrix orders and universal cover inclusion. For the surjective constraint we obtain some partial complexity results.

1 Introduction

Graph homomorphisms, originally obtained as a generalization of graph coloring, have a great deal of applications in computer science and other fields. Beyond these computational aspects they impose an interesting structure on the class of graphs, with many important categorical properties, see e.g. the recent monograph [6]. We focus our attention on graph homomorphisms with local constraints. Originally arising in topological graph theory, these homomorphisms were required to act as a bijection on the neighborhood of each vertex [2].

---

[*] Supported by the Ministry of Education of the Czech Republic as project 1M0021620808.

D. Kratsch (Ed.): WG 2005, LNCS 3787, pp. 115–126, 2005.

We consider further local constraints, namely local injectivity or local surjectivity. Both these kinds of homomorphisms have already been studied due to their applications in models of telecommunication [4] and in social science [3,8].

In related work [5] we have shown that these locally constrained homomorphisms impose an algebraic structure on the class of connected finite graphs. We also extended a necessary condition for the existence of a locally bijective homomorphism between two graphs [7] to a similar but much more sophisticated statement for locally injective or surjective homomorphisms. An important role in this characterization was predicated to matrices that describe the degree structure of a graph, the so-called degree refinement matrices. We gave a characterization of these matrices, and showed that both locally injective and locally surjective graph homomorphisms impose partial orders on degree refinement matrices [5].

New Results

In this paper we continue this work and turn our attention away from categorical questions to focus instead on the following computational questions: Given two degree matrices, are they comparable in the partial orders imposed by local injectivity or surjectivity? It is not obvious that these questions are decidable, and indeed for local surjectivity we must leave this as a major open problem. However, for local injectivity we manage to show an upper bound on the size of the smallest graphs that can possibly justify a positive answer and use this to provide a PSPACE algorithm. The existence of a locally bijective homomorphism between two graphs is conditioned by the equivalence of their degree refinement matrices, which can also be expressed as an isomorphism between their universal covers [7]. For the other two kinds of locally constrained homomorphisms this naturally raises the question, and conjecture, of a similar tight relationship between matrix comparison in the partial order and inclusion of universal covers. However, we apply our PSPACE algorithm to disprove this enticing conjecture. For the surjective constraint we obtain some partial results on the complexity of matrix comparison.

2   Preliminaries

Graphs considered in this paper are simple, i.e. with no loops and multiple edges, connected and, if not stated otherwise, they are also finite. We denote the class of such graphs by $\mathcal{C}$. For any vertex $u \in V_G$ the symbol $N(u)$ denotes the *neighborhood* of u, i.e. the set of all vertices adjacent to u. A k-regular graph is a graph, where all vertices have k neighbors (i.e. are of *degree* k). A (k, l)-regular bipartite graph is a bipartite graph where vertices of one class of the bi-partition are of degree k and the remaining vertices are of degree l. A graph G is a *subgraph* of a graph H if $V_G \subseteq V_H$ and $E_G \subseteq E_H$. This is denoted by $G \subseteq H$.

A *degree partition* of a graph G is a partition of the vertex set V_G into blocks $\mathcal{B} = \{B_1, \ldots, B_k\}$ such that whenever two vertices u and v belong to the same block B_i, then for any $j \in \{1, \ldots, k\}$ we have $|N_G(u) \cap B_j| = |N_G(v) \cap B_j| = m_{i,j}$. The $k \times k$ matrix M such that $(M)_{i,j} = m_{i,j}$ is a *degree matrix*. A graph G can allow several degree matrices. The matrix that corresponds to the partition with the *smallest* number of blocks and where these blocks follow the so-called canonical ordering (just some ordering to provide uniqueness) is called its *degree refinement matrix*. It is denoted by $\mathrm{drm}(G)$ for a graph G and computed in polynomial time by a simple stepwise refinement starting from an initial partition by vertex degrees with blocks ordered by increasing degrees. The refinement of the partition continues until any two nodes in the same block have the same number of neighbors in any other block, see e.g. [5]. (See Fig. 1 for an example.) We denote the class of degree refinement matrices of graphs in $\mathcal{C}$ by $\mathcal{M}$.

A graph homomorphism is an edge-preserving mapping $f : V_G \to V_H$, i.e. $(f(u), f(v))$ is an edge of H whenever $(u, v) \in E_G$. A homomorphism $f : G \to H$ may be further confined to adhere to some local constraints, as in the following definition.

Definition 1. *We call a graph homomorphism* $f : G \to H$ *locally bijective, locally injective or locally surjective if for every vertex* $u \in V_G$ *the restriction of* f *to* $N(u)$ *is a bijection, injection or surjection between* $N(u)$ *and* $N(f(u))$, *respectively. We denote it as* $f : G \overset{B}{\to} H$ *or* $f : G \overset{I}{\to} H$ *or* $f : G \overset{S}{\to} H$, *respectively.*

For each of the three types of local constraints $* = B$ (bijective), $* = I$ (injective) or $* = S$ (surjective), we will in this paper focus on the following three relations on the class of degree refinement matrices $\mathcal{M}$:

$$M \overset{*}{\to} N \iff \text{exist } G, H \in \mathcal{C} : \mathrm{drm}(G) = M, \ \mathrm{drm}(H) = N \text{ and } G \overset{*}{\to} H$$

In [5] we showed that all three relations $(\mathcal{M}, \overset{B}{\to})$, $(\mathcal{M}, \overset{I}{\to})$ and $(\mathcal{M}, \overset{S}{\to})$ are partial orders. Note that $(\mathcal{M}, \overset{B}{\to})$ is in fact a trivial order, since in [7] it has been shown that $\mathrm{drm}(G) = \mathrm{drm}(H)$ is a necessary condition for $G \overset{B}{\to} H$.

For a graph $G \in \mathcal{C}$ the *universal cover* T_G is defined in [1] as the only (possibly infinite) tree that allows $T_G \overset{B}{\to} G$. The vertices of T_G can be represented as walks in G starting in a fixed vertex u that do not traverse the same edge in two consecutive steps. Edges in T_G connect those walks that differ in the presence of the last edge. The mapping $T_G \overset{B}{\to} G$ sending a walk in V_{T_G} to its last vertex is a locally bijective homomorphism. Universal covers are in one-to-one correspondence with degree refinement matrices, hence for $M \in \mathcal{M}$ we can define $T_M = T_G$ for any G with $\mathrm{drm}(G) = M$.

Proposition 1 ([5]). *The relation* $M \overset{I}{\to} N$ *holds if and only if there exist graphs* G *with* $\mathrm{drm}(G) = M$ *and* H *with* $\mathrm{drm}(H) = N$ *such that* $G \subseteq H$.

We use the following relationship between degree refinement matrices and universal covers.

Proposition 2 ([5]). *For any degree refinement matrices* $M, N \in \mathcal{M}$ *it holds that if* $M \overset{I}{\to} N$ *then* $T_M \subseteq T_N$, *and if* $M \overset{S}{\to} N$ *then* $T_N \subseteq T_M$.

For computational complexity purposes $\langle X \rangle$ denotes the size of the instance X (graph, matrix, etc.) in usual binary encoding of numbers. Formally we represent vertices of a graph G by numbers $\{1, 2, \ldots, |V_G|\}$ and its edges as a list of its vertices. A graph with m edges on n vertices hence requires space $\langle G \rangle = \Theta(m \log n)$. For an integral-valued $k \times l$ matrix A let $a^* = 2 + \max\{|A_{i,j}| \mid 1 \leq i \leq k \text{ and } 1 \leq j \leq l\}$. Then the size of A is given by $\langle A \rangle = \Theta(kl \log a^*)$.

We will need the following technical lemma for our PSPACE algorithm.

Lemma 1. *Let A be an integral-valued $k \times l$ matrix with $l > k$. If $Ax = 0$ allows a nontrivial nonnegative solution, then it allows a nontrivial nonnegative integer solution $\mathbf{x}$ with at most $k + 1$ nonzero entries and with $\langle x_i \rangle = O(k \log(ka^*))$ for each entry x_i.*

Proof. If a solution $\mathbf{x}$ with more than $k + 1$ positive coefficients exists, then the columns corresponding to $k + 1$ of these variables are linearly dependent. Let the coefficients of such a linear combination form a vector $\mathbf{x}'$. Obviously $A\mathbf{x}' = \mathbf{0}$, but the entries of $\mathbf{x}'$ may not be necessarily nonnegative.

Without loss of generality we assume that at least one of the entries in $\mathbf{x}'$ is positive. Then, for a suitable value $\alpha = -\min\{\frac{x_i}{x_i'} \mid x_i' > 0\}$ the vector $\mathbf{x} + \alpha\mathbf{x}'$ is also a nontrivial nonnegative solution with more zero entries than $\mathbf{x}$.

Repeating this trimming iteratively we obtain a nontrivial nonnegative solution with at most $k + 1$ nonzero entries. As the other entries are zero, we may restrict the matrix A to columns corresponding to nonzero entries of the solution. It may happen that the rank of the modified matrix decreases. Then we reduce the number of rows until the remaining ones become linearly independent. By repeating the above process we finally get an $k' \times (k' + 1)$ matrix B of rank $k' \leq k$, such that $B\mathbf{y} = \mathbf{0}$ allows a nontrivial nonnegative solution $\mathbf{y}$. Such $\mathbf{y}$ can be extended to a solution $\mathbf{x}$ of the original system by inserting zero entries.

Without loss of generality we assume that the first k' columns of B are linearly independent, and we arrange them in a regular matrix R. Then its inverse can be expressed as $R^{-1} = \frac{adj(R)}{\det(R)}$, where $adj(R)$ is the adjoint matrix of R. By the determinant expansion we have that $\det(R) \leq k'!(a^*)^{k'} \leq k!(a^*)^k \leq k^k(a^*)^k$. Then we find that $\langle \det(R) \rangle = O(k \log(ka^*))$. Each element of $adj(R)$ is a determinant of a minor of R and hence is smaller than $(k - 1)^{k-1}(a^*)^{k-1}$.

Now consider the integral valued matrix $B' = \det(R) \cdot R^{-1}B$. Then

- $\mathbf{y}$ is a solution of $B'\mathbf{y} = \mathbf{0}$ if and only if $B\mathbf{y} = \mathbf{0}$.
- The first k' columns of B' form the matrix $\det(R) \cdot I_{k'}$.
- In the last column the entries $z_1, \ldots, z_l$, are all negative (if $\det(R) > 0$) or all positive (otherwise).

If $\det(R) > 0$ then $\mathbf{y} = (-z_1, \ldots, -z_{k'}, \det(R))$ is a nonnegative nontrivial integral solution to $B\mathbf{y} = \mathbf{0}$. In the other case we swap the sign and choose $\mathbf{y} = (z_1, \ldots, z_{k'}, -\det(R))$. As each $z_i \leq ka^* \max_{ij}(adj(R)_{ij}) \leq k^k(a^*)^k$, we obtain $\langle z_i \rangle = O(k \log(ka^*))$, which concludes the proof. □

3   Matrix Comparison Via Local Injectivity

In this section we consider the problem of deciding whether for given degree refinement matrices M and N the comparison $M \xrightarrow{l} N$ holds.

Observe that according to the definition of the order $(\mathcal{M}, \xrightarrow{l})$, there is no obvious bound on the sizes of graphs G and H with M and N as degree refinement matrices that should justify the comparison $M \xrightarrow{l} N$.

The main result of this paper is the following theorem:

Theorem 1. *Let M, N be degree refinement matrices of order k and l. If $M \xrightarrow{l} N$, then there exist a graph G of size $(klm^*)^{O(k^2 l^2)}$ and a graph H of size $(klm^* n^*)^{O(k^2 l^2)}$ such that $G \xrightarrow{l} H$, $\mathrm{drm}(G) = M$ and $\mathrm{drm}(H) = N$.*

Proof. Throughout this proof we assume that indices i, j, r, s used later always belong to feasible intervals $1 \leq i, r \leq k$ and $1 \leq j, s \leq l$. For clarity we often abbreviate pairs of sub-/super-scripts i, j by ij, so in this notation, ij does not mean multiplication.

The main idea of the construction is as follows. Assume that $M \xrightarrow{l} N$ holds. Then by Proposition 1 there exist a graph H and a subgraph $G \subseteq H$ witnessing $M \xrightarrow{l} N$. Let $\{U_1, \ldots, U_k\}$ be the degree partition of G and $\{V_1, \ldots, V_l\}$ the one for H. We further partition $V_G \subseteq V_H$ as follows. For each pair of indices r and s we define the set

$$W_{rs} = \{v \mid v \in U_r \cap V_s\},$$

and for some vertex $w \in W_{rs}$ we can write a vector describing the distribution of neighbors of w in the classes $W_{11}, \ldots, W_{kl}$.

We first show that for given M and N the set T containing all such vectors is finite. Then, with help of T, we design a set of equations that allows a solution if and only if the desired graphs G and H exist. As the size of T is bounded, we can establish the desired bounds on the size of G and H.

Let $\mathbf{p}^{rs}$ be a vector of length kl whose entries are positive integers and are indexed by pairs ij. If the vector $\mathbf{p}^{rs}$ further satisfies

$$\sum_{j=1}^{l} p_{ij}^{rs} = m_{ri} \qquad \text{for all } 1 \leq i \leq k, \tag{1}$$

$$\sum_{i=1}^{k} p_{ij}^{rs} \leq n_{sj} \qquad \text{for all } 1 \leq j \leq l, \tag{2}$$

then we call $\mathbf{p}^{rs}$ an *injective distribution row for indices r and s*. Note that for given matrices M and N and any feasible choice of r, s the number of all different injective distribution rows for r and s is finite. We denote the set of all injective distribution rows for indices r and s by

$$T(r, s) = \{\mathbf{p}^{rs(1)}, \ldots, \mathbf{p}^{rs(t(rs))}\}.$$

Due to (1), the number of distribution rows for every $\mathbf{p}^{rs}$ is bounded by $t(r,s) \leq \binom{m^*+l-1}{m^*}^k = O((m^*+1)^{kl})$. The total number of distribution rows is then

$$t_0 = \sum_{r,s} t(r,s) = O(kl(m^*+1)^{kl}).$$

Now consider a set of t_0 variables $w^{rs(t)}$ for all feasible r,s and all $1 \leq t \leq t(r,s)$. We claim that the existence of a nontrivial *nonnegative* solution of the following homogeneous system of $k^2 l^2$ equations in t_0 variables:

$$\sum_{t=1}^{t(r,s)} p_{ij}^{rs(t)} w^{rs(t)} = \sum_{t'=1}^{t(i,j)} p_{rs}^{ij(t')} w^{ij(t')} \qquad 1 \leq i,r \leq k,\ 1 \leq j,s \leq l \qquad (3)$$

is a necessary and sufficient condition for the existence of finite graphs G and H witnessing $M \xrightarrow{l} N$.

Necessity: For given G and H we assume without loss of generality that $G \subseteq H$. Firstly determine the sets W_{rs}, and for each vertex $u \in W_{rs} \subseteq V_G$ compute the distribution vector of its neighbors $\mathbf{p}(u) = (|N(u) \cap W_{11}|, \ldots, |N(u) \cap W_{kl}|)$. Then the vector $\mathbf{w}$ with entries $w^{rs(t)} = |\{u : \mathbf{p}(u) = \mathbf{p}^{rs(t)}\}|$ is a nontrivial solution of (3), since in each equation both sides are equal to the number of edges connecting sets W_{rs} and W_{ij}.

Sufficiency: Assume that the system (3) has a nontrivial nonnegative solution. By appropriate scaling we obtain a nonnegative integer solution $\mathbf{w} = (w^{11(1)}, \ldots, w^{kl(t(k,l))})$ with each $w^{rr(t)}$ is even.

We first build a multigraph G_0 upon t_0 sets of vertices $W^{11(1)}, \ldots, W^{kl(t(k,l))}$, where $|W^{rs(t)}| = w^{rs(t)}$ (some sets may be empty) as follows: Denote $W^{rs} = W^{rs(1)} \cup \cdots \cup W^{rs(t(r,s))}$.

Our choice of even values $w^{rr(t)}$ allows us to build an arbitrary $p_{rr}^{rr(t)}$-regular multigraph on each set $W^{rr(t)}$.

As $\mathbf{w}$ satisfies (3), we can easily build a bipartite multigraph between any pair of different sets W^{rs} and W^{ij} such that the number of edges between them is equal to $\sum_{t=1}^{t(r,s)} p_{ij}^{rs(t)} w^{rs(t)} = \sum_{t'=1}^{t(i,j)} p_{rs}^{ij(t')} w^{ij(t')}$.

For any vertex u in $W^{rs(t)}$ with more than $p_{ij}^{rs(t)}$ neighbors in W^{ij} there exists a vertex u^* in some $W^{ij(t^*)}$ with less than $p_{ij}^{rs(t^*)}$ neighbors, and vice versa. Now we remove an edge between u and some neighbor $v \in W^{ij}$ and add the edge (u', v). We repeat this procedure until all vertices of W^{rs} have the right number of neighbors in W^{ij}. Then we do the same for vertices in W^{ij}.

This way we have constructed a bipartite multigraph between W^{rs} and W^{ij} such that each vertex of each $W^{rs(t)}$ is incident with exactly $p_{ij}^{rs(t)}$ edges and each vertex of each $W^{ij(t')}$ is incident with exactly $p_{rs}^{ij(t')}$ edges.

It may happen in some instances that multiple edges are unavoidable. In that case let $d \leq m^*$ be the maximal edge multiplicity in G_0. We obtain the graph G

by taking d copies of the multigraph G_0 and replace each collection of d parallel edges of multiplicity $d' \leq d$ by a simple d'-regular bipartite graph.

Due to the construction, it is straightforward to check that vertices from sets that share the same index r form the r-th block of the degree partition of G and that $\mathrm{drm}(G) = M$.

For the construction of H we first distribute the vertices of G into sets $V_1', \ldots, V_l'$, where

$$V_s' = \bigcup_{r=1}^{k} \bigcup_{t=1}^{t(r,s)} W^{rs(t)}.$$

Since N is a degree refinement matrix, the following homogeneous system whose equations represent the number of edges between two different blocks in N has nontrivial solutions:

$$n_{sj} v_s = n_{js} v_j \qquad 1 \leq j, s \leq l \qquad (4)$$

Then we form sets $V_1, \ldots, V_l$ by further inserting new vertices into $V_1', \ldots, V_l'$ until for each $s, j : |V_s| n_{sj} = |V_j| n_{js}$ and $|V_s| > 0$ is even.

Next we build a multigraph H_0 by constructing an (n_{sj}, n_{js})-regular bipartite multigraph between any two sets V_s and V_j, and an n_{jj}-regular multigraph on each V_j. In case multiple edges cannot be avoided we take sufficient copies of H_0 and make the appropriate reparations. So we perform these steps in the same way as before, however without removing any edges between vertices in (any copy of) G.

Clearly, G is a subgraph of the resulting graph H and H has N as its degree refinement matrix.

To conclude the proof of the theorem we discuss the size of G and H. Note that all coefficients $p_{ij}^{rs(t)}$ of system (3) are at most m^*. Then, by Lemma 1, we find a nontrivial nonnegative integer solution $\mathbf{w}$ whose entry sizes are bounded by $O(k^2 l^2 \log(klm^*))$.

We can use the entries of $2\mathbf{w}^*$ for the sizes of the blocks in the multigraph G_0. Since we take at most m^* copies of G_0 to obtain our final graph G, we find that $\langle G \rangle = (klm^*)^{O(k^2 l^2)}$.

Analogously, the size of each entry of a solution $\mathbf{v}$ of system 4 is bounded by $O(l^2 \log(ln^*))$. Since multigraph H_0 must contain graph G, we use the entries of $\langle G \rangle$ for the block sizes of H_0. We need at most n^* copies of H_0 for graph H. Hence, each block size $|V_i|$ can be chosen within the upper bound $\langle G \rangle \cdot (ln^*)^{O(l^2)}$ implying that $\langle H \rangle = (klm^*n^*)^{O(k^2 l^2)}$. $\qquad \square$

We can now settle the first computational complexity result for the following matrix comparison problem:

MATRIX INJECTIVITY (MI)
Instance: Degree refinement matrices M and N.
Question: Does $M \xrightarrow{I} N$ hold?

Corollary 1. *The* MI *problem is decidable in polynomial space.*

Proof. The proof of Theorem 1 showed that $M \xrightarrow{I} N$ if and only if system (3) has a nontrivial nonnegative solution. Then by Lemma 1 there exists a nontrivial nonnegative integral solution with at most $k^2 l^2 + 1$ nonzero entries, which are each bounded in size by $O(k^2 l^2 \log(klm^*))$.

So we only have to consider vectors of this form. As the size of any such vector is polynomial, we can by brute force sequentially list them all, and test their feasibility for (3). Note that any restriction of (3) to polynomially many columns can be generated in PSPACE as well.   □

As we have discussed in the introduction, the matrix order $(\mathcal{M}, \xrightarrow{I})$ was considered as a nontrivial necessary condition for the decision problem whether $G \xrightarrow{I} H$. As the size of M and N should vary from being independent in the size of the given graphs to be of approximately the same size of G, H, even the exponential time-complexity of the MI problem might be plausible as a precomputation for some instances.

We apply Theorem 1 to disprove the following interesting conjecture on the equivalence between comparison of degree matrices in $\xrightarrow{I}$ and inclusion of universal covers.

Conjecture 1. For any two degree refinement matrices M and N the following equivalence holds: $M \xrightarrow{I} N \iff T_M \subseteq T_N$.

We note here that the affirmative answer for the only if implication was already shown in Proposition 2. The following example acts both as an example for the application of Theorem1, and as an counterexample of Conjecture 1.

Corollary 2. *There exist matrices M and N such that $T_M \subseteq T_N$, but $M \xrightarrow{I\!\!\!/} N$.*

Proof. We first construct graphs G and H such that $H \xrightarrow{S} G$. Denote $M = \mathrm{drm}(G)$ and $N = \mathrm{drm}(H)$. Then according to Proposition 2 we get that $T_M \subseteq T_N$. We will now show that the MI problem for matrices M and N has a negative answer.

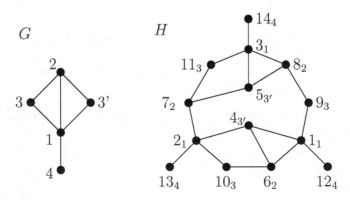

Fig. 1. Graphs G and H, vertices of H are labeled by $u_{f(u)}$ for a $f : H \xrightarrow{S} G$

The graphs G and H together with a mapping $f : H \overset{s}{\rightarrow} G$ are depicted in Fig. 1.

The graph G has 4 classes in its degree refinement and H has 14 classes. Then N is the adjacency matrix of H and the degree refinement matrix of G is

$$M = \begin{pmatrix} 0 & 1 & 2 & 1 \\ 1 & 0 & 2 & 0 \\ 1 & 1 & 0 & 0 \\ 1 & 0 & 0 & 0 \end{pmatrix}.$$

Note that N is the adjacency matrix of H. In order to obtain a contradiction suppose $T_M \overset{I}{\rightarrow} T_N$ holds. By Proposition 1 there exist a graph G' with $\mathrm{drm}(G') = M$ and a graph H' with $\mathrm{drm}(H') = N$ such that $G' \subseteq H'$. Let $\{U_1, \ldots, U_k\}$ be the degree partition of G' and $\{V_1, \ldots, V_l\}$ the one for H'. We define the sets W_{rs} as in proof of Theorem 1.

As we have seen in the proof of Theorem 1 the pair (G', H') corresponds with a nontrivial solution of (3). Below we will show, however, that (3) only allows the trivial solution. For simplicity reasons we will first restrict the length of the injective distribution rows.

A vertex in class U_1 has four neighbors in G'. A vertex in class V_4 has three neighbors in H'. This means that a vertex of U_1 can never be in V_4, i.e., $W_{1,4}$ is empty. Hence the set $T(1, 4)$ is empty. By the same argument we find that the sets $T(r, s)$ with $(r, s) = (1, 5), \ldots, (1, 14), (2, 9), \ldots, (2, 14), (3, 12), \ldots, (3, 14)$ are empty.

A vertex in U_2 has a neighbor of degree four in G'. A vertex in V_1 does not have a neighbor of degree four in H'. Hence the set $T(2, 1)$ is empty. By the same argument we exclude pairs $(2, 2), (2, 3), (3, 1), (3.2), (3, 3), (4, 1), (4, 2), (4, 3)$.

Any vertex in U_4 has degree one in G'. Suppose $u \in U_4$ belongs to V_4. So it does not have degree one in H'. Let $v \in U_1$ be the (only) neighbor of u in G'. Then v has degree four in G' and must belong to $V_1 \cup V_2 \cup V_3$. The other three neighbors of v all have degree greater than one in G'. However, one of these three remaining neighbors of v must have degree one in H'. Hence, the set $T(4, 4)$ is empty. In the same way we may exclude pairs $(4, 4), \ldots, (4, 11)$.

Every vertex in $W_{2,4}$ needs a neighbor in $W_{3,1}$ or $W_{3,2}$. These sets are empty, since both $T(3, 1)$ and $T(3, 2)$ are empty. Hence $T(2, 4)$ is empty, and consequently, by a similar argument, $T(3, 6)$ is empty. Furthermore, $T(2, 4) = \emptyset$ implies that a vertex in $W_{1,2}$ does not have neighbor in $W_{3,7}$. Since every vertex in $W_{3,7}$ must have a neighbor in $W_{1,2}$, the latter implies $T(3, 7) = \emptyset$, and consequently $T(2, 5) = \emptyset$ and $T(3, 8) = \emptyset$.

Only the pairs $(3, 4)$ and $(3, 5)$ allow two distribution rows, the other pairs all allow one. So we have reduced the total number of feasible distribution rows to $4 \cdot 14 - 20 - 9 - 8 - 5 + 2 = 16$.

The equation (3) for $p, q = 1, 1$ and $i, j = 2, 6$ gives $w^{1,1} = w^{2,6}$. Analogously, $w^{1,1} = w^{3,4(1)}$ while $w^{2,6} = w^{3,4(1)} + w^{3,4(2)}$. Hence $w^{3,4(2)} = 0$. Further $w^{3,4(2)} = w^{1,2} = w^{3,10} = w^{2,6}$, and $w^{1,2} = w^{2,7} = w^{3,11} = w^{1,3}$. Consequently, $w^{1,1} = w^{1,2} = w^{1,3} = 0$.

Table 1. The distribution rows for M (only nonzero entries are shown)

i	1	1	1	2	2	2	3	3	3	3	3	4	4	4
j	1	2	3	6	7	8	4	5	9	10	11	12	13	14
$p^{1,1}$				1			1		1			1		
$p^{1,2}$					1		1		1				1	
$p^{1,3}$						1	1		1					1
$p^{2,6}$	1						1		1					
$p^{2,7}$	1						1		1					
$p^{2,8}$		1					1	1						
$p^{3,4(1)}$	1			1										
$p^{3,4(2)}$		1		1										
$p^{3,5(1)}$			1		1									
$p^{3,5(2)}$			1			1								
$p^{3,9}$	1					1								
$p^{3,10}$		1		1										
$p^{3,11}$			1		1									
$p^{4,12}$	1													
$p^{4,13}$		1												
$p^{4,14}$			1											

It can be further shown that (3) allows only trivial solution via values of $w^{r,s}$. However, at this moment we can already claim that no witnesses G, H for $M \xrightarrow{I} N$ exist, since it is impossible to map vertices from the first class of degree partition of G on any vertex of H. $\qquad\square$

4   Matrix Comparison Via Local Surjectivity

In this section we are interested in the following matrix comparison problem:

MATRIX SURJECTIVITY (MS)
Instance: A degree refinement matrix M and a degree refinement matrix N.
Question: Does $M \xrightarrow{S} N$ hold?

We were not able to answer the decidability of this problem. However, we can show some partial results.

Proposition 3. *Let G be a graph with* $\mathrm{drm}(G)$ *of order k and H be a graph on l vertices such that $G \xrightarrow{S} H$. Then there exists a graph G' with* $\mathrm{drm}(G') = \mathrm{drm}(G)$ *such that $G' \xrightarrow{S} H$ and $\langle G' \rangle = (klm^*)^{O(k^2 l^2)}$.*

Proof. Let $f : V_G \to V_H$ be a locally surjective homomorphism from G to H. Let $\{U_1, \ldots, U_k\}$ be the degree partition of G and let $\{v_1, \ldots, v_l\}$ be the vertex set of H. We further partition V_G as follows. For each pair of indices r and s we define the set

$$W_{rs} = \{u \mid u \in U_r \text{ and } f(u) = v_s\},$$

and for some vertex $w \in W_{rs}$ we can write a vector describing the distribution of neighbors of w in the classes $W_{11}, \ldots, W_{kl}$.

Let $\mathbf{p}^{rs}$ be a vector of length kl whose entries are positive integers and are indexed by pairs ij. If the vector $\mathbf{p}^{rs}$ further satisfies

$$\sum_{j=1}^{l} p_{ij}^{rs} = m_{ri} \qquad \text{for all } 1 \leq i \leq k, \qquad (5)$$

$$(v_s, v_j) \in E_H \quad \Rightarrow \quad \sum_{i=1}^{k} p_{ij}^{rs} \geq 1 \qquad \text{for all } 1 \leq j \leq l. \qquad (6)$$

$$(v_s, v_j) \notin E_H \quad \Rightarrow \quad \sum_{i=1}^{k} p_{ij}^{rs} = 0 \qquad \text{for all } 1 \leq j \leq l. \qquad (7)$$

then we call $\mathbf{p}^{rs}$ a *surjective distribution row for indices r and s*. The number of surjective distribution rows is bounded.

We now involve the system of equations (3). We claim that the existence of a nontrivial *nonnegative* solution of (3) is a necessary and sufficient condition for the existence of a finite graph G' with $\mathrm{drm}(G') = M$ and $G \xrightarrow{S} H$. The proof of this claim and the bound on the size of G' is using the same arguments as in the proof of Theorem 1.   □

Now we consider the following decision problem.

MATRIX GRAPH SURJECTIVITY (MGS)
Instance: A degree refinement matrix M and a graph H.
Question: Does there exist a graph G with $\mathrm{drm}(G) = M$ such that $G \xrightarrow{S} H$ holds?

Corollary 3. *The* MGS *problem problem is decidable in polynomial space.*

Proof. We can use Proposition 3 and proceed with a proof analogous to the one in Corollary 1.   □

We can use Corollary 3 to answer decidability of the MS problem for instances (M, N), where N is the degree refinement matrix of a unique graph H. The proposition below shows that this is only the case if H is a tree.

Proposition 4. *A matrix N is a degree refinement matrix of a unique graph H if and only if N is the degree refinement matrix of a tree.*

Proof. Suppose N is the degree refinement matrix of a tree T. Then the universal cover T_N is isomorphic to T itself. Since all graphs that contain a cycle have an infinite universal cover, there can not be another graph H with $\mathrm{drm}(H) = N$.

In order to prove the reverse statement let H be the only graph that has N as a degree refinement matrix. Suppose H is not a tree. Then H contains an edge $e = (u, v)$ such that the graph $H - e$ is still connected. We take a copy H'

of H. Let $e' = (u'v')$ be the copy of e. We remove e in H and e' in H', and we add the edges (u, v') and (u', v). The resulting graph H^* has the same degree refinement matrix as H and is connected. □

We can also use Corollary 3 to answer decidability of the MS problem for instances (M, N), where the $l \times l$ degree refinement matrix N is the adjacency matrix of a graph H. This can be seen as follows. Suppose $M \xrightarrow{S} N$ holds with witnesses G and H'. Since N is an adjacency matrix of graph H, the rows of N are in one-to-one correspondence with vertices of H, i.e., we can say that vertex $v_i \in V_H$ corresponds to row i. Then the function that maps all vertices of H' that belong to block $V_i \subseteq V_{H'}$ to v_i for $1 \leq i \leq l$ is a locally bijective homomorphism from H' to H. The mappings $H' \xrightarrow{B} H$ and $G \xrightarrow{S} H'$ imply $G \xrightarrow{S} H$. So we can restrict ourselves to graph H.

In general, even if we construct a graph G with respect to feasible block sizes, there is no evident rule how to limit the size of some plausible graph H and how to define the locally surjective mapping $G \xrightarrow{S} H$. We leave the general question on decidability of the MS as an open problem.

References

1. ANGLUIN, D. Local and global properties in networks of processors. In *Proceedings of the 12th ACM Symposium on Theory of Computing* (1980), 82–93.
2. BIGGS, N. *Algebraic Graph Theory*. Cambridge University Press, 1974.
3. EVERETT, M. G., AND BORGATTI, S. Role colouring a graph. *Math. Soc. Sci. 21*, 2 (1991), 183–188.
4. FIALA, J., AND KRATOCHVÍL, J. Partial covers of graphs. *Discussiones Mathematicae Graph Theory 22* (2002), 89–99.
5. FIALA, J., PAULUSMA, D., AND TELLE, J. A. Matrix and graph orders derived from locally constrained graph homomorphisms. accepted for MFCS 2005.
6. HELL, P., AND NEŠETŘIL, J. *Graphs and Homomorphisms*. Oxford University Press, 2004.
7. LEIGHTON, F. T. Finite common coverings of graphs. *Journal of Combinatorial Theory B 33* (1982), 231–238.
8. ROBERTS, F. S., AND SHENG, L. How hard is it to determine if a graph has a 2-role assignment? *Networks 37*, 2 (2001), 67–73.

Network Discovery and Verification[*]

Zuzana Beerliova[1], Felix Eberhard[1], Thomas Erlebach[2], Alexander Hall[1],
Michael Hoffmann[2], Matúš Mihaľák[2], and L. Shankar Ram[1]

[1]Department of Computer Science, ETH Zürich
{bzuzana, mhall, lshankar}@inf.ethz.ch
[2]Department of Computer Science, University of Leicester
{te17, mh55, mm215}@mcs.le.ac.uk

Abstract. Consider the problem of discovering (or verifying) the edges
and non-edges of a network, modeled as a connected undirected graph,
using a minimum number of queries. A query at a vertex v discovers (or
verifies) all edges and non-edges whose endpoints have different distance
from v. In the network discovery problem, the edges and non-edges are
initially unknown, and the algorithm must select the next query based
only on the results of previous queries. We study the problem using
competitive analysis and give a randomized on-line algorithm with competitive ratio $O(\sqrt{n \log n})$ for graphs with n vertices. We also show that
no deterministic algorithm can have competitive ratio better than 3. In
the network verification problem, the graph is known in advance and the
goal is to compute a minimum number of queries that verify all edges
and non-edges. This problem has previously been studied as the problem of placing landmarks in a graph or determining the metric dimension
of a graph. We show that there is no approximation algorithm for this
problem with ratio $o(\log n)$ unless $\mathcal{P} = \mathcal{NP}$.

1 Introduction

In recent years, there has been an increasing interest in the study of networks
whose structure has not been imposed by a central authority but arisen from local
and distributed processes. Prime examples of such networks are the Internet and
unstructured peer-to-peer networks such as Gnutella. For these networks, it is
very difficult and costly to obtain a "map" providing an accurate representation
of all nodes and the links between them. Such maps would be useful for many
purposes, e.g., for studying routing aspects or robustness properties.

In order to create maps of the Internet, a commonly used technique is to
obtain local views of the network from various locations (vantage points) and
combine them into a map that is hopefully a good approximation of the real
network [2,13]. More generally, one can view this technique as an approach for
discovering the topology of an unknown network by using a certain type of
queries—a query corresponds to asking for the local view of the network from

---

[*] Research partially supported by the EU within the 6th Framework Programme under
contract 001907 (DELIS).

one specific vantage point. In this paper, we formalize *network discovery* as a combinatorial optimization problem whose goal is to minimize the number of queries required to discover all edges and non-edges of the network. We study the problem as an on-line problem using competitive analysis. Initially, the network is unknown to the algorithm. To decide the next query to ask, the algorithm can only use the knowledge about the network it has gained from the answers of previously asked queries. In the end, the number of queries asked by the algorithm is compared to the optimal number of queries sufficient to discover the network. We consider a query model in which the answer to a query at a vertex v consists of all edges and non-edges whose endpoints have different (graph-theoretic) distance from v.

In the off-line version of the network discovery problem, the network is known to the algorithm from the beginning. The goal is to compute a minimum number of queries that suffice to discover the network. Although an algorithm for this off-line problem would not be useful for network discovery (if the network is known in advance, there is no need to discover it), it could be employed for network verification, i.e., for checking whether a given map is accurate. Thus, we refer to the off-line version of network discovery as *network verification*. Here, we are interested in polynomial-time optimal or approximation algorithms.

Motivation. As mentioned above, the motivation for our research comes from the problem of discovering information about the topology of communication networks such as the Internet or peer-to-peer networks. The query model that we study is motivated by the following considerations. First, notice that our query model can be interpreted in the following way: A query at v yields the shortest-path subgraph rooted at v, i.e., the set of all edges on shortest paths between v and any other vertex. To see that this is equivalent to our definition (where a query yields all edges and non-edges between vertices of different distance from v), note that an edge connects two vertices of different distance from v if and only if it lies on a shortest path between v and one of these two vertices. Furthermore, the shortest-path subgraph rooted at v implicitly confirms the absence of all edges between vertices of different distance from v.

Real-life scenarios where the shortest-path subgraph rooted at a node of the network can be determined arise as follows. With traceroute tools, one can determine the path that packets take in the Internet if they are sent from one's node to some destination. If each traceroute experiment returns a random shortest path to the destination, one could use repeated traceroute experiments to all destinations to discover all edges of the shortest-path subgraph. Making a query at v would mean getting access to node v and running repeated traceroute experiments from v to all other nodes. If we assume that the cost of getting access to a node is much higher than that of running the traceroute-experiments, minimizing the number of queries is a meaningful goal. Along similar lines, in a network that routes all packets along arbitrary shortest paths, one could imagine a routing protocol in which each node stores the shortest-path subgraph rooted at that node. In this case, reading out the routing table at a node would correspond to making a query at that node.

Our model of network discovery is a simplification of reality. In real networks, routing is not necessarily along shortest paths, but may be affected by routing policies, link qualities, or link capacities. Furthermore, routing tables or traceroute experiments will often reveal only a single path (or at most a few different paths) to each destination, but not the whole shortest-path subgraph. Nevertheless, we believe that our model is a good starting point for a theoretical investigation of fundamental issues arising in network discovery.

Related Work. Graph discovery problems have been studied in distributed settings where one or several agents move along the edges of the graph (see, e.g., [3]); the problems arising in such settings appear to require very different techniques from the ones in our setting.

It turns out, however, that the network verification problem has previously been considered as the problem of placing landmarks in graphs [9]. Here, the motivation is to place landmarks in as few vertices of the graph as possible in such a way that each vertex of the graph is uniquely identified by the vector of its distances to the landmarks. The smallest number of landmarks that are required for a given graph G is also called the *metric dimension* of G [8]. For a survey of known results, we refer to [5]. Results for the problem variant where extra constraints are imposed on the set of landmarks (e.g., connectedness or independence) are surveyed in [11].

The problem of determining whether k landmarks suffice (i.e., of determining if the metric dimension is at most k) is $\mathcal{NP}$-complete [6]; see [9] for an explicit proof by reduction from 3-SAT. In [9] it is also shown that the problem of minimizing the number of landmarks admits an $O(\log n)$-approximation algorithm for graphs with n vertices, based on SETCOVER. For trees, they show that the problem can be solved optimally in polynomial time. Furthermore, they prove that one landmark is sufficient if and only if G is a path, and discuss properties of graphs for which 2 landmarks suffice. They also show that if k landmarks suffice for a graph with n vertices and diameter D, we must have $n \leq D^k + k$. For d-dimensional grids they show that d landmarks suffice. For d-dimensional hypercubes, a special case of d-dimensional grids, it was shown in [12] (using an earlier result from [10] on a coin weighing problem) that the metric dimension is asymptotically equal to $2d/\log_2 d$. See also [4] for further results on the metric dimension of Cartesian products of graphs.

Our Results. For network discovery, we give a lower bound showing that no deterministic on-line algorithm can have competitive ratio better than 3, and we present a randomized on-line algorithm with competitive ratio $O(\sqrt{n \log n})$ for networks with n nodes. For the network verification problem, we prove that it cannot be approximated within a factor of $o(\log n)$ unless $\mathcal{P} = \mathcal{NP}$, thus showing that the approximation algorithm from [9] is best possible (up to constant factors). We also give a useful lower bound formula for the optimal number of queries of a given graph. The remainder of the paper is structured as follows. Section 2 gives preliminaries and defines the problems formally. Sections 3 and 4 give our results for network discovery and network verification, respectively. Section 5 points to open problems and promising directions for future research.

2   Preliminaries and Problem Definitions

Throughout this paper, the term *network* refers to a connected, undirected graph. For a given graph $G = (V, E)$, we denote the number of nodes by $n = |V|$ and the number of edges by $m = |E|$. For two distinct nodes $u, v \in V$, we say that $\{u, v\}$ is an *edge* if $\{u, v\} \in E$ and a *non-edge* if $\{u, v\} \notin E$. The set of non-edges of G is denoted by $\bar{E}$. We assume that the set V of nodes is known in advance and that it is the presence or absence of edges that needs to be discovered or verified.

A *query* is specified by a vertex $v \in V$ and called a *query at v*. The query at v is also denoted by v. The answer of a query at v consists of a set E_v of edges and a set $\bar{E}_v$ of non-edges. These sets are determined as follows. Label every vertex $u \in V$ with its distance (number of edges on a shortest path) from v. We refer to sets of vertices with the same distance from v as *layers*. Then E_v is the set of all edges connecting vertices in different layers, and $\bar{E}_v$ is the set of all non-edges whose endpoints are in different layers. Because the query result can be seen as a layered graph, we refer to this query model as the *layered-graph* query model.

A set $Q \subseteq V$ of queries discovers (all edges and non-edges of) a graph $G = (V, E)$ if $\bigcup_{q \in Q} E_q = E$ and $\bigcup_{q \in Q} \bar{E}_q = \bar{E}$. In the off-line case, we also say "verifies" instead of "discovers". The network verification problem is to compute, for a given network G, a smallest set of queries that verifies G. The network discovery problem is the on-line version of the network verification problem. Its goal is to compute a smallest set of queries that discovers G. Here, the edges and non-edges of G are initially unknown to the algorithm, the queries are made sequentially, and the next query must always be determined based only on the answers of previous queries.

We denote by $OPT(G)$, for a given graph G, the cardinality of an optimal query set for verifying G, and by $A(G)$ the cardinality of the query set produced by an algorithm A. The quality of an algorithm is measured by the worst possible ratio $A(G)/OPT(G)$ over all networks G. In the off-line case, an algorithm is a ρ-approximation algorithm (and achieves approximation ratio ρ) if it runs in polynomial time and satisfies $A(G)/OPT(G) \leq \rho$ for all networks G. In the on-line case, an algorithm is ρ-competitive (and achieves competitive ratio ρ) if $A(G)/OPT(G) \leq \rho$ for all networks G. It is weakly ρ-competitive if $A(G) \leq \rho \cdot OPT(G) + c$ for some constant c. If the on-line algorithm is randomized, $A(G)$ is replaced by $\mathbb{E}[A(G)]$ in these definitions. We do not require on-line algorithms to run in polynomial time.

We use LG–ALL–DISCOVERY to refer to the network discovery problem with the layered-graph query model and the goal of discovering all edges and non-edges, and we use LG–ALL–VERIFICATION to refer to its off-line version.

3   Network Discovery

We consider the on-line scenario. Clearly, any algorithm that does not repeat queries has competitive ratio at most $n-1$, since $n-1$ queries are always sufficient to discover a network. Furthermore, the inapproximability result that we will derive in Section 4 (Theorem 3) shows that we cannot hope for a polynomial-time on-line algorithm with competitive ratio $o(\log n)$; it may still be possible to obtain such a ratio using exponential-time on-line algorithms, however. We present a lower bound on the competitive ratio of all deterministic on-line algorithms.

Theorem 1. *No deterministic on-line algorithm for* LG-ALL-DISCOVERY *can have weak competitive ratio* $3 - \varepsilon$ *for any* $\varepsilon > 0$.

Proof. Let A be any deterministic algorithm for LG–ALL–DISCOVERY. We first give a simpler proof that A cannot be better than 2-competitive. Consider Fig. 1(a). We refer to the subgraph induced by the vertices labeled r, x, y, and z as a 2-*gadget*. Assume that the given graph G consists of a global root g and k, $k \geq 2$, disjoint copies of the 2-gadget, with the r-vertex of each 2-gadget connected to the global root g. One can easily verify that $OPT(G) = k$ for this graph, and that the set of all x-vertices of the 2-gadgets constitutes an optimal query set. On the other hand, algorithm A can be forced to make the first query at g (as, initially, the vertices are indistinguishable to the algorithm). This will not discover any information about edges or non-edges between vertices x, y and z of each 2-gadget. The only queries that can discover this information are queries at x, y and z. In fact, a query at x or y suffices to discover the edge between x and y and the non-edges between x and z and between y and z. When A makes the first query among the vertices in $\{x, y, z\}$ of a 2-gadget, we can force it to make that query at z, since the three vertices are indistinguishable to the algorithm. The query at z does not discover the edge between x and y. The algorithm must make a second query in the 2-gadget to discover that edge. In total, the algorithm must make at least $2k+1$ queries. As the construction works for arbitrary values of k, this shows that no deterministic on-line algorithm can guarantee weak competitive ratio $2 - \varepsilon$ for any constant $\varepsilon > 0$.

To get a stronger lower bound of 3, we create a new gadget, called the 3-*gadget*, as shown in Fig. 1(b). The 3-gadget is the subgraph induced by all vertices except g in the figure. We claim that A can be forced to make 6 queries in each 3-gadget, whereas the optimum query set consists of only 2 vertices in each 3-gadget (drawn shaded in the figure). If we construct a graph with k, $k \geq 2$, disjoint copies of the 3-gadget, the s-vertex in each of them connected to the global root g as indicated in the figure, we get a graph G for which we claim that $OPT(G) = 2k$ and the algorithm A can be forced to make at least $6k + 1$ queries. This shows that no deterministic on-line algorithm can guarantee weak competitive ratio $3 - \varepsilon$ for any constant $\varepsilon > 0$.

To see that $OPT(G) = 2k$, let Q be the set of queries consisting of the two shaded vertices from each copy of the 3-gadget as shown in Fig. 1(b). We claim that Q discovers G. This can be verified manually as follows: For every vertex in a 3-gadget Π, consider the 3-tuple whose components are the distances from that vertex to the two query vertices in Π and the distance to an arbitrary query vertex from Q outside Π. One finds that each vertex in Π has a unique 3-tuple, showing that all edges and non-edges of Π are discovered by Q. Each non-edge between two different 3-gadgets is discovered by one of the queries inside these

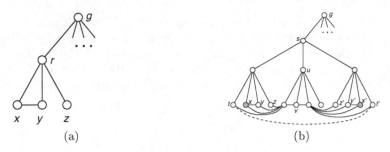

Fig. 1. Lower bound constructions

two 3-gadgets. The edges and non-edges between g and each 3-gadget are also discovered. Hence, $OPT(G) \leq 2k$. We have $OPT(G) \geq 2k$, because each of the edges $\{x, y\}$ and $\{x', y'\}$ (see Fig. 1(b)) of a 3-gadget requires a separate query.

To show that $A(G) \geq 6k + 1$, we argue as follows. First, we can force A to make the first query at g. This will not reveal any information about edges within the same layer of any of the 3-gadgets. We view each 3-gadget as consisting of s and a left part, a middle part, and a right part. The left part consists of the left child of s and its four adjacent vertices below (these four vertices are called *bottom vertices*, and the left child of s is called the *root* of that part); the middle and right part are defined analogously. The three parts of a 3-gadget Π are indistinguishable to A until it makes its first query inside Π. A query at s would not discover any new information about Π, so we can ignore queries that A might make at s in the following arguments. When A makes its first query inside Π, we can force this query to be in the middle part, and we can force it to be at u or v. In both cases, the query does not discover any information about the edges and non-edges between the bottom vertices of the left part, nor does it discover any information about the edges and non-edges between the bottom vertices of the right part, nor does it discover the edge drawn dashed. When A chooses its second query in Π, it could be in the left part, in the middle part, or in the right part. Assume that A chooses the left part; since the bottom vertices of the left part are still indistinguishable to A, we can force A to make the query either at the root of the left part or at the bottom vertex t. Similarly, in the right part we can force A to make the query at its root or at t'. In the middle part, A can make the query anywhere. In any case, the second query made by A does not discover any information about edges and non-edges between vertices in the set $\{x, y, z\}$ and in the set $\{x', y', z'\}$. Similarly as in the case of Fig. 1(a), for each of these sets we can force A to make the first query at z (at z') and thus require a second query at x or y (at x' or y') to discover everything about these groups. In total, A must make at least 6 queries in each 3-gadget. □

With the gadget of Fig. 1(a) one can prove easily that no randomized on-line algorithm for LG-ALL-DISCOVERY can have weak competitive ratio $4/3 - \varepsilon$ for any $\varepsilon > 0$; just observe that we can force a randomized algorithm to make the first query at z with probability at least $1/3$. Note that all lower bounds on the

```
E ← ∅;    /* discovered edges */
N ← ∅;    /* discovered non-edges */
A ← (V₂);    /* all pairs of distinct nodes */
/* Phase 1 */
for i = 1 to 3√(n ln n) do
        v ← randomly chosen node from V;
        (Eᵥ, Nᵥ) ← query(v);
        E ← E ∪ Eᵥ;
        N ← N ∪ Nᵥ;
od;
/* Phase 2 */
while E ∪ N ≠ A do
        {u, v} ← an arbitrary element of A \ (E ∪ N);
        (Eᵤ, Nᵤ) ← query(u);
        (Eᵥ, Nᵥ) ← query(v);
        E ← E ∪ Eᵤ ∪ Eᵥ;
        N ← N ∪ Nᵤ ∪ Nᵥ;
        S ← set of nodes from which the (non-)edge {u, v} is discovered;
        foreach x ∈ S \ {u, v} do
            (Eₓ, Nₓ) ← query(x);
            E ← E ∪ Eₓ;
            N ← N ∪ Nₓ;
        od;
od;
```

Fig. 2. On-line algorithm for LG-ALL-Discovery

weak competitive ratio also hold for the (standard) competitive ratio where no additive constant c is allowed.

Theorem 2. *There is a randomized on-line algorithm that achieves competitive ratio $O(\sqrt{n \log n})$ for* LG-ALL-Discovery.

Proof. The on-line algorithm is shown in Fig. 2. In the first phase, it makes $3\sqrt{n \ln n}$ queries at nodes chosen uniformly at random. In the second phase, as long as node pairs with unknown status exist, it picks an arbitrary such pair $\{u, v\}$ and proceeds as follows. First, it queries u and v in order to determine the distance of all nodes to u and v. From this it can deduce the set S of nodes from which the edge or non-edge between u and v can be discovered; these are simply the nodes for which the distance to u differs from the distance to v. Then, it queries all remaining nodes in S.

To analyze the algorithm, it is helpful to view LG-ALL-Discovery as a HittingSet problem. For every edge or non-edge $\{u, v\}$, let S_{uv} be the set of nodes from which a query discovers $\{u, v\}$. The task of the LG-ALL-Discovery problem translates into the task of computing a subset of V that hits all sets S_{uv}. The goal of the first phase is to hit all sets that have size at least $\sqrt{n \ln n}$ with high probability. If this succeeds, the problem remaining for the second phase is a HittingSet problem where all sets have size at most $\sqrt{n \ln n}$. The algorithm of the second phase repeatedly picks an arbitrary set that is not yet hit, and includes all its elements in the solution. As the sets have size at most $\sqrt{n \ln n}$, the number of queries made in the second phase is at most a factor of $\sqrt{n \ln n}$ away from the optimum.

Let us make this analysis precise. Consider a node pair $\{u, v\}$ for which the set S_{uv} has size at least $\sqrt{n \ln n}$. In each query of the first phase, the probability that S_{uv} is not hit is at most $1 - \frac{\sqrt{n \ln n}}{n} = 1 - \frac{\sqrt{\ln n}}{\sqrt{n}}$. Thus, the probability that S_{uv} is not hit throughout the first phase is at most $\left(1 - \frac{\sqrt{\ln n}}{\sqrt{n}}\right)^{3\sqrt{n \ln n}} \leq e^{-3 \ln n} = \frac{1}{n^3}$. There are at most $\binom{n}{2}$ sets S_{uv} of cardinality at least $\sqrt{n \ln n}$. The probability that at least one of them is not hit in the first phase is at most $\binom{n}{2} \cdot \frac{1}{n^3} \leq \frac{1}{n}$.

Now consider the second phase, conditioned on the event that the first phase has hit all sets S_{uv} of size at least $\sqrt{n \ln n}$. In each iteration of the while-loop of the second phase, the algorithm asks at most $\sqrt{n \ln n}$ queries. Let ℓ be the number of iterations. It is clear that the optimum must make at least ℓ queries, because no two unknown pairs $\{u, v\}$ considered in different iterations of the second phase can be resolved by the same query.

Since $OPT(G) \geq 1$ and $OPT(G) \geq \ell$, the number of queries made by the algorithm is at most $3\sqrt{n \ln n} + \ell \sqrt{n \ln n} = O(\sqrt{n \log n}) \cdot OPT(G)$.

With probability at least $1 - \frac{1}{n}$, the first phase succeeds and the algorithm makes $O(\sqrt{n \log n}) \cdot OPT(G)$ queries. If the first phase fails, the algorithm makes at most n queries. This case increases the expected number of queries made by the algorithm by at most $\frac{1}{n} \cdot n = 1$. Thus, the expected number of queries is at most $O(\sqrt{n \log n}) \cdot OPT(G) + \frac{1}{n} \cdot n = O(\sqrt{n \log n}) \cdot OPT(G)$.   □

4   Network Verification

Theorem 3. *It is $\mathcal{NP}$-hard to approximate* LG–ALL–VERIFICATION *within ratio $o(\log n)$.*

Proof. We prove the inapproximability result using an approximation-preserving reduction from the *test collection problem (TCP)*:

Problem TCP
Input: ground set S and collection $\mathcal{C}$ of subsets of S
Feasible solution: subset $\mathcal{C}' \subseteq \mathcal{C}$ such that for each two distinct elements x
    and y of S, there exists a set $C \in \mathcal{C}'$ such that exactly one of x and y is in C.
Objective: minimize the cardinality of $\mathcal{C}'$

In the original application for TCP, S is a set of diseases and $\mathcal{C}$ is a collection of tests. A test $C \in \mathcal{C}$, applied to a patient, will give a positive result if the patient is infected by a disease in C. If a patient is known to be infected by exactly one of the diseases in S, the goal of TCP is to compute a minimum number of tests that together can uniquely identify that disease.

Without loss of generality, we can restrict ourselves to instances of TCP in which any two elements of the ground set can be separated by at least one of the sets in $\mathcal{C}$; instances without this property do not have any feasible solutions.

Halldórsson et al. [7] prove that TCP cannot be approximated with ratio $o(\log|S|)$ unless $\mathcal{P} = \mathcal{NP}$. Their proof uses an approximation-preserving reduction from SETCOVER; the latter problem was shown $\mathcal{NP}$-hard to approximate within $o(\log n)$, where n is the cardinality of the ground set, by Arora and Sudan [1]. The proof by Arora and Sudan establishes the inapproximability result for SETCOVER even for instances in which the size of the ground set and the number of sets are polynomially related. The reduction from SETCOVER to TCP maintains this property. Hence, we know that it is $\mathcal{NP}$-hard to approximate TCP with ratio $o(\log|S|)$ even for instances satisfying $|\mathcal{C}| \leq |S|^g$ for some positive constant g.

Let an instance $(S, \mathcal{C})$ of TCP be given. Let $n_{\text{TCP}} = |S|$ and $m_{\text{TCP}} = |\mathcal{C}|$. By the remark above, we can assume that $m_{\text{TCP}} = n_{\text{TCP}}^{O(1)}$. We construct an instance $G = (V, E)$ of LG–ALL–VERIFICATION as follows. First, we add $n_{\text{TCP}} + m_{\text{TCP}}$ vertices to V: an *element vertex* v_s for every element $s \in S$ and a *test vertex* u_C for every $C \in \mathcal{C}$. We initially add the following edges to E: Any two element vertices are joined by an edge, and every test vertex u_C is joined to all element vertices v_s with $s \in C$. The idea behind this construction is that queries at test vertices verify all edges in the clique of element vertices if and only if the corresponding tests form a test cover. We have to extend the construction slightly since, in LG–ALL–VERIFICATION, the edges and non-edges incident to the test vertices need to be verified as well. We add $h = 2(\lceil \log m_{\text{TCP}} \rceil + 2)$ auxiliary vertices $w_1, \ldots, w_h$ to take care of this. For each i, $1 \leq i \leq h/2$, the auxiliary vertices w_{2i-1} and w_{2i} are said to form a *pair*. In addition, we add one extra node z. We add the following edges:

- The two auxiliary vertices in each pair are joined by an edge.
- Number the m_{TCP} test vertices arbitrarily from 0 to $m_{\text{TCP}} - 1$. Both auxiliary vertices in the i-th pair, $1 \leq i \leq h/2 - 2$, are joined to those of the m_{TCP} test vertices whose number has a 1 in the i-th position of its binary representation.
- Both auxiliary vertices in the last two pairs are joined to all test vertices.
- The extra node z is joined to all other vertices of the graph.

The graph constructed in this way is denoted by $G = (V, E)$. See Fig. 3 for an illustration. We prove two claims:

Claim 1. Given a solution $\mathcal{C}'$ to the TCP instance $(S, \mathcal{C})$, there is a solution Q of the constructed instance $G = (V, E)$ of LG–ALL–VERIFICATION satisfying $|Q| = |\mathcal{C}'| + \lceil \log m_{\text{TCP}} \rceil + 2$.

Proof (of Claim 1). Let a solution $\mathcal{C}'$ to the TCP instance $(S, \mathcal{C})$ be given. Let Q contain all test vertices corresponding to sets $C \in \mathcal{C}'$ as well as the first vertex of every pair of auxiliary vertices. Obviously, we have $|Q| = |\mathcal{C}'| + \lceil \log m_{\text{TCP}} \rceil + 2$. It is not difficult to verify that Q discovers all edges and non-edges of G.   □

Claim 2. Given a solution Q to the constructed instance $G = (V, E)$ of LG–ALL–VERIFICATION, one can construct in polynomial time a solution $\mathcal{C}'$ of the original TCP instance $(S, \mathcal{C})$ satisfying $|\mathcal{C}'| \leq |Q| - \lceil \log m_{\text{TCP}} \rceil - 2$.

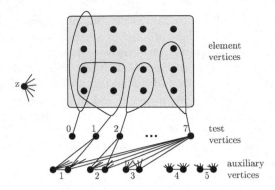

Fig. 3. Illustration of the construction of the graph $G = (V, E)$ that is an instance of LG–ALL–VERIFICATION. The auxiliary vertices in pairs 4 and 5 are adjacent to all test vertices. The auxiliary vertices in pair i, $1 \leq i \leq 3$, are adjacent to the test vertices whose number has a 1 in position i of the binary representation. For example, the auxiliary vertices in pair 2 are adjacent to test vertices $2, 3, 6$ and 7.

Proof (of Claim 2). Observe that Q must contain at least one vertex from each pair of auxiliary vertices; otherwise, the edge joining this pair would not be discovered. The queries at these vertices do not discover any edges between element vertices (all element vertices are at distance 2 from any auxiliary vertex because of the extra vertex z). Let Q' be the vertices in Q that are not auxiliary vertices. We have $|Q'| \leq |Q| - \lceil \log m_{\text{TCP}} \rceil - 2$. Now, Q' is a set of element vertices and test vertices that, in particular, discovers all edges between element vertices.

Let Q_S be the set of element vertices in Q' and let Q_C be the set of test vertices in Q'. If Q_S is empty, the queries at the vertices in Q_C discover all edges of the clique of element vertices. In particular, this means that for any two distinct element vertices v_s and v_t in V, there must be a query at a vertex adjacent to one of v_s, v_t but not to the other. This shows that the set $C' = \{C \in C \mid u_C \in Q'\}$ is a solution of the original TCP instance of the required size.

Now assume Q_S is nonempty. The set of edges between element vertices that are not discovered by Q_C is a disjoint union of cliques. The queries in Q_S must discover all edges in these cliques. As the only edges between element vertices that a query at an element vertex discovers are the edges incident to that vertex, a clique of size k requires $k-1$ queries. Assume that there are p cliques and denote the number of vertices in these cliques by $k_1, \ldots, k_p$. Then Q_S contains at least $\sum_{i=1}^{p}(k_i - 1)$ vertices. All edges in a clique of size k can always be discovered by $k - 1$ queries at test vertices: simply select these queries greedily by choosing, as long as there is an edge $\{u, v\}$ in the clique that has not yet been discovered, any test vertex that is adjacent to one of u, v but not the other. Hence, we can replace the queries in Q_S by at most $\sum_{i=1}^{p}(k_i - 1)$ queries at test vertices and add these to Q_C, obtaining a set of queries at test vertices that discovers all edges between element vertices. As in the previous paragraph, this set of test vertices gives a solution to the original TCP instance of cardinality at most $|Q'|$.   □

Assume there is an approximation algorithm A for LG–ALL–VERIFICATION that achieves ratio $o(\log n)$, where $n = |V|$. Consider the algorithm B for TCP that, given an instance of TCP, constructs an instance of LG–ALL–VERIFICATION as described above, applies A to this instance, and transforms the result into a solution to the TCP instance following Claim 2. Recall that $m_{\text{TCP}} = n_{\text{TCP}}^{O(1)}$. We claim that B achieves ratio $o(\log n_{\text{TCP}})$ for TCP. Let OPT_{TCP} be the optimum objective value for the given TCP instance and OPT_{LG} be the optimum objective value for the constructed instance of LG–ALL–VERIFICATION. Let B_{TCP} and A_{LG} denote the objective values of the solutions computed by B and A, respectively. Note that $OPT_{\text{TCP}} \geq \log n_{\text{TCP}}$ always holds, since n_{TCP} elements cannot be separated by fewer than $\log n_{\text{TCP}}$ test sets.

Claims 1 and 2 imply that $OPT_{\text{TCP}} = OPT_{\text{LG}} - \lceil \log m_{\text{TCP}} \rceil - 2$. We have $OPT_{\text{LG}} = OPT_{\text{TCP}} + \lceil \log m_{\text{TCP}} \rceil + 2 \leq OPT_{\text{TCP}} + O(\log n_{\text{TCP}}) = O(OPT_{\text{TCP}})$. Claim 2 implies $B_{\text{TCP}} \leq A_{\text{LG}}$ and thus we get $B_{\text{TCP}} \leq o(\log n) \cdot OPT_{\text{LG}} = o(\log n) \cdot O(OPT_{\text{TCP}}) = o(\log n_{\text{TCP}}) \cdot O(OPT_{\text{TCP}})$, where the last equality follows from $n = n_{\text{TCP}} + m_{\text{TCP}} + 2(\lceil \log m_{\text{TCP}} \rceil + 2) + 1 = n_{\text{TCP}}^{O(1)}$. This shows $B_{\text{TCP}} \leq o(\log n_{\text{TCP}}) \cdot OPT_{\text{TCP}}$ and completes the proof of Theorem 3.   □

Theorem 4. *If a graph $G = (V, E)$ contains a subgraph H of diameter D_H with n_H vertices, then $OPT(G) \geq \log_{D_H+1} n_H$.*

Proof. Imagine the queries being performed sequentially. At any instant, the unknown edges and non-edges induce disjoint cliques, which we call *unknown groups*. Two vertices are in the same unknown group if and only if they were in the same layer of all queries made so far. Consider the n_H vertices of subgraph H. Initially, all vertices form an unknown group. For each query, the n_H vertices of H will be in at most $D_H + 1$ consecutive layers of the layered graph returned by the query. Therefore, after the first query, at least $n_H/(D_H + 1)$ vertices of H will still be in the same unknown group. Similarly, after k queries, at least $n_H/(D_H + 1)^k$ vertices of H will be in an unknown group together. If k queries suffice to verify all edges and non-edges, the unknown groups must be singletons in the end. So we must have $n_H/(D_H + 1)^k \leq 1$. This proves the theorem.   □

This theorem implies that a graph containing a clique on k vertices requires at least $\log_2 k$ queries, and a graph with maximum degree Δ at least $\log_3(\Delta + 1)$ queries. For the former, take H to be the clique on k vertices, and for the latter, take H to be the subgraph induced by a vertex of degree Δ and its neighbors.

5   Directions for Future Work

In this paper, we have considered network discovery and network verification problems in the layered-graph query model. The goal was to discover or verify all edges and non-edges of a network. For network discovery, the major problem left open by our work is to close the gap between our randomized upper bound of $O(\sqrt{n \log n})$ and the small constant lower bounds.

The subject of our study is an example of a family of problem settings in which the goal is to discover or verify information about a graph using queries. Different problems are obtained if the query model is varied, or if the objective is changed. Other natural query models are, e.g., that a query at v returns only the distances from v to all other vertices of the graph; that a query is specified by two vertices u and v, and returns the set of all edges on shortest paths between u and v; or that a query returns an arbitrary shortest-path tree rooted at v. Concerning the objective, the goal could be to discover or verify a certain graph parameter such as diameter, average path length, or independence number. One could also relax the requirement and only ask for an approximate answer, e.g., one could consider the problem of minimizing the number of queries required to approximate the average path length within a factor of $1 + \varepsilon$. We believe that the study of such problems could be a fruitful area of research with applications in the monitoring and analysis of communication networks such as the Internet.

References

1. S. Arora and M. Sudan. Improved low-degree testing and its applications. In *Proc. 29th Ann. ACM Symp. on Theory of Computing (STOC'97)*, pages 485–495, 1997.
2. P. Barford, A. Bestavros, J. Byers, and M. Crovella. On the marginal utility of deploying measurement infrastructure. In *Proc. ACM SIGCOMM Internet Measurement Workshop 2001*, November 2001.
3. M. A. Bender and D. K. Slonim. The power of team exploration: Two robots can learn unlabeled directed graphs. In *Proc. 35th Ann. IEEE Symp. on Foundations of Computer Science (FOCS'94)*, pages 75–85, 1994.
4. J. Cáceres, C. Hernando, M. Mora, I. M. Pelayo, M. L. Puertas, C. Seara, and D. R. Wood. On the metric dimension of Cartesian products of graphs. Manuscript, 2005.
5. G. Chartrand and P. Zhang. The theory and applications of resolvability in graphs: A survey. *Congr. Numer.*, 160:47–68, 2003.
6. M. R. Garey and D. S. Johnson. *Computers and Intractability. A Guide to the Theory of NP-Completeness*. W. H. Freeman and Company, New York, 1979.
7. B. V. Halldórsson, M. M. Halldórsson, and R. Ravi. On the approximability of the minimum test collection problem. In *Proc. 9th Ann. European Symposium on Algorithms (ESA'01)*, LNCS 2161, pages 158–169. Springer-Verlag, 2001.
8. F. Harary and R. Melter. The metric dimension of a graph. *Ars Combin.*, 2:191–195, 1976.
9. S. Khuller, B. Raghavachari, and A. Rosenfeld. Landmarks in graphs. *Discrete Appl. Math.*, 70:217–229, 1996.
10. B. Lindström. On a combinatory detection problem I. *Magyar Tud. Akad. Mat. Kutató Int. Közl.*, 9:195–207, 1964.
11. V. Saenpholphat and P. Zhang. Conditional resolvability in graphs: A survey. *Int. J. Math. Math. Sci.*, 38:1997–2017, 2004.
12. A. Sebő and E. Tannier. On metric generators of graphs. *Math. Oper. Res.*, 29(2):383–393, 2004.
13. L. Subramanian, S. Agarwal, J. Rexford, and R. Katz. Characterizing the Internet hierarchy from multiple vantage points. In *INFOCOM'02*, June 2002.

Complete Graph Drawings
Up to Triangle Mutations

Emeric Gioan

LIRMM, CNRS Montpellier

Abstract. The logical structure we introduce here to describe a (topological) graph drawing, called subsketch, is intermediate between the map (determining the drawing when it is planar), and the sketch introduced by Courcelle (determining the drawing in general but assuming we know the order of the crossings on each edge). For a complete graph drawing, the subsketch is determined, through first order logic formulas, by the size, a corner of the drawing and the crossings of the edges.

We prove, constructively, that two complete graph drawings have the same subsketch if and only if they can be transformed into each other by a sequence of triangle mutations - or triangle switches. This construction generalizes Ringel's theorem on uniform pseudoline arrangements. Moreover, it applies to plane projections of spatial graphs encoded by rank 4 uniform oriented matroids.

Keywords: Graph drawing, logical structure, triangle switch, mutation, pseudoline arrangement, oriented matroid, spatial graph visualization.

1 Introduction

Three subjects meet in this paper: first the dynamical structure of geometrical objects with triangle mutations (or triangle switches), secondly axiomatics of graph drawings using logical structures as concise as possible, and thirdly the combinatorial study of visualization of spatial graphs encoded by oriented matroids.

In the whole paper, graph drawing is understood in the sense of topological graph drawing, that is drawing of which edges are represented by Jordan arcs (not supposed to be straight), whereas a graph drawing is called geometrical when its edges are represented by (straight) line segments. We consider graph drawings of a graph on a plane where two edges cross at most once and where the unbounded region is defined by the choice of two given adjacent edges called a *corner* (equivalently, we could consider drawings on a sphere, but we would have then to choose a particular point "infinity" so that the region containing it would be considered as the "unbounded" one).

From an axiomatic point of view, a general setting is introduced by Courcelle in [2], allowing both logical and geometrical points of view on graph drawings, and leading to applications of monadic second order logic to graph drawings. In this setting, a graph drawing is determined by its sketch, that is: its underlying graph, the circular ordering of the edges at each vertex, the pairs of edges that

D. Kratsch (Ed.): WG 2005, LNCS 3787, pp. 139–150, 2005.
© Springer-Verlag Berlin Heidelberg 2005

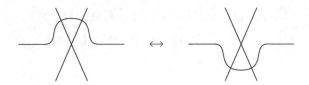

Fig. 1. Triangle mutation (or triangle switch)

cross, and the order of crossings on each edge. If the last data is removed, we get the subsketch of the graph drawing. Hence the subsketch is intermediate between the sketch and the so-called map of the drawing (which determines the drawing if it is planar, see for instance [5]). We prove in Section 3 that, for a complete graph drawing, the subsketch and other useful information, are determined through first order logic formulas by its number of vertices, a corner, and the pairs of edges that cross.

A triangle mutation - or triangle switch - in a graph drawing is passing an edge over the crossing of two other edges, when no obstruction occurs. This local transformation is shown on Figure 1. Obviously a triangle mutation does not change the subsketch. We consider the problem of finding a logical structure for graphs drawings defined up to a sequence of triangle mutations.

We prove constructively in Section 4 that, for a complete graph drawing, the subsketch structure plays this part: it determines the drawing, up to a sequence of triangle mutations and orientation preserving homeomorphisms.

Note that, if one considers a complete graph drawing with an even number of vertices, all of them being drawn on the same circle, then the pairs of opposite vertices define a pseudoline arrangement in a neighbourhood of the centre of the circle, see Figure 2. In fact, the above result generalizes Ringel's theorem on uniform pseudoline arrangements [7] (see Section 5.1).

A consequence of the above result - the original purpose of this paper - is that two projections of complete spatial graphs, defined by finite sets of points in general position representing the same rank 4 uniform oriented matroid [1], are equivalent up to homeomorphism and a sequence of mutations. Hence the com-

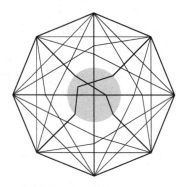

Fig. 2. Complete graph drawing and pseudoline arrangement

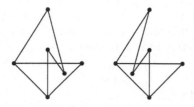

Fig. 3. Two 2-connected graph drawings with same subsketch but no triangle

binatorial structure of the oriented matroid together with the logical structure of the projected drawing form the two levels of a modelization of perspectives in spatial graph visualization (see Section 5.2).

In a graph which is not complete, the subsketch is no more sufficient to determine the drawing up to triangle mutations. In general, additional data would be necessary. In this paper, it is an open question. As an example Figure 3 below represents two graph drawings with same crossings and same circular orderings around each edge, but which cannot be tranformed into each other with triangle mutations, since they simply have no triangle.

NB: All proofs of this paper have been removed or shortened in order to fit the requested size for papers of this WG05 proceedings. A full version is forthcoming.

2   Preliminaries

In this paper, a *graph* is always a finite, directed, loop-free, connected graph. The set of vertices of a graph G is denoted V_G, or simply V, and its set of edges is denoted $\overrightarrow{E}_G$, or simply $\overrightarrow{E}$. The underlying undirected set of edges is denoted E_G, or simply E. In fact, the direction of an edge will be used only to define an order of the points on a geometrical representation of this edge. So, for $a, b \in V_G$ and $(a, b) \in \overrightarrow{E}_G$, we will denote $[a, b] = [b, a] \in E_G$.

A *(topological) drawing* of a graph G in the real oriented affine plane is a set of points representing V_G together with a set of drawn edges representing E_G satisfying the following properties:

D1 - a drawn edge is a Jordan arc (i. e. homeomorphic to a closed segment) between the two extremities representing the vertices ; a drawn edge contains no other representation of a vertex of the graph than its extremities.

D2 - two edges having extremities in common (two in the case of multiple edges) meet only at these extremities ; when two edges with no common extremity meet, they cross at this intersection point ; two edges with no common extremity cross at most once.

D3 - no three edges meet at the same point, except if this point is an extremity of the three edges.

Note that if Jordan arcs were replaced by line segments in axiom D1, we would define *geometrical* graph drawings, for which various properties would become trivial (for instance the two Lemmas in Section 3).

With a drawing D of the graph G various pieces of information are associated, encoding the drawing at different levels of abstraction. We call *drawn* element the topological representation of this element in the given drawing.

First the relation $inc_G \subseteq \overrightarrow{E}_G \times V_G \times V_G$ is defined by $(e, x, y) \in inc_G$ if and only if the edge e is directed from the vertex x to the vertex y. Then inc_G describes the structure of the graph G.

Secondly the relation $sig_D \subseteq V_G \times E_G \times E_G$ is defined by $(x, e, f) \in sig_D$ if and only if x is an extremity of e and f, and f is the next edge in the circular ordering around x in the trigonometric sense of rotation, which is well defined by definition of a drawing (property D2).

A *corner* of D is an element $(P, \beta, \alpha) \in sig_D$ such that the drawn vertex P is in the topological boundary of the infinite region of the plane delimited by D, and the intersections of the drawn edges β and α with this boundary are homeomorphic to line segments (containing P). Note that if the graph is complete then β and α are entirely contained in this boundary.

The set of relations inc_G, sig_D define the *map* associated with the drawing D of the graph G. It is well known (see for example [5]) that if D is a drawing with no edge crossing (except for common extremities), and thus G planar, then D is determined up to an orientation preserving homeomorphism of the plane by its map and a corner.

Thirdly, in [2], the relation $dcross_D \subseteq \overrightarrow{E}_G \times \overrightarrow{E}_G$ is defined by $(e, f) \in dcross_D$ if and only if the drawn edges e and f have no extremity in common, the drawn edges e and f have one intersection point and f goes from the left of e to its right when e is directed from bottom to top. Of course $(e, f) \in dcross_D$ implies $(f, e) \notin dcross_D$. In this paper we do not need directed edges for the crossing relation, it is sufficient to consider the relation $cross_D \subseteq E_G \times E_G$, defined by $(e, f) \in cross_D$ if and only if the drawn edges e and f have no extremity in common and the drawn edges e and f have one intersection point. Of course $(e, f) \in cross_D$ implies $(f, e) \in cross_D$. Then we say that $e \in E_G$ and $f \in E_G$ *cross* in D.

The set of relations $inc_G, sig_D, cross_D$ define the *subsketch* of the drawing D.

Fourthly, in [2], the relation $before_D \subseteq \overrightarrow{E}_G \times E_G \times E_G$ is defined by $(e, f, g) \in before_D$ if and only if $f \neq g$, e and f cross in D, e and g cross in D, and the intersection point of e and f is before the intersection point of e and g on the directed drawn edge e. Note that if e crosses f and g then either $before_D(e, f, g)$ or $before_D(e, g, f)$ but not both. The set of relations $inc_G, sig_D, dcross_D, before_D$ define the *sketch* associated with the drawing D, as introduced in [2]. By definition of a drawing, the relation $before_D$ induces, for any edge e, a linear ordering on the elements that cross e. A result of [2] is that the drawing D is determined up to an orientation preserving homeomorphism of the plane by its sketch and its corner.

In view of this result, we will assume from now on that drawings are always given with a certain corner, and are considered up to orientation preserving homeomorphisms (that is an homeomorphism of the plane which preserves the orientation of one - or equivalently any - triangle of the plane). Then *we can*

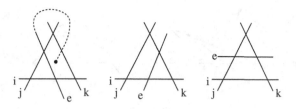

Fig. 4. Triangle $[i, j, k]$ cut twice by e

identify drawings and sketches, and the following definitions about drawings or sketches can be made equivalently for one of these two objects, depending on the point of view: geometrical, or logical. When the context is not ambiguous, we may omit the suffix $_D$ referring to the drawing.

Let D be a drawing of a graph G. We call *triangle* of D an element $(e, f, g) \in E_G \times E_G \times E_G$ such that e and f cross in D, e and g cross in D, and f and g cross in D. The order of the elements in the triplet have no importance, and we denote the triangle $[e, f, g]$.

The *segments* of a triangle $[e, f, g]$ are the subsets of the drawn elements e, f, or g which are delimited by the intersection with the two other elements of the triangle. The *interior* of a triangle $[e, f, g]$ is the bounded region of the plane delimited by its segments and containing these segments. A triangle is *contained* in another triangle if the two triangles are not equal, they have two common elements, and the interior of the first one is contained in the interior of the second one. We say that $h \in E_G$ *cuts the triangle* $[e, f, g]$, resp. *cuts the triangle* $[e, f, g]$ *twice*, if, geometrically, the drawn element h has a non empty intersection with at least one, resp. two, segment(s) of $[e, f, g]$. The following easy Lemma 1 is illustrated by Figure 4.

Lemma 1. *If $[i, j, k]$ is a triangle cut twice by e, then one and only one triplet in $\{\ \{i, j, e\}, \{i, k, e\}, \{j, k, e\}\ \}$ defines a triangle contained in $[i, j, k]$.* $\square$

Let D and D' be two drawings of the graph G with same subsketch. As D and D' have the same *cross* relation, they have same triangles. We say that a triangle $[e, f, g]$ is *permuted between D and D'* if the ordering of crossings between its edges along each of its three edges is different in the two drawings, that is if $before_D(e, f, g) = \neg before_{D'}(e, f, g)$, $before_D(f, e, g) = \neg before_{D'}(f, e, g)$, and $before_D(g, e, f) = \neg before_{D'}(g, e, f)$,

We call *free* a triangle of which interior has an intersection with the drawing reduced to the segments of the triangle. In particular it is not cut by any element, but not that the converse is false as show the triangle $[e, k, i]$ in the left Figure 4 when j is removed.

Given a drawing D of a graph G and a free triangle $[e, f, g]$ of D, the *mutation of $[e, f, g]$ from D* is the sketch D' of G for which all relations are the same as in D, except that e and f, and resp. e and g, and resp. f and g, are permuted on the drawn edge g, and resp. f, and resp. e. In other words all relations are the

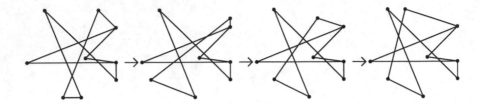

Fig. 5. A sequence of mutations

same in D' as in D except that the triangle $[e, f, g]$ is permuted between D and D'. We denote $D \to D'$, and call $[e, f, g]$ the *mutated* triangle from D to D'.

Hence, a triangle $[e, f, g]$, which is free in D, is permuted between D and its mutation from D. But, of course, a triangle may be permuted between two drawings D and D', without being free in D nor in D'.

A *sequence of mutations* from the sketch of a drawing D is a sequence of sketches, each one being the mutation of a free triangle from the previous one. On the example of Figure 5, the triangle containing a vertex cannot be mutated, but the three other triangles can be mutated triangles in a sequence of mutations.

3   Logical Structure of Complete Graph Drawings

In this section, we prove that, for a complete graph drawing with given number of vertices and given corner, the *cross* relation is sufficient to determine, through first order logic formulas, not only the the *sig* relation and thus the subsketch of the drawing, but also an *ins* relation which states if a vertex of a graph is inside the triangle formed by three other vertices. This is not true for general graph drawings (see Figure 3). We shall see that these relations determine also several other relations and finally determine the sketch of the drawing except the *before* relations for edges of triangles containing no vertex.

Let D be a graph drawing, with corner (P, β, α). The vertex P is called *vertex at the corner*, and the other extremities of α and β are denoted respectively A and B.

For three vertices $e, f, g \in V_G$, we denote $[e, f, g]$ the bounded region of the plane delimited by the drawn edges $[e, f]$, $[f, g]$ and $[g, e]$, containing these drawn edges. Thus this region does not contain the vertex at the corner P when $P \notin \{e, f, g\}$. Not that by definition, such a region is equivalent to a closed ball up to homeomorphism. The relation $ins_D \subseteq V_G \times V_G \times V_G \times V_G$ is defined by $(x, e, f, g) \in ins_D$ if and only if $x \notin \{e, f, g\}$ and the drawn vertex x is inside the region $[e, f, g]$.

For the construction of the next theorem, we introduce a relation $bet_D \subseteq V_G \times E_G \times E_G \times E_G$ called *between* relation for the drawing D, such that $(x, e, f, g) \in bet_D$ if the edges e, f, g all have extremity x, and f is between e and g in the circular order of the edges around x (note that the order is essential in the sentence: f is not between g and e).

The *size* of a complete graph drawing is the number of vertices of the underlying complete graph.

Theorem 1. *The subsketch and the inside relation of a complete graph drawing are determined, through first order logic formulas, by its size, its crossing relation and its corner.*

Proof. The construction is step by step and uses extensively the topological definition of the corner and properties (D1) (D2) (D3) of a drawing. The proof is not difficult and is about two pages long. However the ordering of the steps is important. Briefly: begin with the inside relations for triplets containing P, then for general triplets, then consider the between relations around P, and then the between relations around any vertex. □

Since the *sig* relations are determined, we easily get the following corollary by using the restrictions to 4 vertices subdrawings.

Corollary 1. *Let D be a complete graph drawing. Its dcross relation is determined with first order logic formulas by its size, crossing relation and corner.* □

The following results are trivial in the geometrical case. They generalize to topological graph drawings, quite technically but easily, using Theorem 1 and the axioms (D1), (D2), (D3), by considering the several possible representations.

Lemma 2. *Let D be a complete graph drawing with given size, crossing relation and corner. Let f and g be two edges such that either f and g have same extremity, or f and g do not cross. If f and g both cross an edge e, then the $before(e, f, g)$ relation is determined by first order logic formulas.* □

Corollary 2. *Let D and D' be two complete graph drawings with same size, crossing relation and corner. Then $D \neq D'$ if and only if there exists a permuted triangle between D and D'.* □

We say that a drawn triangle T *contains* a drawn vertex a, if the drawn vertex a is inside the bounded region of the plane delimited by drawn edges of T

Lemma 3. *Let D be a complete graph drawing, with given size, crossing relation and corner. Let $T = [e, f, g]$ be a triangle, and a a vertex of D. The property that the drawn triangle T contains the drawn vertex a is expressible by a first order logic formula. Moreover, when this property is true for some a, the $before(e, f, g)$ relation is also determined by a first order logic formula.* □

Corollary 3. *If two complete graph drawings have same size, crossing relation and corner, then a drawn triangle permuted between the two sketches contains no drawn vertex of the graph.* □

4   Triangle Mutations in Complete Graph Drawings

In the previous Section we saw that two complete graph drawings with same corner and subsketch have the same *before* relations except for triangles containing no drawn vertex. The aim of this Section is to prove that two complete graph drawings with same corner have same subsketch if and only if they can be transformed into each other by a sequence of mutations. The "if" way is obvious since a mutation does not change the subsketch, the "only if" way is made by an algorithm.

For a drawing D of a graph G, and a drawn edge e of D, we denote $D - e$ the drawing obtained by removing the drawn edge e except the intersection points with other edges. Note that if $G - e$ is not connected, then an extremity a of e is isolated in $G - e$, and by definition is not represented in $D - e$.

Let G be a complete graph with vertices $\{a_1, ..., a_n\}$, the (undirected) edges of G are denoted $e_{i,j} = [a_i, a_j]$, $1 \le i < j \le n$. For a drawing D of G, we denote $D_n = D$ and, for $1 \le i < n$, $D_i = D - \{e_{i,n}, e_{i+1,n}, ..., e_{n-1,n}\}$. In particular, D_1 is a drawing of the complete graph on $n - 1$ vertices $a_1, ..., a_{n-1}$. When D is given with a corner (P, β, α), we choose to numerate vertices so that $P = a_1$, $\beta = [a_1, a_2]$ and $\alpha = [a_1, a_3]$, so that it remains a corner of the considered subdrawings.

Lemma 4. *Let $1 \le i < n$, and let D and D' be two complete graph drawings, with same size, crossing relation and corner, such that $D_i = D'_i$. Then there exists a permuted triangle between D_{i+1} and D'_{i+1}, and a sequence of mutations from D_{i+1} to D'_{i+1} containing only permuted triangles between D_{i+1} and D'_{i+1}.*

Proof. The proof is about one page long and consists in a sweeping of e_i. □

Theorem 2. *Let D and D' be two complete graph drawings with same size, crossing relation and corner. There exists a sequence $S(D, D')$ of mutations $D = D^{(0)} \to D^{(1)} \to ... \to D^{(k-1)} \to D^{(k)} = D'$ from D to D'. Moreover this sequence can be chosen such that, for any intermediate sketch $D^{(i)}$, $1 \le 0 \le k-1$ the mutated triangle from $D^{(i)}$ to $D^{(i+1)}$ is contained in a permuted triangle between $D^{(i)}$ and D'. It is given by the following algorithm.*

Computation of the first triangle $T(D_i, D'_i)$ from D_i to D'_i

if $n \le 3$ or $D_i = D'_i$ then $T(D_i, D'_i) = \emptyset$
if $n > 3$ and $1 < i \le n$ then let $T = T(D_{i-1}, D'_{i-1})$
    if $T \ne \emptyset$ then
        if T is free in D_i then $T(D_i, D'_i) := T$
        otherwise T is cut by $e_{i,n}$ in D_i then there exists (by lemma 1) a unique T'
            contained in T, free in D_i, with $e_{i,n} \in T'$, and $T(D_i, D'_i) := T'$
    if $T = \emptyset$ then there exists (by lemma 4) T', free in D_i, with $e_{i,n} \in T'$,
        permuted between D_i and D'_i, and $T(D_i, D'_i) := T'$ (arbitrary choice)

Computation of $S(D, D')$

if $T(D, D') = \emptyset$ then $S(D, D') := D$
otherwise D'' being obtained by mutation of $T(D, D')$ from D
$$S(D, D') := D \to S(D'', D')$$

Proof (sum up). We prove Theorem 2 by induction on n and $1 < i \le n$, using the previous algorithms. Recall that D_1 is a drawing of the complete graph on

$n - 1$ vertices, hence $T(D_1, D_1')$ and $S(D_1, D_1')$ are built for drawings of K_{n-1}. Note that, by Corollary 2, for all $1 < i \leq n$, we have $D_i \neq D_i'$ if and only if there exists a permuted triangle between D_i and D_i'.

The direct computation of $S(D_i, D_i')$ can be done the following way: first build $S(D_i - e_{i,n}, D_i - e_{i,n}) = S(D_{i-1}, D_{i-1}')$. The key point is that any triangle in this sequence at level $i - 1$ is contained by induction hypothesis in a triangle which is permuted between the current sketch and the final one. Hence it cannot contain a vertex of the graph according to Corollary 3. So free triangles used in the sequence of mutations at level $i - 1$ which are not cut by $e_{i,n}$, remain free triangles at level i.

Then add the mutations built in the algorithm when $T \neq \emptyset$ and T is cut by $e_{i,n}$ using Lemma 1. These added mutations all contain $e_{i,n}$. The sequence obtained here is denoted S'', and the arrangement obtained from D_i by S'' is D_i''. Then $D_{i-1}'' = D_{i-1}'$ and by Lemma 4 there exists a sequence S'' from D_i' to D_i' using only mutations containing $e_{i,n}$. Then $S = S' \to S''$ is a sequence of mutations from D_i to D_i'.

At last, $T(D_i, D_i')$ is contained in a permuted triangle between D_i and D_i': either $T(D_{i-1}, D_{i-1}') = \emptyset$ and it is a permuted triangle between D_i and D_i', or it is contained in $T(D_{i-1}, D_{i-1}')$, which is contained in a permuted triangle between D_{i-1} and D_{i-1}' (by induction hypothesis), and so between D_i and D_i'. □

5   Examples and Applications

5.1   Triangle Mutations in Pseudoline Arrangements

A *pseudoline arrangement* may be defined as a finite set of curves in the affine plane, each one being homeomorphic to a line, and such that any two pseudolines cross each other exactly once. We will always consider *uniform* pseudoline arrangements, i. e. no three pseudolines can meet at the same point. We consider that a pseudoline arrangement is labelled and given with the *circular ordering* of the pseudolines at infinity, and is defined up to an orientation preserving homeomorphism. Pseudoline arrangements (equivalent to rank 3 oriented matroids) are well studied objects, see [1] chapter 4. They satisfy simple axiomatics with the *before* relation [1], and even first order axiomatics [3].

Here, a pseudoline arrangement can be considered as a structure similar to a sketch of which *inc* and *sig* relations are not useful, of which crossing relation is trivial (each element crosses each other element once), and determined, when each pseudoline is directed, by the linear ordering of the crossings on each pseudoline, that is by a *before* relation. Hence all definitions about triangles and mutations can be done exactly the same way in pseudoline arrangements. So the previous result and algorithm apply naturally: for an arrangement A on $E = \{e_1, ..., e_n\}$, we denote A_k, $1 \leq k \leq n$, the arrangement on $E_k = \{e_1, ..., e_k\}$ obtained by restriction from A, and we replace D_i with A_i and $e_{i,n}$ with e_i in Theorem 2. Note that a similar natural inductive construction for a sequence of mutations has been used for pseudoline arrangements by Roudneff in [8].

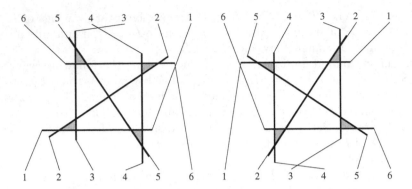

Fig. 6. Two arrangements with no permuted free triangle

The well known Ringel's theorem on pseudoline arrangements [7] states that if A and A' are two uniform pseudoline arrangements with same number of elements and same circular ordering at infinity then there exists a sequence of mutations from A to A'. Hence Theorem 2 gives a slight strengthening of this theorem, which allows to transform A into A' avoiding mutations of triangles not contained in a permuted triangle. Indeed, in the generalization to graph drawings, we want to avoid mutations of triangles containing drawn vertices.

The very important point is that it is not possible in general to transform a configuration into another one using only mutations of permuted free triangles, as it would mean there is always a permuted free triangle between two different configurations, which is false as shown on the example below. This has been mentioned in [4] from which Figure 6 is taken and made straight. Note that one of these two arrangements had already been a significant example for another problem in [1] Figure 1.11.2.

Example. The sequences of triangles built by the previous algorithm applied to the arrangements of Figure 6 are the following. We separate the two built subsequences: the first one (S' in the proof of Theorem 2) built from the previous level, and the second one when only the last pseudoline has to be moved (S'' in the proof of Theorem 2).

- at level 3: $\emptyset$ (triangles 123 are the same in both arrangements)
- at level 4: $(\emptyset) \rightarrow (234 \rightarrow 134 \rightarrow 124)$ (only 4 has to be moved)
- at level 5: $(235 \rightarrow 234 \rightarrow 135 \rightarrow 134 \rightarrow 125 \rightarrow 124) \rightarrow (\emptyset)$ (the first is sufficient)
- at level 6: $(\mathbf{356} \rightarrow 235 \rightarrow 346 \rightarrow 234 \rightarrow 135 \rightarrow 134 \rightarrow 125 \rightarrow 124) \rightarrow (236 \rightarrow 126 \rightarrow 136 \rightarrow 146 \rightarrow 156 \rightarrow 456 \rightarrow 256 \rightarrow \mathbf{356})$

This example shows two pseudoline arrangements having all their free triangles (123, 145, 356 and 246) in the same position. Then a sequence of mutations from one to the other must begin with the mutation of a non permuted triangle. Hence the minimal number of mutations needed in the sequence may be strictly larger than the number of permuted triangles. For instance in the above sequence, we used twice the mutation of 356. The problem of building a minimal sequence of mutations in general is open.

5.2   Visualization of Spatial Graphs Encoded by Oriented Matroids

Consider a set E of $n + 1$ points in the 3-dimensional real (or rational) space in general position, a plane in general position with this configuration, and $a \in E$ the extremal point in E with respect to the plane (i. e. the distance from a to the plane is maximal). Then the projections, from a to the plane, of the segments formed by all pairs of vertices is a complete (geometrical) graph drawing on n vertices (see Figure 7).

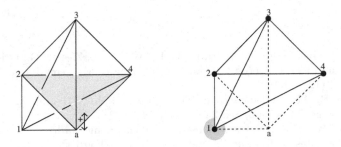

Fig. 7. Perspective on a spatial graph

Theorem 3. *The rank 4 oriented matroid defined by E determines a corner and the cross relations of the drawing obtained by projection from the extremal point $a \in E$. Hence it determines the drawing up to a sequence of triangle mutations.*

Proof. With the oriented matroid, we know for each triplet in E, and for each pair of other points, if these two points are on the same side or the opposite sides of the plane spanned by the triplet, i. e. we know the relative signs of elements in a cocircuit defined by the triplet. Then we easily get a corner of the drawing and its *cross* relations (but not all the drawing). We end using Theorem 2.   □

With theorem 3 we know that if two such configurations of points define the same oriented matroid up to a bijection of the ground set, then their projections, from extremal points being in bijection, are the same up to a sequence of triangle mutations and orientation preserving homeomorphisms.

Note that this application uses mainly particular cases of the constructions of the paper because: first, the graph drawing obtained by projection is a geometrical graph drawing, that is a drawing with straight edges, and secondly, the oriented matroid structure may determine directly the inside and map relations on the drawing.

Note nevertheless that the obtained result is not trivial since it is impossible in general to transform the first point configuration into the second by an isotopy of the space preserving the oriented matroid structure (which would have been, if true, an immediate way to build the required sequence of mutations). This fact is known in oriented matroid theory [1] as the *Universality Theorem of Mnëv*, stating that realization spaces of oriented matroid are not connected, and in fact are birationally equivalent to semi-algebraic varieties. For some other spatial transformation problems related to spatial graphs, see [6].

Finally, the point a plays the part of a point of view. When a moves in a region delimited by the planes formed by other points of the configuration, the oriented matroid data, and the subksetch, are unchanged, but the drawing, and its sketch, change with a sequence of triangle mutations. When a crosses a plane, the oriented matroid data changes (a sign changes in some cocircuit). Thus, it is a certain modelization, using two structural levels, of spatial graph visualization.

References

[1] Björner, A., Las Vergnas, M., Sturmfels, B., White, N., Ziegler, G.: Oriented matroids. Cambridge Univertisty Press. Encyclopedia of Mathematics ans its Applications **46** (1993, 1999)
[2] Courcelle, B.: The monadic second-order logic of graphs XIII: graph drawings with edge crossings. Th. Comp. Sci. **244** (2000) 63–94
[3] Courcelle, B., Olive, F.: Une axiomatisation au premier ordre des arrangements de pseudodroites euclidiennes. Annales de l'Institut Fourier (Universit J. Fourier, Grenoble, France) **49** (1999) 883–903
[4] Felsner, S., Weil, H.: A theorem on higher Bruhat order. Disc. Comp. Geom. **23** (2000) 121–127
[5] Mohar, B., Thomassen, C.: Graphs on surfaces. John Hopkins University Press, Baltimore, MD (2000).
[6] Ramírez Alfonsín, J.L.: Knots and links in spatial graphs: A Survey. Disc. Math., to appear.
[7] Ringel, G.: Über Geraden in allgemeiner Lage. Elemente der Math. **12** (1957) 75–82
[8] Roudneff, J-P.: Tverberg-type theorems for pseudoconfigurations of points in the plane. Europ. J. Comb. **9** (1988) 189–198

Collective Tree 1-Spanners for Interval Graphs

Derek G. Corneil[1], Feodor F. Dragan[2], Ekkehard Köhler[3], and Chenyu Yan[2]

[1] Department of Computer Science, University of Toronto, Toronto, Ontario, Canada
dgc@cs.toronto.edu
[2] Department of Computer Science, Kent State University, Kent, Ohio, U.S.A
{dragan, cyan}@cs.kent.edu
[3] Institut für Mathematik, Technische Universität Berlin, Berlin, Germany
ekoehler@math.TU-Berlin.DE

Abstract. In this paper we study the existence of a small set $\mathcal{T}$ of spanning trees that collectively "1-span" an interval graph G. In particular, for any pair of vertices u, v we require a tree $T \in \mathcal{T}$ such that the distance between u and v in T is at most one more than their distance in G. We show that:

- there is no constant size set of collective tree 1-spanners for interval graphs (even unit interval graphs),
- interval graph G has a set of collective tree 1-spanners of size $O(\log D)$, where D is the diameter of G,
- interval graphs have a 1-spanner with fewer than $2n - 2$ edges.

Furthermore, at the end of the paper we state other results on collective tree c-spanners for $c > 1$ and other more general graph classes.

1   Introduction

A spanning subgraph H of G is called a *spanner* of G if H provides a "good" approximation of the distances in G. More formally, for $c \geq 1$, H is called an *additive c-spanner* of G if for any pair of vertices u and v their distance in H is at most c plus their distance in G [10]. (A similar definition can be given for multiplicative c-spanners [1,14,13]; however since we are only concerned with additive spanners, we will often omit "additive".) In this paper, we continue the approach taken in [5,4,7] of studying *collective tree spanners*. We say that a graph $G(V, E)$ *admits a system of μ collective additive tree c-spanners* if there is a system $\mathcal{T}(G)$ of at most μ spanning trees of G such that for any two vertices u, v of G a spanning tree $T \in \mathcal{T}(G)$ exists such that the distance in T between x and v is at most c plus their distance in G. We say that system $\mathcal{T}(G)$ collectively c-spans the graph G. Clearly, if G admits a system of μ collective additive tree c-spanners, then G admits an additive c-spanner with at most $\mu \times (n - 1)$ edges (take the union of all those trees), and if $\mu = 1$ then G admits an additive tree c-spanner. Note also that any graph on n vertices admits a system of at most $n - 1$ collective additive tree 0-spanners (take $n - 1$ Breadth-First-Search–trees (also known as *shortest path trees*) rooted at different vertices of G).

D. Kratsch (Ed.): WG 2005, LNCS 3787, pp. 151–162, 2005.
© Springer-Verlag Berlin Heidelberg 2005

One of the motivations to introduce this new concept steams from the problem of designing compact and efficient routing schemes in graphs. In [6,15], a shortest path routing labeling scheme for trees is described that assigns each vertex of an n-vertex tree a $O(\log^2 n/\log\log n)$-bit label. Given the label of a source vertex and the label of a destination, it is possible to compute in constant time, based solely on these two labels, the neighbor of the source that heads in the direction of the destination. Clearly, if an n-vertex graph G admits a system of μ collective additive tree r-spanners, then G admits a routing labeling scheme of deviation (i.e., additive stretch) r with addresses and routing tables of size $O(\mu \log^2 n/\log\log n)$ bits per vertex. Once computed by the sender in μ time (by choosing for a given destination an appropriate tree from the collection to perform routing), headers of messages never change, and the routing decision is made in constant time per vertex (for details see [4,5]).

Previously, collective tree spanners of particular classes of graphs were considered in [4,5,7]. Paper [5] showed that any chordal graph, chordal bipartite graph or cocomparability graph admits a system of at most $\log_2 n$ collective additive tree 2–spanners. These results were complemented by lower bounds, which say that any system of collective additive tree 1–spanners must have $\Omega(\sqrt{n})$ spanning trees for some chordal graphs and $\Omega(n)$ spanning trees for some chordal bipartite graphs and some cocomparability graphs. Furthermore, it was shown that any k-chordal graph admits a system of at most $\log_2 n$ collective additive tree $(2\lfloor k/2 \rfloor)$–spanners and any circular-arc graph admits a system of two collective additive tree 2–spanners. Paper [4] showed that any AT-free graph (graph without asteroidal triples) admits a system of two collective additive tree 2-spanners, any graph having a dominating shortest path admits a system of two collective additive tree 3-spanners and a system of five collective additive tree 2-spanners, and any graph with asteroidal number $\mathsf{an}(G)$ admits a system of $\mathsf{an}(G)(\mathsf{an}(G) - 1)/2$ collective additive tree 4-spanners and a system of $\mathsf{an}(G)(\mathsf{an}(G) - 1)$ collective additive tree 3-spanners. Collective multiplicative tree spanners of planar graphs were investigated in [7]. It was shown that any weighted n–vertex planar graph admits a system of $O(\sqrt{n})$ collective multiplicative tree 1-spanners (equivalently, additive tree 0-spanners) and a system of at most $2\log_{3/2} n$ collective multiplicative tree 3–spanners.

In this paper we study collective tree 1-spanners for interval graphs. In Section 2, we show that no constant number of trees can collectively 1-span interval graphs (even unit interval graphs). Surprisingly there is, as shown in Section 4, an additive 1-spanner that uses fewer than $2n - 2$ edges, the number of edges required for two disjoint spanning trees. In Section 3, we present a polynomial time algorithm to find a set of $O(\log D)$ trees that collectively 1-span a given interval graph G, where D is the diameter of G. In the final section we briefly list other results on families of graphs that strictly contain interval graphs. First we present the definitions used in this paper.

Notation and Definitions: All graphs occurring in this paper are connected, finite, undirected, loopless and without multiple edges. In a graph $G(V, E)$ ($n = |V|, m = |E|$) the *length* of a path from a vertex v to a vertex u is the number

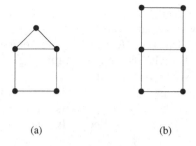

Fig. 1. (a) A house, (b) A domino

of edges in the path. The *distance* $d_G(u, v)$ between the vertices u and v is the length of a shortest path connecting u and v. The *eccentricity* $ecc(v)$ of a vertex v of G is $\max_{u \in V} d_G(u, v)$. The *diameter* $diam(G)$ of G is $\max_{v \in V} ecc(v)$. The *i*th *neighborhood* of a vertex v of G is the set $N_i(v) = \{u \in V : d_G(v, u) = i\}$. For a vertex v of G, the sets $N(v) = N_1(v)$ and $N[v] = N(v) \cup \{v\}$ are called the *open neighborhood* and the *closed neighborhood* of v, respectively. For a set $S \subseteq V$, by $N[S] = \bigcup_{v \in S} N[v]$ we denote the *closed neighborhood* of S and by $N(S) = N[S] \setminus S$ the *open neighborhood* of S. A set $D \subseteq V$ is called a *dominating set* of a graph $G = (V, E)$ if $N[D] = V$.

An independent set of three vertices such that each pair is joined by a path that avoids the neighborhood of the third is called an *asteroidal triple* (AT). A graph G is an *AT-free graph* if it does not contain any asteroidal triples [2]. A graph is *chordal* if it does not contain any induced cycles of length greater than 3. A graph is an *interval graph* if one can associate with each vertex an interval on the real line such that two vertices are adjacent if and only if the corresponding intervals have a nonempty intersection. Furthermore, an interval graph is a *unit interval graph* if all intervals are of the same length. Unit interval graphs are equivalent to *proper interval graphs* where no interval can properly contain any other interval. It is well known that a graph is an interval graph if and only if it is both a chordal graph and an AT-free graph [9].

A graph is *weakly chordal* (also called *weakly triangulated*) if neither G nor its complement $\overline{G}$ contain an induced hole (cycle of size at least 5). A graph G is *house-hole-domino-free (HHD-free)* if it does not contain the house, the domino, and holes as induced subgraphs (see Fig. 1). Clearly, chordal graphs are strictly contained in both weakly chordal and HHD-free graphs.

2 Lower Bound

Independently McKee [12] and Kratsch et al. [8] showed that no single tree can c-span a chordal graph for any constant c. We now show a similar result for collectively 1-spanning a unit interval graph.

Theorem 1. *No constant number of trees can collectively 1-span a unit interval graph.*

Proof. First we will show that two trees do not suffice and then show how to extend this result to any constant number of trees.

The general "gadget" will be a K_3 with two independent universal vertices x and y (i.e. we have a K_5 with the edge xy missing). The vertices of the K_3 will be labelled 1, 2, 3. Now make a sufficiently long chain of these gadgets by identifying the y vertex of a gadget with the x vertex of its right neighbor. It is straightforward to confirm that this graph G is a unit interval graph. Consider two trees T_1 and T_2 that supposedly collectively 1-span G. By making the chain sufficiently long, by the "pigeonhole principle", we are guaranteed that there are three gadgets in G namely, A, B and C where A is left of B which is left of C such that:

- T_1 restricted to A, B and C is exactly the same spanning tree for all three gadgets. Exactly the same means from the labelled vertex point of view,
- T_2 restricted to A, B and C is also exactly the same spanning tree for all three gadgets. Note that T_1 restricted to $\{A, B, C\}$ is not necessarily the same as T_2 restricted to $\{A, B, C\}$.

The vertices in A, B and C will be denoted A_x, B_3, C_y, where, for example, A_x refers to the x-vertex of A. We say that a tree provides a 1-approximating path between two vertices if the distance between the vertices in the tree is at most 1 more than their distance in G. We now show that in order for T_1 or T_2 to provide such an approximating path, certain edges of G must be present in the tree.

Claim. Let i be an element of $\{1, 2, 3\}$. If either T_1 or T_2 provides a 1-approximating path between A_i and C_i, then it must contain the xi and yi edges in all of A, B and C.

Proof. Without loss of generality, assume that T_1 provides the 1-approximating path between A_i and $C_i, i \in \{1, 2, 3\}$. Such a path requires either A_i to be adjacent to A_y and/or C_i to be adjacent to C_x. Without loss of generality, assume C_i is adjacent to C_x; thus since T_1 when restricted to A, B and C is exactly the same, A_i is adjacent to A_x and B_i is adjacent to B_x as well. We now show that in all three of A, B and C, i is also adjacent to y. Suppose not; now in each gadget, the distance between i and y is at least 2 which means that the tree path between A_i and C_i must be at least 2 greater than the distance in G (since in T_1 the distance between B_x and B_y must be at least 3 by following the edge $B_x B_i$ and the path between B_i and B_y). □

From the claim, it is clear that each of T_1 and T_2 can provide at most one path between A_1, C_1 or A_2, C_2 or A_3, C_3 and thus at least three trees are required to 1-approximate G.

To generalize this argument, i.e. to show that at least k trees are required, merely replace the K_3 in the gadget by a K_k. The same use of the claim shows that $k - 1$ trees are not enough. □

A straightforward analysis (that will be presented in the journal version of the paper) shows that the size of the collective tree 1-spanners is $\Omega(\sqrt{\log n})$.

3   Upper Bound

In light of $\Omega(\sqrt{\log n})$ spanning trees being needed to collectively 1-span an interval graph G, we now show that $2\log_2(D-1)+4$ spanning trees suffice, where D is the diameter of G.

Let P be a shortest path of a graph G. If every vertex z of G belongs to the neighborhood $N[P]$ of P, then we say that P is a *dominating shortest path (DS-path)* of G. It is known that any AT-free graph has a DS-path which can be found in linear time [2]. In what follows we will need a slightly stronger result from [8].

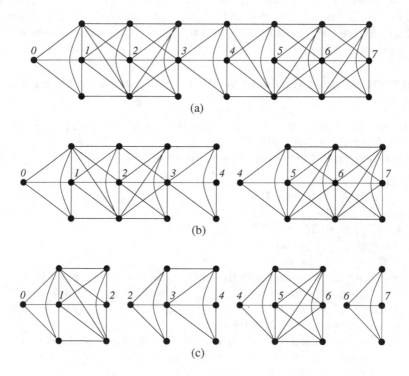

Fig. 2. (a) Graph $G = G_{0,7}$, (b) graphs $G_{0,4}$ and $G_{4,7}$, (c) graphs $G_{0,2}$, $G_{2,4}$, $G_{4,6}$ and $G_{6,7}$. Graphs $G_{0,1}$, $G_{1,2}$, $G_{2,3}$, $G_{3,4}$, $G_{4,5}$ and $G_{5,6}$ are not shown.

Lemma 1. *[2,8] Any AT-free graph G admits a DS-path $(x_0, x_1, \cdots, x_{ecc(x_0)})$ such that for every $i = 1, 2, \cdots, ecc(x_0)$, every vertex $z \in N_i(x_0)$ is adjacent to x_i or x_{i-1}. Moreover, such a DS-path can be constructed in linear time.*

Now let G be an interval graph and let $(x_0, x_1, \cdots, x_{ecc(x_0)})$ be such a DS-path of G described by Lemma 1. The following lemma is important for our future discussion.

Lemma 2. *For any two adjacent vertices $u \in N_i(x_0)$ and $v \in N_{i+1}(x_0)$, $u, v \in N[x_i]$ or $u, v \in N[x_{i+1}]$. Moreover, if $u \neq x_i$, then $ux_i \in E$.*

Proof. If $u = x_i$ or $v = x_{i+1}$, then the lemma is trivially true. Hence, we may assume that $u \neq x_i$ and $v \neq x_{i+1}$. If $ux_i \notin E$ then, by Lemma 1, $ux_{i-1} \in E$. If now $vx_i \in E$, then u, v, x_i, x_{i-1} give an induced cycle of length 4 in G, which is impossible for an interval graph. If $vx_i \notin E$ then, by Lemma 1, $vx_{i+1} \in E$. Then, we obtain either an induced cycle of length 5 or induced cycle of length 4, depending on whether or not ux_{i+1} is in E. So, if $u \neq x_i$, then ux_i must be in E. If now $vx_{i+1} \in E$ but neither ux_{i+1} nor vx_i is in E, then x_i, u, v, x_{i+1} form an induced cycle of length 4 in G, which is impossible. $\qquad\square$

Let l denote $ecc(x_0)$. For any two integers i, j, $0 \leq i < j \leq l$, we define $G_{i,j}$ to be the subgraph of G induced by vertices $\{x_i\} \cup N_{i+1}(x_0) \cup \cdots \cup N_j(x_0)$ (see Fig. 2 for an illustration). In view of Lemma 1, obviously, $G_{i,j}$ is connected and $G = G_{0,l}$. We use the following procedure to construct a system of *local shortest path trees* of G.

PROCEDURE 1. A system of local shortest path trees for an interval graph G.

Input: An interval graph G, a DS-path $(x_0, \cdots, x_l)$ and the layering $\{x_0\}, N_1(x_0), \cdots, N_l(x_0)$ of G.
Output: A system of local shortest path trees of G.
Method:
    set $k := 0$; $\mathcal{G}_k := \{G_{0,l}\}$; $\mathcal{T} := \emptyset$;
    while $\mathcal{G}_k \neq \emptyset$ **do**
        set $\mathcal{G}_{k+1} := \emptyset$; $T_k' := \emptyset$; $T_k'' := \emptyset$;
        for each $G_{i,j} \in \mathcal{G}_k$ **do**
          if $j = i + 1$ **then**
              construct a shortest path tree of $G_{i,j}$ rooted at x_i and put it in T_k';
              construct a shortest path tree of $G_{i,j}$ rooted at x_j and put it in T_k'';
          else /* if $j > i + 1$ */
              set $s := \lceil (j - i)/2 \rceil + i + 1$;
              construct a shortest path tree of $G_{i,j}$ rooted at x_{s-1} and put it in T_k';
              construct a shortest path tree of $G_{i,j}$ rooted at x_s and put it in T_k'';
              set $\mathcal{G}_{k+1} := \mathcal{G}_{k+1} \cup \{G_{i,s-1}, G_{s-1,j}\}$;
        set $\mathcal{T} := \mathcal{T} \cup T_k' \cup T_k''$;
        set $k := k + 1$;
    return $\mathcal{T}$.

Note that the while loop in the procedure above will be executed at most $\log_2(l-1)+2$ times. Let $G_{i,j}$ be an arbitrary subgraph generated by the procedure with $j > i + 1$. Let also $s = \lceil (j - i)/2 \rceil + i + 1$ and $a \in N_r(x_0), b \in N_t(x_0)$ be two arbitrary vertices in $G_{i,j}$, where $r \leq t$ are two integers between i and j inclusive. Let $T_s, T_{s-1} \in \mathcal{T}$ be the two shortest path trees of $G_{i,j}$ rooted at x_s, x_{s-1}, respectively. Clearly, both spanning trees span all the vertices of $G_{i,j}$ and the subgraphs $G_{i,s-1}$ and $G_{s-1,j}$ of $G_{i,j}$ have only one common vertex x_{s-1}. The following lemmata hold.

Lemma 3. *If $r=t=s$, then $d_{T_s}(a,b) \leq d_G(a,b)+1$ or $d_{T_{s-1}}(a,b) \leq d_G(a,b)+1$.*

Proof. Since T_s and T_{s-1} are shortest path trees, using Lemma 1, one can easily show that $d_{T_s}(a,b) \leq 3$ or $d_{T_{s-1}}(a,b) \leq 3$. So, if $ab \notin E$ or $a,b \in N[x_s]$ or $a,b \in N[x_{s-1}]$, then the lemma holds. If now $ab \in E$ and, without loss of generality, $ax_s, bx_{s-1} \in E$ and $bx_s, ax_{s-1} \notin E$, then the vertices a, b, x_{s-1}, x_s form an induced cycle of length 4 in G, which is impossible.   □

In a similar way one can show the following.

Lemma 4. *If a and b are vertices of a graph $G_{i,i+1}$ then $d_{T'}(a,b) \leq d_G(a,b)+1$ or $d_{T''}(a,b) \leq d_G(a,b)+1$, where $T', T'' \in \mathcal{T}$ are shortest path trees of $G_{i,i+1}$ rooted at x_i and x_{i+1}, respectively.*

Lemma 5. *If $i \leq r < s \leq t \leq j$, then $d_{T_s}(a,b) \leq d_{G_{i,j}}(a,b)+1$ or $d_{T_{s-1}}(a,b) \leq d_{G_{i,j}}(a,b)+1$.*

Proof. Using Lemma 1, it is easy to show that $d_{T_s}(a,b) \leq t-r+3$ or $d_{T_{s-1}}(a,b) \leq t-r+3$. So, when $d_{G_{i,j}}(a,b) \geq t-r+2$, the lemma clearly holds. Therefore, we may assume that $d_{G_{i,j}}(a,b)$ is $t-r+1$ or $t-r$. Let first $d_{G_{i,j}}(a,b) = t-r$ and $(z_r = a, z_{r+1}, \cdots, z_t = b)$ be a shortest path between a and b in $G_{i,j}$. Consider vertices z_{s-1} and z_s. According to Lemma 2, they both belong to $N[x_s]$ or to $N[x_{s-1}]$. Without loss of generality, assume $z_s, z_{s-1} \in N[x_s]$. Since T_s is a shortest path tree, $d_{T_s}(x_s, a) \leq s-r$ and $d_{T_s}(x_s, b) \leq t-s+1$. So, $d_{T_s}(a,b) \leq d_{T_s}(x_s, a) + d_{T_s}(x_s, b) \leq t-r+1 = d_{G_{i,j}}(a,b)+1$.

Now assume that $d_{G_{i,j}}(a,b) = t-r+1$. Let $z_s z_{s-1} \in E$ be an edge on the shortest path between a and b in $G_{i,j}$ such that $z_s \in N_s(x_0)$ and $z_{s-1} \in N_{s-1}(x_0)$. Obviously, such an edge must exist, and we have $d_{G_{i,j}}(a,b) = d_{G_{i,j}}(a, z_{s-1}) + d_{G_{i,j}}(b, z_s) + 1$. According to Lemma 2, both z_s and z_{s-1} belong to $N[x_s]$ or to $N[x_{s-1}]$. Without loss of generality, assume they belong to $N[x_s]$. Then, since T_s is a shortest path tree of $G_{i,j}$, $d_{T_s}(x_s, a) \leq 1 + d_{G_{i,j}}(z_{s-1}, a)$ and $d_{T_s}(x_s, b) \leq 1 + d_{G_{i,j}}(z_s, b)$. Hence, $d_{T_s}(a,b) \leq 2 + d_{G_{i,j}}(z_s, b) + d_{G_{i,j}}(z_{s-1}, a) = 1 + d_{G_{i,j}}(a,b)$. This concludes our proof.   □

Lemma 6. *If $d_{G_{i,j}}(a,b) \neq d_G(a,b)$, then $a \in N_{i+1}(x_0)$ or $b \in N_{i+1}(x_0)$.*

Proof. Without loss of generality, assume that $a \in N_r(x_0), b \in N_t(x_0)$ and $i+1 \leq r \leq t \leq j$. We claim that there always exists a shortest path $P^G(a,b)$ between a and b in G such that $P^G(a,b) \cap N_{j+1}(x_0) = \emptyset$. If this is not the case, then there must exist vertices $c, d \in P^G(a,b) \cap N_j(x_0)$ and $c', d' \in N_{j+1}(x_0) \cap P^G(a,b)$ such that cc' and dd' are edges of $P^G(a,b)$. Obviously, $cd \notin E$. According to Lemma 2(second part), $cx_j, dx_j \in E$. Then, if we replace the part of $P^G(a,b)$ between c and d with the path (c, x_j, d), obviously we will get a shortest path between a and b that does not intersect $N_{j+1}(x_0)$. So, we may assume that $P^G(a,b) \cap N_{j+1}(x_0) = \emptyset$.

If neither $a \in N_{i+1}(x_0)$ nor $b \in N_{i+1}(x_0)$, then $i+1 < r \leq t$. Since $d_{G_{i,j}}(a,b) \neq d_G(a,b)$, we must be able to find four vertices $e, f \in N_{i+2}(x_0) \cap P^G(a,b)$ and $e', f' \in N_{i+1}(x_0)$ such that ee' and ff' are edges of $P^G(a,b)$. If

$e'f' \in E$ or $e' = f'$, then $P^G(a,b)$ is in $G_{i,j}$, i.e., $d_{G_{i,j}}(a,b) = d_G(a,b)$. Hence, one may assume that $e'f' \notin E$ and $e' \neq f'$. Then, according to Lemma 2(second part), $e'x_{i+1}, f'x_{i+1} \in E$ and we can choose another shortest path between a and b that does not intersect $N_i(x_0)$ and get $d_{G_{i,j}}(a,b) = d_G(a,b)$ again. Thus, if neither $a \in N_{i+1}(x_0)$ nor $b \in N_{i+1}(x_0)$, then $d_{G_{i,j}}(a,b) = d_G(a,b)$. ☐

We are ready to prove the following main lemma of this section.

Lemma 7. *For any two vertices* $a, b \in V(G)$*, there exists a local shortest path tree* $T \in \mathcal{T}$ *such that* $d_T(a,b) \leq d_G(a,b) + 1$.

Proof. Let $G_{i,j}$ be a subgraph of G, generated by Procedure 1, which contains both vertices a and b and has the minimum difference $j - i$. If $j - i = 1$ then we are done by Lemma 4. Therefore, in what follows we assume that $j > i+1$, and let $s = \lceil (j-i)/2 \rceil + i + 1$ and $a \in N_r(x_0), b \in N_t(x_0)$, where $i \leq r \leq t \leq j$. By minimality of $j - i$, $r < s \leq t$ (if $t < s$ then $G_{i,s-1}$ contains both a and b, and if $r \geq s$ then $G_{s-1,j}$ contains both a and b).

The case $i \leq r < s \leq t \leq j$ when $d_{G_{i,j}}(a,b) = d_G(a,b)$ is handled by Lemma 5. Assume now that $d_{G_{i,j}}(a,b) \neq d_G(a,b)$. Let $P^G(a,b)$ be an arbitrary shortest path between a and b in G. By Lemma 6, $r = i + 1$. We claim that $d_G(a,b) = t - r + 2$. Indeed, since $d_{G_{i,j}}(a,b) \leq t - r + 3$ (recall that $a \in N[x_{i+1}] \cup N(x_i)$ and $b \in N[x_t] \cup N(x_{t-1})$ by Lemma 1) and $d_{G_{i,j}}(a,b) \neq d_G(a,b)$, we must have $d_G(a,b) \leq t - r + 2$. On the other hand, if $d_G(a,b) \leq t - r + 1$, then we can easily show that all the vertices of $P^G(a,b)$ are in $G_{i,j}$, and thus $d_{G_{i,j}}(a,b) = d_G(a,b)$.

Consider now the local shortest path tree $T_s \in \mathcal{T}$ of $G_{i,j}$ rooted at x_s, where $s = \lceil (j-i)/2 \rceil + i + 1$. It is easy to show that $d_{T_s}(x_s, a) \leq s - r + 2$ and $d_{T_s}(x_s, b) \leq t - s + 1$. Combining the two inequalities, we get $d_{T_s}(a,b) \leq t - r + 3$. Since $d_G(a,b) = t - r + 2$, the lemma holds. ☐

We can group the local shortest path trees from $\mathcal{T}$ into at most $2 \log_2(l-1)+4$ spanning trees of G. Consider Procedure 1. At the beginning, $G_{0,l} = G$ and we

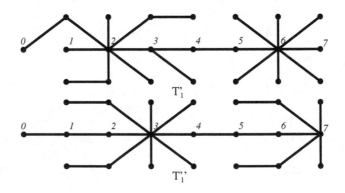

Fig. 3. Spanning trees T_1' and T_1'' of an interval graph G from Fig. 2

construct only two spanning trees of G, i.e., $T_0' = \{T_0'\}$, $T_0'' = \{T_0''\}$. In the second iteration, G is decomposed into two subgraphs $G_{0,s-1}$ and $G_{s-1,l}$ where $s = \lceil l/2 \rceil + 1$. For each of the two subgraphs, the algorithm constructs two local shortest path trees, i.e., $T_1' = \{T'^1_1, T'^2_1\}$, $T_1'' = \{T''^1_1, T''^2_1\}$. Since $G_{0,s-1}$ and $G_{s-1,l}$ have only vertex x_{s-1} in common, we conclude $T_1' := T'^1_1 \cup T'^2_1$ and $T_1'' := T''^1_1 \cup T''^2_1$ are two spanning trees of G (see Fig. 3). In general, during the iteration k of Procedure 1, for each of the 2^{k-1} subgraphs $G_{0,j_1}, G_{j_1,j_2}, \cdots, G_{j_{2^{k-1}-1},j}$ of G, we construct two local shortest path trees, i.e., $T_k' = \{T'^1_k, T'^2_k, \cdots, T'^{2^{k-1}}_k\}$, $T_k'' = \{T''^1_k, T''^2_k, \cdots, T''^{2^{k-1}}_k\}$, where T'^γ_k and T''^γ_k are the local shortest path trees constructed for $G_{j_{\gamma-1},j_\gamma}$ ($\gamma = 1, \cdots, 2^{k-1}$). Again, for any $\gamma = 1, \cdots, 2^{k-1} - 1$, $G_{j_{\gamma-1},j_\gamma}$ and $G_{j_\gamma,j_{\gamma+1}}$ have only vertex x_{j_γ} in common. Therefore, $T_k' := \bigcup_{1 \le \gamma \le 2^{k-1}} T'^\gamma_k$ and $T_k'' := \bigcup_{1 \le \gamma \le 2^{k-1}} T''^\gamma_k$ are two spanning trees of G. Since the number of iterations is bounded by $\alpha \le \log_2(l-1) + 2$, in this way we will create a system $\mathcal{ST} := \{T_0', T_0'', T_1'.T_1'', \cdots, T_\alpha', T_\alpha''\}$ of at most 2α spanning trees of G. Furthermore, each local shortest path tree from $\mathcal{T}$ will be contained in one of the spanning trees from $\mathcal{ST}$ as a subtree. Thus, we proved the following result.

Theorem 2. *Any interval graph of diameter D admits a system of $2\log_2(D-1)+4$ collective additive tree 1-spanners. Moreover, these trees can be constructed in $O(m \log D)$ total time.*

4   Sparse Spanner

Given the result in Theorem 1 that no constant number of trees can collectively 1-span a unit interval graph, it is somewhat surprising that there is a sparse 1-spanner of an interval graph that has fewer than $2n - 2$ edges (i.e. the number of edges in two disjoint spanning trees). To see this, we first present an algorithm to produce a subgraph H of interval graph G. We then show that H has the required number of edges and is in fact a 1-spanner of G.

PROCEDURE 2. Construction of a sparse 1-spanner for an interval graph G.

Input: An interval graph G, and an interval ordering $\prec$ of V where for all $x \prec y \prec z$ if $xz \in E$, then $xy \in E$. Let D be the diameter of the graph G.
Output: A sparse 1-spanner H of G.
Method:
　　　let x_D be the last vertex in the ordering $\prec$; set $E_H := \emptyset$;
　　　add the edge from x_D to its leftmost neighbor to E_H;
　　　for i from D downto 1 do
　　　　　let x_{i-1} be the left most neighbor of x_i;
　　　　　add to E_H all edges from x_{i-1} to vertices to the right of x_{i-1} up to x_i;
　　　　　if $i > 1$ then add to E_H all edges in G from x_{i-1} to vertices to the left of x_{i-1}.

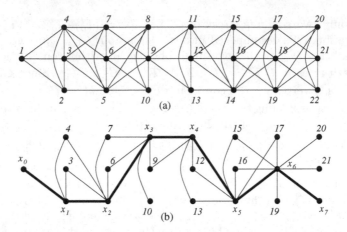

Fig. 4. (a) Graph G and its interval ordering $\prec$. (b) Sparse 1-spanner H with the edges of P bold.

As an example of Procedure 2, consider Fig. 4(a) where the interval graph of Fig. 2 is repeated together with an interval numbering. The 1-spanner H is shown in Fig. 4(b) and the bold edges denote P, the path induced on $\{x_i, 0 \leq i \leq D\}$.

We now show that H is a sparse 1-spanner of G.

Lemma 8. H is a 1-spanner of G with at most $2n - D - 2$ edges.

Proof. First we show that H has at most $2n - D - 2$ edges. To see this note:

- all vertices to the right of x_{D-1} have degree 1 in H and there is at least one vertex here;
- all vertices to the left of x_{D-1} that are not on P have degree at most 2 in H (by the interval ordering property);
- there are $D - 1$ edges joining the $P \setminus \{x_D\}$ vertices.

Thus the total number of edges in H is at most $1 + 2(n - (D+1)) + D - 1 = 2n - D - 2$, as required.

To see that H is a 1-spanner, consider arbitrary vertices x and y where $x \prec y$ in the interval ordering. We now show that $d_H(x, y) \leq d_G(x, y) + 1$. This is clearly true if x is in P, so we assume that x is not in P. Now, suppose x is between x_i and x_{i+1} for $i \geq 0$ and y satisfies $x_j \prec y \preceq x_{j+1}$, where $i < j$. (Note that if $i = j$, then immediately $d_H(x, y) \leq 2$.)

Claim. $d_G(x, y) \geq j - i$.

Proof. Suppose to the contrary that there is a path Q in G of length less than $j - i$. It is easy to see that the number of P vertices strictly between x and y is $j - i$ and thus some edge uv (where $u \prec v$) of Q surrounds two P vertices x_k and x_{k+1} (i.e. $u \prec x_k \prec x_{k+1} \prec v$). Since $uv \in E_G$, $ux_{k+1} \in E_G$ contradicting x_k being the left most neighbor of x_{k+1}. □

Now suppose $d_G(x,y) = j - i$ as witnessed by path $Q := (q_0 = x, q_1, \cdots, q_{j-i} = y)$. Using the same argument as in the claim, for each $k \in \{0, 1, \cdots, j-1\}$, q_k must lie between x_{i+k} and x_{i+k+1}. Since $x = q_0$ is adjacent to q_1, and x_{i+1} is between q_0 and q_1, we know that x is adjacent to x_{i+1}. Now consider the path in H from x to $x_{i+1}, \cdots x_j, y$. This path has length $j - i + 1$.

Thus we may assume that $d_G(x,y) > j - i$. But the path in H from x to $x_i, \cdots x_j, y$ has length $j - i + 2$ and we are finished.   □

Thus we have the following result:

Theorem 3. *Any interval graph G of diameter D admits a sparse additive 1-spanner with at most $2n - D - 2$ edges. Moreover, this spanner can be constructed in $O(n + m)$ time.*

Proof. Given Lemma 4, we only have to establish the time complexity. There are many linear time interval graph recognition algorithms that can be used to determine an interval ordering of the given graph (for example see [3]). Using this ordering, a straightforward implementation of Procedure 2 can be achieved in linear time.   □

Furthermore, in the journal version of the paper we will show that the sparse spanner returned by Procedure 2 can be used for efficient routing.

5   Concluding Remarks

The most obvious open question in this paper is to tighten the gap between the lower and upper bounds for the size of a collective tree 1-spanner for interval graphs.

The results stated in this paper also raise questions about additive c-spanners for $c > 1$ for graph classes containing interval graphs. (Recall that interval graphs have a single tree that 2-spans the graph [11,8].) In the journal version of the paper, we will present proofs of the following theorems.

Theorem 4. *No constant number of trees can collectively additively c-span chordal graphs for $c \leq 3$.*

Theorem 5. *No constant number of trees can collectively additively c-span weakly chordal graphs for all constants c.*

Theorem 6. *Any HHD-free graph admits a system of at most $2 \log_2 n$ collective additive tree 2-spanners. Moreover, such a set of trees can be constructed in $O(m \log n)$ time.*

For the proof of Theorem 6 we show an auxiliary result of independent interest that any n-vertex HHD-free graph G has a separator $S \subseteq V$ such that

- any connected component of $G \setminus S$ has no more than $n/2$ vertices and
- $S \subseteq (N[x] \cup N[y])$ for some vertices $x, y \in S$.

Moreover, S and such two vertices x and y can be found in linear time.

Acknowledgements. DGC wishes to thank the Natural Sciences and Engineering Research Council of Canada for financial assistance in the support of this research.

References

1. L.P. CHEW, There are planar graphs almost as good as the complete graph, *J. of Computer and System Sciences*, 39 (1989), 205–219.
2. D.G. CORNEIL, S. OLARIU and L. STEWART, Asteroidal Triple–free Graphs, *SIAM J. Discrete Math.*, 10 (1997), 399–430.
3. D.G. CORNEIL, S. OLARIU and L. STEWART, The LBFS structure and recognition of interval graphs, under revision.
4. F.F. DRAGAN, C. YAN and D.G. CORNEIL, Collective Tree Spanners and Routing in AT-free Related Graphs (Extended Abstract), Proceedings of *30th International Workshop Graph-Theoretic Concepts in Computer Science (WG '04)*, June 2004, Bad Honnef, Germany, Springer, *Lecture Notes in Computer Science* 3353, 68-80.
5. F.F. DRAGAN, C. YAN and I. LOMONOSOV, Collective tree spanners of graphs, Proc. of the *9th Scandinavian Workshop on Algorithm Theory (SWAT'04)*, 8-10 July, 2004, Humlebæk, Denmark, Springer, Lecture Notes in Computer Science 3111, pp. 64-76.
6. P. FRAIGNIAUD and C. GAVOILLE, Routing in Trees, *Proceedings of the 28th Int. Colloquium on Automata, Languages and Programming (ICALP 2001)*, Lecture Notes in Computer Science 2076, 2001, pp. 757–772.
7. A. GUPTA, A. KUMAR and R. RASTOGI, Traveling with a Pez Dispenser (or, Routing Issues in MPLS), *SIAM J. Comput.*, 34 (2005), pp. 453-474.
8. D. KRATSCH, H.-O. LE, H. MÜLLER, E. PRISNER AND D. WAGNER Additive tree spanners *SIAM J. Discrete Math.* 17 (2003), 332-340.
9. C. LEKKERKERKER AND J. BOLAND Representation of a finite graph by a set of intervals on the real line *Fund. Math.*, 51 (1962), 45-64.
10. A.L. LIESTMAN AND T. SHERMER, Additive graph spanners, *Networks*, 23 (1993), 343-364.
11. M.S. MADANLAL, G. VENKATESAN, and C. PANDU RANGAN, Tree 3-spanners on interval, permutation and regular bipartite graphs, *Inform. Process. Lett.*, 59 (1996), 97-102.
12. T.A. MCKEE, personal communication to E. Prisner, 1995.
13. D. PELEG, and A.A. SCHÄFFER, Graph Spanners, *J. Graph Theory*, 13 (1989), 99-116.
14. D. PELEG AND J.D. ULLMAN, An optimal synchronizer for the hypercube, *in Proc. 6th ACM Symposium on Principles of Distributed Computing*, Vancouver, 1987, 77-85.
15. M. THORUP and U. ZWICK, Compact routing schemes, *Proceedings of the 13th Ann. ACM Symp. on Par. Alg. and Arch. (SPAA 2001)*, ACM 2001, pp. 1–10.

On Stable Cutsets in Claw-Free Graphs and Planar Graphs

Van Bang Le[1], Raffaele Mosca[2], and Haiko Müller[3]

[1] Institut für Informatik, Universität Rostock, 18051 Rostock, Germany
le@informatik.uni-rostock.de
[2] Dipartimento di Scienze, Universitá degli Studi "G.D'Annunzio",
Viale Pindaro 42, Pescara 65127, Italy
r.mosca@unich.it
[3] School of Computing, University of Leeds, Leeds, LS2 9JT, UK
hm@comp.leeds.ac.uk

Abstract. To decide whether a line graph (hence a claw-free graph) of maximum degree five admits a stable cutset has been proven to be an **NP**-complete problem. The same result has been known for K_4-free graphs. Here we show how to decide this problem in polynomial time for (claw, K_4)-free graphs and for a claw-free graph of maximum degree at most four. As a by-product we prove that the stable cutset problem is polynomially solvable for claw-free planar graphs, and for planar line graphs. Now, the computational complexity of the stable cutset problem restricted to claw-free graphs and claw-free planar graphs is known for all bounds on the maximum degree.

Moreover, we prove that the stable cutset problem remains **NP**-complete for K_4-free planar graphs of maximum degree five.

1 Introduction

In a graph, a *clique* (*stable set*) is a set of pairwise (non-)adjacent vertices. A *cutset* (or *separator*) of a graph G is a set S of vertices such that $G - S$ is disconnected. A *clique cutset* (*stable cutset*) is a cutset which is also a clique (stable set).

Clique cutsets are a well-studied kind of separators in the literature, and have been used in divide-and-conquer algorithms for various graph problems, such as graph colouring and finding maximum stable sets; see [18,22]. Applications of clique cutsets in algorithm design use the fact that these cutsets (in arbitrary graphs) can be found in polynomial time [18,21,22].

The importance of stable cutsets has been demonstrated first in [6,20] in connection to perfect graphs. Tucker [20] proved that if S is a stable cutset in G and if no induced cycle of odd length at least five in G has a vertex in S then the colouring problem on G can be reduced to the same problem on the smaller subgraphs induced by S and the components of $G - S$.

Later, the papers [2,3,4,10,13,15] discussed the computational complexity of the problem STABLE CUTSET ("Does a given graph admit a stable cutset?").

D. Kratsch (Ed.): WG 2005, LNCS 3787, pp. 163–174, 2005.
© Springer-Verlag Berlin Heidelberg 2005

Stable cutsets (in line graphs) have been also studied under other notion. A graph is *decomposable* (cf. [11]) if its vertices can be coloured red and blue in such a way that each colour appears on at least one vertex but each vertex v has at most one neighbour having a different colour from v. In other words, a graph is decomposable if its vertices can be partitioned into two nonempty parts such that the edges connecting vertices of different parts form an induced matching, a *matching-cut*. It turns out that matching-cuts in a graph correspond to stable cutsets in its line graphs. Matching-cuts have been studied in [1,5,8,9,15,16,17]. The papers [7,17] point out an application in graph drawing.

The relationship between decomposability and a stable cutset is (cf. [2]): If $L(G)$ has a stable cutset, then G is decomposable. If G is decomposable and has minimum degree at least two, then $L(G)$ has a stable cutset.

Chvátal [5] proved that recognising decomposable graphs is **NP**-complete, even for graphs with maximum degree four. Thus, in terms of stable cutsets in line graphs, Chvátal's result may be reformulated and improved as follows.

Theorem 1 (Chvátal [5]). STABLE CUTSET *is* **NP**-*complete, even if the input is restricted to line graphs with maximum degree six.*

Theorem 2 ([15]). STABLE CUTSET *remains* **NP**-*complete if restricted to line graphs with maximum degree five, and is polynomial solvable for line graphs of maximum degree at most four.*

Hence, the computational complexity of STABLE CUTSET for line graphs is completely characterised with respect to maximum degree constraints.

In particular, STABLE CUTSET is **NP**-complete for claw-free graphs with maximum degree five. In [15], it is shown that STABLE CUTSET is solvable in linear time for arbitrary graphs with maximum degree at most three. The complexity of STABLE CUTSET for graphs with maximum degree 4 is still open.

In this paper we will improve the second part of Theorem 2 to the larger class of claw-free graphs as follows: STABLE CUTSET becomes polynomial for claw-free graphs of maximum degree at most four. Thus the computational complexity of STABLE CUTSET for claw-free graphs is completely characterised with respect to maximum degree constraints.

STABLE CUTSET for K_3-free graphs is trivial. In [2] it was shown that STABLE CUTSET is **NP**-complete for K_4-free graphs. Our second result is that STABLE CUTSET can be solved in polynomial time for (claw, K_4)-free-graphs. As a by-product, we will show that STABLE CUTSET is polynomially solvable for claw-free planar graphs, and in particular for planar line graphs.

Finally, we show that STABLE CUTSET remains **NP**-complete on planar K_4-free graphs with maximum degree five.

2   Preliminaries

Let G be a graph. The vertex set and the edge set of G are denoted by $V(G)$ and $E(G)$, respectively. The neighbourhood of a vertex v in G, denoted by $N(v)$, is the set of all vertices in G adjacent to v. Let $\deg(v) = |N(v)|$ be the degree of the vertex v, and $\Delta(G) = \max\{\deg(v) \mid v \in V(G)\}$ the maximum degree of G. For a subset $W \subseteq V(G)$, $G[W]$ is the subgraph of G induced by W.

Let $\mathrm{scs}(G)$ denote the minimum size of a stable cutset of G. If G has no stable cutset we write $\mathrm{scs}(G) = \infty$.

When discussing the computational complexity of STABLE CUTSET we may assume that G is connected. Moreover we assume that no vertex v of G has a stable neighbourhood $N(v)$. Otherwise $N(v)$ or $\{v\}$ would be a stable cutset in G, or G has at most two vertices, and we are done. Thus, we have (cf. [15]):

Lemma 1. *If* $\mathrm{scs}(G) < \infty$, *then* $\mathrm{scs}(G - v) < \infty$ *for all* $v \in V(G)$.

Lemma 2. *Let* C *be a clique cutset in a graph* G, $|C| \geqslant 2$. *Then* $\mathrm{scs}(G) < \infty$ *if and only if there is a component* $G[A]$ *of* $G - C$ *such that* $\mathrm{scs}(G[A \cup C]) < \infty$.

Since a clique cutset can be found in polynomial time ([18,21]), and singletons are stable, Lemma 2 allows us to assume that G has no clique cutset.

3   Rigid Sets

A set $R \subseteq V$ is said to be *rigid* in $G = (V, E)$ if, for every stable set $S \subseteq V$, there is a connected component $G[A]$ of $G - S$ with $R \setminus S \subseteq A$. Rigid sets naturally come in because G has a stable cutset if and only if V is not rigid.

Clearly, every clique of G is rigid. Moreover, if Q and R are rigid sets such that $Q \cap R$ contains a pair of adjacent vertices, then $Q \cup R$ is rigid. However, further rigid sets exist, see Fig. 1 for examples.

By definition, a *chordal graph* has no induced cycle of length four or more.

 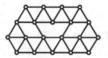

Fig. 1. Graphs without stable cutset

Lemma 3. *Let* $H = (R, F)$ *be a 2-connected chordal subgraph of* $G = (V, E)$. *Then* R *is rigid in* G.

Proof. The base step of the inductive proof is for complete H. In the inductive step we consider a minimal separator of H and use that it is a clique in G.   □

4   Claw-Free Graphs of Maximum Degree Four

We are going to improve the second part of Theorem 2. We will show that STABLE CUTSET is polynomial solvable for claw-free graphs with maximum degree four by reducing the problem to line graphs.

Recall that the *line graph* $L(G)$ of a graph G has the edges of G as its vertices, and two distinct edges of G are adjacent in $L(G)$ if they are incident in G. Line graphs have been characterised in terms for forbidden induced subgraphs as follows: A graph is a line graph if and only if it does not contain any of the nine graphs listed in Fig. 2 as an induced subgraph (cf. [12]).

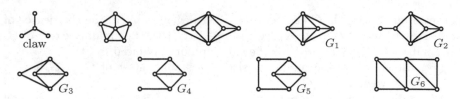

Fig. 2. Forbidden induced subgraphs for line graphs

Lemma 4. *Let G be a claw-free graph without clique cutset and $\Delta(G) = 4$.*

  (i) *If G contains an induced G_1, then $G = G_1$ or $\mathrm{scs}(G) \leqslant 2$.*
 (ii) *If G contains an induced G_2, then $|V(G)| \leqslant 8$ or $\mathrm{scs}(G) \leqslant 3$.*
(iii) *If G contains an induced G_3, then $|V(G)| \leqslant 8$ or $\mathrm{scs}(G) \leqslant 3$.*
(iv) *If $\mathrm{scs}(G) < \infty$ and G contains an induced G_4 or G_5, then $\mathrm{scs}(G) \leqslant 3$.*
 (v) *If $\mathrm{scs}(G) < \infty$ and G contains an induced G_6 then $\mathrm{scs}(G) \leqslant 2$.*

Theorem 3. *Let G be a claw-free graph with $\Delta(G) = 4$ and without clique cutset. Assume that G is not a line graph and has at least 9 vertices. Then $\mathrm{scs}(G) < \infty$ if and only if $\mathrm{scs}(G) \leqslant 3$.*

Proof. As G is not a line graph, G must contain one of the nine forbidden induced subgraphs listed in Fig. 2. As G is claw-free and has maximum degree four, G therefore must contain one of the graphs $G_1, \ldots, G_6$ in Fig. 2 as an induced subgraph. Now the Theorem follows from Lemma 4. • □

Theorem 4. STABLE CUTSET *can be solved in polynomial time for claw-free graphs with maximum degree at most four.*

    Thus the computational complexity of STABLE CUTSET for claw-free graphs is completely characterised with respect to maximum degree constraints.

5   (Claw, K_4)-Free Graphs

This section shows that STABLE CUTSET can be solved efficiently for (claw, K_4)-free graphs by reducing the problem to claw-free graphs with maximum degree at most four. We observe first:

Lemma 5. *The maximum degree in a (claw, K_4)-free graph is at most five.*

Proof. Let v be a vertex of degree at least six in any graph G. By a Ramsey-argument, $G[N(v)]$ contains either a triangle or the complement thereof. That is, G contains a K_4 including v, or there is a claw with central vertex v. □

    Let G be a (claw, K_4)-free graph on at least 11 vertices that contains neither clique cutsets nor vertices with stable neighbourhood. We will show that, for all vertices v of G with $\deg(v) = 5$, G has a stable cutset if and only if $G - v$ has a stable cutset. By Theorem 4, STABLE CUTSET is then solvable in polynomial time for (claw, K_4)-free graphs.

Let v be a vertex of degree five in G. By Lemma 1 it remains to show that if $G - v$ has a stable cutset then G has a stable cutset.

Assume to the contrary that G has no stable cutset, and consider an inclusion-minimal stable cutset S in $G - v$. By the minimality of S, every vertex in S has at least one neighbour in each connected component of $(G - v) - S$. Hence $(G - v) - S$ has exactly two connected components, otherwise there would be a claw in G. Moreover, $1 \leqslant |N(v) \cap S| \leqslant 2$, otherwise $S \cup \{v\}$ would be a stable cutset in G (if $N(v) \cap S = \varnothing$) or there would be a claw in G (if $|N(v) \cap S| \geqslant 3$).

Let A and B induce connected components of $(G - v) - S$. Then for all $u \in S \cup \{v\}, N(u) \cap A$ and $N(u) \cap B$ are cliques, each containing one or two vertices.

If $|N(v) \cap A| = 2 = |N(v) \cap B|$ then, as G is K_4-free, no vertex in $N(v) \cap S$ is adjacent to a vertex in $N(v) \cap A$ or $N(v) \cap B$. But then G admits a claw, a contradiction. Thus, $|N(v) \cap A| = 1$ or $|N(v) \cap B| = 1$, hence $|N(v) \cap S| = 2$. Let, without loss of generality, $N(v) \cap A = \{a_1, a_2\}$, $N(v) \cap B = \{b\}$, and $N(v) \cap S = \{s_1, s_2\}$.

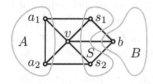

Fig. 3. Minimal stable cutset S in $G - v$ and the neighbourhood of v in $G - v$

Recall that a_1 and a_2 are adjacent. As G is K_4-free, we may assume $s_1 a_2 \notin E(G)$. Then s_2 and a_2 are adjacent (otherwise, v, s_1, s_2, and a_2 would form a claw) and hence s_2 and a_1 are nonadjacent, implying $s_1 a_1 \in E(G)$ (otherwise, v, s_1, s_2, and a_1 would form a claw). Finally, s_1 and s_2 both must be adjacent to b (otherwise there would be a claw), see also Fig. 3.

We complete the proof by case analysis according to the number of neighbours of s_i in A and B.

Theorem 5. STABLE CUTSET *is polynomial on (claw, K_4)-free graphs.*

6   Claw-Free Planar Graphs

In [4], it was shown that every graph with n vertices and $2n - 4$ edges contains a stable cutset (and, by the proof given there, such one can be found in polynomial time). Consequently one might ask the computational complexity of STABLE CUTSET in graphs with few edges. A natural candidate in this direction is the class of planar graphs. In this section we show that STABLE CUTSET can be solved efficiently for claw-free planar graphs.

It is well-known that planar graphs do not contain a K_5-minor.

Lemma 6. *Let G be a graph without clique cutset. If G contains no K_5-minor, then $G = K_4$ or G is K_4-free.*

Proof. We show that G cannot properly contain a K_4. Assume the contrary and consider four pairwise adjacent vertices a, b, c, and d in G. Then $H :=$

$G - \{a, b, c, d\}$ is non-empty and connected (otherwise, $\{a, b, c, d\}$ would be a clique cutset in G). Moreover, for each vertex $v \in \{a, b, c, d\}$, $N(v) \cap H \neq \varnothing$, otherwise $\{a, b, c, d\} \setminus \{v\}$ would be a clique cutset in G separating v and H. Thus, $\{a\}, \{b\}, \{c\}, \{d\}$, and H form a K_5-minor, a contradiction.   □

Theorem 6. STABLE CUTSET *is polynomial on claw-free planar graphs.*

Proof. Theorem 6 directly follows from Lemma 6 and Theorem 5 since we may assume that our graphs do not contain any clique cutset.   □

Corollary 1. STABLE CUTSET *becomes polynomial on planar line graphs.*

7   Planar Graphs of Degree at Most Five

In this section we prove that STABLE CUTSET remains **NP**-complete when restricted to partial subgraphs of the triangular grid without vertices of degree six. Since these graphs are K_4-free, this substantially improves the **NP**-completeness result in [2]. We use a reduction from a restricted version of planar 3SAT [14].

Let $\varphi = \bigwedge_{j=1}^{m} c_j$ be the conjunction of clauses. Each clause is the disjunction of literals. The literals are boolean variables or their negations. By X and C we denote the set of variables and clauses. For $x \in X$ and $c \in C$, $x \in c$ means that x or its negation $\overline{x}$ is a literal in c. We may assume the following restrictions:

- each variable appears (as x or $\overline{x}$) in at least three and at most four clauses,
- each clause consists of exactly three literals, and
- the graph $G = (V, E)$ is planar, where $V = X \cup C$ and $E = \{xc : x \in c\}$.

Note that these conditions ensure $|X| \leqslant |C| \leqslant \frac{4}{3}|X|$, i.e. $|V|$ is linear in $|X|$.

7.1   Construction

Let G' be a partial subgraph of a square grid such that each edge of G corresponds with a path in G', and the vertices having degree three or four in G' are in one-to-one correspondence with the vertices of G. Such an embedding G' of G into an $n \times n$-grid, $n = \mathcal{O}(|X|)$, can be constructed in quadratic time [19]. For each $e \in E$ let $\ell(e)$ be the number of horizontal edges on the path representing e in G'. We compute a ℓ-minimum spanning tree $T = (V, F)$ of G. Then each edge in $E \setminus F$ is represented by a path containing a horizontal edge because we cannot make a cycle of vertical edges only.

Starting from the embedding G', we construct a reduction graph as follows:

- each vertex in X is replaced by a truth assignment component,
- each vertex in C is replaced by a satisfaction test component, and
- each path corresponding with an edge in E is replaced by a channel.

Channels consist of three strips. The outer ones are banks and appear as double lines in the subsequent figures. The inner strip is the water, depicted in

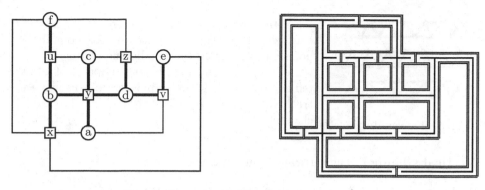

Fig. 4. Planar embedding and channel map

bold. Unlike edges in F, those in $E \setminus F$ contain a bridge in a horizontal part. The bridge interrupts the water and connects the two banks.

The water component is still connected because T is connected. Similarly, the bank component becomes connected via the bridges because all the water is surrounded by banks.

For example, let $X = \{u, v, x, y, z\}$ and $C = \{a, b, c, d, e, f\}$, where $\varphi = a \wedge b \wedge c \wedge d \wedge e \wedge f$ and

$$a = v \vee x \vee y \qquad b = u \vee \overline{x} \vee y \qquad c = u \vee \overline{y} \vee z$$
$$d = v \vee \overline{y} \vee \overline{z} \qquad e = \overline{v} \vee \overline{x} \vee \overline{z} \qquad f = \overline{u} \vee x \vee z$$

An grid embedding G' of the graph G corresponding with φ is shown on the left hand side of Fig. 4. A spanning tree T is indicated by bold edges. The right-hand side of this figure maps the channels and shows the bridges.

Now we are ready to describe building blocks in more detail. All the vertices are either bold (water) or double (bank), except four black vertices in the satisfaction test component. Edges are double (if both endpoints are double), bold (if both endpoints are bold), dotted (a double and a bold endpoint) for the reed between bank and water, and black (if one endpoint is black). A *monochrome component* is a maximal connected set of vertices of the same style (double or bold). All building blocks have the following properties:

- they are partial subgraphs of the triangular grid,
- they do not contain vertices of degree six (or more), and
- all monochrome components are rigid.

In the entire reduction graph, all double vertices (bank) will form one monochrome component, and all bold vertices (water) will form another one. If this graph has a stable cutset at all, then it separates bank from water. That is, each stable cutset will contain exactly one endpoint from each dotted edge.

Horizontal Channel. The horizontal channel is depicted in Fig. 5.

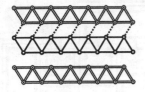

Fig. 5. Horizontal channel

Note that exactly two different stable cutsets exist which separate the upper monochrome component (bank) from the middle one (water). These cutsets are disjoint. That is, one endpoint of a dotted edge fixes the entire stable cutset. This way the truth values are propagated through the horizontal channel.

Vertical Channel. The vertical channel is depicted in Fig. 6.

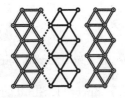

Fig. 6. Vertical channel

As in the horizontal channel, exactly two different stable cutsets exist which separate the left monochrome component (bank) from the middle one (water). Again, these cutsets are disjoint, and one endpoint of a dotted edge fixes the entire stable cutset. The truth values are propagated through the vertical channel in a similar way.

Bends. Two mini-bends are depicted in Fig. 7. At the hart of each bend in the channel we have one of them, or a reflection thereof.

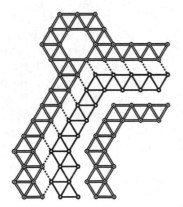

Fig. 7. Mini-bends

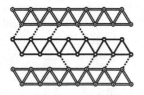

Fig. 8. Across

While the vertical part of a mini-bend always fits to a vertical channel, this is not the case for the horizontal part. The gadgets depicted in Fig. 8 and 9 their reflections will rectify. Note that all these building blocks propagate the truth values as the straight channels do.

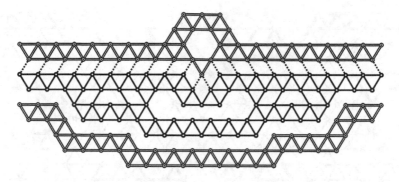

Fig. 9. Change of tilt

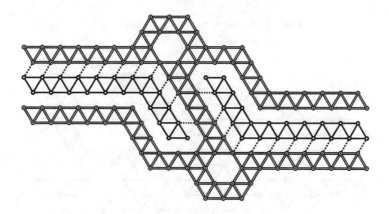

Fig. 10. Channel with bridge

Channel with Bridge. The bridge is depicted in Fig. 10.

The essential part in the centre resembles the idea of Fig. 8 with interchanged styles. The rest keeps the monochrome components rigid.

Truth Assignment Component. We give a mini-version with four horizontal outlets in Fig. 11. For a variable appearing in only three clauses cap one outlet.

The central part is known from Fig. 8, serving four outlets rather than two. The remaining parts are struts to keep the monochrome parts rigid.

Satisfaction Test Component. A mini-version of this component is given in Fig. 12. It has three inlets, on the top right, on the left, and bottom right. Let x, y and z be the literals whose truth values are fed in at these positions.

On the left we first split the y-channel into two, as in Figure 11. What follows is a strut to keep the water component rigid. The interesting part follows further to the right. The two black houses really test whether the clause is satisfied.

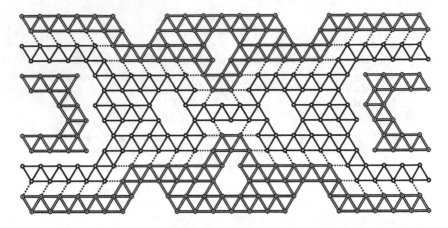

Fig. 11. Truth assignment component

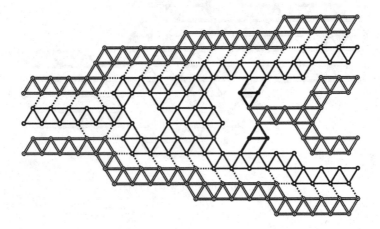

Fig. 12. Satisfaction test component

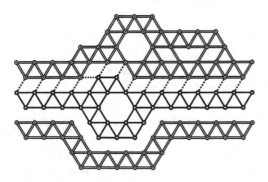

Fig. 13. Negator

The upper house tests $x \vee y$, the lower one $y \vee z$. Both houses together test $(x \vee y) \vee (y \vee z)$.

Each inlet of the satisfaction test component is directly connected to an outlet of a truth assignment component if the corresponding variable x is a positive literal in the clause, i.e. x appears unnegated as x. Otherwise ($\overline{x}$ appears as negative literal in the clause) we include the negator from Fig. 13 into the channel.

7.2 Equivalence

Let $a : X \rightarrow \{0, 1\}$ be a truth assignment of the variables in φ such that $a(\varphi) = 1$. We describe a stable cutset in the reduction graph.

The truth assignment component with caps at all four outlets allows exactly two stable cutsets, which are disjoint. These correspond with the truth values 0 (false) and 1 (true). For each variable $x \in X$ we choose the stable set in the truth assignment component that is given by $\varphi(x)$. These stable sets are extended along the channels into the satisfaction test components.

Because $a(\varphi) = 1$, for each clause there is at least one true literal. If literal x is true (upper right inlet), we choose two nonadjacent vertices in the four-cycle of the upper black house, and the bank vertex in the lower house. Whatever the truth value of the literals y (left inlet) and z (lower right inlet) is, this set of vertices extends to a stable cutset in the satisfaction test component. If z is true we swap the roles of upper and lower house. Finally, if y is true we can choose nonadjacent vertices in the four-cycles of both houses because the stable cutset enforced by the left inlet contains vertices both in the lower and upper branch of the component. Since this works in every satisfaction test component, we constructed a stable cutset of the reduction graph.

Now assume a stable cutset S of the reduction graph R is given. Then there is a bank component of $R - S$ containing all double vertices not in S, and a water component of $R - S$ containing all remaining bold vertices. We claim that S, restricted to the truth assignment components, defines a satisfying truth assignment $a : X \rightarrow \{0, 1\}$ for φ. Because the channels propagate the truth values between the truth assignment components and satisfaction test component, it remains to be shown that for each clause there is a true literal.

Each satisfaction test component contains two adjacent bank vertices incident with black edges. Clearly at most one of them belongs to S. This vertex separates its black house from the bank component. The other black house belongs to the bank component, and is separated from the water component by two nonadjacent vertices in its four-cycle. One of these vertices is bold. It marks a true literal in clause corresponding with this satisfaction test component.

8   Conclusion

While it has been shown that deciding whether or not a claw-free graph with maximum degree five [15], or a graph without 4-clique [2] contains a stable cutset is an **NP**-complete problem, we have proved in this paper that it can be decided in polynomial time whether or not

- a claw-free graph with maximum degree at most four,
- a claw-free graph without 4-clique, or
- a claw-free planar graph

contains a stable cutset.

In contrast, it is **NP**-complete to decide whether or not a planar graph with maximum degree five contains a stable cutset. The computational complexity of the stable cutset problem still remains open for graphs with maximum degree four, and even for planar graphs with maximum degree at most four.

References

1. P. Bonsma, *The complexity of the matching-cut problem for planar graphs and other graph classes*, Proc. WG 2003, LNCS 2880 (2003) 93–105.
2. A. Brandstädt, F. Dragan, V.B. Le, T. Szymczak, *On stable cutsets in graphs*, Discr. Appl. Math. 105 (2000) 39–50.
3. G. Chen, R.J. Faudree, M.S. Jacobson, *Fragile graphs with small independent cuts*, J. Graph Theory 41 (2002) 327–341.
4. G. Chen, X. Yu, *A note on fragile graphs*, Discrete Math. 249 (2002) 41–43.
5. V. Chvátal, *Recognizing decomposable Graphs*, J. Graph Theory 8, (1984) 51–53.
6. D.G. Corneil, J. Fonlupt, *Stable set bonding in perfect graphs and parity graphs*, J. Combin. Theory (B) 59 (1993) 1–14.
7. G. di Battista, M. Patrignani, F. Vargiu, *A Split&Push approach to 3D orthogonal drawing*, J. Graph Algorithms Appl. 1 (2000) 105–133.
8. A.M. Farley, A. Proskurowski, *Networks immune to isolated line failures*, Networks 12 (1982) 393–403.
9. A.M. Farley, A. Proskurowski, *Extremal graphs with no disconnecting matching*, Congressus Nummerantium 41 (1984) 153–165.
10. T. Feder, P. Hell, S. Klein, R. Motwani, *List partitions*, SIAM J. Discrete Math. 16 (2003) 449–478.
11. R.L. Graham, *On primitive graphs and optimal vertex assigments*, Ann. N.Y. Acad. Sci. 175 (1970) 170–186.
12. R.L. Hemminger, L.W. Beineke, *Line graphs and line digraphs*, In: Selected Topics in Graph Theory I, L.W. Beineke, R.T. Wilson, eds., Academic Press, London, (1978) 271–305.
13. S. Klein, C.M.H. de Figueiredo, *The NP-completeness of multi-partite cutset testing*, Congressus Numerantium 119 (1996) 217–222.
14. D. Lichtenstein, *Planar formulae and their uses*, SIAM Journal on Computing 11 (1982) 320–343.
15. V.B. Le, B. Randerath, *On stable cutsets in line graphs*, Theor. Comput. Sci. 301 (2003) 463–475.
16. A.M. Moshi, *Matching cutsets in graphs*, J. Graph Theory 13, (1989) 527–536.
17. M. Patrignani, M. Pizzonia, *The complexity of the matching-cut problem*, Proc. WG 2001, LNCS 2204 (2001) 284–295.
18. R.E. Tarjan, *Decomposition by clique separators*, Discr. Math. 55 (1985) 221–232.
19. I. Tollis, G. di Battista, P. Eades, R. Tamassia, *Graph drawing. Algorithms for the visualization of graphs*, Prentice Hall, Upper Saddle River, NJ, 1999.
20. A. Tucker, *Coloring graphs with stable cutsets*, J. Combin. Theory (B) 34 (1983) 258–267.
21. S.H. Whitesides, *An algorithm for finding clique cut-sets*, Inf. Process. Lett. 12 (1981) 31–32.
22. S.H. Whitesides, *An method for solving certain graph recognition and optimization problems, with applications to perfect graphs*, Ann. Discr. Math. 21 (1984) 281–297.

Induced Subgraphs of Bounded Degree
and Bounded Treewidth*

Prosenjit Bose[1], Vida Dujmović[1], and David R. Wood[2]

[1] School of Computer Science,
Carleton University, Ottawa, Canada
{jit, vida}@scs.carleton.ca
[2] Departament de Matemàtica Aplicada II,
Universitat Politècnica de Catalunya, Barcelona, Spain
david.wood@upc.edu

Abstract. We prove that for all $0 \leq t \leq k$ and $d \geq 2k$, every graph G with treewidth at most k has a 'large' induced subgraph H, where H has treewidth at most t and every vertex in H has degree at most d in G. The order of H depends on t, k, d, and the order of G. With $t = k$, we obtain large sets of bounded degree vertices. With $t = 0$, we obtain large independent sets of bounded degree. In both these cases, our bounds on the order of H are tight. For bounded degree independent sets in trees, we characterise the extremal graphs. Finally, we prove that an interval graph with maximum clique size k has a maximum independent set in which every vertex has degree at most $2k$.

1 Introduction

The 'treewidth' of a graph has arisen as an important parameter in the Robertson/Seymour theory of graph minors and in algorithmic complexity. See Bodlaender [2] and Reed [7] for surveys on treewidth. The main result of this paper, proved in Section 5, states that every graph G has a large induced subgraph of bounded treewidth in which every vertex has bounded degree in G. The order of the subgraph depends on the treewidth of G, the desired treewidth of the subgraph, and the desired degree bound. Moreover, we prove that the bound is best possible in a number of cases.

Before that, in Sections 2 and 3 we consider two relaxations of the main result, firstly without the treewidth constraint, and then without the degree constraint. That is, we determine the minimum number of vertices of bounded degree in a graph of given treewidth (Section 2), and we determine the minimum number of vertices in an induced subgraph of bounded treewidth, taken over all graphs of given treewidth (Section 3). This latter result is the first ingredient in the proof of the main result. The second ingredient is in Section 4, where we

---

* Research of P. Bose and V. Dujmović is supported by NSERC. Research of D. Wood is supported by the Government of Spain grant MEC SB2003-0270, and by the projects MCYT-FEDER BFM2003-00368 and Gen. Cat 2001SGR00224.

D. Kratsch (Ed.): WG 2005, LNCS 3787, pp. 175–186, 2005.

prove that the subgraph of a k-tree induced by the vertices of bounded degree has surprisingly small treewidth.

A graph with treewidth 0 has no edges. Thus our results pertain to independent sets for which every vertex has bounded degree in G. Here our bounds are tight, and in the case of trees, we characterise the extremal trees. Furthermore, by exploiting some structural properties of interval graphs that are of independent interest, we prove that every interval graph with no $(k + 2)$-clique has a maximum independent set in which every vertex has degree at most $2k$. These results are presented in Section 6.

1.1   Preliminaries

Let G be a graph. All graphs considered are finite, undirected, and simple. The vertex-set and edge-set of G are denoted by $V(G)$ and $E(G)$, respectively. The number of vertices of G is denoted by $n = |V(G)|$. The subgraph *induced* by a set of vertices $S \subseteq V(G)$ has vertex set S and edge set $\{vw \in E(G) : v, w \in S\}$, and is denoted by $G[S]$.

A k-*clique* ($k \geq 0$) is a set of k pairwise adjacent vertices. Let $\omega(G)$ denote the maximum number k such that G has a k-clique. A *chord* of a cycle C is an edge not in C whose endpoints are both in C. G is *chordal* if every cycle on at least four vertices has a chord. The *treewidth* of G is the minimum number k such that G is a subgraph of a chordal graph G' with $\omega(G') \leq k + 1$.

A vertex is *simplicial* if its neighbourhood is a clique. For each vertex $v \in V(G)$, let $G \setminus v$ denote the subgraph $G[V(G) \setminus \{v\}]$. The family of graphs called k-*trees* ($k \geq 0$) are defined recursively as follows. A graph G is a k-*tree* if either (a) G is a $(k+1)$-clique, or (b) G has a simplicial vertex v whose neighbourhood is a k-clique, and $G \setminus v$ is a k-tree.

By definition, the graph obtained from a k-tree G by adding a new vertex v adjacent to each vertex of a k-clique C is also a k-tree, in which case we say v is *added onto* C. For every k-tree G on n vertices, $\omega(G) = k + 1$; G has minimum degree k; and G has $kn - \frac{1}{2}k(k + 1)$ edges, and thus G has average degree $2k - k(k + 1)/n$. It is well known that the treewidth of a graph G equals the minimum number k such that G is a spanning subgraph of a k-tree.

We will express our results using the following notation. Let G be a graph. Let $V_d(G) = \{v \in V(G) : \deg_G(v) \leq d\}$ denote the set of vertices of G with degree at most d. Let $G_d = G[V_d(G)]$. A subset of $V_d(G)$ is called a *degree-d* set. For an integer $t \geq 0$, a t-*set* of G is a set S of vertices of G such that the induced subgraph $G[S]$ has treewidth at most t. Let $\alpha^t(G)$ be the maximum number of vertices in a t-set of G. Let $\alpha_d^t(G)$ be the maximum number of vertices in a degree-d t-set of G. Observe that $\alpha_d^t(G) = \alpha^t(G_d)$.

Let $\mathcal{G}$ be a family of graphs. Let $\alpha^t(\mathcal{G})$ be the minimum of $\alpha^t(G)$, and let $\alpha_d^t(\mathcal{G})$ be the minimum of $\alpha_d^t(G)$, taken over all $G \in \mathcal{G}$. Let $\mathcal{G}_{n,k}$ be the family of n-vertex graphs with treewidth k. Note that every graph in $\mathcal{G}_{n,k}$ has at least $k+1$ vertices. These definitions imply the following. Every graph $G \in \mathcal{G}$ has $\alpha_d^t(G) \geq \alpha_d^t(\mathcal{G})$ and $\alpha^t(G) \geq \alpha^t(\mathcal{G})$. Furthermore, there is at least one graph G for which $\alpha_d^t(G) = \alpha_d^t(\mathcal{G})$, and there is at least one graph G for which $\alpha^t(G) = \alpha^t(\mathcal{G})$. Thus the lower bounds we derive in this paper are universal and the upper bounds are existential.

As described above, our main result is a lower bound on $\alpha_d^t(\mathcal{G}_{n,k})$ that is tight in many cases. Here, lower and upper bounds are 'tight' if they are equal when ignoring the terms independent of n. Many of our upper bound constructions are based on the k-th power of an n-vertex path P_n^k. This graph has vertex set $\{v_1, v_2, \ldots, v_n\}$ and edge set $\{v_i v_j : |i - j| \leq k\}$. Obviously P_n^k is a k-tree.

For $t = k$, a degree-d t-set in a graph G with treewidth k is simply a set of vertices with degree at most d. Thus in this case, $\alpha_d^k(G) = |V_d(G)|$. At the other extreme, a graph has treewidth 0 if and only if it has no edges. A set of vertices $I \subseteq V(G)$ is *independent* if $G[I]$ has no edges. Thus a 0-set of G is simply an independent set of vertices of G. As is standard, we abbreviate $\alpha^0(G)$ by $\alpha(G)$, $\alpha_d^0(G)$ by $\alpha_d(G)$, etc. An independent set I of G is *maximum* if $|I| \geq |J|$ for every independent set J of G. Thus $\alpha(G)$ is the cardinality of a maximum independent set of G.

2   Large Subgraphs of Bounded Degree

In this section we prove tight lower bounds on the number of vertices of bounded degree in graphs of treewidth k. We will use the following result of Bose *et al.* [3].

Lemma 1 ([3]). *Let G be a graph on n vertices, with minimum degree δ, and with average degree α. Then for every integer $d \geq \delta$,*

$$|V_d(G)| \geq \left(\frac{d + 1 - \alpha}{d + 1 - \delta}\right) n \ .$$

Theorem 1. *For all integers $k \geq 0$ and $d \geq 2k - 1$,*

$$\lim_{n \to \infty} \frac{\alpha_d^k(\mathcal{G}_{n,k})}{n} = \frac{d - 2k + 1}{d - k + 1} \ .$$

Proof. First we prove a lower bound on $\alpha_d^k(\mathcal{G}_{n,k})$. Let G be a graph in $\mathcal{G}_{n,k}$ with $\alpha_d^k(G) = \alpha_d^k(\mathcal{G}_{n,k})$. If a vertex v of G has degree at most d in a spanning supergraph of G, then v has degree at most d in G. Thus we can assume that G is a k-tree. Hence G has minimum degree k and average degree $2k - k(k+1)/n$. By Lemma 1,

$$\alpha_d^k(\mathcal{G}_{n,k}) = |V_d(G)| \geq \left(\frac{d + 1 - 2k + k(k+1)/n}{d + 1 - k}\right) n$$

$$= \left(\frac{d - 2k + 1}{d - k + 1}\right) n + \frac{k(k+1)}{d - k + 1} \ . \tag{1}$$

Now we prove an upper bound on $\alpha_d^k(\mathcal{G}_{n,k})$ for all $n \equiv 2k \pmod{d-k+1}$, and for all $k \geq 0$ and $d \geq 2k - 1$. Let s be the integer such that $n - 2k = s(d - k + 1)$. Then $s \geq 0$. We now construct a graph $G \in \mathcal{G}_{n,k}$. Initially let $G = P_{(s+2)k}^k$ be the k-th power of the path $(v_1, v_2, \ldots, v_{(s+2)k})$. Let $r = d - 2k + 1$. Then $r \geq 0$. Add r vertices onto the clique $(v_{ik+1}, v_{ik+2}, \ldots, v_{ik+k})$ for each $1 \leq i \leq s$. Thus G is a k-tree, as illustrated in Figure 1. The number of vertices in G is

$$(s + 2)k + sr = (s + 2)k + s(d - 2k + 1) = s(d - k + 1) + 2k = n . \qquad (2)$$

Each vertex v_i, $k + 1 \leq i \leq (s + 1)k$, has degree $2k + r = d + 1$. Hence such a vertex is not in a degree-d set. The remaining vertices all have degree at most d. Thus

$$\alpha_d^k(\mathcal{G}_{n,k}) \leq \alpha_d^k(G) = |V_d(G)| = rs + 2k = \left(\frac{d - 2k + 1}{d - k + 1} \right) n + \frac{2k^2}{d - k + 1} . \qquad (3)$$

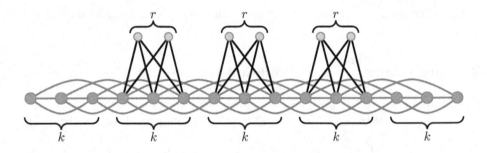

Fig. 1. The graph G with $k = 3$, $d = 7$, and $s = 3$ (and thus $r = 2$)

Observe that the difference between the lower and upper bounds in (1) and (3) is only

$$\frac{2k^2 - k(k + 1)}{d + 1 - k} = \frac{k(k - 1)}{d + 1 - k} \leq k - 1 .$$

It is easily seen that for all $\epsilon > 0$, there is an n_0 such that for all $n \geq n_0$,

$$0 \leq \frac{\alpha_d^k(\mathcal{G}_{n,k})}{n} - \frac{d - 2k + 1}{d - k + 1} \leq \epsilon .$$

Therefore the sequence $\{\alpha_d^k(\mathcal{G}_{n,k})/n : n \geq 2k\}$ converges to $\frac{d-2k+1}{d-k+1}$. □

3   Large Subgraphs of Bounded Treewidth

We now prove a tight bound on the maximum order of an induced subgraph of bounded treewidth in a graph of treewidth k.

Theorem 2. *For all integers n and $0 \leq t \leq k$,*

$$\alpha^t(\mathcal{G}_{n,k}) = \left(\frac{t + 1}{k + 1} \right) n .$$

Proof. First we prove the lower bound. Let G be a graph in $\mathcal{G}_{n,k}$. First suppose that G is a k-tree. By definition, $V(G)$ can be ordered $(v_1, v_2, \ldots, v_n)$ so that for each vertex v_i, the *predecessors* $\{v_j : j < i, v_i v_j \in E(G)\}$ of v_i are a clique of $\min\{k, i-1\}$ vertices. Now colour G greedily in this order. That is, for $i = 1, 2, \ldots, n$, assign to v_i the minimum positive integer (a colour) not already assigned to a neighbour of v_i. Clearly $k+1$ colours suffice. Let S be the union of the $t+1$ largest colour classes (monochromatic set of vertices). Thus $|S| \geq (t+1)n/(k+1)$. For each vertex v_i in S, the predecessors of v_i that are in S and v_i itself form a clique, and thus have pairwise distinct colours. Thus v_i has at most t predecessors in S, and they form a clique in $G[S]$. Hence $G[S]$ has treewidth at most t, and S is the desired t-set. Now suppose that G is not a k-tree. Then G is a spanning subgraph of a k-tree G'. Thus G' has a t-set S with at least $(t+1)n/(k+1)$ vertices. Now $G[S]$ is a subgraph of $G'[S]$. Thus $G[S]$ also has treewidth at most t.

For the upper bound, we now show that every t-set of P_n^k has at most $(t+1)n/(k+1)$ vertices. First suppose that $t = 0$. A 0-set is an independent set. Clearly every independent set of P_n^k has at most $n/(k+1)$ vertices. Now consider the case of general t. Let S be a t-set of P_n^k. By the above bound, $P_n^k[S]$ has an independent set I of at least $|S|/(t+1)$ vertices. Now I is also an independent set of P_n^k. Thus $|I| \leq n/(k+1)$. Hence $|S|/(t+1) \leq n/(k+1)$, and $|S| \leq (t+1)n/(k+1)$. □

4   Structure of Bounded Degree Subgraphs

In this section we study the structure of the subgraph of a k-tree induced by the vertices of bounded degree. We first prove that in a k-tree with sufficiently many vertices, not all the vertices of a clique have low degree. A clique $C = (v_1, v_2, \ldots, v_k)$ of a graph G is said to be *ordered by degree* if $\deg_G(v_i) \leq \deg_G(v_{i+1})$ for all $1 \leq i \leq k-1$.

Theorem 3. *Let G be a k-tree on $n \geq 2k+1$ vertices. Let $(u_1, u_2, \ldots, u_q)$ be a clique of G ordered by degree. Then $\deg_G(u_i) \geq k+i-1$ for all $1 \leq i \leq q$.*

Note that Theorem 3 is not true if $n \leq 2k$, as the statement would imply that a $(k+1)$-clique has a vertex of degree n. Thus the difficulty in an inductive prove of Theorem 3 is the base case. Theorem 3 follows from the following stronger result with $n \geq 2k+1 \geq k+q$.

Lemma 2. *Let G be a k-tree on n vertices. Let $C = (u_1, u_2, \ldots, u_q)$ be a clique of G ordered by degree. If $n \geq k+q$ then*

$$\deg_G(u_i) \geq k+i-1, \ 1 \leq i \leq q \ ; \tag{4}$$

otherwise $n \leq k+q-1$, and

$$\deg_G(u_i) \geq \begin{cases} k+i-1 & \text{if } 1 \leq i \leq n-k-1 \ , \\ n-1 & \text{if } n-k \leq i \leq q \ . \end{cases} \tag{5}$$

Proof. We proceed by induction on n. In the base case, G is a $(k + 1)$-clique, and every vertex has degree k. The claim follows trivially. Assume the result holds for k-trees on less than n vertices. Let C be a q-clique of a k-tree G on $n \geq k + 2$ vertices. Since every k-tree on at least $k + 2$ vertices has two non-adjacent simplicial vertices [4], at least one simplicial vertex v is not in C. Since $n \geq k + 2$ and v is simplicial, the graph $G_1 = G \setminus v$ is a k-tree on $n - 1$ vertices. Now C is a q-clique of G_1. Let $C = (u_1, u_2, \ldots, u_q)$ be ordered by degree in G_1. By induction, if $n \geq k + q + 1$ then

$$\deg_{G_1}(u_i) \geq k + i - 1, \ 1 \leq i \leq q; \tag{6}$$

otherwise $n \leq k + q$, and

$$\deg_{G_1}(u_i) \geq \begin{cases} k + i - 1 & \text{if } 1 \leq i \leq n - k - 2 , \\ n - 2 & \text{if } n - k - 1 \leq i \leq q . \end{cases} \tag{7}$$

First suppose that $n \geq k+q+1$. Then by (6), $\deg_G(u_i) \geq \deg_{G_1}(u_i) \geq k+i-1$, and (4) is satisfied. Otherwise $n \leq k + q$. Let $B = \{u_{n-k-1}, u_{n-k}, \ldots, u_q\}$. Then $|B| \geq 2$, and by (7), every vertex in B has degree $n-2$ in G_1. That is, each vertex in B is adjacent to every other vertex in G_1. Let X be the set of neighbours of v. Since v is simplicial, X is a k-clique. At most one vertex of B is not in X, as otherwise $X \cup B$ would be a $(k + 2)$-clique of G_1. Without loss of generality, this exceptional vertex in B, if it exists, is u_{n-k-1}. The other vertices in B are adjacent to one more vertex, namely v, in G than in G_1. Thus $\deg_G(u_i) \geq k+i-1$ for all $1 \leq i \leq n - k - 1$, and $\deg_G(u_i) = n - 1$ for all $n - k \leq i \leq q$. Hence (5) is satisfied. □

We can now prove the main result of this section.

Theorem 4. *For all integers $1 \leq k \leq \ell \leq 2k$, and for every k-tree G on $n \geq \ell+2$ vertices, the subgraph G_ℓ of G induced by the vertices of degree at most ℓ, has treewidth at most $\ell - k$.*

Proof. Let $C = (u_1, u_2, \ldots, u_q)$ be a clique of G ordered by degree. Suppose, for the sake of contradiction, that there are at least $\ell - k + 2$ vertices of C with degree at most ℓ. Let $j = \ell - k + 2$. Since C is ordered by degree, $\deg(u_j) \leq \ell$. Since $n \geq \ell + 2$, we have $j \leq n - k$. By Lemma 2, $\deg(u_j) \geq k + j - 1$ (unless $j = n-k$, in which case $\deg(u_j) = n-1 \geq \ell+1$, which is a contradiction). Hence $k+j-1 \leq \ell$. That is, $k + (\ell - k + 2) - 1 \leq \ell$, a contradiction. Thus C contributes at most $\ell - k + 1$ vertices to G_ℓ, and $\omega(G_\ell) \leq \ell - k + 1$. Now, G_ℓ is an induced subgraph of G, which is chordal. Thus G_ℓ is chordal. Since $\omega(G_\ell) \leq \ell - k + 1$, G_ℓ has treewidth at most $\ell - k$. □

Note the following regarding Theorem 4:

- There are graphs of treewidth $k \geq 2$ for which the theorem is not true. For example, for any $p \geq k + 1$, consider the graph G consisting of a $(k + 1)$-clique C and a p-vertex path with one endpoint v in C. Then G has at least

$2k + 1$ vertices, has treewidth k, and every vertex of G has degree at most k, except for v which has $\deg(v) = k + 1$. For $\ell = k$, G_ℓ is comprised of two components, one a k-clique and the other a path, in which case G_ℓ has treewidth $k - 1 > \ell - k = 0$. For $k + 1 \leq \ell \leq 2k - 1$, $G_\ell = G$ has treewidth $k > \ell - k$.

– The theorem is not true if $k \leq n \leq \ell + 1$. For example, for any $1 \leq k \leq \ell \leq 2k - 1$, the k-tree obtained by adding $\ell + 1 - k$ vertices onto an initial k-clique has $\ell + 1$ vertices, maximum degree ℓ, and treewidth $k > \ell - k$.

– The case of $\ell = k$ is the well-known fact that in a k-tree with at least $k + 2$ vertices, distinct simplicial vertices are not adjacent. Put another way, the set of simplicial vertices of a k-tree with at least $k + 2$ vertices is a 0-set.

5 Large Subgraphs of Bounded Treewidth and Bounded Degree

The following theorem is the main result of the paper.

Theorem 5. *For all integers* $0 \leq t \leq k$, $d \geq 2k$, *and* $n \geq 2k + 1$,

$$\alpha_d^t(\mathcal{G}_{n,k}) \geq \left(\frac{d - 2k + 1}{d - \frac{3}{2}k + 1 + \frac{t(t+1)}{2(k+1)}} \right) \left(\frac{t + 1}{k + 1} \right) n + \frac{k(t + 1)}{d - \frac{3}{2}k + 2 + \frac{t(t+1)}{2(k+1)}}$$

Proof. Let G be a graph in $\mathcal{G}_{n,k}$ with $\alpha_d^t(G) = \alpha_d^t(\mathcal{G}_{n,k})$. A degree-$d$ t-set of a spanning supergraph of G is a degree-d t-set of G. Thus we can assume that G is a k-tree.

Consider ℓ with $k + t \leq \ell \leq 2k$. By Theorem 4, G_ℓ has treewidth at most $\ell - k$. Since $t \leq \ell - k$, by Theorem 2,

$$\alpha^t(G_\ell) \geq \left(\frac{t + 1}{\ell - k + 1} \right) |V_\ell(G)| \ .$$

Since $\ell \leq d$, $\alpha^t(G_\ell) \leq \alpha_d^t(G)$, which implies that

$$|V_\ell(G)| \leq \left(\frac{\ell - k + 1}{t + 1} \right) \alpha_d^t(G) \ . \tag{8}$$

Now, G has $kn - \frac{1}{2}k(k + 1)$ edges and minimum degree k. Let n_i be the number of vertices of G with degree exactly i. Thus,

$$\sum_{i \geq k} i \cdot n_i = 2|E(G)| = 2kn - k(k + 1) = -k(k + 1) + \sum_{i \geq k} 2k \cdot n_i \ .$$

Thus,

$$\sum_{i \geq 2k+1} (i - 2k)n_i = -k(k + 1) + \sum_{i=k}^{2k-1} (2k - i)n_i = -k(k + 1) + \sum_{i=k}^{2k-1} |V_i(G)| \ ,$$

and

$$\sum_{i \geq 2k+1} (i - 2k)n_i = -k(k+1) + \sum_{i=k}^{k+t-1} |V_i(G)| + \sum_{i=k+t}^{2k-1} |V_i(G)| .$$

By (8),

$$\sum_{i \geq 2k+1} (i - 2k)n_i \leq -k(k+1) + t \cdot |V_{k+t}(G)| + \sum_{i=k+t}^{2k-1} \frac{(i - k + 1) \cdot \alpha_d^t(G)}{t+1}$$

$$\leq -k(k+1) + t \cdot \alpha_d^t(G) + \frac{\alpha_d^t(G)}{t+1} \sum_{i=t+1}^{k} i$$

$$= -k(k+1) + \alpha_d^t(G) \left(t + \frac{1}{t+1} \left(\frac{k(k+1) - t(t+1)}{2} \right) \right)$$

$$= -k(k+1) + \alpha_d^t(G) \left(\frac{t(t+1) + k(k+1)}{2(t+1)} \right) .$$

Since $d \geq 2k$,

$$-k(k+1) + \alpha_d^t(G) \left(\frac{t(t+1) + k(k+1)}{2(t+1)} \right) \geq \sum_{i \geq d+1}(i-2k)n_i \geq (d-2k+1) \sum_{i \geq d+1} n_i .$$

Hence,

$$|V_d(G)| = n - \sum_{i \geq d+1} n_i \geq n + \frac{k(k+1)}{d - 2k + 1} - \alpha_d^t(G) \left(\frac{t(t+1) + k(k+1)}{2(t+1)(d - 2k + 1)} \right) .$$

By Theorem 2,

$$\alpha_d^t(G) = \alpha^t(G_d)$$

$$\geq \frac{t+1}{k+1} |V_d(G)|$$

$$\geq \frac{(t+1)n}{k+1} + \frac{k(t+1)}{d - 2k + 1} - \alpha_d^t(G) \left(\frac{t(t+1) + k(k+1)}{2(k+1)(d - 2k + 1)} \right) .$$

It follows that

$$\alpha_d^t(G) \geq \frac{(d - 2k + 1)(t + 1)n + k(k+1)(t+1)}{(d - \frac{3}{2}k + 1)(k+1) + \frac{1}{2}t(t+1)} .$$

The result follows.   □

A number of notes regarding Theorem 5 are in order:

- Theorem 5 with $t = k$ is equivalent to the lower bound in Theorem 1.
- For $d < 2k$, no result like Theorem 5 is possible, since $\alpha_d^t(P_n^k) = 2(t+1)$.
- The proof of Theorem 5 is similar to a strategy developed by Biedl and Wilkinson [1] for finding bounded degree independent sets in planar graphs.

We now prove an existential upper bound on the cardinality of a degree-d t-set.

Theorem 6. *For all integers $k \geq 1$ and $d \geq 2k - 1$ such that $2(d - 2k + 1) \equiv 0$ (mod $k(k+1)$), there are infinitely many values of n, such that for all $0 \leq t < k$,*

$$\alpha_d^t(\mathcal{G}_{n,k}) \leq \left(\frac{d - 2k + 1}{d - \frac{3}{2}k + 1}\right)\left(\frac{t + 1}{k + 1}\right) n + \frac{(k - 1)(t + 1)(d - 2k + 1) + k(t + 1)(k + 1)}{(d - \frac{3}{2}k + 1)(k + 1)} \ .$$

Proof. Our construction employs the following operation. Let G be a k-tree containing an ordered k-clique $C = (v_1, v_2, \ldots, v_k)$. A *block* at C consists of $k+1$ new vertices $\{x_1, x_2, \ldots, x_{k+1}\}$ where x_1 is added onto the k-clique $\{v_1, v_2, \ldots, v_k\}$; x_2 is added onto the k-clique $\{v_1, v_2, \ldots, v_{k-1}, x_1\}$; x_3 is added onto the k-clique $\{v_1, v_2, \ldots, v_{k-2}, x_1, x_2\}$; and so on, up to x_{k+1} which is added onto the k-clique $\{x_1, x_2, \ldots, x_k\}$. Clearly the graph obtained by adding a block to a k-clique of a k-tree is also a k-tree

Our graph is parameterised by the positive integer $n_0 \geq 2k + 3$. Initially let G be the k-th power of a path $(v_1, v_2, \ldots, v_{n_0})$. Note that any $k + 1$ consecutive vertices in the path form a clique. Let r be the non-negative integer such that $2(d - 2k + 1) = rk(k + 1)$. Add r blocks to G at $(v_i, v_{i+1}, \ldots, v_{i+k-1})$ for each $3 \leq i \leq n_0 - k - 1$, as illustrated in Figure 2.

G is a k-tree with $n = n_0 + r(k + 1)(n_0 - (k + 3))$ vertices. Let S be a maximum degree-d t-set of G. Consider a vertex v_i for $k + 2 \leq i \leq n_0 - k - 1$. Since $n_0 \geq 2k + 3$ there is such a vertex. The degree of v_i is

$$2k + r\sum_{i=1}^{k} i \ = \ 2k + \tfrac{1}{2}rk(k + 1) \ = \ d + 1 \ .$$

Thus $v_i \notin S$. Since each block $\{x_1, x_2, \ldots, x_{k+1}\}$ is a clique, and treewidth-t graphs have no $(t + 2)$-clique, at most $t + 1$ vertices from each block are in S. Similarly, since $\{v_1, v_2, \ldots, v_{k+1}\}$ and $\{v_{n_0-k}, v_{n_0-k+1}, \ldots, v_{n_0}\}$ are cliques, at most $t + 1$ vertices from each of these sets are in S. Thus

$$\alpha_d^t(\mathcal{G}_{n,k}) \ \leq \ \alpha_d^t(G) \ = \ |S| \ \leq \ (t + 1)\big(r(n_0 - (k + 3)) + 2\big) \ . \tag{9}$$

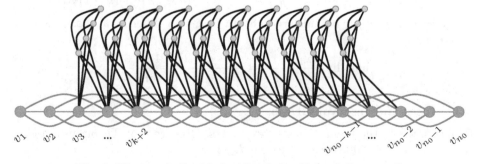

Fig. 2. The graph G with $k = 3$ and $d = 11$ (and thus $r = 1$)

Substituting the equality $n_0 = \frac{n+r(k+1)(k+3)}{1+r(k+1)}$ into (9),

$$\frac{\alpha_d^t(\mathcal{G}_{n,k})}{t+1} \leq \frac{r(n+k-1)+2}{1+r(k+1)} \ . \tag{10}$$

The claimed bound on $\alpha_d^t(\mathcal{G}_{n,k})$ follows by substituting the equality $r = \frac{2(d-2k+1)}{k(k+1)}$ into (10). Observe that n is a function of n_0 and n_0 is independent of d. Thus there are infinitely many values of n for each value of d. ☐

6   Bounded Degree Independent Sets

Intuitively, one would expect that a maximum independent set would not have vertices v of high degree, as this would prevent the many neighbours of v from being in the independent set. In this section, we explore the accuracy of this intuition in the case of k-trees. Recall that $\alpha_d(G)$ is the maximum cardinality of a degree-d independent set in a graph G.

Motivated by applications in computational geometry, the previously known results regarding bounded degree independent sets have been for planar graphs [6,5,8,1]. The best results were obtained by Biedl and Wilkinson [1], who proved tight bounds (up to an additive constant) on $\alpha_d(G)$ for planar G with $d \leq 15$. For $d \geq 16$ there is a gap in the bounds.

Theorem 2 with $t = 0$ proves that every n-vertex graph G with treewidth k has $\alpha(G) \geq n/(k+1)$, and that this bound is tight for P_n^k. Theorem 5 with $t = 0$ gives the following lower bound on the size of a degree-d independent set in a graph of treewidth k (for all $k \geq 1$ and $d \geq 2k$):

$$\alpha_d(\mathcal{G}_{n,k}) \geq \left(\frac{d-2k+1}{d-\frac{3}{2}k+1}\right)\left(\frac{n}{k+1}\right) + \frac{k}{d-\frac{3}{2}k+1} \ .$$

Note that such a bound is not possible for $d < 2k$ since $\alpha_d(P_n^k) = 2$ for $d < 2k$.

Theorem 6 proves the corresponding upper bound. In particular, for all $k \geq 1$, there are infinitely many values of d, and for each such d, there are infinitely many values of n for which

$$\alpha_d(\mathcal{G}_{n,k}) \leq \left(\frac{d-2k+1}{d-\frac{3}{2}k+1}\right)\left(\frac{n}{k+1}\right) + \frac{(k-1)(d-2k+1)+k(k+1)}{(d-\frac{3}{2}k+1)(k+1)} \ .$$

These lower and upper bounds are tight. In fact, they differ by at most one. We conclude that

$$\lim_{n\to\infty}\lim_{d\to\infty}\frac{\alpha_d(\mathcal{G}_{n,k})}{n} = \frac{d-2k+1}{(d-\frac{3}{2}k+1)(k+1)} \ .$$

6.1   Trees and Interval Graphs

$\mathcal{G}_{n,1}$ is precisely the family of n-vertex forests. Observe that Theorems 5 and 6 with $k = 1$ and $t = 0$ prove that for all $d \geq 1$,

$$\alpha_d(\mathcal{G}_{n,1}) = \frac{(d-1)n+2}{2d-1} \ .$$

A tree T for which $\alpha_d(T) = \frac{(d-1)n+2}{2d-1}$ is called α_d-*extremal*. We omit the proof of the following characterisation of the α_d-extremal trees. A tree is d-*regular* if every vertex has degree 1 or d, and there is at least one vertex of degree d.

Theorem 7. *Let d be a positive integer. A tree T on $n \geq 5$ vertices is α_d-extremal if and only if T is obtained from a $(d+1)$-regular tree by subdividing every leaf-edge once.*

A graph G is an *interval graph* if one can assign to each vertex $v \in V(G)$ a closed interval $[L_v, R_v] \subseteq \mathbb{R}$ such that $vw \in E(G)$ if and only if $[L_v, R_v] \cap [L_w, R_w] \neq \emptyset$. An interval graph G has tree-width equal to $\omega(G) + 1$. (In fact, it has path-width equal to $\omega(G) + 1$.) Thus the previous results of this paper apply to interval graphs. However, for bounded degree independent sets in interval graphs, we can say much more, as we show in this section. In an interval graph, it is well known that we can assume that the endpoints of the intervals are distinct. We say a vertex w is *dominated* by a vertex v if $L(v) < L(w) < R(w) < R(v)$.

Lemma 3. *Let G be an interval graph with $\omega(G) \leq k + 1$. Suppose G has a vertex v with $\deg(v) \geq 2k + 1$. Then there is a vertex w that is dominated by v and $\deg(w) \leq 2k - 1$.*

Proof. For each vertex $y \in V(G)$, let $A(y) = \{x \in V(G) : L(x) < L(y) < R(x)\}$ and $B(y) = \{x \in V(G) : L(x) < R(y) < R(x)\}$. Observe that x is dominated by y if and only if $xy \in E(G)$ but $x \notin A(y) \cup B(y)$. Also $|A(y)| \leq k$ as otherwise $A(y) \cup \{y\}$ would be a clique of at least $k+2$ vertices. Similarly $|B(y)| \leq k$. Thus $|A(y) \cup B(y)| \leq 2k$.

Now consider the given vertex v. Since $\deg(v) \geq 2k + 1$, v has a neighbour $u \notin A(v) \cup B(v)$. Thus u is dominated by v. Let w be a vertex with the shortest interval that is dominated by v. That is, if u and w are dominated by v, then $R(w) - L(w) \leq R(u) - L(u)$. Thus w does not dominate any vertex, and every neighbour of w is in $A(w) \cup B(w)$. Now $|A(w)| \leq k$, $|B(w)| \leq k$, and $v \in A(w) \cap B(w)$. Thus $\deg(w) \leq 2k - 1$. □

Note that Lemma 3 with $k = 1$ is the obvious statement that a vertex of degree at least three in a caterpillar is adjacent to a leaf.

Theorem 8. *Every interval graph G with $\omega(G) \leq k + 1$ has a degree-$2k$ maximum independent set. That is, $\alpha_{2k}(G) = \alpha(G)$.*

Proof. Let I be a maximum independent set of G. If I contains a vertex v with $\deg(v) \geq 2k + 1$, apply Lemma 3 to obtain a vertex w dominated by v such that $\deg(w) \leq 2k - 1$. Replace v by w in I. The obtained set is still independent, since every neighbour of w is also adjacent to v, and is thus not in I. Apply this step repeatedly until every vertex in I has degree at most $2k$. Thus $\alpha_{2k}(G) \geq |I| = \alpha(G)$. By definition, $\alpha_{2k}(G) \leq \alpha(G)$. □

The bound of $2k$ in Theorem 8 is best possible, since P_n^k is an interval graph with $\omega(G) \leq k + 1$ and only $2k$ vertices of degree at most $2k - 1$. Thus $\alpha(P_n^k) = \lceil n/(k+1) \rceil \gg \alpha_{2k-1}(P_n^k)$.

References

1. THERESE BIEDL AND DANA F. WILKINSON. Bounded-degree independent sets in planar graphs. *Theory Comput. Syst.*, in press. In PROSENJIT BOSE AND PAT MORIN, eds., *Proc. 13th International Conf. on Algorithms and Computation* (ISAAC '02), vol. 2518 of *Lecture Notes in Comput. Sci.*, pp. 416–427. Springer, 2002.

2. HANS L. BODLAENDER. A partial k-arboretum of graphs with bounded treewidth. *Theoret. Comput. Sci.*, 209(1-2):1–45, 1998.

3. PROSENJIT BOSE, MICHIEL SMID, AND DAVID R. WOOD. Light edges in degree-constrained graphs. *Discrete Math.*, 282(1-3):35–41, 2004.

4. GABRIEL A. DIRAC. On rigid circuit graphs. *Abh. Math. Sem. Univ. Hamburg*, 25:71–76, 1961.

5. HERBERT EDELSBRUNNER. *Algorithms in Combinatorial Geometry*. Springer, 1988.

6. DAVID KIRKPATRICK. Optimal search in planar subdivisions. *SIAM J. Comput.*, 12(1):28–35, 1983.

7. BRUCE A. REED. Algorithmic aspects of tree width. In BRUCE A. REED AND CLÁUDIA L. SALES, eds., *Recent Advances in Algorithms and Combinatorics*, pp. 85–107. Springer, 2003.

8. JACK SNOEYINK AND MARC VAN KREVELD. Linear-time reconstruction of Delaunay triangulations with applications. In RAINER E. BURKHARD AND GERHARD J. WOEGINGER, eds., *Proc. 5th Annual European Symp. on Algorithms* (ESA '97), vol. 1284 of *Lecture Notes in Comput. Sci.*, pp. 459–471. Springer, 1997.

Optimal Broadcast Domination of Arbitrary Graphs in Polynomial Time*

Pinar Heggernes and Daniel Lokshtanov

Department of Informatics, University of Bergen, N-5020 Bergen, Norway
pinar.heggernes@ii.uib.no, daniel.lokshtanov@student.uib.no

Abstract. Broadcast domination was introduced by Erwin in 2002, and it is a variant of the standard dominating set problem, such that vertices can be assigned various domination powers. Broadcast domination assigns a power $f(v) \geq 0$ to each vertex v of a given graph, such that every vertex of the graph is within distance $f(v)$ from some vertex v having $f(v) \geq 1$. The optimal broadcast domination problem seeks to minimize the sum of the powers assigned to the vertices of the graph. Since the presentation of this problem its computational complexity has been open, and the general belief has been that it might be $\mathcal{NP}$-hard. In this paper, we show that optimal broadcast domination is actually in $\mathcal{P}$, and we give a polynomial time algorithm for solving the problem on arbitrary graphs, using a non standard approach.

1 Introduction

A *dominating set* in a graph is a subset of the vertices of the graph such that every vertex of the graph either belongs to the dominating set or has a neighbor in the dominating set. A vertex outside of the dominating set is said to be *dominated* by one of its neighbors in the dominating set. The standard optimal domination problem seeks to find a dominating set of minimum cardinality. Since the introduction of this problem [2], [12], many domination related graph parameters have been introduced and studied, and domination in graphs is one of the most well known and widely studied subjects within graph algorithms [7], [8].

The standard dominating set problem can be seen as to represent a set of cities having broadcast stations, where every city can hear a broadcast station placed in it or in a neighboring city [11]. In 2002 Erwin [5] introduced the *broadcast domination* problem, which is more realistic in the sense that the various broadcast stations are allowed to transmit at different powers. FM radio stations are distinguished both by their transmission frequency and by their ERP (Effective Radiated Power). A transmitter with a higher ERP can transmit further, but it is more expensive to build and to operate. Consequently, the optimal broadcast domination problem asks to compute an integer valued power function

* This work is supported by the Research Council of Norway through the SPECTRUM project grant.

D. Kratsch (Ed.): WG 2005, LNCS 3787, pp. 187–198, 2005.

f on the vertices, such that every vertex of the graph is at distance at most $f(v)$ from some vertex v having $f(v) \geq 1$, and the sum of the powers are minimized.

Since the introduction of this problem, its computational complexity has been open [4], [10]. The standard optimal domination problem is $\mathcal{NP}$-hard [6], and so are some variants that might resemble broadcast domination: optimal r-domination asks for a dominating set of minimum cardinality where every vertex of the graph is within distance r from some vertex of the dominating set for a given r [9], [13], and the (k,r)-center problem asks to find an r-dominating set containing at most k vertices, where one parameter is given and the other is to be minimized [1], [6]. Since most of the interesting domination problems are $\mathcal{NP}$-hard on general graphs, this gave some indication that optimal broadcast domination might also be $\mathcal{NP}$-hard for general graphs. Following this, in 2003 Blair et al. gave polynomial time algorithms for optimal broadcast domination of trees, interval graphs, and series-parallel graphs [3].

In this paper, we show that, quite surprisingly, optimal broadcast domination is in $\mathcal{P}$. We first prove that every graph has an optimal broadcast domination in which the subsets of vertices dominated by the same vertex are ordered in a path or a cycle. Using this, we give a polynomial time algorithm for computing optimal broadcast dominations of arbitrary graphs. Our algorithm computes minimum weight paths in an auxiliary graph, and thus differs from standard methods of proving polynomial time bounds, like reductions to 2-SAT or 2-dimensional matching.

This paper is organized as follows. In the next section, we give the necessary background. In Section 3, we prove the necessary results on the structure of optimal broadcast dominations. In Section 4, we use this result to develop a polynomial time algorithm for all graphs. We conclude with a few remarks in Section 5.

2   Definitions and Terminology

In this paper we work with unweighted, undirected, connected, and simple graphs as input graphs to our problem. Let $G = (V, E)$ be a graph with $n = |V|$ and $m = |E|$. For any vertex $v \in V$, the *neighborhood* of v is the set $N_G(v) = \{u \mid uv \in E\}$. Similarly, for any set $S \subseteq V$, $N_G(S) = \cup_{v \in S} N(v) - S$. We let $G(S)$ denote the subgraph of G induced by S.

The *distance* between two vertices u and v in G, denoted by $d_G(u, v)$, is the minimum number of edges on a path between u and v. The *eccentricity* of a vertex v, denoted by $e(v)$, is the largest distance from v to to any vertex of G. The *radius* of G, denoted by $rad(G)$, is smallest eccentricity in G. The *diameter* of G, denoted by $diam(G)$, is the largest distance between any pair of vertices in G.

A function $f : V \rightarrow \{0, 1, \cdots, diam(G)\}$ is a *broadcast* on G. The set of *broadcast dominators* defined by f is the set $V_f = \{v \in V \mid f(v) \geq 1\}$. A broadcast is *dominating* if for every vertex $u \in V$ there is a vertex $v \in V_f$ such that $d(u, v) \leq f(v)$. In this case f is also called a *broadcast domination*. The *cost*

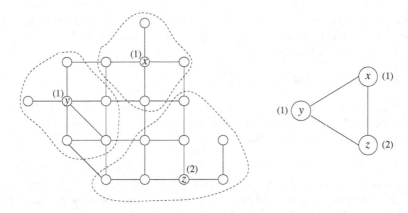

Fig. 1. On the left hand side, a graph G with an efficient broadcast domination f is shown. For vertices v with $f(v) \geq 1$, the broadcast powers $f(v)$ are shown in parentheses, and the dashed curves indicate the balls $B(v, f(v))$. For all other vertices w, $f(w) = 0$. On the right hand side, the corresponding domination graph G_f is given, and the weight of each vertex is shown in parentheses.

of a broadcast f incurred by a set $S \subseteq V$ is $c_f(S) = \sum_{v \in S} f(v)$. Thus, $c_f(V)$ is the total cost incurred by broadcast function f on G.

For a vertex $v \in V$ and an integer $p \geq 1$, we define the *ball* $B_G(v, p)$ to be the set of vertices that are at distance $\leq p$ from v in G. Thus $B_G(v, f(v))$ is the set of all vertices that are dominated by v (including v itself) if $f(v) \geq 1$. We will omit the subscript G in the notation for balls, since a ball will always refer to the input graph G. A broadcast domination f on G is *efficient* if $B(u, f(u)) \cap B(v, f(v)) = \emptyset$ for all pairs of distinct vertices $u, v \in V$.

For an efficient broadcast domination f on G, we define the *domination graph* $G_f = (V_f, \{uv \mid N_G(B(u, f(u))) \cap B(v, f(v)) \neq \emptyset\})$. Hence the domination graph can be seen as a modification of G in which every ball $B(v, f(v))$ is contracted to the single vertex v, and neighborhoods are preserved. Since G is connected and f is dominating, G_f is always connected. An example is given in Figure 1.

The *optimal broadcast domination problem* on a given graph G asks to compute a broadcast domination on G with the minimum cost. Note that if f is an optimal broadcast domination on $G = (V, E)$, then $c_f(V) \leq rad(G)$ since one can always choose a vertex v of smallest eccentricity and dominate all other vertices with $f(v) = e(v) = rad(G)$. If $c_f(V) = rad(G) = f(v)$ for a single vertex v in G, then f is called a *radial* broadcast domination.

3   The Structure of an Optimal Broadcast Domination

In [4], Dunbar et al. show that every graph has an optimal broadcast domination that is efficient. In particular, the following lemma is implicit from the proof of this result.

Lemma 1. (Dunbar et al. [4]) *For any non efficient broadcast domination f on a graph $G = (V, E)$, there is an efficient broadcast domination f' on G such that $|V_{f'}| < |V_f|$ and $c_{f'}(V) = c_f(V)$.*

We now add the following results.

Lemma 2. *Let f be an efficient broadcast domination on $G = (V, E)$. If the domination graph G_f has a vertex of degree > 2, then there is an efficient broadcast domination f' on G such that $|V_{f'}| < |V_f|$ and $c_{f'}(V) = c_f(V)$.*

Proof. Let v be a vertex with degree > 2 in G_f, and let x, y, and z be three of the neighbors of v in G_f. By the way the domination graph G_f is defined, v, x, y, and z are also vertices in G, and they all have broadcast powers ≥ 1 in f. Since f is efficient, $d_G(v, x) = f(v) + f(x) + 1$. Similarly, $d_G(v, y) = f(v) + f(y) + 1$ and $d_G(v, z) = f(v) + f(z) + 1$. Assume without loss of generality that $f(x) \leq f(y) \leq f(z)$.

If $f(x) + f(y) > f(z)$ then we construct a new broadcast f' on G with $f'(u) = f(u)$ for all vertices $u \in V \setminus \{v, x, y, z\}$. Furthermore, we let $f'(v) = f(v) + f(x) + f(y) + f(z)$, and $f'(x) = f'(y) = f'(z) = 0$. The new broadcast f' is dominating since every vertex that was previously dominated by one of v, x, y, or z is now dominated by v. To see this, let u be any vertex that was dominated by x, y, or z in f. Thus $d_G(v, u) \leq f(v) + 2f(z) + 1$ by our assumptions. Since $f'(v) > f(v) + 2f(z)$, vertex u is now dominated by v in f'. The cost of f' is the same as that of f, and the number of broadcast dominators in f' is smaller.

Let now $f(x) + f(y) \leq f(z)$. As we mentioned above, there is a path P in G between v and z of length $f(v) + f(z) + 1$. Let w be a vertex on P such that the number of edges between w and z on P is $f(v) + f(x) + f(y)$. Since f is efficient, $f(w) = 0$. We construct a new broadcast f' on G such that $f'(u) = f(u)$ for all vertices $u \in V \setminus \{v, w, x, y, z\}$. Furthermore, we let $f'(w) = f(v) + f(x) + f(y) + f(z)$, and $f'(v) = f'(x) = f'(y) = f'(z) = 0$. By the way $d_G(z, w)$ is defined, any vertex that was dominated by z or v in f is now dominated by w, since $d_G(v, w) < f(z)$. Let u be a vertex that was dominated by y in f. The distance between u and w in G is $\leq 2f(y) + 2f(v) + f(z) + 2 - f(v) - f(x) - f(y) = f(y) + f(v) + f(z) + 2 - f(x) \leq f(y) + f(v) + f(z) + f(x) = f'(w)$. Thus u is now dominated by w. The same is true for any vertex that was dominated by x in f since we assumed that $f(x) \leq f(y)$. Thus f' is a broadcast domination. Clearly, the costs of f' and f are the same, and f' has fewer broadcast dominators.

Thus we have shown how to compute a new broadcast domination f' as desired. If f' is not efficient, then by Lemma 1 there exists an efficient broadcast domination with the same cost and fewer broadcast dominators, so the lemma follows.   ∎

We are now ready to state the main result of this section, on which our algorithm will be based.

Theorem 1. *For any graph G, there is an efficient optimal broadcast domination f on G such that the domination graph G_f is either a path or a cycle.*

Proof. Let f be any efficient optimal broadcast domination on $G = (V, E)$. If G_f has a vertex of degree > 2 then by Lemma 2, an efficient broadcast domination f' on G with $|V_{f'}| < |V_f|$ and $c_{f'}(V) = c_f(V)$ exists. The proofs of both Lemmas 1 and 2 are constructive, so we know how to obtain f'. As long as there are vertices of degree > 2 in the domination graph, this process can be repeated. Since we always obtain a new domination graph with a strictly smaller number of vertices, the process has to stop after $< n$ steps. Since domination graphs are connected, the theorem follows.   ■

Note that a path can be a single edge or a single vertex. If G_f is a single vertex then f is a radial broadcast.

Corollary 1. *For any graph $G = (V, E)$, there is an efficient optimal broadcast domination f on G such that removing the vertices of $B(v, f(v))$ from G results in at most two connected components, for every $v \in V_f$.*

Proof. Since there is always an efficient optimal broadcast domination f on G such that the balls $B(v, f(v))$ with $v \in V_f$ are ordered in a path or a cycle by Theorem 1, it suffices to observe that $B(v, f(v))$ induces a connected subgraph in G for each $v \in V_f$.   ■

Corollary 2. *For any graph $G = (V, E)$, there is an efficient optimal broadcast domination f on G such that a vertex $x \in V_f$ satisfies the following: $G' = G(V \setminus B(x, f(x)))$ is connected (or empty), and G' has an efficient optimal broadcast domination f' such that $G'_{f'}$ is a path (or empty).*

Proof. By Theorem 1, let f be an efficient optimal broadcast domination of G such that G_f is a path or a cycle. Let x be any vertex of G_f if G_f is a cycle, any of the two endpoints of G_f if G_f is a path with at least two vertices, or G_f itself if G_f is a single vertex. Let $f'(v) = f(v)$ for all $v \in V \setminus \{x\}$. Since f is efficient on G, f' is an efficient dominating broadcast on G', and $G'_{f'}$ is the result of removing x from G_f. Thus $G'_{f'}$ is a path or empty.   ■

4   Computing an Optimal Broadcast Domination

By Theorem 1 we know that an efficient optimal broadcast f on G must exist such that G_f is a path or a cycle. We will first give an algorithm for handling the case when G_f is a path.

4.1   Optimal Broadcast Domination When the Domination Graph Is a Path

In this subsection, we want to find an efficient broadcast domination of minimum cost over all broadcast dominations f on $G = (V, E)$ such that G_f is a path. Our approach will be as follows: for each vertex u of G, we will compute a new graph $\mathcal{G}_u$, and use this to find the best possible broadcast domination f such that G_f

is a path and u belongs to a ball corresponding to one of the endpoints of G_f. We will repeat this process for every u in G, and choose at the end the best f ever computed.

Given a vertex $u \in V$, we define a directed graph $\mathcal{G}_u$ with weights assigned to its vertices as follows: For each $v \in V$ and each $p \in [1, ..., rad(G)]$, there is a vertex (v, p) in $\mathcal{G}_u$ if and only if one of the following is true:

- $G(V \setminus B(v, p))$ is connected or empty and $u \in B(v, p)$
- $G(V \setminus B(v, p))$ has at most two connected components and $u \notin B(v, p)$.

Thus $\mathcal{G}_u$ has a total of at most $n \cdot rad(G)$ vertices. Following Corollaries 1 and 2, each vertex (v, p) represents the situation that $f(v) = p$ in the broadcast domination f that we are aiming to compute. We define the *weight* of each vertex (v, p) to be p.

The role of u is to define the "left" endpoint of the path that we will compute. All edges will go from "left" to "right". We partition the vertex set of $\mathcal{G}_u$ into four subsets:

- $A_u = \{(v, p) \mid G(V \setminus B(v, p))$ is connected and $u \in B(v, p)\}$
- $B_u = \{(v, p) \mid G(V \setminus B(v, p))$ has two connected components$\}$
- $C_u = \{(v, p) \mid G(V \setminus B(v, p))$ is connected and $u \notin B(v, p)\}$
- $D_u = \{(v, p) \mid B(v, p) = V\}$

For each vertex (v, p), let $L_u(v, p)$ be the connected component of $G(V \setminus B(v, p))$ that contains u (i.e., the component to the "left" of $B(v, p)$), and let $R_u(v, p)$ be the connected component of $G(V \setminus B(v, p))$ that does not contain u (i.e., the component to the "right" of $B(v, p)$). Thus $L_u(v, p) = \emptyset$ for every $(v, p) \in A_u \cup D_u$, and $R_u(v, p) = \emptyset$ for every $(v, p) \in C_u \cup D_u$.

The edges of $\mathcal{G}_u$ are directed and defined as follows: A directed edge $(v, p) \rightarrow (w, q)$ is an edge of $\mathcal{G}_u$ if and only if all of the following three conditions are satisfied:

- $B(v, p) \cap B(w, q) = \emptyset$ in G
- $R_u(v, p) \neq \emptyset$ and $L_u(w, q) \neq \emptyset$
- $(N_G(B(w, q)) \cap L_u(w, q)) \subset B(v, p)$ and $(N_G(B(v, p)) \cap R_u(v, p)) \subset B(w, q)$ in G.

To restate the last requirement in plain text: $B(v, p)$ must contain all neighbors of $B(w, q)$ in $L_u(w, q)$, and $B(w, q)$ must contain all neighbors of $B(v, p)$ in $R_u(v, p)$.

Observation 1. *Given the first two requirements that an edge of $\mathcal{G}_u$ must satisfy, the two conditions of the last requirement are equivalent.*

Proof. Note first that $(N_G(B(w, q)) \cap L_u(w, q)) \neq B(v, p)$ and $(N_G(B(v, p)) \cap R_u(v, p)) \neq B(w, q)$ since $B(v, p) \cap B(w, q) = \emptyset$ and thus v has no neighbor in $B(w, q)$ and w has no neighbor in $B(v, p)$ in G. Let now $(N_G(B(w, q)) \cap L_u(w, q)) \subset B(v, p)$. Observe that $B(v, p) \subseteq L_u(w, q)$, since $B(v, p) \cap B(w, q) = \emptyset$ and each ball induces a connected subgraph of G. Furthermore, since $B(w, q) \cup$

$R_u(w,q)$ is connected and does not intersect with $B(v,p)$, and since there is no path from $B(w,q)$ to u that avoids $B(v,p)$, we can also conclude that $B(w,q) \subseteq R_u(v,p)$. Assume now, for a contradiction, that $(N_G(B(v,p)) \cap R_u(v,p)) \not\subseteq B(w,q)$. Thus $B(v,p)$ has a neighbor z in $R_u(v,p)$ and $z \notin B(w,q)$. Since $R_u(v,p)$ is connected there is a path between z and a vertex of $B(w,q)$ in $R_u(v,p)$, and in particular, this path contains a vertex y of $R_u(w,q)$. But this means that there is a path between u and y in $G(V \setminus B(w,q))$, which contradicts that $y \in R_u(w,q)$. The proof in the other direction is analogous.   ∎

By the way we have defined the edges of $\mathcal{G}_u$, all vertices belonging to A_u have indegree 0 and all vertices belonging to C_u have outdegree 0. Hence, any path in $\mathcal{G}_u$ can contain at most one vertex from A_u (which must be the starting point of the path) and at most one vertex from C_u (which must be the ending point of the path). The vertices of D_u are isolated, and every vertex of D_u defines a radial broadcast domination on its own.

Lemma 3. *Given $G = (V, E)$ and a vertex u in G, let $(v_1, p_1) \rightarrow (v_2, p_2) \rightarrow \ldots \rightarrow (v_k, p_k)$ be a directed path in $\mathcal{G}_u$ with $(v_1, p_1) \in A_u \cup D_u$ and $(v_k, p_k) \in C_u \cup D_u$. Then for $1 \le i \le k$, the following is true: $\bigcup_{j=1}^{i-1} B(v_j, p_j) = L_u(v_i, p_i)$ and $\bigcup_{j=i+1}^{k} B(v_j, p_j) = R_u(v_i, p_i)$.*

Proof. Observe that $k = 1$ if and only if the path contains a vertex of D_u, in which case the lemma follows trivially. Let us for the rest of the proof assume that $k \ge 2$.

We first show that $\bigcup_{j=1}^{i-1} B(v_j, p_j) = L_u(v_i, p_i)$ by induction on i, starting from $i = 1$ and continuing to $i = k$.

Let us consider the base cases $i = 1$ and $i = 2$. When $i = 1$, we must show that $L_u(v_1, p_1) = \emptyset$, which follows trivially since $(v_1, p_1) \in A_u \cup D_u$. When $i = 2$, we need to show that $B(v_1, p_1) = L_u(v_2, p_2)$. Since $(v_1, p_1) \rightarrow (v_2, p_2)$ is an edge of $\mathcal{G}_u$ and $L_u(v_1, p_1) = \emptyset$, we know that $N_G(B(v_1, p_1)) \subset B(v_2, p_2)$. By the definition of an edge of $\mathcal{G}_u$, we also know that $N_G(B(v_2, p_2)) \cap L_u(v_2, p_2) \subset B(v_1, p_1)$. Thus there cannot exist a path between a vertex of $B(v_2, p_2)$ and a vertex of $B(v_1, p_1)$ that avoids $B(v_1, p_1)$ and the result follows since $L_u(v_2, p_2)$ is connected.

For the induction step, assume that $\bigcup_{j=1}^{i-1} B(v_j, p_j) = L_u(v_i, p_i)$, and we will show that $\bigcup_{j=1}^{i} B(v_j, p_j) = L_u(v_{i+1}, p_{i+1})$. Because of the edge $(v_i, p_i) \rightarrow (v_{i+1}, p_{i+1})$, by the proof of Observation 1, we know that $B(v_i, p_i) \subseteq L_u(v_{i+1}, p_{i+1})$ and $B(v_{i+1}, p_{i+1}) \subseteq R_u(v_i, p_i)$. Thus, by the induction assumption, $B(v_{i+1}, p_{i+1})$ does not intersect with $\bigcup_{j=1}^{i} B(v_j, p_j)$. Again by the induction assumption, $\bigcup_{j=1}^{i} B(v_j, p_j)$ is connected and contains u. As a consequence, we can conclude that $\bigcup_{j=1}^{i} B(v_j, p_j) \subseteq L_u(v_{i+1}, p_{i+1})$. Now, if $L_u(v_{i+1}, p_{i+1})$ contains a vertex x that does not belong to $\bigcup_{j=1}^{i} B(v_j, p_j)$ then due to the induction assumption, there must be a path (possibly a single edge) between x and a vertex of $B(v_i, p_i)$ whose vertices are all outside of $\bigcup_{j=1}^{i-1} B(v_j, p_j)$. Consequently, $B(v_i, p_i)$ must have a neighbor y in $R_u(v_i, p_i)$ such that that $x \notin$

$B(v_{i+1}, p_{i+1})$, which contradicts the existence of the edge $(v_i, p_i) \to (v_{i+1}, p_{i+1})$. Thus $\bigcup_{j=1}^{i} B(v_j, p_j) = L_u(v_{i+1}, p_{i+1})$, and the proof of this part is complete.

Showing that $\bigcup_{j=i+1}^{k} B(v_j, p_j) = R_u(v_i, p_i)$ for $1 \le i \le k$ is completely analogous, and we skip this part. ∎

Lemma 4. *Given $G = (V, E)$, there is a vertex $u \in V$ such that $(v_1, p_1) \to (v_2, p_2) \to \dots \to (v_k, p_k)$ is a directed path in $\mathcal{G}_u$ with $(v_1, p_1) \in A_u \cup D_u$ and $(v_k, p_k) \in C_u \cup D_u$ if and only if G has an efficient broadcast domination f such that G_f is the undirected path $v_1 - v_2 - \dots - v_k$ and $f(v_i) = p_i$ for $1 \le i \le k$.*

Proof. Let f be an efficient broadcast on $G = (V, E)$ with broadcast dominators $V_f \subset V$ such that G_f is a path. Let $V_f = \{v_1, v_2, \dots, v_k\}$ so that $v_1 - v_2 - \dots - v_k$ is the path equivalent to G_f, and let u be any vertex of $B(v_1, f(v_1))$. If $k = 1$ then $V = B(v_1, p_1)$, and the lemma trivially follows since $\mathcal{G}_u$ contains a vertex $(v_1, p_1) \in D_u$. Let $k \ge 2$. By the proofs of Corollaries 1 and 2, removing $B(v_1, f(v_1))$ or $B(v_k, f(v_k))$ from G results in a connected graph, and removing $B(v_i, f(v_i))$ from G results in exactly two connected components for $2 \le i \le k-1$ (if $k \ge 3$). Consequently, for each $v_i \in V_f$, $(v_i, f(v_i))$ is a vertex of $\mathcal{G}_u$. In $\mathcal{G}_u$, $(v_1, f(v_1))$ belongs to A_u, $(v_k, f(v_k))$ belongs to C_u, vertices $(v_i, f(v_i))$ belong to B_u for $2 \le i \le k-1$ (for $k \ge 3$), and $(v_1, f(v_1)) \to (v_2, f(v_2)) \to \dots \to (v_k, f(v_k))$ is a path by the definition of the edges in $\mathcal{G}_u$.

In the other direction, let u be a vertex of G, and let $P = (v_1, p_1) \to (v_2, p_2) \to \dots \to (v_k, p_k)$ be a directed path in $\mathcal{G}_u$ such that $(v_1, p_1) \in A_u \cup D_u$ and $(v_k, p_k) \in C_u \cup D_u$. Define a function f so that $f(v_i) = p_i$ for $1 \le i \le k$, and $f(v) = 0$ for all other vertices of G. By Lemma 3, $\bigcup_{i=1}^{k} B(v_i, p_i) = V$, and $B(v_i, p_i) \cap B(v_j, p_j) = \emptyset$, for $1 \le i < j \le k$. Thus f is an efficient broadcast domination on G. ∎

Now the idea is to find a directed path P_u in $\mathcal{G}_u$ from a vertex of $A_u \cup D_u$ to a vertex of $C_u \cup D_u$ such that the sum of the weights of the vertices of P_u (including the endpoints) is minimized. Let us call this sum $W(P_u)$. Then we will compute $\mathcal{G}_u$ for each vertex u in G, and repeat this process, and at the end choose a path with the minimum total weight. Our algorithm for the path case is given in Figure 2.

Theorem 2. *Given a graph $G = (V, E)$, Algorithm MPBD computes an efficient broadcast domination f on G of minimum cost such that G_f is a path.*

Proof. We compute a minimum weight path in $\mathcal{G}_u$ for every $u \in V$, and among all these paths we choose a path P with the lowest $W(P)$. By Lemma 4, P corresponds to a broadcast domination f of G such that G_f is a path, and by the way each $\mathcal{G}_u$ is constructed, $W(P) = c_f(V)$. Assume that there is a broadcast domination f' on G with $c_{f'}(V) < c_f(V)$ such that $G_{f'}$ is a path. Lemma 4 guarantees the existence of a path P' in $\mathcal{G}_v$ for some vertex $v \in V$ such that $W(P') < W(P)$, which is a contradiction. ∎

Corollary 3. *Let $G = (V, E)$ be a graph such that there is an efficient optimal broadcast domination on G where the domination graph is a path. Algorithm MPBD computes such a broadcast domination on G.*

Algorithm Minimum Path Broadcast Domination - MPBD
Input: A graph $G = (V, E)$.
Output: An efficient broadcast domination function f of minimum cost on G,
such that G_f is a path.
begin
    for each vertex v in G **do**
        $f(v) = 0$;
    Let P be a dummy path with $W(P) = rad(G) + 1$;
    for each vertex u in G **do**
        Compute $\mathcal{G}_u$ with vertex sets A_u, B_u, C_u, and D_u;
        Find a minimum weight path P_u starting in a vertex of $A_u \cup D_u$ and
        ending in a vertex of $C_u \cup D_u$;
        if $W(P_u) < W(P)$ **then**
            $P = P_u$;
    end-for
    for each vertex (v, p) on P **do**
        $f(v) = p$;
end

Fig. 2. The algorithm for computing the best path broadcast domination

4.2   Optimal Broadcast Domination for All Cases

Now we want to compute an optimal broadcast domination for any given graph
G. Our approach will be as follows. Let x be any vertex of G. For each k between
1 and $rad(G)$ such that $G' = G(V \setminus B(x, k))$ is connected or empty, we run the
minimum path broadcast domination algorithm MPBD on G'. Our algorithm
for the general case is given in Figure 3.

In this way, we consider all broadcast dominations f whose corresponding
domination graphs are paths or cycles. The advantage of this approach is its
simplicity. The disadvantage is that we also consider many cases that do not
correspond to a path or a cycle, which we could have detected with a longer and
more involved algorithm. However, these unnecessary cases do not threaten the
correctness of the algorithm, and detecting them does not decrease the asymptotic time bound.

Theorem 3. *Algorithm OBD computes an optimal broadcast domination of any
given graph.*

Proof. Let $G = (V, E)$ be the input graph. By Theorem 1 and Corollary 2, there
is a vertex x in V and an integer $k \in [1, rad(G)]$ such that the graph $G' = G(V \setminus
B(x, k))$ has an efficient optimal broadcast domination f' where the domination
graph $G'_{f'}$ is a path, and that f' can be extended to an optimal broadcast
domination f for G with $f(x) = k$, $f(v) = 0$ for $v \in B(x, k)$ with $x \neq v$, and
$f(v) = f'(v)$ for all other vertices v. Algorithm MPBD computes an optimal
broadcast domination of G', and since Algorithm OBD tries all possibilities for
(x, k), the result follows.   ∎

Algorithm Optimal Broadcast Domination - OBD
Input: A graph $G = (V, E)$.
Output: An efficient optimal broadcast domination function f on G.
begin
    $opt = rad(G) + 1$;
    for each vertex x in G **do**
        for $k = 1$ **to** $rad(G)$ **do**
            if $G' = G(V \setminus B(x, k))$ is connected or empty **then**
                $f = \text{MPBD}(G')$;
                if $c_f(V \setminus B(x, k)) + k < opt$ **then**
                    $opt = c_f(V \setminus B(x, k)) + k$;
                    $f(x) = k$;
                    for each vertex v in $B(x, k) \setminus \{x\}$ **do**
                        $f(v) = 0$;
            end-if
        end-if
end

Fig. 3. The algorithm for computing an optimal broadcast domination

Note that although there is always an efficient optimal broadcast domination f such that G_f is a cycle or a path, there can of course exist other optimal broadcast dominations f' with $c_{f'}(V) = c_f(V)$ such that $G_{f'}$ is not a path or a cycle, and such that f' is not efficient. The optimal broadcast domination returned by algorithm OBD does not necessarily correspond to a path or a cycle, since we do not force the endpoints (or forbid the interior points) of the path for G' to be neighbors of $B(x, k)$. Nor is the returned broadcast necessarily efficient, as some ball $B(v, p)$ might have an outreach outside of G' and might overlap with $B(x, k)$.

4.3   Time Complexity

We explain a straight forward implementation of our algorithms to justify the polynomial time complexity. In Algorithm MPBD, given a graph G, for each vertex u in G, we create an auxiliary graph $\mathcal{G}_u$ with at most $n \cdot rad(G)$ vertices. Thus we create n graphs with $O(n^2)$ vertices and $O(n^4)$ edges each. In each such graph, we first compute vertex sets A_u, B_u, C_u, and D_u, which requires a breadth first search for each vertex of $\mathcal{G}_u$, thus $O(n^6)$ time for each auxiliary graph. The edges of $\mathcal{G}_u$ can also be computed within this bound. Then we compute a path of minimum weight in each such graph. A shortest path in a connected graph $H = (U, D)$ can be computed in time $O(|D| \log |U|)$ by well-known algorithms like the one by Dijkstra. Minimum weight paths can be computed by simple modifications of such algorithms within the same time bound. Thus in each graph that we create, it takes $O(n^4 \log n)$ time to find a minimum weight path. Consequently the total time required for each $\mathcal{G}_u$ is $O(n^6)$, giving a total of $O(n^7)$ time for Algorithm MPBD. In Algorithm OBD, we repeat this process $n \cdot rad(G) = O(n^2)$ times to find an optimal broadcast domination. As a result, the overall time complexity of a straight forward implementation is $O(n^9)$ and thus polynomial.

It can be shown that each auxiliary graph $\mathcal{G}_u$ is acyclic and has $O(n^3)$ edges. Therefore minimum weight paths can be computed by topological sort in time $O(n^3)$ in each such graph. With a preprocessing step of computing all pairs of shortest paths on G in $O(n^3)$ time, and using an $O(n^3)$ space data structure to store the information about the edges between all possible balls in all auxiliary graphs, we can actually reduce the total running time to $O(n^3 r^2 m)$, where $r = rad(G)$. We leave the details of this implementation to the full paper, due to limited space here.

5   Concluding Remarks

In this paper we have shown that the broadcast domination problem is solvable in polynomial time on all graphs. Our focus has been on polynomial time and not the best possible time bound. Our algorithm can be enhanced to run substantially faster, as explained. For further research, more efficient algorithms for this problem should be of interest.

The optimal broadcast domination problem studies the cost $c_f(V) = \sum_{v \in V} f(v)$ of a broadcast domination f on a graph $G = (V, E)$. Other definitions of the cost of a broadcast may be appropriate depending on the application, since the cost of a broadcast can be different from the value of a broadcast. To be more precise, one could define a cost function $c(i)$, and let the total cost be $c_f(V) = \sum_{v \in V} c(f(v))$. Thus in our case $c(i) = i$ for all i. Our polynomial time algorithm can be used for all cost functions c, where $c(i) + c(j) \geq c(i+j)$ for all integers i and $j \geq 0$. For general cost functions the problem becomes $\mathcal{NP}$-hard, because we can let $c(0) = 0$, $c(1) = 1$, and $c(i) > n$ for all $i > 1$, which gives a direct reduction from the standard dominating set problem.

References

1. J. Bar-Ilan, G. Kortsarz, and D. Peleg. How to allocate network centers. *J. Algorithms*, 15:385–415, 1993.
2. C. Berge. *Theory of Graphs and its Applications*. Number 2 in Collection Universitaire de Mathematiques. Dunod, Paris, 1958.
3. J. R. S. Blair, P. Heggernes, S. Horton, and F. Manne. Broadcast domination algorithms for interval graphs, series-parallel graphs, and trees. *Congressus Numerantium*, 169:55 – 77, 2004.
4. J. E. Dunbar, D. J. Erwin, T. W. Haynes, S. M. Hedetniemi, and S. T. Hedetniemi. Broadcasts in graphs. 2002. Submitted.
5. D. J. Erwin. Dominating broadcasts in graphs. *Bull. Inst. Comb. Appl.*, 42:89–105, 2004.
6. M. R. Garey and D. S. Johnson. *Computers and Intractability*. W. H. Freeman and Co., 1978.
7. T. W. Haynes, S. T. Hedetniemi, and P. J. Slater. *Domination in Graphs: Advanced Topics*. Marcel Dekker, New York, 1998.

8. T. W. Haynes, S. T. Hedetniemi, and P. J. Slater. *Fundamentals of Domination in Graphs*. Marcel Dekker, New York, 1998.

9. M. A. Henning. Distance domination in graphs. In T. W. Haynes, S. T. Hedetniemi, and P. J. Slater, editors, *Domination in Graphs: Advanced Topics*, pages 321–349. Marcel Dekker, New York, 1998.

10. S. B. Horton, C. N. Meneses, A. Mukherjee, and M. E. Ulucakli. A computational study of the broadcast domination problem. Technical Report 2004-45, DIMACS Center for Discrete Mathematics and Theoretical Computer Science, 2004.

11. C. L. Liu. *Introduction to Combinatorial Mathematics*. McGraw-Hill, New York, 1968.

12. O. Ore. *Theory of Graphs*. Number 38 in American Mathematical Society Publications. AMS, Providence, 1962.

13. P. J. Slater. *R*-domination in graphs. *J. Assoc. Comput. Mach.*, 23:446–450, 1976.

Ultimate Generalizations of LexBFS and LEX M

Anne Berry[1], Richard Krueger[2], and Genevieve Simonet[3]

[1] LIMOS, UMR, bat. ISIMA, 63173 Aubière cedex, France
berry@isima.fr
[2] Department of Computer Science, University of Toronto,
Toronto, Ontario M5S 3G4, Canada
krueger@cs.toronto.edu
[3] LIRMM, 161, Rue Ada, F-34392 Montpellier, France
simonet@lirmm.fr

Abstract. Many graph search algorithms use a labelling of the vertices to compute an ordering of the vertices. We examine such algorithms which compute a peo (perfect elimination ordering) of a chordal graph, and corresponding algorithms which compute an meo (minimal elimination ordering) of a non-chordal graph.

We express all known peo-computing search algorithms as instances of a generic algorithm called MLS (Maximal Label Search) and generalize Algorithm MLS into CompMLS, which can compute any peo.

We then extend these algorithms to versions which compute an meo, and likewise generalize all known meo-computing search algorithms. We show the surprising result that all these search algorithms compute the same set of minimal triangulations, even though the computed meos are different.

1 Introduction

Graph searching plays a fundamental role in many algorithms, particularly using Breadth-First or Depth-First searches and their many variants. One important application is to compute special graph orderings related to the chordality of a graph. When the input graph is chordal, one wants to find an ordering of the vertices called a *peo* (perfect elimination ordering), which repeatedly selects a vertex whose neighbourhood is a clique (called a *simplicial vertex*), and removes it from the graph. This is a certificate of chordality, as, given an ordering of the vertices, one can determine in linear time whether it is a peo of the graph.

When the input graph fails to be chordal, it is often interesting to embed it into a chordal graph by adding an inclusion-minimal set of edges, a process called *minimal triangulation*. One of the ways of accomplishing this is to use an ordering of the vertices called an *meo* (minimal elimination ordering), and use this to simulate a peo by repeatedly adding any edges whose absence would violate the simplicial condition.

Though some earlier work had been done on these problems ([11], [10]), the seminal paper is that of Rose, Tarjan and Lueker [12], which presented two very efficient algorithms to compute a peo or an meo. They introduced the concept

D. Kratsch (Ed.): WG 2005, LNCS 3787, pp. 199–213, 2005.

of *lexicographic order* (which roughly speaking is a dictionary order), and used this for graph searches which at each step choose an unnumbered vertex of maximal label. With this technique, they introduced Algorithm LEX M, which for a non-chordal graph computes an meo in a very efficient $O(nm)$ time, and then streamlined this for use on a chordal graph, introducing what is now called Algorithm LexBFS, a Breadth-First Search which runs in optimal $O(n+m)$ time and computes a peo if the input graph is chordal.

Later work has been done on computing peos. Tarjan and Yannakakis [14] presented Algorithm MCS (Maximal Cardinality Search) which is similar to LexBFS but uses a simplified labelling and order (a cardinality choice criterion is used instead of a lexicographic one). MCS also computes in linear time a peo if the input graph is chordal.

Shier [13] remarks that neither LexBFS nor MCS is capable of computing all peos. He proposes a generalization of both LexBFS and MCS, Algorithm MEC, which can compute any peo of a chordal graph.

Recently, Corneil and Krueger [6] introduced Algorithm LexDFS as a Depth-First analogue to LexBFS. They also introduced Algorithm MNS (Maximal Neighbourhood Search) as a generalization of LexBFS, LexDFS and MCS, which simply chooses at each step a vertex whose set of numbered neighbours is inclusion-maximal. They gave characterizations of the orderings computed by these search Algorithms and observed, from a result of Tarjan and Yannakakis [14] on the property characterizing MNS orderings, every MNS ordering yields a peo if the input graph is chordal.

Berry, Blair, Heggernes and Peyton [1] recently introduced Algorithm MCS-M, which computes an meo. MCS-M is extended from MCS in the same fashion LEX M can be extended from LexBFS. The sets of meos defined by LEX M and by MCS-M are different, but Villanger [15] recently showed that the same sets of minimal triangulations were obtained.

In this paper, we address natural questions which arise about peos and meos: how can the existing algorithms be generalized? Do these new algorithms compute all peos of a chordal graph? Can they all be extended to compute meos? What sets of minimal triangulations are obtained?

We show that LexBFS, MCS, LexDFS and MNS can be described as instances of a generic algorithm called MLS (Maximal Label Search) which computes a peo of a chordal graph, but cannot compute every peo of every chordal graph. In order to obtain all possible peos, we extend MLS to CompMLS, which uses Shier's idea of working on the connected components of the subgraph induced by the unnumbered vertices. We show that every instance of gereric CompMLS is capable of computing any peo of a chordal graph.

We then go on to examine the issues pertaining to meos and minimal triangulations. We show that MNS, MLS and CompMLS can all be extended to compute an meo, in the same way that LEX M is extended from LexBFS. We show the very strong result that all the sets of minimal triangulations computed are the same, independent of the meo-computing algorithm which is used, and that not all minimal triangulations can be computed by this new family of algorithms.

Because of space limitations, we give either abridged proofs or no proof for our results; the proofs may be skipped without hindering comprehension. The reader is referred to the journal version of this paper for the full proofs.

The paper is organized as follows: Section 2 gives some definitions and notations, in Section 3 we discuss peos, and in Section 4 we discuss meos.

2   Preliminaries

All graphs in this work are undirected and finite. A graph is denoted $G = (V, E)$, with $n = |V|$, and $m = |E|$. The *neighbourhood* of a vertex x in G is denoted $N_G(x)$, or simply $N(x)$ if the meaning is clear. An ordering on V is a one-to-one mapping from $\{1, 2, ..., n\}$ to V. In every figure in this paper showing an ordering α on V, α is defined by giving on the figure the number $\alpha^{-1}(x)$ for every vertex x. We denote by $\mathbb{Z}^+$ the set of positive integers $\{1, 2, 3, \dots\}$.

A *chordal* (or *triangulated*) graph is a graph with no chordless cycle of length greater or equal to 4. To recognize chordal graphs efficiently, Fulkerson and Gross [9] used a greedy elimination scheme on simplicial vertices: "A graph is chordal iff one can repeatedly find a simplicial vertex and delete it from the graph, until no vertex is left." This defines an ordering on the vertices which is called a *perfect elimination ordering (peo)* of the graph.

When a graph G fails to be chordal, any ordering α on the vertices can be used to embed G into a chordal graph (called a *triangulation* of G) by repeatedly choosing the next vertex x, adding any edges necessary to make it simplicial, and removing x. If F is the set of added edges, graph $H = (V, E + F)$ is chordal and is denoted G_α^+.

If $H = (V, E + F)$ is a triangulation of $G = (V, E)$, and if for every proper subset $F' \subset F$, graph $(V, E + F')$ fails to be chordal, H is called a *minimal triangulation* of G. If moreover $H = G_\alpha^+$, α is called a *minimal elimination ordering (meo)* of G.

In [12], two very important characterizations are given:

Path Lemma
For any graph $G = (V, E)$, any ordering α on V and any x, y in V such that $\alpha^{-1}(y) < \alpha^{-1}(x)$, xy is an edge of G_α^+ iff there is a path μ in G from x to y such that $\forall t \in \mu \setminus \{x, y\}$, $\alpha^{-1}(t) < \alpha^{-1}(y)$.

Unique Chord Property
For any graph $G = (V, E)$ and any triangulation $H = (V, E + F)$ of G, H is a minimal triangulation of G iff each edge in F is the unique chord of a 4-cycle of H.

3   Computing Peos

Algorithm MNS, as defined by [6], works in the following fashion: start with a graph where all vertices are unnumbered and all labels of all vertices are empty. Repeatedly choose an unnumbered vertex x whose label is maximal (with respect to set inclusion), give x the following number i, and add (by a union operation) i to the label of all as-yet unnumbered neighbours of x. Note that the graph search algorithms in [6] number vertices from 1 to n. In this paper, our algorithms compute peos and meos directly, and thus number vertices from n down to 1. MNS is a generalization of both LexBFS and LexDFS, where labels are lists and maximality is defined using lexicographic order, and it is also a generalization of MCS, where maximality is decided using the cardinality of the MNS labels. Corneil and Krueger [6] observed that MNS computes a peo of a chordal graph, and that any LexBFS, LexDFS or MCS ordering of a graph is an MNS one.

We will now extend MNS, using a *labelling structure*:

Definition 1. *A **labelling structure** is a structure $(L, \preceq, l_0, Inc)$, where:*

- L *is a set (the set of labels),*
- $\preceq$ *is a partial order on L (which may be total or not, with $\prec$ denoting the corresponding strict order), which will be used to choose a vertex of maximal label,*
- l_0 *is an element of L, which will be used to initialize the labels,*
- *Inc is a mapping from $L \times \mathbb{Z}^+$ to L, which will be used to increment a label, and such that for any integer n in $\mathbb{Z}^+$, any integer i from 1 to n and any labels l and l' in L_i^n, the following properties hold:*
  (ls1) $l \prec Inc(l, i)$
  (ls2) if $l \prec l'$ then $Inc(l, i) \prec Inc(l', i)$
  where L_i^n is the subset of L defined by induction on i by:
  - *$L_n^n = \{l_0\}$, and*
  - *$L_{i-1}^n = L_i^n \cup \{l = Inc(l', i) \mid l' \in L_i^n\}$, for any i from n down to 2.*

The corresponding algorithm, which we introduce as MLS (Maximal Label Search), is given by Figure 1. MLS iteratively selects a vertex to add to the ordering and increments the labels of its unselected neighbours. We will refer to the iteration of the loop that defines $\alpha(i)$ as Step i of the algorithm. We observe that L_i^n corresponds to the set of labels that could possibly be assigned to an unselected vertex at the beginning of Step i.

Algorithm MLS (Maximal Label Search)
input : A graph $G = (V, E)$ and a labelling structure $(L, \preceq, l_0, Inc)$.
output: An ordering α on V.
Initialize all labels as l_0; $G' \leftarrow G$;
for $i = n$ **downto** 1 **do**
  Choose a vertex x of G' of maximal label;
  $\alpha(i) \leftarrow x$;
  foreach y in $N_{G'}(x)$ **do**
    $\lfloor$ $label(y) \leftarrow Inc(label(y), i)$;
  Remove x from G';

Fig. 1. Algorithm MLS

LexBFS, MCS, LexDFS and MNS are all special cases of MLS, with the following labelling structures $(L, \preceq, l_0, Inc)$:

<u>LexBFS</u> (Structure S_1): L is the set of lists of elements of $\mathbb{Z}^+$, $\preceq$ is lexicographical order (a total order), l_0 is the empty list, $Inc(l, i)$ is obtained from l by adding i to the end of the list.

<u>MCS</u> (Structure S_2): $L = \mathbb{Z}^+ \cup \{0\}$, $\preceq$ is $\leq$ (a total order), $l_0 = 0$, $Inc(l, i) = l + 1$.

<u>LexDFS</u> (Structure S_3): L is the set of lists of elements of $\mathbb{Z}^+$, $\preceq$ is lexicographical order (a total order), l_0 is the empty list, $Inc(l, i)$ is obtained from l by adding $n + 1 - i$ to the beginning of the list.

<u>MNS</u> (Structure S_4): L is the power set of $\mathbb{Z}^+$, $\preceq$ is $\subseteq$ (not a total order), $l_0 = \emptyset$, $Inc(l, i) = l \cup \{i\}$.

In our proofs, we will use the following notations: for any graph $G = (V, E)$, any execution of our algorithms on G computing some ordering α on V, and any integer i from 1 to n, we say V_i is the set of still unnumbered vertices at the beginning of Step i (i.e. $V_i = \{\alpha(j), 1 \leq j \leq i\}$), G'_i is graph G' at the beginning of Step i (i.e. $G'_i = G[V_i]$), and, for each $y \in V_i$, $label_i(y)$ is the value of $label(y)$ at the beginning of Step i and $Num_i(y) = \{j \in \{i+1, i+2, ..., n\} \mid label(y)$ has been incremented at Step $j\}$.

We can view MLS as a generic algorithm with parameter S. For every labelling structure S, we denote by S-MLS the instance of generic Algorithm MLS using this particular labelling structure S and by "S-MLS ordering of a graph G" any ordering that can be computed by S-MLS on input graph G. Thus, LexBFS is S_1-MLS, MCS is S_2-MLS, LexDFS is S_3-MLS, and MNS is S_4-MLS.

The set of S-MLS orderings of a given graph depends on S. An MLS ordering of a graph G is an ordering that can be computed by MLS on G, i.e. by S-MLS for some labelling structure S. Thus, the set of MLS orderings of G is the union of the sets of S-MLS orderings of G for all labelling structures S.

The following theorem shows that MNS can compute every S-MLS ordering of a given graph for every labelling structure S, so that any graph has the same MNS and MLS orderings. This theorem can easily be proved using the MNS characterization presented in [6] or by a proof similar to that of Theorem 4 given in Section 4.

Theorem 1. *For any graph $G = (V, E)$ and any labelling structure S, any S-MLS ordering of G is an MNS ordering of G.*

A corollary of Theorem 1 is that any instance of MLS computes a peo of a chordal graph, since this is true for MNS [6]. Another consequence is that any LexBFS, MCS or LexDFS ordering of a graph is also an MNS ordering, which already follows from the characterizations given in [6]. However, for arbitrary labelling structures S and S', an ordering computed by S-MLS need not be computable by S'-MLS. For instance, Figure 2(a) shows a LexBFS ordering which is not an MCS ordering, while Figure 2(b) shows an MCS ordering which is not a LexBFS ordering. There also exist graphs with MNS orderings that are neither LexBFS nor MCS orderings.

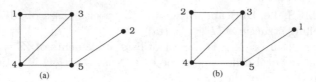

Fig. 2. A chordal graph with different (a) LexBFS and (b) MCS orderings

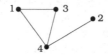

Fig. 3. α is a CompMNS ordering of G but not an MNS one

It is interesting to remark that even though MLS, or equivalently MNS, is more general than LexBFS and MCS, it still is not powerful enough to compute every possible peo of a given chordal graph. This is shown by the simple counterexample in Figure 3: no MLS execution on this graph will find the ordering indicated, although it is clearly a peo.

In order to make it possible to find *any* peo, we further generalize MLS using Shier's idea [13] of using the *connected components* of the subgraph G' induced by the unnumbered vertices. We thus introduce Algorithm CompMLS, defined from Algorithm MLS by replacing:

"Choose a vertex x of G' of maximal label;"
with
"Choose a connected component C of G';
Choose a vertex x of C of maximal label in C;".

This generalizes the whole family of peo-computing algorithms discussed in this paper: for any X in {LexBFS, MCS, LexDFS, MNS, MLS}, Algorithm CompX is a generalization of X, and computes a peo if the graph is chordal. The CompMLS algorithms also generalize Algorithms MEC and MCC defined by Shier [13]: MEC is CompMNS and MCC is CompMCS.

Algorithm CompMNS can compute the peo of Figure 3. In fact, Shier proved in [13] that CompMNS and even CompMCS compute all peos of a chordal graph. We show that this result holds for every instance of Algorithm CompMLS, as for any labelling structure S, a chordal graph has the same CompMNS and S-CompMLS orderings. This follows from Theorem 6 in Section 4.

Theorem 2. *For any chordal graph G and any labelling structure S, the S-CompMLS orderings of G are exactly its peos.*

Let us conclude this section by some remarks on running the MLS family of algorithms on *non-chordal* graphs. LexBFS has been used on AT-free graphs [7] and has been shown to have very interesting invariants even on an arbitrary graph ([2], [3]). Likewise, MCS has also been used on various graph classes ([8], [5]).

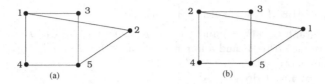

Fig. 4. A non-chordal graph with different (a) CompLexBFS and (b) CompMCS orderings

Unlike a chordal graph, a non-chordal graph does not necessarily have the same CompLexBFS, CompMCS, CompLexDFS, and CompMNS orderings. Figure 4(a) shows a CompLexBFS ordering which is not a CompMCS one, while Figure 4(b) shows a CompMCS ordering which is not a CompLexBFS one.

4   Computing Meos

We will now introduce the extensions of Algorithms MNS, MLS and CompMLS into their meo-computing counterparts.

To extend LexBFS into LEX M, at each step choosing a vertex x of maximum label $label(x)$, an edge is added between x and any unnumbered vertex y whenever there is a path from x to y in the subgraph induced by the unnumbered vertices such that all internal vertices on the path have a label strictly smaller than the label of y. This approach has been used recently in [1] to extend MCS into meo-computing Algorithm MCS-M; here, we extend MLS into MLSM, as given by Figure 5. Thus LEX M is S_1-MLSM, MCS-M is S_2-MLSM, LexDFS-M is defined as S_3-MLSM, and MNSM is defined as S_4-MLSM. We will see in the next section that Algorithm MNSM is in fact as general as MLSM: every MLSM ordering of a graph is an MNSM ordering.

4.1   The MLSM Family of Algorithms

Theorem 3. *For any execution of MLSM, $H = G_\alpha^+$, and α is a meo of G.*

To prove this, we will need several technical lemmas. The proofs of Lemmas 1 and 2 are straightforward, and that of Lemma 3 is given in the Appendix.

Lemma 1. *For any graph $G = (V, E)$, any execution of MLS (resp. MLSM) on G computing ordering α, any integers i, j such that $1 \leq i < j \leq n$ and any y in V_i, the following propositions are equivalent:*

1. *$label_j(y) \neq label_{j-1}(y)$,*
2. *$label_{j-1}(y) = Inc(label_j(y), j)$,*
3. *$label_j(y) \prec label_{j-1}(y)$,*
4. *$j \in Num_i(y)$,*
5. *$\alpha(j)y$ is an edge of G (resp. H),*
6. *(for MLSM) there is a path μ in G'_j from $\alpha(j)$ to y such that $\forall t \in \mu \setminus \{\alpha(j), y\}$, $label_j(t) \prec label_j(y)$.*

Algorithm MLSM (Maximal Label Search for Meo)

input : A graph $G = (V, E)$ and a labelling structure $(L, \preceq, l_0, Inc)$.
output: An meo α on V and a minimal triangulation $H = G_\alpha^+$ of G.
Initialize all labels as l_0; $E' \leftarrow \emptyset$; $G' \leftarrow G$;
for $i = n$ **downto** 1 **do**
> Choose a vertex x of G' of maximal label;
> $\alpha(i) \leftarrow x$;
> **foreach** *vertex y of G' different from x* **do**
> > **if** *there is a path from x to y in G' such that every internal vertex on the path has a label strictly smaller than $label(y)$* **then**
> > > $E' \leftarrow E' \cup \{xy\}$;
>
> **foreach** y *in* V *such that* $xy \in E'$ **do**
> > $label(y) \leftarrow Inc(label(y), i)$;
> Remove x from G';

$H \leftarrow (V, E')$;

Fig. 5. Algorithm MLSM

Lemma 2. *For any graph $G = (V, E)$, any execution of MLS or MLSM on G, any integer i from 1 to n and any x, y in V_i,*
(i) if $Num_i(x) = Num_i(y)$ then $label_i(x) = label_i(y)$, and
(ii) if $Num_i(x) \subset Num_i(y)$ then $label_i(x) \prec label_i(y)$.

Lemma 3. *For any graph G, any execution of MLSM on G computing ordering α, any integer i from 1 to n and any path μ in G'_i ending in some vertex y,*
a) $\forall t \in \mu \setminus \{y\}$, $label_i(t) \prec label_i(y)$ iff $\forall t \in \mu \setminus \{y\}$, $Num_i(t) \subset Num_i(y)$,
b) if $\forall t \in \mu \setminus \{y\}$, $label_i(t) \prec label_i(y)$ then $\forall t \in \mu \setminus \{y\}$, $\alpha^{-1}(t) < \alpha^{-1}(y)$,
c) if $\forall t \in \mu \setminus \{y\}$, $\alpha^{-1}(t) < \alpha^{-1}(y)$ then $\forall t \in \mu \setminus \{y\}$, $Num_i(t) \subseteq Num_i(y)$.

Proof (of Theorem 3). We first show that for any execution of MLSM, $H = G_\alpha^+$. Let $x, y \in V$ such that $\alpha^{-1}(y) < \alpha^{-1}(x) = i$. Let us show that xy is an edge of H iff it is an edge of G_α^+.

If xy is an edge of H then, by Lemma 1, there is a path μ in G'_i from x to y such that $\forall t \in \mu \setminus \{x, y\}$, $label_i(t) \prec label_i(y)$. By Lemma 3 b), $\forall t \in \mu \setminus \{x, y\}$, $\alpha^{-1}(t) < \alpha^{-1}(y)$ and, by the Path Lemma, xy is an edge of G_α^+.

Conversely, let xy be an edge of G_α^+. Let us show that xy is an edge of H. By the Path Lemma, there is a path μ in G from x to y such that $\forall t \in \mu \setminus \{x, y\}$, $\alpha^{-1}(t) < \alpha^{-1}(y) < i$, so $\mu \setminus \{x\} \subseteq V_{i-1}$. By Lemma 3 c), $\forall t \in \mu \setminus \{x, y\}$, $Num_{i-1}(t) \subseteq Num_{i-1}(y)$. Let t_1 be the neighbour of x in μ. xt_1 is an edge of H, so, by Lemma 1, $i \in Num_{i-1}(t_1)$, hence $i \in Num_{i-1}(y)$, and by Lemma 1 xy is an edge of H.

We now show that G_α^+ is a minimal triangulation of G. Let $H = G_\alpha^+ = (V, E + F)$. As G_α^+ is a triangulation of G, by the Unique Chord Property, it is sufficient to show that each edge in F is the unique chord of a cycle in H of length 4. Let xy be an edge in F with $\alpha^{-1}(y) < \alpha^{-1}(x) = i$. xy is an edge of H

so by Lemma 1 there is a path μ in G'_i from x to y such that $\forall t \in \mu \setminus \{x,y\}$, $label_i(t) \prec label_i(y)$, and also $\alpha^{-1}(t) < \alpha^{-1}(y)$ by Lemma 3 b). $\mu \setminus \{x,y\} \neq \emptyset$ since xy is not an edge in G. Let t_1 be the vertex in $\mu \setminus \{x,y\}$ such that $\alpha^{-1}(t_1)$ is maximum. By the Path Lemma, xt_1 and t_1y are edges of G^+_α and therefore of H. As $label_i(t_1) \prec label_i(y)$, by Lemma 2 (i) and (ii) $Num_i(y) \not\subseteq Num_i(t_1)$. Let $j \in Num_i(y) \setminus Num_i(t_1)$, and $z = \alpha(j)$. $j > i$ and by Lemma 1 yz is an edge of H (and therefore of G^+_α) but t_1z is not. Since yx and yz are edges of G^+_α with $\alpha^{-1}(y) < \alpha^{-1}(x) = i < j = \alpha^{-1}(z)$, by definition of G^+_α xz is an edge of G^+_α, and therefore of H. Hence xy is the unique chord of cycle (x, t_1, y, z, x) in H of length 4.

Thus MLSM (and also LEX M, MCS-M, LexDFS-M and MNSM) computes an meo and a minimal triangulation of the input graph. MLSM has the same behaviour (same labelling and numbering) on the input graph as MLS on the output minimal triangulation. As a result, MLS and MLSM have the same behavior on chordal graphs and the following property holds for any graph.

Property 1. For any graph G and any labelling structure S, any S-MLSM ordering α of G is an S-MLS ordering of G^+_α.

Let us remark that for two given structures S and S', the sets of orderings computed by S-MLSM and S'-MLSM may be different, as is the case for S-MLS and S'-MLS on chordal graphs. LEX M and MCS-M for example are different, as shown in Figure 2 (since MLS and MLSM compute the same orderings on a chordal graph). In the same way that MNS is as general as MLS, it turns out that MNSM is as general as MLSM, thus every graph has the same MLSM and MNSM orderings.

Theorem 4. *For any graph $G = (V, E)$ and any labelling structure S, any S-MLSM ordering of G is an MNSM ordering of G.*

Proof. Let S be a labelling structure and let α be a S-MLSM ordering of G. Let us show that α is a MNSM ordering of G. Let e be an execution of MLSM on G and S computing α, and let e' be an execution of MNSM on G numbering at each step the same vertex as e, provided that this vertex is still unnumbered and of maximal label among the unnumbered vertices. We rename $label_i$ and Num_i into $label'_i$ and Num'_i in e'. For any i from 1 to n, let $P(i)$ be the property: e' numbers successively $\alpha(n)$, $\alpha(n-1)$, ..., $\alpha(i+1)$ and for any vertex $y \in V_i$, $Num_i(y) = Num'_i(y)$. Let us show $P(i)$ by induction on i from n down to 1. $P(n)$ obviously holds. Assume $P(i)$ holds for some i, $n \geq i > 1$. Let us show $P(i-1)$. We first show that e' numbers $\alpha(i)$ at step i, i.e. that there is no vertex $y \in V_i$ such that $label'_i(\alpha(i)) \subset label'_i(y)$. If there was such a vertex y then we would have by definition of MNSM labelling $Num'_i(\alpha(i)) = label'_i(\alpha(i)) \subset label'_i(y) = Num'_i(y)$, and also $Num_i(\alpha(i)) \subset Num_i(y)$ by induction hypothesis, hence by Lemma 2 $label_i(\alpha(i)) \prec label_i(y)$, which contradicts the maximality of $label_i(\alpha(i))$ in V_i in e. Thus e' numbers $\alpha(i)$ at step i, and, when processing $\alpha(i)$, labels of the same vertices are increased in executions e and e' by Lemma 3 a) and induction hypothesis. So $P(i-1)$ holds, which completes the proof of $P(i)$ by induction. From $P(1)$, α is a MNSM ordering of G.

4.2   The CompMLSM Family of Algorithms

We define CompMLSM from MLSM in the same way as we defined CompMLS from MLS. Properties extend readily from an MLSM algorithm to its CompMLSM version: at step i, our proofs only compare the label of $\alpha(i)$ to labels of vertices along paths in the graph G'_i of unnumbered vertices, so $\alpha(i)$ needs only be maximal within the connected component of G'_i containing it.

We thus have similar results:

Theorem 5. *For any input graph G and any X in $\{LEX\ M,\ MCS\text{-}M,\ LexDFS\text{-}M,\ MNSM,\ MLSM\}$, CompX computes a meo α of G and the associated minimal triangulation G^+_α of G.*

We also easily extend results such as Property 1.

An important difference between MLSM and CompMLSM is that the set of orderings CompMLSM can find is independent of the labelling structure used, and is a superset of the set of orderings obtainable by any algorithm of the MLSM family.

Theorem 6. *Any graph has the same S-CompMLSM orderings for all labelling structures S.*

We need the following technical lemma.

Lemma 4. *For any graph G, any labelling structure S, any execution of S-CompMLSM on G computing ordering α and any integer i from 1 to n, $Num_i(\alpha(i))$ is inclusion-maximum in the connected component of G'_i containing $\alpha(i)$.*

Proof. Let C be the connected component of G'_i containing $\alpha(i)$ and y be a vertex of C. Let us show that $Num_i(y) \subseteq Num_i(\alpha(i))$. Let $z \in Num_i(y)$. Let us show that $z \in Num_i(\alpha(i))$. Let μ_1 be a path from $\alpha(i)$ to y in the subgraph of G'_i induced by C. By Lemma 1 and Theorem 5, yz is an edge of G^+_α, so by the Path Lemma there is a path μ_2 in G from y to z such that $\forall t \in \mu_2 \setminus \{y, z\}$, $\alpha^{-1}(t) < \alpha^{-1}(y) \leq i$. The path μ from $\alpha(i)$ to z obtained by concatenation of μ_1 and μ_2 is such that $\forall t \in \mu \setminus \{\alpha(i), z\}$, $\alpha^{-1}(t) < i < \alpha^{-1}(z)$. By the Path Lemma, $\alpha(i)z$ is an edge of G^+_α and, by Lemma 1 and Theorem 5, $z \in Num_i(\alpha(i))$.

Proof (of Theorem 6). Let G be a graph and S, S' be labelling structures. Let us show that any S-CompMLSM ordering of G is a S'-CompMLSM ordering. The proof is similar to that of Theorem 4, replacing S-MLSM by S-CompMLSM and MNSM by S'-CompMLSM. We need only revise our argument that e' numbers $\alpha(i)$ at step i as follows:

Let C be the connected component of G'_i containing $\alpha(i)$, which is the same in e and e' by induction hypothesis. By Lemma 4, $Num_i(\alpha(i))$ is inclusion maximum in C, so by induction hypothesis $Num'_i(\alpha(i))$ is inclusion maximum in C, and by Lemma 2 $label'_i(\alpha(i))$ is maximum in C, so e' numbers $\alpha(i)$ at step i.

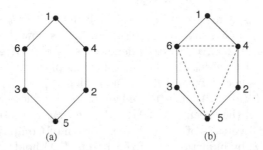

Fig. 6. (a) Graph G and an meo α of G. (b) The corresponding minimal triangulation G_α^+ of G. No version of CompMLSM or MLSM can compute this meo, and the corresponding minimal triangulation is not obtainable by any of these algorithms.

If the input graph is chordal, since the S-CompMLSM execution is the same as the S-CompMLS one, the same orderings are found. By [13], CompMCS computes all peos, so we can deduce that S-CompMLS also computes all peos. Thus Theorem 2 is a corollary of Theorem 6.

Computing all peos does not extend to meos for the MLSM family of algorithms: Figure 6 shows an meo which CompMLSM is not capable of computing.

This raises the question of which minimal triangulations can be obtained by various algorithms of this family. Villanger in [15] proved the surprising result that the sets of minimal triangulations obtainable by LEX M and MCS-M are the same. Upon investigation, it turns out that, given one of these algorithms, using its Comp version does not enlarge the set of computable triangulations, although the set of computable meos may be larger.

Theorem 7. *For any graph G and any given labelling structure S, G has the same sets of S-MLSM and of S-CompMLSM minimal triangulations.*

Proof. Let $G = (V, E)$ be a graph and let S be a labelling structure. Clearly, any S-MLSM minimal triangulation of G is a S-CompMLSM one. Conversely, let H be a S-CompMLSM minimal triangulation of G and let us show that it is a S-MLSM one. Let α be the ordering on V computed by some execution of S-CompMLSM computing H, and, for any i from 1 to n, let C_i be the connected component of G_i' chosen at step i of this execution. Let α' be the ordering on V and H' be the minimal triangulation of G computed by an execution of S-MLSM choosing, for any i from 1 to n, $\alpha'(i)$ at step i in the following way (every variable v is denoted v in the execution of CompMLSM and v' in that of MLSM):

1) Choose a connected component C_i' of G_i'' containing a vertex of maximal label in G_i''.
2) If there is some j from 1 to n such that $C_i' = C_j$ and $label_i'(\alpha(j))$ is maximum in C_i' then choose $\alpha'(i) = \alpha(j)$ else choose for $\alpha'(i)$ any vertex of C_i' of maximal label in G_i''.

Note that there is at most one integer j such that $C'_i = C_j$ since for any j, k such that $j < k$, $C_j \neq C_k$ since $\alpha(k) \in C_k \setminus C_j$. Let us now show that $H' = H$. For any subset J of $\{1, 2, ..., n\}$, let $\alpha(J)$ denote the set of vertices $\{\alpha(j) \mid j \in J\}$, and, for any i from 1 to n, let $P(i)$ be the following property.

$P(i)$: if there is some j from 1 to n such that $C'_i = C_j$ and $\forall y \in C'_i$, $\alpha'(Num'_i(y)) = \alpha(Num_j(y))$ then the edges of H' produced when processing the vertices of C'_i (in the execution of MLSM) are exactly those of H produced when processing the vertices of C_j (in the execution of CompMLSM).

Let us show $P(i)$ by induction on i from 1 to n. $P(1)$ holds since C'_1 contains a single vertex and processing this vertex produces no edge of H (or H'). We suppose $P(i-1)$ for some i, $1 < i \leq n$. Let us show $P(i)$. We suppose that there is some j from 1 to n such that $C'_i = C_j$ and $\forall y \in C'_i, \alpha'(Num'_i(y)) = \alpha(Num_j(y))$. By Lemma 4 $Num_j(\alpha(j))$, and therefore $\alpha(Num_j(\alpha(j)))$, is inclusion-maximum in C_j, so $\alpha'(Num'_i(\alpha(j)))$, and therefore $Num'_i(\alpha(j))$, is inclusion-maximum in C'_i and by Lemma 2 $label'_i(\alpha(j))$ is maximum in C'_i. By definition of α', $\alpha'(i) = \alpha(j)$. The edges of H' produced when processing $\alpha'(i)$ are exactly those of H produced when processing $\alpha(j)$ by Lemma 3 a), and the connected components of G''_{i-1} obtained from C'_i by removing $\alpha'(i)$ are exactly those of G'_{j-1} obtained from C_j by removing $\alpha(j)$ with $\forall y \in C'_i \setminus \{\alpha'(i)\}$, $\alpha'(Num'_{i-1}(y)) = \alpha(Num_{j-1}(y))$. For each such connected component C, there is some $k < i$ and some $l < j$ such that $C = C'_k = C_l$ and $\forall y \in C'_k$, $\alpha'(Num'_k(y)) = \alpha'(Num'_{i-1}(y)) = \alpha(Num_{j-1}(y)) = \alpha(Num_l(y))$, so by induction hypothesis the edges of H' produced when processing the vertices of C are exactly those of H produced when processing the vertices of C. Hence the edges of H' produced when processing the vertices of C'_i are exactly those of H produced when processing the vertices of C_j. So $P(i)$ holds, which completes the induction on i. Now, for each connected component C of G there are some i and j from 1 to n such that $C = C'_i = C_j$ and $\forall y \in C$, $Num'_i(y) = Num_j(y) = \emptyset$, so by $P(i)$ the edges of H' produced when processing the vertices of C are exactly those of H produced when processing the vertices of C. Hence $H' = H$.

Theorem 7, together with Theorem 6, yields the following interesting result:

Theorem 8. *For any graph G, whichever meo-computing algorithm of the MLSM and CompMLSM families is used (e.g., LEX M, MCS-M, LexDFS-M, MNSM, or their Comp extensions), the set of computable minimal triangulations is the same.*

These minimal triangulations fail to cover all possible minimal triangulations: Figure 6(b) shows a minimal triangulation which is obtainable by none of our graph search meo-computing algorithms.

5   Conclusion

We have extended Algorithm LexBFS into Algorithm MLS by defining a general labelling structure, and shown how to extend this further to CompMLS to enable any possible peo to be computed. We have also extended all these algorithms to meo-computing versions. Our work yields alternate (and often simpler) proofs for the results of several papers, as [1,12,13,14,15].

However, we have shown that these new meo-computing algorithms fail to enhance the possibility for finding a wider range of minimal triangulations. LEX M has been studied experimentally, and shown to be very restrictive ([4]), yielding triangulations which are far from minimum. This problem remains with the enlarged family of new meo-computing algorithms we present here, and appears to be a fundamental limitation of graph search.

As for the complexity of Algorithms MLS and MLSM, implementations of LexBFS and MCS in $O(n+m)$ time and of LEX M in $O(nm)$ time are well known [12,14] and serve as valid implementations of MNS and MNSM respectively. In the journal version of this paper we will present a complexity analysis of any algorithm of the MLS and MLSM families. We will study the complexity of the problem of determining whether a given ordering on the vertex set of a given graph can be computed or not by a given algorithm of these families. We will also show that the conjunction of conditions (ls1) and (ls2) on a labelling structure is sufficient, but not necessary, to ensure that the corresponding algorithm of the MLS (resp. MLSM) family computes a peo of every chordal graph (resp. meo of every graph). We will define a weaker condition that is both necessary and sufficient.

As mentioned in the Introduction, LexBFS and MCS, though designed for chordal graphs, have been used for graph classes other than chordal graphs. The more general peo-finding algorithms discussed in this paper could also prove useful on non-chordal graphs, on a wider variety of graph classes and problems.

Acknowledgments

We are very grateful to Derek Corneil for his suggestions, corrections and inspiring work on graph search algorithms.

References

1. A. Berry, J. Blair, P. Heggernes and B. Peyton. Maximum Cardinality Search for computing minimal triangulations of graphs. *Algorithmica*, 39(4):287–298, 2004.
2. A. Berry and J.-P. Bordat. Separability generalizes Dirac's theorem. *Discrete Applied Mathematics*, 84:43–53, 1998.
3. A. Berry and J.-P. Bordat. Moplex elimination orderings. *Electronic Notes in Discrete Mathematics* , Volume 8, 2001, Proceedings of First Cologne-Twente Workshop on Graphs and Combinatorial Optimization.
4. J. R. S. Blair, P. Heggernes, and J. A. Telle. A practical algorithm for making filled graphs minimal. *Theoretical Computer Science A*, 250-1/2: 125–141, 2001.
5. H. L. Bodlaender, A. M. C. A. Koster. On the Maximum Cardinality Search Lower Bound for Treewidth. *Proceedings WG 2004*, 81–92, 2004.
6. D. G. Corneil and R. Krueger. A unified view of graph searching. Submitted.

7. D. G. Corneil, S. Olariu, and L. Stewart. Linear Time Algorithms for Dominating Pairs in Asteroidal Triple-free Graphs. *SIAM Journal on Computing*, 28:1284–1297, 1999.
8. E. Dahlhaus, P. L. Hammer, F. Maffray, and S. Olariu. On Domination Elimination Orderings and Domination Graphs. *Proceedings of WG 1994*, 81-92, 1994.
9. D.R. Fulkerson and O.A. Gross. Incidence matrixes and interval graphs. *Pacific Journal of Math.*, 15:835–855, 1965.
10. T. Ohtsuki. A fast algorithm for finding an optimal ordering in the vertex elimination on a graph. *SIAM Journal on Computing*, 5:133–145, 1976.
11. T. Ohtsuki, L. K. Cheung, and T. Fujisawa. Minimal triangulation of a graph and optimal pivoting order in a sparse matrix. *Journal of Math. Analysis and Applications*, 54:622–633, 1976.
12. D. Rose, R.E. Tarjan, and G. Lueker. Algorithmic aspects of vertex elimination on graphs. *SIAM Journ. Comput*, 5:146–160, 1976.
13. D. R. Shier. Some aspects of perfect elimination orderings in chordal graphs. *Discrete Applied Mathematics*, 7:325–331, 1984.
14. R. E. Tarjan and M. Yannakakis. Simple linear-time algorithms to test chordality of graphs, test acyclicity of hypergraphs, and selectively reduce acyclic hypergraphs. *SIAM J. Comput.*, 13:566–579, 1984.
15. Y. Villanger. Lex M versus MCS-M. *to appear in a Special Issue of Discrete Mathematics*, 2004.

Appendix

We give here the proof of Lemma 3. It uses the following technical Lemmas 5 and 6. For any path μ containing vertices x and y, $\mu[x, y]$ denotes the subpath of μ between x and y.

Lemma 5. *For any graph G, any execution of MLSM on G, any integer i from 1 to n and any path μ in G'_{i-1} ending in some vertex y, if $\forall t \in \mu \setminus \{y\}$, $Num_i(t) \subset Num_i(y)$ then $\forall t \in \mu \setminus \{y\}$, $Num_{i-1}(t) \subset Num_{i-1}(y)$.*

Proof. We suppose that $\forall t \in \mu \setminus \{y\}$, $Num_i(t) \subset Num_i(y)$ (and therefore $label_i(t) \prec label_i(y)$ by Lemma 2). Let $t \in \mu \setminus \{y\}$ and let us show that $Num_{i-1}(t) \subset Num_{i-1}(y)$. It is sufficient to show that if $i \in Num_{i-1}(t)$ then $i \in Num_{i-1}(y)$. We suppose that $i \in Num_{i-1}(t)$. By Lemma 1 there is a path λ in G'_i from $\alpha(i)$ to t such that $\forall t' \in \lambda \setminus \{\alpha(i), t\}$, $label_i(t') \prec label_i(t)$. Let μ' be the path obtained by concatenation of λ and $\mu[t, y]$. Then μ' is a path in G'_i from $\alpha(i)$ to y such that $\forall t' \in \mu' \setminus \{\alpha(i), y\}$, $label_i(t') \prec label_i(y)$. Hence by Lemma 1 $i \in Num_{i-1}(y)$.

Lemma 6. *For any graph G, any execution of MLSM on G, any integer i from 1 to n and any path μ in G'_i ending in some vertex y, if $\exists t \in \mu \setminus \{y\} \mid Num_i(t) \not\subset Num_i(y)$ then $\exists t_1 \in \mu \setminus \{y\} \mid \forall t \in \mu[t_1, y] \setminus \{t_1\}$, $Num_i(t) \subset Num_i(t_1)$.*

Proof. We suppose that $\exists t \in \mu \setminus \{y\} \mid Num_i(t) \not\subset Num_i(y)$. Let j be the largest integer such that $\exists t \in \mu \setminus \{y\} \mid Num_{j-1}(t) \not\subset Num_{j-1}(y)$ and let t_1 be the vertex of μ closest to y such that $Num_{j-1}(t_1) \not\subset Num_{j-1}(y)$. So $j -$

$1 \geq i,\ j \in Num_{j-1}(t_1)$ and $\forall t \in \mu[t_1, y] \setminus \{t_1\}$, $j \notin Num_{j-1}(t)$. Let us show that $Num_j(t_1) = Num_j(y)$. We assume for contradiction that $Num_j(t_1) \neq Num_j(y)$. Let t_2 be the vertex of $\mu[t_1, y]$ closest to t_1 such that $Num_j(t_2) = Num_j(y)$. By the choice of j, $\forall t \in \mu[t_1, t_2] \setminus \{t_2\}$, $Num_j(t) \subset Num_j(t_2)$ and by Lemma 5 $Num_{j-1}(t_1) \subset Num_{j-1}(t_2)$. So $j \in Num_{j-1}(t_2)$ with $t_2 \in \mu[t_1, y] \setminus \{t_1\}$, a contradiction.

So $\forall t \in \mu[t_1, y] \setminus \{t_1\}$, $Num_{j-1}(t) = Num_j(t) \subseteq Num_j(y) = Num_j(t_1) \subset Num_j(t_1) \cup \{j\} = Num_{j-1}(t_1)$. As $j - 1 \geq i$, by Lemma 5 $\forall t \in \mu[t_1, y] \setminus \{t_1\}$, $Num_i(t) \subset Num_i(t_1)$.

Lemma 3. *For any graph G, any execution of MLSM on G computing ordering α, any integer i from 1 to n and any path μ in G'_i ending in some vertex y,*
a) $\forall t \in \mu \setminus \{y\}$, $label_i(t) \prec label_i(y)$ iff $\forall t \in \mu \setminus \{y\}$, $Num_i(t) \subset Num_i(y)$,
b) if $\forall t \in \mu \setminus \{y\}$, $label_i(t) \prec label_i(y)$ then $\forall t \in \mu \setminus \{y\}$, $\alpha^{-1}(t) < \alpha^{-1}(y)$,
c) if $\forall t \in \mu \setminus \{y\}$, $\alpha^{-1}(t) < \alpha^{-1}(y)$ then $\forall t \in \mu \setminus \{y\}$, $Num_i(t) \subseteq Num_i(y)$.

Proof. a) For the forward direction, we assume for contradiction that $\forall t \in \mu \setminus \{y\}$, $label_i(t) \prec label_i(y)$ and $\exists t \in \mu \setminus \{y\} \mid Num_i(t) \not\subset Num_i(y)$ (and therefore $Num_i(t) \not\subseteq Num_i(y)$ since, by Lemma 2 (i), $Num_i(t) \neq Num_i(y)$). By Lemma 6, $\exists t_1 \in \mu \setminus \{y\} \mid Num_i(y) \subset Num_i(t_1)$ and by Lemma 2 $label_i(y) \prec label_i(t_1)$, a contradiction.

The reverse direction follows immediately from Lemma 2.

b) We suppose that $\forall t \in \mu \setminus \{y\}$, $label_i(t) \prec label_i(y)$. Let $k = max\{\alpha^{-1}(t),\ t \in \mu\}$. By a) and Lemma 5, $\forall t \in \mu \setminus \{y\}$, $label_k(t) \prec label_k(y)$. So $\alpha(k) = y$, which completes the proof.

c) We assume for contradiction that $\forall t \in \mu \setminus \{y\}$, $\alpha^{-1}(t) < \alpha^{-1}(y)$ and $\exists t \in \mu \setminus \{y\} \mid Num_i(t) \not\subseteq Num_i(y)$. By Lemma 6 $\exists t_1 \in \mu \setminus \{y\} \mid \forall t \in \mu[t_1, y] \setminus \{t_1\}$, $Num_i(t) \subset Num_i(t_1)$ and by a) and b), $\alpha^{-1}(y) < \alpha^{-1}(t_1)$, a contradiction.

Adding an Edge in a Cograph

Stavros D. Nikolopoulos and Leonidas Palios

Department of Computer Science,
University of Ioannina, GR-45110 Ioannina, Greece
{stavros, palios}@cs.uoi.gr

Abstract. In this paper, we establish structural properties of cographs which enable us to present an algorithm which, for a cograph G and a non-edge xy (i.e., two non-adjacent vertices x and y) of G, finds the minimum number of edges that need to be added to the edge set of G such that the resulting graph is a cograph and contains the edge xy. The motivation for this problem comes from algorithms for the dynamic recognition and online maintenance of graphs; the proposed algorithm could be a suitable addition to the algorithm of Shamir and Sharan [13] for the online maintenance of cographs. The proposed algorithm runs in time linear in the size of the input graph and requires linear space.

Keywords: Perfect graphs, cographs, cotrees, connected components, co-connected components, optimization problems.

1 Introduction

In this paper, we study the following problem:

(*Cograph,*+1)*-MinEdgeAddition*: Given a cograph G and a non-edge xy (i.e., a pair of non-adjacent vertices x and y) of G, find the minimum number of non-edges of G that need to be added to G so that the resulting graph is a cograph and contains xy as an edge.

This problem is an instance of a more general $(\Pi, +k)$-MinEdgeAddition problem in which we deal with a class Π of graphs and we want to have k given non-edges added. Similarly, we can define the $(\Pi, -k)$-MinEdgeAddition problem: we are given a graph G from a class Π and k edges of G which we want removed; since the removal of these edges yields a graph G' which may not necessarily belong to Π, we want to find the minimum number of non-edges of G which when added to G' give a graph in Π (note that the fact that we add non-edges of G prevents the addition of an edge of G which we want removed). Further extensions lead to the $(\Pi, \pm k)$-MinEdgeDeletion problem, in which we remove the minimum number of edges of G (instead of adding non-edges) in order to obtain a graph in Π.

The above problems are motivated by the dynamic recognition problem on (or on-line maintenance of) graphs: a series of requests for the addition or the deletion of an edge or a vertex (potentially incident on a number of edges) are submitted and each is executed only if the resulting graph remains in the same

D. Kratsch (Ed.): WG 2005, LNCS 3787, pp. 214–226, 2005.

class of graphs. Several authors have studied this problem for different classes of graphs and have given algorithms supporting some or all the above operations; we mention the edges-only fully dynamic algorithm of Ibarra [8] for chordal graphs, the fully dynamic algorithm of Hell *et al.* [7] for proper interval graphs, and the fully dynamic algorithm of Shamir and Sharan [13] for cographs.

The *cographs*, short for *complement reducible graphs*, are defined as the class of graphs formed from a single vertex under the closure of the operations of union and complementation, namely: (i) a single-vertex graph is a cograph; (ii) the disjoint union of cographs is a cograph; (iii) the complement of a co-graph is a cograph. Cographs were introduced in the early 1970s by Lerchs [11] who studied their structural and algorithmic properties. Along with other prop-erties, Lerchs has shown that they admit a unique tree representation, up to isomorphism, called a *cotree*. Cographs have arisen in many disparate areas of applied mathematics and computer science and have been independently redis-covered by various researchers under various names such as D^*-graphs [10], P_4 restricted graphs [4,5], 2-parity graphs and Hereditary Dacey graphs or *HD*-graphs [15]. They are perfect and in fact form a proper subclass of permutation graphs and distance hereditary graphs; they contain the class of quasi-threshold graphs and, thus, the threshold graphs [1,9]. Furthermore, they are precisely the graphs which contain no induced subgraph isomorphic to a P_4 (i.e., a chordless path on four vertices).

The study of cographs led naturally to constructive characterizations that implied several linear-time recognition algorithms that also enabled the con-struction of the cotree in linear time [1,14]. Surprisingly, despite the structural simplicity of cographs, constructing linear-time recognition algorithms has been challenging. The first linear-time recognition and cotree-construction algorithm was proposed by Corneil, Perl, and Stewart [5] in 1985. Recently, Bretscher *et al.* [2] presented a simple linear-time recognition algorithm which uses a multi-sweep LexBFS approach; their algorithm either produces the cotree of the input graph or identifies an induced P_4. Additionally, since the cographs are perfect, many interesting optimization problems in graph theory, which are NP-complete in general graphs, have polynomial sequential solutions [1,9]; for example, for the problem of determining the minimum path cover for a cograph, Lin *et al.* [12] presented a linear-time algorithm, which can be used to produce a Hamiltonian cycle or path, if such a structure exists.

In this paper, we solve the (Cograph,+1)-MinEdgeAddition problem. We consider (what we call) the component-partition of a graph G with respect to any of its vertices v: this is related to the partition of the subgraph of G induced by the neighbors of v in G into co-components and to the partition of the sub-graph induced by the non-neighbors of v into components. By taking advantage of the fact that a cograph contains no induced subgraph isomorphic to a P_4 [11], we establish structural properties for the component-partition of a cograph with respect to any of its vertices. These properties and the use of dynamic program-ming enable us to describe an algorithm for the above problem which runs in time linear in the size of the input graph.

2   Theoretical Framework

We consider finite undirected graphs with no loops or multiple edges. For a graph G, we denote by $V(G)$ and $E(G)$ the vertex set and edge set of G, respectively. Let S be a subset of the vertex set $V(G)$ of a graph G; the subgraph of G induced by S is denoted by $G[S]$.

The *neighborhood* $N(x)$ of a vertex x of the graph G is the set of all the vertices of G which are adjacent to x. The *closed neighborhood* of x is defined as $N[x] := N(x) \cup \{x\}$. The neighborhood of a subset S of vertices is defined as $N(S) := \left(\bigcup_{x \in S} N(x) \right) - S$ and its closed neighborhood as $N[S] := N(S) \cup S$. If two vertices x and y are adjacent in G, we say that x *sees* y; otherwise we say that x *misses* y. We extend this notion to vertex sets: $V_i \subseteq V(G)$ sees (misses) $V_j \subseteq V(G)$ if and only if every vertex $x \in V_i$ sees (misses) every vertex $y \in V_j$.

If the graph G contains a path from a vertex x to a vertex y, we say that x *is connected to* y. The *connected components* (or *components*) of G are the equivalence classes of the "is connected to" relation on the vertex set $V(G)$ of G. The *co-connected components* (or *co-components*) of G are the connected components of the complement $\overline{G}$ of G.

3   The Component-Partition

Let us consider for a vertex v of a graph G the partition of the subgraphs $G[N(v)]$ and $G[V(G) - N[v]]$ into co-components and connected components, respectively; then, we define:

Definition 1. *Let G be a graph and v a vertex of G. We define the* component-partition *of G with respect to v, denoted by $(v; \widehat{C}_{1..\ell}; C_{1..k})$, as the partition of the vertex set $V(G)$*

$$V(G) \;=\; \{v\} \;+\; \widehat{C}_1 + \widehat{C}_2 + \ldots + \widehat{C}_\ell \;+\; C_1 + C_2 + \ldots + C_k,$$

where $\widehat{C}_1, \widehat{C}_2, \ldots, \widehat{C}_\ell$ are the co-connected components of $G[N(v)]$ and $C_1, C_2, \ldots, C_k$ are the connected components of $G[V(G) - N[v]]$.

In particular, we restrict our attention to component-partitions such that there are no P_4s with vertices in both $N(v)$ and $V(G) - N[v]$; thus, we define:

Definition 2. *Let G be a graph, v a vertex of G, and $(v; \widehat{C}_{1..\ell}; C_{1..k})$ the component-partition of G with respect to v. We say that this component-partition is* good *if and only if G contains no P_4 with a vertex in some $\widehat{C}_i$ $(1 \le i \le \ell)$ and a vertex in some C_j $(1 \le j \le k)$.*

Our interest in good component-partitions comes from the property described in the following observation:

Observation 1. *Suppose that the component-partition $(v; \widehat{C}_{1..\ell}; C_{1..k})$ of a graph G with respect to a vertex v is good. If G contains a P_4, then all the* *vertices of the P_4 belong to the same co-component $\widehat{C}_i$ or to the same component C_j.*

Observation 1 follows from the fact that the vertices of any P_4 in the subgraph $G[N[v]]$ all belong to the same co-component of $G[N(v)]$, and the vertices of any P_4 in the subgraph $G[V(G) - N[v]]$ all belong to the same component of $G[V(G) - N[v]]$. Additionally, the definition of a good component-partition and the fact that the cographs do not contain P_4s clearly imply:

Observation 2. *If G is a cograph, then the component-partition of any induced subgraph of G with respect to any of its vertices is good.*

In Lemma 1 we establish necessary and sufficient conditions for a component-partition to be good.

Lemma 1. *Let G be a graph, v a vertex of G, and $(v; \widehat{C}_{1..\ell}; C_{1..k})$ the component-partition of G with respect to v. Then, the component-partition of G with respect to v is good if and only if the following two conditions hold:*

(i) for each co-component $\widehat{C}_i$ and each component C_j, $\widehat{C}_i$ either sees or misses C_j;

(ii) if, for each co-component $\widehat{C}_i$, $1 \leq i \leq \ell$, we define the set $\widehat{I}_i = \{ j \mid \widehat{C}_i \text{ sees } C_j \}$, then the co-components of $G[N(v)]$ have the following monotonicity property: $|\widehat{I}_i| \leq |\widehat{I}_j|$ implies that $\widehat{I}_i \subseteq \widehat{I}_j$.

Condition (ii) of Lemma 1 can be phrased in another equivalent way, as given in the following corollary.

Corollary 1. *Let G be a graph, v a vertex of G, and $(v; \widehat{C}_{1..\ell}; C_{1..k})$ the component-partition of G with respect to v. Then, the component-partition of G with respect to v is good if and only if the following two conditions hold:*

(i) for each co-component $\widehat{C}_i$ and each component C_j, $\widehat{C}_i$ either sees or misses C_j;

(ii) Suppose that the ordering of the co-components $\widehat{C}_1, \widehat{C}_2, \ldots, \widehat{C}_\ell$ corresponds to their ordering by non-decreasing $|\widehat{I}_i|$, where $\widehat{I}_i = \{ j \mid \widehat{C}_i \text{ sees } C_j \}$. If we associate each component C_i, $1 \leq i \leq k$, with the set $I_i = \{ j \mid C_i \text{ sees } \widehat{C}_j \}$, then the components of $G[V(G) - N[v]]$ have the following property: if $I_i \neq \emptyset$ and h is the minimum element of I_i, then $I_i = \{h, h+1, \ldots, \ell\}$.

We also note that because the co-components of the neighbors of a vertex and the components of its non-neighbors trade places in the complement of the graph, then properties similar to those described in conditions (ii) of Lemma 1 and Corollary 1 hold for the sets I_i and $\widehat{I}_i$, respectively.

Let us assume that the component-partition $(v; \widehat{C}_{1..\ell}; C_{1..k})$ of the graph G is good. We partition the set of co-components $\{\widehat{C}_1, \widehat{C}_2, \ldots, \widehat{C}_\ell\}$ of the subgraph $G[N(v)]$ into a collection of sets $\widehat{S}_1, \widehat{S}_2, \ldots, \widehat{S}_{\ell'}$ defined as follows:

Definition 3. *Consider the equivalence relation R on the set of co-components $\{\widehat{C}_1, \widehat{C}_2, \ldots, \widehat{C}_\ell\}$ such that $(\widehat{C}_i, \widehat{C}_j) \in R$ if and only if $\widehat{I}_i = \widehat{I}_j$, i.e., $\widehat{C}_i$ and $\widehat{C}_j$ see the same components of the subgraph $G[V(G) - N[v]]$. We define the sets $\widehat{S}_1, \widehat{S}_2, \ldots, \widehat{S}_{\ell'}$ as the equivalence classes of the relation R where, for every i, j such that $1 \leq i < j \leq \ell'$, and every $\widehat{C}_r \in \widehat{S}_i$ and $\widehat{C}_s \in \widehat{S}_j$, it holds that $\widehat{I}_r \subset \widehat{I}_s$.*

The value ℓ' is equal to the number of distinct sets $\widehat{I}_i$, and thus each set $\widehat{S}_j$ is nonempty. It is not difficult to see that the partition sets $\widehat{S}_1, \widehat{S}_2, \ldots, \widehat{S}_{\ell'}$ have the following properties:

- If a connected component $\mathcal{C}$ of the subgraph $G[V(G) - N[v]]$ sees a co-component $\widehat{\mathcal{C}}_i \in \widehat{S}_j$, then $\mathcal{C}$ sees all the co-components in $\widehat{S}_j$.
- Let us consider the ordering of the co-components $\{\widehat{\mathcal{C}}_1, \widehat{\mathcal{C}}_2, \ldots, \widehat{\mathcal{C}}_\ell\}$ consisting of an arbitrary ordering of the elements of the set $\widehat{S}_1$ followed by an arbitrary ordering of the elements of $\widehat{S}_2$ and so on up to the set $\widehat{S}_{\ell'}$. In this ordering, the co-components $\widehat{\mathcal{C}}_i$, $1 \leq i \leq \ell$, are ordered by non-decreasing value of $|\widehat{I}_i|$.

In light of these properties and the fact that the component-partition $(v; \widehat{\mathcal{C}}_{1..\ell}; \mathcal{C}_{1..k})$ is good (thus condition (ii) of Corollary 1 holds), we have:

Definition 4. *We define the following partition of the set of components $\{\mathcal{C}_1, \mathcal{C}_2,$ $\ldots, \mathcal{C}_k\}$ of the subgraph $G[V(G) - N[v]]$:*

$$
\begin{aligned}
S_1 &= \{\, \mathcal{C}_j \mid \forall \widehat{\mathcal{C}} \in \widehat{S}_1,\; \mathcal{C}_j \text{ sees } \widehat{\mathcal{C}} \,\} \\
S_i &= \{\, \mathcal{C}_j \mid \forall \widehat{\mathcal{C}} \in \widehat{S}_i \text{ and } \widehat{\mathcal{C}}' \in \widehat{S}_{i-1},\; \mathcal{C}_j \text{ sees } \widehat{\mathcal{C}} \text{ but misses } \widehat{\mathcal{C}}' \,\} \quad (2 \leq i \leq \ell') \\
S_{\ell'+1} &= \{\, \mathcal{C}_1, \mathcal{C}_2, \ldots, \mathcal{C}_k \,\} - \bigcup_{i=1,\ldots,\ell'} S_i
\end{aligned}
$$

The definition of the sets $\widehat{S}_j$, $j = 1, 2, \ldots, \ell'$, implies that $S_i \neq \emptyset$ for all $i = 2, 3, \ldots, \ell'$. However, $S_{\ell'+1}$ and S_1 may be empty. In particular, $S_{\ell'+1}$ is empty if and only if the graph G is connected; in fact, $S_{\ell'+1}$ contains the connected components of G except for the component to which v belongs. Figure 1 illustrates the partitions of the set of co-components and of the set of components described above and their adjacencies in a good component-partition of the graph G with respect to vertex v; the dotted ovals indicate the partition sets, and the circles inside the ovals indicate the components or co-components belonging to the partition set.

In terms of the partitions into sets $\widehat{S}_1, \widehat{S}_2, \ldots, \widehat{S}_{\ell'}$ and $S_1, S_2, \ldots, S_{\ell'}, S_{\ell'+1}$, the cotree of a cograph G has a very special structure, which is described in

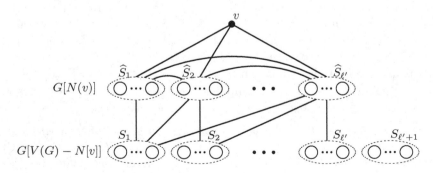

Fig. 1

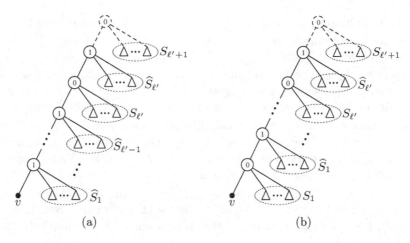

Fig. 2

the following observation (because of Observation 2, the sets $\widehat{S}_1, \widehat{S}_2, \ldots, \widehat{S}_{\ell'}$ and $S_1, S_2, \ldots, S_{\ell'}, S_{\ell'+1}$ are well defined).

Observation 3. *Let G be a cograph, v a vertex of G, and $\widehat{S}_1, \widehat{S}_2, \ldots, \widehat{S}_{\ell'}$ and $S_1, S_2, \ldots, S_{\ell'}, S_{\ell'+1}$, respectively, the partitions of the co-connected components of $G[N(v)]$ and of the connected components of $G[V(G) - N[v]]$ as defined above. Then,*

(i) if $S_1 = \emptyset$, the cotree of G has the general form depicted in Figure 2(a);

(ii) if $S_1 \neq \emptyset$, the cotree of G has the general form depicted in Figure 2(b).

In either case, the dashed part[1] appears in the tree if and only if $S_{\ell'+1} \neq \emptyset$.

The circular nodes labeled with a 0 or a 1 in Figure 2 are 0-nodes and 1-nodes, respectively, whereas the shaded node is a leaf node; the triangles denote the cotrees of the corresponding connected components or co-components.

4   Adding an Edge in a Cograph

Let G be a cograph and let x, y be two vertices of G which are not adjacent. We want to solve the (Cograph,+1)-MinEdgeAddition problem for G, x, y, i.e., we wish to make x and y adjacent, while adding the minimum number of non-edges of G so that the resulting graph is a cograph. Instrumental in the algorithm that we will be presenting is the component-partition of the graph G with respect to a vertex of G (see Definition 1) and in particular the partitions into sets $\widehat{S}_i$ and

---

[1] Lerchs' definition required that the root of a cotree be a 1-node [11]; here, we relax this condition and allow the root to be a 0-node as well, thus obtaining cotrees whose internal nodes all have at least two children, and whose root is a 1-node if and only if the corresponding cograph is connected.

S_j (see Definitions 3 and 4); since G is a cograph, Observation 2 holds and thus the adjacencies between the $\widehat{S}_i$s and S_js are as shown in Figure 1.

In particular, let $\widehat{S}_1(x)$, $\widehat{S}_2(x), \ldots,$ $\widehat{S}_{\ell'_x}(x)$ and $S_1(x), S_2(x), \ldots,$ $S_{\ell_x}(x)$, $S_{\ell_x+1}(x)$ be the sets of the co-components of the subgraph $G[N(x)]$ and of the connected components of the subgraph $G[V(G) - N[x]]$, respectively. Since x and y are non-adjacent, then y belongs to a set, say, $S_{k_x}(x)$; in particular, let $\mathcal{C}_y$ be the component in $S_{k_x}(x)$ to which y belongs. Similarly, let $\widehat{S}_1(y)$, $\widehat{S}_2(y), \ldots,$ $\widehat{S}_{\ell'_y}(y)$ and $S_1(y), S_2(y), \ldots,$ $S_{\ell_y}(y), S_{\ell_y+1}(y)$ be the sets of the co-components of $G[N(y)]$ and of the connected components of $G[V(G) - N[y]]$, respectively, and suppose that x belongs to the component $\mathcal{C}_x$ of the set $S_{k_y}(y)$. Because the elements of the sets $\widehat{S}_{k_x+i}(x)$ and $\widehat{S}_{k_y+i}(y)$, $i \geq 0$, and $S_{k_x+i}(x)$ and $S_{k_y+i}(y)$, $i \geq 1$, correspond to the subtrees of the cotree of G hanging from the nodes in the path from the parent of the least common ancestor of x and y to the root (see Figure 2), it holds that

$$\widehat{S}_{k_x+i}(x) = \widehat{S}_{k_y+i}(y) \quad \text{for all } i \geq 0$$
$$\text{and} \qquad S_{k_x+i}(x) = S_{k_y+i}(y) \quad \text{for all } i \geq 1$$

which also implies that $\ell'_x - k_x = \ell'_y - k_y$. Moreover, from any subtrees, other than those containing x and y, hanging from the least common ancestor of x and y, we have:

$$S_{k_x}(x) - \{\mathcal{C}_y\} = S_{k_y}(y) - \{\mathcal{C}_x\}.$$

For the sake of simplicity of the notation, let us define

$$V_i(x) = \bigcup_{1 \leq t \leq i} \left(\widehat{S}_t(x) \cup S_t(x)\right) \quad \text{and} \quad V_i(y) = \bigcup_{1 \leq t \leq i} \left(\widehat{S}_t(y) \cup S_t(y)\right).$$

Note that $V_0(x) = \emptyset$ and $V_0(y) = \emptyset$. Then, the properties of a good component-partition (in light of Observation 2) imply (see also Figure 1):

P1: the common neighbors of x and y are precisely the vertices in $\widehat{S}_{k_x}(x) \cup \widehat{S}_{k_x+1}(x) \cup \ldots \cup \widehat{S}_{\ell'_x}(x) = \widehat{S}_{k_y}(y) \cup \widehat{S}_{k_y+1}(y) \cup \ldots \cup \widehat{S}_{\ell'_y}(y)$;

P2: $\mathcal{C}_y = \{y\} \cup V_{k_y-1}(y)$ and similarly, $\mathcal{C}_x = \{x\} \cup V_{k_x-1}(x)$.

In order to show Property P2, we note that $\mathcal{C}_y$ is the connected component to which y belongs after all the common neighbors of x and y have been removed; then, Property P2 follows from considering the removal of the vertices in $\widehat{S}_{k_y}(y) \cup \widehat{S}_{k_y+1}(y) \cup \ldots \cup \widehat{S}_{\ell'_y}(y)$ (see Property P1) in the component-partition of the graph G with respect to y.

Let G' be an optimal solution to the (Cograph,+1)-MinEdgeAddition problem, i.e., G' is a cograph for which $V(G') = V(G)$, $E(G) \cup \{xy\} \subseteq E(G')$, and $|E(G')|$ is minimum. Clearly, Observation 2 holds for G'; the properties of G' are described in the following two lemmata.

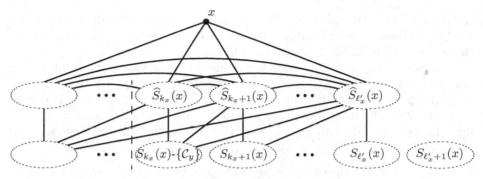

Fig. 3. $S_{k_x}(x) - \{C_y\} \neq \emptyset$: the rightmost sets $\widehat{S}'_i(x)$ and $S'_i(x)$, $i = r_x - (\ell'_x - k_x), \ldots, r_x$, in the partitions of the subgraphs $G'[N_{G'}(x)]$ and $G'[V(G') - N_{G'}[x]]$

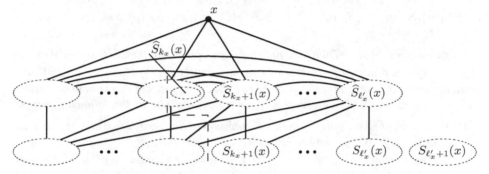

Fig. 4. $S_{k_x}(x) = \{C_y\}$: the rightmost sets $\widehat{S}'_i(x)$ and $S'_i(x)$, $i = r_x - (\ell'_x - k_x) + 1, \ldots, r_x$, and $\widehat{S}'_{r_x - (\ell'_x - k_x)}(x)$ in the partitions of $G'[N_{G'}(x)]$ and $G'[V(G') - N_{G'}[x]]$

Lemma 2. *Let G be a cograph, x, y be two non-adjacent vertices of G, and let*

- $\widehat{S}_1(x), \widehat{S}_2(x), \ldots, \widehat{S}_{\ell_x}(x)$ *and* $S_1(x), S_2(x), \ldots, S_{\ell'_x}(x), S_{\ell'_x+1}(x)$,
- $\widehat{S}_1(y), \widehat{S}_2(y), \ldots, \widehat{S}_{\ell_y}(y)$ *and* $S_1(y), S_2(y), \ldots, S_{\ell'_y}(y), S_{\ell'_y+1}(y)$,
- $k_x, k_y, C_y, V_i(x),$ *and* $V_i(y)$

be as described above. Then, for the partition of the subgraphs $G'[N_{G'}(x)]$ and $G'[V(G') - N_{G'}[x]]$ of an optimal graph G' into sets of co-components $\widehat{S}'_1(x)$, $\widehat{S}'_2(x), \ldots, \widehat{S}'_{r_x}(x)$, and connected components $S'_1(x), S'_2(x), \ldots, S'_{r_x}(x), S'_{r_x+1}(x)$ respectively, the following properties hold:

(i) $\widehat{S}'_{r_x-i}(x) = \widehat{S}_{\ell_x-i}(x) = \widehat{S}_{\ell_x-i}(y)$ *for all* $i = 0, 1, 2, \ldots, \ell'_x - k_x - 1$, *and* $S'_{r_x+1-i}(x) = S_{\ell'_x+1-i}(x) = S_{\ell'_y+1-i}(y)$ *for all* $i = 0, 1, 2, \ldots, \ell'_x - k_x$ *(see Figures 3 and 4);*

(ii) *if $S_{k_x}(x)$ contains at least one connected component in addition to C_y, then* $\widehat{S}'_{r_x-(\ell'_x-k_x)}(x) = \widehat{S}_{k_x}(x) = \widehat{S}_{k_y}(y)$ *and* $S'_{r_x-(\ell'_x-k_x)}(x) = S_{k_x}(x) - \{C_y\}$ *(see Figure 3);*

(iii) *if $S_{k_x}(x)$ contains just the connected component C_y, then all the co-components in $\widehat{S}_{k_x}(x)$ form co-components in $\widehat{S}'_{r_x-(\ell'_x-k_x)}(x)$ (see Figure 4).*

In terms of the cotree of the graph G, Lemma 2 implies that all changes that need to be done in order to obtain the cotree of the graph G' are restricted in the subtree rooted at the least common ancestor of x and y.

The remaining sets of co-components and components in the component-partition of the optimal graph G' with respect to x are obtained from an optimal "grouping" of the vertices in $V_{k_x-1}(x) \cup C_y = V_{k_x-1}(x) \cup \{y\} \cup V_{k_y-1}(y)$ (see Property P2). The following lemma gives the possible cases of such a "grouping." It does not take into account case (iii) of Lemma 2; if this case applies, then the set $\widehat{S}'_{r_x-(\ell'_x-k_x)}(x)$, in addition to the co-components that result by the "grouping," contains the co-components in $\widehat{S}_{k_x}(x)$ as well (see Figure 4).

Lemma 3. *Let G be a cograph, x, y be two non-adjacent vertices of G, and let*

- $\widehat{S}_1(x), \widehat{S}_2(x), \ldots, \widehat{S}_{\ell'_x}(x)$ *and* $S_1(x), S_2(x), \ldots, S_{\ell'_x}(x), S_{\ell'_x+1}(x),$
- $\widehat{S}_1(y), \widehat{S}_2(y), \ldots, \widehat{S}_{\ell'_y}(y)$ *and* $S_1(y), S_2(y), \ldots, S_{\ell'_y}(y), S_{\ell'_y+1}(y),$
- $V_i(x)$ *and* $V_i(y)$

be as described above. Then, an optimal "grouping" of the vertices in $V_i(x) \cup \{y\} \cup V_j(y)$ in order to form sets of co-components $\widehat{S}'_1(x), \widehat{S}'_2(x), \ldots, \widehat{S}'_k(x)$ and sets of connected components $S'_1(x), S'_2(x), \ldots, S'_k(x)$ in the component-partition of an optimal graph G' with respect to vertex x is of one of the following forms:

(a) *$k = 1$, $\widehat{S}'_k(x)$ contains a single co-component induced by all the vertices in $V_i(x) \cup \{y\} \cup V_j(y)$, and $S'_k(x) = \emptyset$ (see Figure 5(a));*

(b) *provided that $V_j(y) \neq \emptyset$, $k = i + 1$, $\widehat{S}'_k(x)$ consists of a single co-component involving just vertex y, $S'_k(x)$ consists of a single connected component induced by the vertices in $V_j(y)$, whereas the remaining sets $\widehat{S}'_1(x), \widehat{S}'_2(x), \ldots, \widehat{S}'_{k-1}(x)$ and $S'_1(x), S'_2(x), \ldots, S'_{k-1}(x)$ are identical to $\widehat{S}_1(x), \widehat{S}_2(x), \ldots, \widehat{S}_i(x)$ and $S_1(x), S_2(x), \ldots, S_i(x)$, respectively (see Figure 5(b));*

(c) *provided that $j \geq 2$ or $j = 1$ and $S_1(y) \neq \emptyset$, $\widehat{S}'_k(x) = \widehat{S}_j(y)$ and $S'_k(x) = S_j(y)$ (see Figure 5(c));*

(d) *provided that $i \geq 2$ or $i = 1$ and $S_1(x) \neq \emptyset$, $\widehat{S}'_k(x) = \widehat{S}_i(x)$ and $S'_k(x) = S_i(x)$ (see Figure 5(d)).*

Lemma 2 and the fact that the vertices in $\widehat{S}_{k_x}(x) \cup \widehat{S}_{k_x+1}(x) \cup \ldots \cup \widehat{S}_{\ell'_x}(x)$ see all the vertices in $\{x, y\} \cup V_{k_x-1}(x) \cup V_{k_y-1}(y)$ (see Property P1) imply that new edges are added only as a result of the "grouping;" it is important to note that we do not need to add new edges connecting vertices in the same set $\widehat{S}'_t(x)$ or $S'_t(x)$ as the vertices in each such set induce subgraphs not containing any P_4s. Then, in light of Lemma 3, we get a recursive expression for the number of additional edges that such an optimal "grouping" requires; this is given in Lemma 4.

Lemma 4. *Suppose that the conditions of Lemma 2 hold and let $cost(i, j)$ denote the number of edges with both endpoints in $\{x\} \cup V_i(x) \cup \{y\} \cup V_j(y)$ which need to be added to G in an optimal "grouping" of the vertices in $V_i(x) \cup \{y\} \cup V_j(y)$ to form sets $\widehat{S}'_t(x)$ and $S'_t(x)$. Then,*

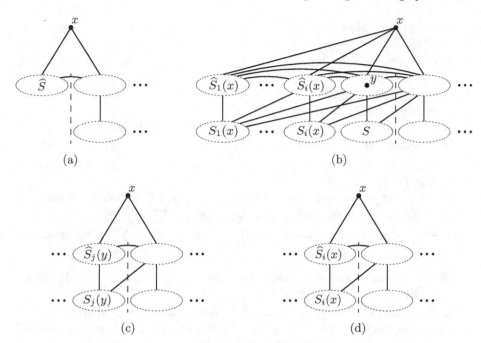

Fig. 5. Cases (a)-(d) of Lemma 3 where $\widehat{S}$ contains a single co-component induced by $V_i(x) \cup \{y\} \cup V_j(y)$ and S contains a single component induced by $V_j(y)$

(i)   the number of additional edges in the graph G' (optimal solution for the problem (Cograph,+1)-MinEdgeAddition for G, x, y) is $cost(k_x - 1, k_y - 1)$;

(ii) the value $cost(i, j)$ is the minimum among the costs of the cases below provided that they apply:

(a) $1 + \sum_{t=1}^{i} |S_t(x)| + \sum_{t=1}^{j}(|\widehat{S}_t(y)| + |S_t(y)|)$;

(b) $1 + \sum_{t=1}^{i}(|\widehat{S}_t(x)| + |S_t(x)|) + \sum_{t=1}^{j} |S_t(y)|$, provided that $j \geq 1$;

(c) $cost(i, j - 1) + |\widehat{S}_j(y)| \cdot \left(1 + \sum_{t=1}^{i}(|\widehat{S}_t(x)| + |S_t(x)|)\right)$, provided that $j \geq 2$ or $j = 1$ and $S_1(y) \neq \emptyset$;

(d) $cost(i - 1, j) + |\widehat{S}_i(x)| \cdot \left(1 + \sum_{t=1}^{j}(|\widehat{S}_t(y)| + |S_t(y)|)\right)$, provided that $i \geq 2$ or $i = 1$ and $S_1(x) \neq \emptyset$.

It is important to note the symmetry between cases (a) and (b) and between cases (c) and (d) with respect to x and y, as it is expected. Let us now consider some special cases.

○   If $i = 0$ and $j = 0$, then only case (a) applies and $cost(0, 0) = 1$.

○   If $i = 0$ and $j = 1$, then case (d) does not apply while the cost in case (a) is no smaller that the cost in case (b), thus, $cost(0, 1)$ is the minimum of $1 + \sum_{t=1}^{1} |S_t(y)|$ and of $cost(0, j - 1) + |\widehat{S}_j(y)|$ assuming that case (c) applies: if $S_1(y) \neq \emptyset$ then case (c) applies and since $cost(0, 0) = 1$, we have that $cost(0, 1) = \min\{1 + |S_1(y)|, 1 + |\widehat{S}_1(y)|\}$; if $S_1(y) = \emptyset$ then case (c) does not apply and thus $cost(0,1)$ is $1 + |S_1(y)| = 1$. In either case, $cost(0,1) = \min\{1 + |S_1(y)|, 1 + |\widehat{S}_1(y)|\}$.

○ If $i = 1$ and $j = 0$, then cases (b) and (c) do not apply, so that $cost(1,0)$ is the minimum of $1 + \sum_{t=1}^{1} |S_t(x)|$ and of $cost(0,0) + |\widehat{S}_i(x)|$ assuming that case (d) applies. The case is symmetric to the previous case so that $cost(1,0) = \min\{1 + |S_1(x)|, 1 + |\widehat{S}_1(x)|\}$.

○ If $i = 0$ and $j \geq 2$, then case (d) does not apply, while the cost in case (a) is no smaller than the cost in case (b) since $1 + \sum_{t=1}^{j}(|\widehat{S}_t(y)| + |S_t(y)|) < 1 + \sum_{t=1}^{j} |S_t(y)|$; thus, $cost(0,j)$ is the minimum of $1 + \sum_{t=1}^{j} |S_t(y)|$ and of $cost(0, j-1) + |\widehat{S}_j(y)|$.

○ If $i \geq 2$ and $j = 0$, then case (c) does not apply, while the cost in case (b) is no smaller than the cost in case (a) since $1 + \sum_{t=1}^{i}(|\widehat{S}_t(x)| + |S_t(x)|) < 1 + \sum_{t=1}^{i} |S_t(x)|$; thus, $cost(i,0)$ is the minimum of $1 + \sum_{t=1}^{i} |S_t(x)|$ and of $cost(i - 1, 0) + |\widehat{S}_i(x)|$.

Based on Lemma 4 and the above discussion, we give below our algorithm. The algorithm uses four matrices $A_x[]$, $B_x[]$, $A_y[]$, and $B_y[]$, such that $A_v[i] = \sum_{t=1}^{i} |\widehat{S}_t(v)|$ and $B_v[i] = \sum_{t=1}^{i} |S_t(v)|$. It also uses a 2-dimensional array $cost[\,,\,]$, where it saves the values of $cost(\,,\,)$. The algorithm receives as input a cograph G on n vertices and two non-adjacent vertices x, y of G, and outputs the minimum number of edges that need to be added to G so that x, y become adjacent and the resulting graph is a cograph (we note that the algorithm can be easily modified to produce the set of edges that need to be added, instead of their number only, within the same time and space complexity). In detail, it works as follows:

Algorithm ADD-EDGE-IN-COGRAPH

1. Compute the sets
   $\widehat{S}_1(x), \widehat{S}_2(x), \ldots, \widehat{S}_{\ell_x}(x)$ of co-components of $G[N(x)]$ and
   $S_1(x), S_2(x), \ldots, S_{\ell'_x}(x), S_{\ell'_x+1}(x)$ of conn. components of $G[V(G) - N[x]]$;
   find the set $S_{k_x}(x)$ to which y belongs;
   compute the sets
   $\widehat{S}_1(y), \widehat{S}_2(y), \ldots, \widehat{S}_{\ell'_y}(y)$ of co-components of $G[N(y)]$ and
   $S_1(y), S_2(y), \ldots, S_{\ell'_y}(y), S_{\ell'_y+1}(y)$ of conn. components of $G[V(G) - N[y]]$;
   find the set $S_{k_y}(y)$ to which x belongs;

2. $A_x[0] \leftarrow 0; \qquad B_x[0] \leftarrow 0;$
   for $i = 1, 2, \ldots, k_x - 1$ **do**
   $\qquad A_x[i] \leftarrow A_x[i-1] + |\widehat{S}_i(x)|;$
   $\qquad B_x[i] \leftarrow B_x[i-1] + |S_i(x)|;$
   $A_y[0] \leftarrow 0; \qquad B_y[0] \leftarrow 0;$
   for $i = 1, 2, \ldots, k_y - 1$ **do**
   $\qquad A_y[i] \leftarrow A_y[i-1] + |\widehat{S}_i(y)|;$
   $\qquad B_y[i] \leftarrow B_y[i-1] + |S_i(y)|;$

3. $cost[0,0] \leftarrow 1;$
   for $j = 1, 2, \ldots, k_y - 1$ **do**
   $\qquad cost[0,j] \leftarrow \min\{1 + B_y[j], cost[0, j-1] + A_y[j] - A_y[j-1]\};$

```
for  i = 1, 2, ..., k_x − 1 do
   cost[i, 0] ← min{1 + B_x[i], cost[i − 1, 0] + A_x[i] − A_x[i − 1]};
   for  j = 1, 2, ..., k_y − 1 do
      val1 ← 1 + B_x[i] + A_y[j] + B_y[j];              {case (a)}
      val2 ← 1 + A_x[i] + B_x[i] + B_y[j];              {case (b)}
      if  j ≥ 2 or (j = 1 and B_y[1] ≠ 0)              {case (c)}
      then  val3 ← cost[i, j − 1] + (A_y[j] − A_y[j − 1]) · (1 + A_x[i] + B_x[i])
      else  val3 ← n²;
      if  i ≥ 2 or (i = 1 and B_x[1] ≠ 0)              {case (d)}
      then  val4 ← cost[i − 1, j] + (A_x[i] − A_x[i − 1]) · (1 + A_y[j] + B_y[j])
      else  val4 ← n²;
      cost[i, j] ← min{val1, val2, val3, val4};
return(cost[k_x − 1, k_y − 1]).
```

Note that whenever a value $cost[\,,\,]$ is needed for another cost-computation, it has already been computed. The correctness of Algorithm ADD-EDGE-IN-COGRAPH follows from Lemma 4, the discussion of the special cases, and the definitions of the arrays $A_x[\,]$, $B_x[\,]$, $A_y[\,]$, and $B_y[\,]$, which also imply that $A_x[i] - A_x[i-1] = |\widehat{S}_i(x)|$, $B_x[j] - B_x[j-1] = |S_j(x)|$, and similarly for $A_y[\,]$ and $B_y[\,]$.

Time and Space Complexity: Suppose that the input cograph G has n vertices and m edges. Then, the sets $\widehat{S}_1(x), \widehat{S}_2(x), \ldots, \widehat{S}_{\ell'_x}(x)$ and $S_1(x), S_2(x), \ldots,$ $S_{\ell'_x + 1}(x)$ can be computed in $O(n + m)$ time and space either by computing the cotree of G [5], or by computing the co-components of $G[N(x)]$ [3,6] and the connected components of $G[V(G) - N[x]]$ and then by placing them in the appropriate $\widehat{S}_i(x)$ or $S_i(x)$ based on their number of incident edges to vertices in $V(G) - N[x]$ and in $N(x)$ respectively. Finding the set $S_{k_x}(x)$ can be done in constant time. Similarly, the computation of the corresponding sets $\widehat{S}_i(y)$ and $S_i(y)$, and finding $S_{k_y}(y)$ takes $O(n + m)$ time and space. For the complexity of Steps 2 and 3, we observe that ℓ'_x and ℓ'_y are $O(\sqrt{m})$: since every vertex in any co-component of $\widehat{S}_i$ $(1 \leq i \leq \ell'_x)$ sees every vertex in the co-components of $\widehat{S}_j$ for $j \neq i$, there exist at least $\ell'_x(\ell'_x - 1)/2$ edges connecting vertices in different co-components of $G[N(x)]$; since G contains a total of m edges and there are at least ℓ'_x edges connecting x to its neighbors, we conclude that $m \geq \ell'_x + \ell'_x(\ell'_x - 1)/2 > {\ell'_x}^2/2$, from which the result for ℓ'_x follows; a similar argument holds for ℓ'_y. Step 2 takes $O(\sqrt{m}) = O(n)$ time, since $k_x \leq \ell'_x$ and $k_y \leq \ell'_y$. Step 3 takes $O(k_x \cdot k_y) = O(\ell'_x \cdot \ell'_y) = O(m)$ time. The space needed by Algorithm ADD-EDGE-IN-COGRAPH is equal to the space needed for the representation of the input graph G and the space taken by the arrays $A_x[\,]$, $B_x[\,]$, $A_y[\,]$, $B_y[\,]$, and $cost[\,,\,]$; hence, it is $O(n + m + k_x \cdot k_y) = O(n + m)$. Therefore, Algorithm ADD-EDGE-IN-COGRAPH takes $O(n + m)$ time and space.

5   Concluding Remarks

In this paper, we described a linear-time algorithm for the (Cograph,+1)-Min-EdgeAddition problem; instrumental in our construction are the properties of the component-partition of a cograph that we establish. Since the cographs are complement-invariant, the approach we used when applied on the complement of the given graph gives a solution to the (Cograph,−1)-MinEdgeDeletion problem.

It would be interesting to obtain efficient algorithms for the (Cograph,−1)-MinEdgeAddition and the (Cograph,+1)-MinEdgeDeletion problems as well as for the extensions of all these problems in which k edges or non-edges are involved. Finally, it would also be interesting to study the problems for other classes of graphs; an obvious immediate next step would be to consider the class of P_4-sparse graphs, a superclass of the class of cographs.

Acknowledgment. The authors would like to thank A. Brandstädt for proposing the four variants $(\Pi, \pm 1)$-MinEdgeAddition/Deletion problems, D. Corneil for suggesting a solution to the (Cograph,−1)-MinEdgeDeletion problem, and D. Kratsch for constructive suggestions and comments.

References

1. A. Brandstädt, V.B. Le, and J.P. Spinrad, *Graph Classes: A Survey*, SIAM Monographs on Discrete Mathematics and Applications, 1999.
2. A. Bretscher, D. Corneil, M. Habib, and C. Paul, A simple linear time LexBFS cograph recognition algorithm, *Proc. 29th Int'l Workshop on Graph Theoretic Concepts in Comput. Sci. (WG'03)*, LNCS 2880 (2003) 119–130.
3. K.W. Chong, S.D. Nikolopoulos, and L. Palios, An optimal parallel co-connectivity algorithm, *Theory Comput. Systems* **37** (2004) 527–546.
4. D.G. Corneil, H. Lerchs, and L. Stewart-Burlingham, Complement reducible graphs, *Discrete Appl. Math.* **3** (1981) 163–174.
5. D.G. Corneil, Y. Perl, and L.K. Stewart, A linear recognition algorithm for cographs, *SIAM J. Comput.* **14** (1985) 926–934.
6. E. Dahlhaus, J. Gustedt, and R.M. McConnell, Partially Complemented Representations of Digraphs, *Discrete Math. & Theoret. Comput. Sci.* **5** (2002) 147–168.
7. P. Hell, R. Shamir, and R. Sharan, A fully dynamic algorithm for recognizing and representing proper interval graphs, *SIAM J. Comput.* **31** (2002) 289–305.
8. L. Ibarra, Fully dynamic algorithms for chordal graphs, *Proc. 10th Annual ACM-SIAM Symp. on Discrete Algorithms (SODA'99)*, (1999) 923–924.
9. M.C. Golumbic, *Algorithmic Graph Theory and Perfect Graphs*, Academic Press, Inc., 1980.
10. H.A. Jung, On a class of posets and the corresponding comparability graphs, *J. Combin. Theory Ser. B* **24** (1978) 125–133.
11. H. Lerchs, On cliques and kernels, *Technical Report*, Department of Computer Science, University of Toronto, March 1971.
12. R. Lin, S. Olariu, and G. Pruesse, An optimal path cover algorithm for cographs, *Computers Math. Applic.* **30** (1995) 75–83.
13. R. Shamir and R. Sharan, A fully dynamic algorithm for modular decomposition and recognition of cographs, *Discrete Appl. Math.* **136** (2004) 329–340.
14. J.P. Spinrad, *Efficient Graph Representations*, American Mathematical Society, 2003.
15. D.P. Sumner, Dacey graphs, *J. Austral. Math. Soc.* **18** (1974) 492–502.

The Computational Complexity
of Delay Management
Extended Abstract

Michael Gatto[1], Riko Jacob[1], Leon Peeters[1], and Anita Schöbel[2]

[1] Institute of Theoretical Computer Science, ETH Zurich
{gattom, rjacob, peetersl}@inf.ethz.ch
[2] Institute for Numerical and Applied Mathematics, University of Göttingen
schoebel@math.uni-goettingen.de

Abstract. Delay management for public transport consists of deciding whether vehicles should wait for delayed transferring passengers, with the objective of minimizing the overall passenger discomfort.

This paper classifies the computational complexity of delay management problems with respect to various structural parameters, such as the maximum number of passenger transfers, the graph topology, and the capability of trains to reduce delays. Our focus is to distinguish between polynomially solvable and NP-complete problem variants. To that end, we show that even fairly restricted versions of the delay management problem are hard to solve.

1 Introduction

Even a carefully planned railway system will once in a while have to deal with delayed trains due to unforeseeable events. In such a case, the railway operator can react by maintaining some connections and modifying the schedule accordingly.

This paper considers the impact of such modifications on the overall passenger delay. The problem of managing delayed trains is still not well understood, even though the first research on railway delays started as early as two decades ago (see, for example, [HK81]). In particular, no efficient exact algorithms are known so far for any general problem setting. We present an explanation for this situation by showing that several restricted versions of the delay management problem are NP-complete. We identify various combinatorial aspects that cause the problem to be difficult to solve, and complementarily extend some of the polynomial time algorithms of [GGJ$^+$04]. Thus, we establish a fairly precise complexity boundary that depends on structural parameters of the problem instance.

The delay management problem considers a trade-off that is best explained by an example. Consider a passenger in an on-time train, which decides to wait for a delayed feeder train. Although the passenger was traveling on-time, she now faces a delay because of this decision. Moreover, she herself may later miss a connecting train in a subsequent station. Alternatively, had the train not waited, then the

D. Kratsch (Ed.): WG 2005, LNCS 3787, pp. 227–238, 2005.

connecting passengers in the feeder train would have missed their connection. In particular, they would have had to wait for the next train, thus facing a large delay each. Delay management consists of deciding which connecting trains should wait for which delayed feeder trains, with the objective of minimizing the sum of the delays faced by the passengers.

We model the railway network as a graph, and passenger flows as fixed paths in this graph. In this network, unforeseen events may occur that result in the late arrival of connecting passengers at transfer stations. Given these so-called source delays, our goal is to decide which connecting trains wait for delayed transfer passengers, such that the sum of all passenger delays is minimized. In our opinion, this model captures the key aspects of delay management, such as the propagation of delays through the network. Because of its abstractness, the model is also applicable to other modes of scheduled public transport. Still, some important real-life aspects are not included, such as the availability of track capacity to accommodate the adjusted schedule.

1.1   Contribution of the Paper

This paper identifies a boundary between NP-completeness and polynomial solvability for various natural problem parameters of the delay management problem. In particular, we focus here on the case where all non-zero source delays are of equal size, which we refer to as binary source delays.

We first show that the binary delay management problem is strongly NP-complete if trains cannot catch up on their delay, already on a railway network with series-parallel topology. As the complexity reduction requires the passengers to transfer at most three times, the result complements our earlier finding that the problem is polynomially solvable when passengers transfer at most twice [GGJ+04]. We also extend the latter result to the case of unbounded number of transfers, in which initially on-time passengers are not allowed to miss a connecting train (though they are allowed to be delayed).

Next, we study the binary delay management problem with slack times, meaning that trains can catch up on their delay. We show that this variant is already NP-complete on a railway network with a line structure. Again, this contrasts an earlier result on the polynomial solvability of such a line network without slack times [GGJ+04]. Further, a slightly different NP-completeness reduction yields passengers that transfer twice, on a more general network that is series-parallel. As an ingredient for one of our proofs, we establish that the maximum unweighted directed cut problem on directed acyclic graphs is NP-complete.

Without slack times, all source delays must contribute to the objective. For this setting, we also investigate the objective function without this offset, and show that it is NP-hard to approximate to a certain constant factor.

Finally, we describe a polynomial time "pedal-to-the-metal" algorithm for the delay management problem with slack times, under the restriction that all passengers travel to the same destination station on a network with a tree-like structure.

Table 1. Classification of the binary delay management problem, with implied entries grayed out. Contributions of the paper are in italic.

Network topology	No slack times			With slack times	
	≤ 2 transfers	≤ 3 transfers	< ∞ transfers	≤ 2 transfers	< ∞ transfers
General	min cut	*NP-complete*	*NP-complete*	*NP-complete*	*NP-complete*
Series-parallel	min cut	*NP-complete*	*NP-complete*	*NP-complete*	*NP-complete*
Line	min cut	dynamic prog	dynamic prog	?	*NP-complete*
Tree, single destination				*"pedal-to-the-metal"*	

Given our interest in complexity aspects, we focus on fairly simple versions of the delay management problem, which are perhaps not too realistic. Still, our findings give insight into the structure of more complex and more realistic models. Thus, the main contribution of the paper is to identify combinatorial aspects that are crucial for the problem's complexity. Table 1 summarizes the results. Naturally, the unrestricted delay management problem is NP-complete as well.

Due to space limitations, this paper only summarizes our results and omits most of the proofs. The detailed exposition of our results is available as [GJPS04].

1.2   Related Research

The above described delay management problem was introduced by [Sch01], who proposed a Mixed Integer Programming formulation for a model that is similar to ours. Schöbel [Sch03] also showed that, when no two delayed vehicles meet in an optimal solution to this model, its constraint matrix is totally unimodular. In that case, an optimal solution can be obtained in polynomial time by Linear Programming. Further, [GGJ+04] described a minimum cut reduction for passengers that transfer at most twice, and a polynomial-time dynamic program for railway networks with a path topology.

In spite of these algorithmic results, no strong NP-completeness results were known so far for delay management. [Sch03] showed that the bi-criteria problem of concurrently minimizing the weighted passenger delay and the number of missed connections is weakly NP-complete. For the same bi-criteria problem, [Meg04] provides a slightly different complexity proof and some further theoretical observations.

[GJPW04] provides a first competitive analysis for the on-line version of delay management. A series of papers by Suhl et al., most recently [BS04], evaluates different deterministic delay policies by simulation.

2   Problem Statement

This section describes the delay management model analyzed in the paper. First, we describe the general model, which is similar to the model in [Sch01]. Next, we specify the considered restrictions of the model.

General Model Definition

Let $G = (V, E)$ be a directed acyclic graph. Each vertex $v \in V$ represents a station, and each edge $e = (u, v) \in E$ represents a single direct train only operating the connection between u and v. Trains do not have intermediate stops in this model. At each station v, the outgoing edges (v, w) represent the connecting trains for the passengers traveling on the incoming trains (u, v). A directed path in G then corresponds to a journey a passenger can undertake by transferring between trains. We assume transfers to happen instantaneously. Thus, passengers arriving at a station with a delay can only board the connecting train if it waits for the entire delay of the feeder train. Alternatively, one could model the transfers by additional edges in the graph. We omit this construction for simplicity, but point out that these additional edges do not influence our results.

A train $e = (u, v) \in E$ can reduce a possible delay by $\mathcal{S}(e) \geq 0$ time units on its trip from u to v, for example by driving faster than scheduled. We refer to $\mathcal{S}(e)$ as the *slack time* of the train. Trains must not arrive earlier than scheduled, so slack times can only be used if a train waits for some passengers, and thus departs with a delay.

Passenger flows in the railway system are modeled by a set of directed paths $\mathcal{P}$ in the graph. Such a path P induces transfers at every internal vertex of P. A path $P \in \mathcal{P}$ has an associated weight $w(P)$ representing the number of passengers, or the importance of the path in a more abstract sense. As a direct consequence of an unforeseen event, some passengers may arrive at a transfer station with a delay. In our model, such passengers are represented by a passenger path $P \in \mathcal{P}$ with a source delay $\mathcal{D}(P) > 0$, starting at that transfer station and ending at the passengers' destination. Thus, our model defines source delays on paths rather than on trains. We refer to paths with $\mathcal{D}(P) = 0$ as *source punctual paths*, and to paths with $\mathcal{D}(P) > 0$ as *source delayed paths*.

A passenger path $P \in \mathcal{P}$ misses a connection if it arrives at a transfer station with a delay, and its connecting train does not wait long enough. We assume that trains are operated according to a periodic timetable with period T, and that delays do not propagate to the next period. Although the latter may happen, we do not consider such cases for the sake of simplicity. Hence, a passenger path $P \in \mathcal{P}$ has an arrival delay $\delta_P = T$ if it misses a connection. If all connections on path P are maintained, its arrival delay δ_P equals the arrival delay of its last train. We refer to paths with $\delta_P = 0$, arriving as scheduled at their destination, as *punctual paths*, to those arriving delayed as *delayed paths*, and to those missing a connection as *dropped paths*. Further, we refer to paths not missing a connection as *maintained paths*. The possibility to drop paths is a key aspect of this setting. Indeed, dropping a path P effectively removes P from the network, such that no train is influenced any more by P.

An instance is completely defined by the tuple $(G, \mathcal{S}, \mathcal{P}, \mathcal{D}, w, T)$. For such an instance, a delay policy π specifies which trains wait, for how long, and how much slack time they use. We wish to find a delay policy π^* that minimizes the *total passenger delay* defined as the weighted sum of arrival delays $\sum_{P \in \mathcal{P}} w_P \delta_P$.

Problem Restricting Parameters

Our complexity results consider several restricted versions of the general model. These restrictions include the basic cases of limiting the maximum number of passenger transfers (maximum passenger path length), restricting the source delays to one single non-zero value (binary source delays), and not allowing for trains to catch up on delays (availability of slack times).

As for the network topology, we consider lines, trees, and series-parallel graphs. Series-parallel graphs have treewidth two, which intuitively means they are almost trees. Many NP-complete problems, such as Independent Set and Vertex Cover, become polynomially solvable on bounded-treewidth graphs. Hence, an NP-hardness result for series-parallel graphs in some sense complements a polynomial time algorithm for trees and line graphs.

Our results also apply to the setting where all source delayed paths originate from a single delayed train [GJPS04]. Additionally, one can consider dynamic path choices, i.e., passengers can react dynamically to the chosen delay policy. Such choices become irrelevant if only unique origin-destination paths exist, as can be obtained in most of our constructions. Hence, our hardness results extend to this setting [GJPS04].

3   Delay Management Without Slack Times

Our first analysis considers the restricted setting of binary source delays and no slack times, that is, $\mathcal{D}(P) \in \{0, \delta\}$ for all $P \in \mathcal{P}$, and $\mathcal{S}(e) = 0$ for all $e \in E$. In this setting, an optimal delay policy π^* describes which trains depart on-time, and which ones wait for time δ. We refer to this restricted model as the binary delay management problem, and write an instance as $(G, \mathcal{P}, \mathcal{D}, w, T)$.

3.1   Proof of Hardness with Three Transfers

Here, we show that the binary delay management problem is NP-complete already for unweighted passenger paths on a series-parallel train-network. To that end, we first prove a weaker theorem.

Definition: Decision binary delay management problem.
Instance: A binary delay management instance $(G, \mathcal{P}, \mathcal{D}, w, T)$, $d \in \mathbb{N}$.
Question: Is there a delay policy such that the total passenger delay is less than or equal to d?

Theorem 1. *The decision binary delay management problem with passenger paths changing at most four times is NP-complete.*

Proof. It is easy to see that the problem is in NP, as the weighted delay of the paths induced by a delay policy π can be computed in polynomial time, and the size of π is polynomial as well. We show that the problem is NP-hard by

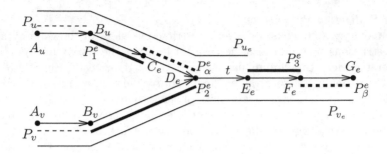

Fig. 1. The construction for an extended edge (u, u_e, v_e, v) in G_2. Directed edges represent the trains in the network. Undirected lines represent the paths. Thick paths are the paths of weight M. Dashed paths represent paths with source delay δ.

reduction from Maximum Independent Set [GJ79–Problem GT20]. Let the undirected graph $G = (V, E)$, $|V| = n$, $|E| = m$ be a Maximum Independent Set instance asking for an independent set of size K. Consider its 2-subdivision [Pol74] $G_2 = (V_2, E_2)$, i.e., the graph obtained by inserting the vertices u_e, v_e for each undirected edge $e = (u, v)$ and splitting the edge into three undirected edges $(u, u_e), (u_e, v_e), (v_e, v)$. We refer to this construction for an edge $e = (u, v)$ of the original graph as the extended edge in the 2-subdivision, symbolized by (u, u_e, v_e, v). The graph G has a maximum independent set of size K if and only if its 2-subdivision G_2 has a maximum independent set of size $K + m$.

In the following, we construct gadgets for every extended edge of the 2-subdivision graph. In the resulting delay management instance, certain paths that are maintained in an optimal delay policy π^* correspond to the vertices in the maximum independent set of the 2-subdivision.

For each vertex q in G_2 we construct a path P_q in the delay management instance, such that two vertices q, r can be in the same independent set if and only if the corresponding paths P_q and P_r can both be maintained in the same optimal delay policy. A maximum independent set in G_2 hence corresponds to an optimal set of maintained paths.

For this construction, consider an extended edge (u, u_e, v_e, v). For the vertices u, v we have paths $P_u, P_v \in \mathcal{P}$, both with unit weight and unit source delay. These paths exist once for every $u \in V$. Further, we introduce paths P_{u_e}, P_{v_e} for u_e and v_e, both with unit weight and no source delay. The exact configuration of all these paths is shown in Figure 1.

For each extended edge (u, u_e, v_e, v) of the 2-subdivision we introduce five paths in the delay management instance, $P_1^e, P_2^e, P_3^e, P_\alpha^e, P_\beta^e$, each with weight $w(P_i^e) = M, i \in \{1, 2, 3, \alpha, \beta\}$, where M is a sufficiently large value. The paths P_α^e and P_β^e have source delay δ, the other paths have no source delay. Because of the large weight M, the source delayed paths P_α^e, P_β^e will never be dropped in an optimal delay policy π^*. For the same reason, the paths $P_i^e, i \in \{1, 2, 3\}$ will always be kept punctual. We refer to these paths as M-paths, and Figure 1 shows their exact configuration.

Let π^* be an optimal delay management policy for the constructed instance, meaning that no M-paths are dropped. In π^*, the paths corresponding to vertices of G_2 interact by sharing edges. Because of the M-paths, π^* cannot maintain two such interacting paths, since one requires the shared edge to be delayed, whereas the other requires it to be on-time. Indeed, P_u and P_{u_e} share the edge (A_u, B_u), and P_{u_e} and P_{v_e} share (D_e, E_e). Note that this construction enforces that each maintained unit weight path arrives at its destination with a delay of δ.

Hence, the unit weight paths that are maintained in π^* correspond to an independent set I in G_2. Since every maintained path reduces the cost of the delay policy, I is a maximum independent set.

More precisely, set $\delta = 1$, $T = 2$, $M = m + 2$, and $d = 2mM\delta + (2m + n)T - (m + K)(T - \delta)$. Now G_2 has an independent set of size $m + K$ if and only if the binary delay management instance has a delay policy π with cost at most d which maintains $m + K$ unit weight paths.

Finally, observe that the longest constructed passenger path requires only four changes. $\qquad\square$

Theorem 1 can be strengthened by constructing an even simpler instance of the delay management problem.

Theorem 2. *The decision binary delay management problem is NP-complete on a series-parallel graph with passenger paths changing at most three times and unweighted passenger paths.*

The proof is given in [GJPS04]. The main idea to obtain a series-parallel network is to contract the seven node classes $A, \cdots, G$ to a single node each, which does not change the interaction of paths.

3.2   Approximating the Additional Delay

As stated in Section 2, our objective is to minimize the total weighted passenger delay. Alternatively, it also makes sense to minimize only the weighted delay that paths face in addition to their source delay. Indeed, as there are no slack times, a source delayed path can never do better than arrive at its destination with a delay of δ. This portion of the delay cannot be optimized, so it is reasonable to omit it from the objective function. We refer to this alternative objective function as the additional weighted delay, which should be minimized.

As Independent Set and Vertex Cover are complementary problems, the results from [Hås01] provide an inapproximability result for the delay management problem with the additional delay objective function. The proof involves a different reduction from independent set, and generally needs more than three passenger transfers. The proof can be found in [GJPS04].

Lemma 1. *For any $\epsilon > 0$, it is NP-hard to approximate the binary decision delay management problem with the objective of minimizing the additional delay within a factor $\frac{15}{14} - \epsilon$.*

3.3   Polynomially Solvable Cases

Several special cases of the delay management problem can be solved by reduction to a minimum directed cut problem [GGJ+04]. This section extends these results for the case in which no delay policy is allowed to drop source punctual paths. As dropping source punctual paths is in some sense unfair, this restricted case may very well be reasonable from a practical point of view.

A minimum directed s-t-cut is a partition of the vertex set into two disjoint sets $S, \bar{S}$, with $s \in S, t \in \bar{S}$, such that the sum of the costs of the edges traversing the cut from S to $\bar{S}$ is minimal. We construct a new graph $G' = (V', E', c)$, with a cost function $c : E \to \mathbb{N}$. A minimum directed cut in G' with respect to c corresponds to an optimal delay policy on $(G, \mathcal{P}, \mathcal{D}, w, T)$.

We map the trains E to vertices in G', and add two new vertices s and t. The idea of the reduction is that trains in S wait, whereas trains in $\bar{S}$ depart on time. For each source punctual passenger path we introduce infinite weight edges between every two subsequent trains the path uses. Further, for each such path we introduce a new vertex and infinite weight edges from each train used by the path to the new vertex, such that it must be in S if the path is delayed. An appropriately weighted edge connected to $t \in \bar{S}$ accounts for the delay occurring if the path is delayed. For each source delayed path, we also introduce a new vertex. This vertex is connected to all trains used by the path with infinite weight edges, such that the vertex is in $\bar{S}$ if one of these trains is on-time. An appropriately weighted edge connects $s \in S$ with this vertex, accounting for the dropping costs. An additional weighted edge (s, t) accounts for the delay of the source delayed path. The detailed construction and the proof of the following lemma are given in [GJPS04].

Lemma 2. *Given that source punctual passenger paths cannot be dropped, the minimum total passenger delay for $(G, \mathcal{D}, \mathcal{P}, w, T)$ is equal to the cost of the minimum directed s-t-cut $[S, \bar{S}]$ in G'. In such an optimal delay policy all trains corresponding to vertices in S wait and all trains corresponding to vertices in $\bar{S}$ depart on-time.*

Now, if the source punctual paths are short enough that they cannot be dropped, the above construction works in general.

Corollary 1. *The delay management problem with an unrestricted delay policy can be solved by reduction to a minimum cut problem if each source punctual path uses one train only.*

Moreover, a slightly different construction than above for the source punctual paths yields the following more general theorem. The proof can be found in [GJPS04].

Theorem 3. *The delay management problem with an unrestricted delay policy can be solved by reduction to a minimum cut problem if each source punctual path transfers at most twice.*

4   Delay Management with Slack Times

In this section, we analyze the case where trains can have some slack time. A train e having slack time $\mathcal{S}(e)$ is able to catch up $\mathcal{S}(e)$ time units on its delay. As stated earlier, we do not allow trains to catch up more time than they are delayed, implying that they can never arrive early. Also for this case, the never-meet-property allows for a polynomial-time algorithm [Sch03].

4.1   Proof of Hardness

In Section 3.1, we showed that the delay management problem is NP-complete already on series-parallel networks. Here, we show that by including slack times the delay management problem becomes NP-complete already on a network with the topology of a line. In contrast, the delay management problem without slack times is still polynomially solvable in this case [GGJ$^+$04]. Further, we show that by including slack times the delay management problem becomes NP-complete already with passenger paths transferring twice. This variant as well is still polynomial-time solvable without slack times [GGJ$^+$04].

We reduce from Maximum Directed Acyclic Cut. As the hardness proof of Maximum Directed Cut in [PY91] does not create an acyclic graph, we first show that the Maximum Directed Acyclic Cut problem is NP-complete.

Definition: Maximum Directed Acyclic Cut
Instance: Directed acyclic unweighted graph $G = (V, E)$, $K \in \mathbb{N}$.
Question: Does a partition of V exist into two disjoint sets $V_1, V_2, V = V_1 \cup V_2$, such that the number of edges traversing the partition from V_1 to V_2 is greater than or equal to K?

Theorem 4. *Maximum Directed Acyclic Cut is NP-complete.*

Proof. Clearly, the problem is in NP. We prove it to be NP-hard by reduction from Maximum Unweighted Directed Cut [GJ79–Problem ND16]: given a directed graph $G = (V, E), |V| = n, |E| = m$ and a positive integer $K \in N$, is there a partition of V into two disjoint sets $V_1, V_2, V = V_1 \cup V_2$, such that the number of edges traversing the cut from V_1 to V_2 is at least K?

We first build a maximum directed acyclic cut instance $G' = (V', E')$ using edge weights c' as follows. For each vertex $v_i \in V$, we build a gadget of five vertices, $\{v_i^1, v_i^2, v_i^3, v_i^4, v_i^5\}$, connected by four edges $(v_i^j, v_i^{j+1}), j \in \{1, \dots, 4\}$, with weight $c'(v_i^j, v_i^{j+1}) = m$. At most two non-consecutive edges of each gadget can traverse the cut. By setting their weights to m we enforce that two of these edges actually do traverse the cut. For each edge $e = (v_i, v_j) \in E$, we insert the edge (v_i^2, v_j^4) in E' with weight $c'(v_i^2, v_j^4) = 1$.

The reduction is polynomial in space and time: we have $5n$ vertices and $4n + m$ edges, and the graph can be constructed efficiently. The graph G has a maximum cut V_1, V_2 of size K if and only if G' has a maximum cut V_1', V_2' of size $2nm + K$.

The crucial observation for the reduction's correctness is that a node gadget cannot have both $v_i^2 \in V_1'$ and $v_i^4 \in V_2'$, and at the same time contribute $2m$ from gadget-internal edges.

Finally, the above reduction also works for unweighted graphs G'. In that case we introduce, for each gadget, m parallel paths of length two between even-numbered vertices instead of the edges of weight m, multiplying the odd-numbered vertices. Still, the cut consistently separates the vertices as above. As we introduce $4m$ edges for each vertex in G, the construction remains polynomial.   □

In the following, we show that fairly restricted versions of the delay management problem with slack times are already NP-complete.

Definition: Decision delay management problem with slack times.
Instance: A delay management instance $(G, \mathcal{S}, \mathcal{P}, \mathcal{D}, w, T)$, $d \in \mathbb{N}$.
Question: Does a delay policy exist, such that the total passenger delay does not exceed d?

Theorem 5. *The decision delay management problem with slack times is NP-complete with binary delays, binary slack times, unweighted passenger paths, and passengers transferring at most twice.*

Proof. The proof is by reduction from Maximum Directed Acyclic Cut. It is clear that the problem is in NP, as the delay of each path can be efficiently computed from a delay policy.

Given a maximum directed acyclic cut instance $G = (V, E)$, we build a delay management problem $(G', \mathcal{P}, \mathcal{D}, w, \mathcal{S}, T)$, with $G' = (V', E')$, as follows. For every $v \in V$, we introduce an edge $f_v \in E'$ without slack. For each edge $e = (u, v) \in E$, we introduce an edge g_e from f_u to f_v having slack time equal to δ. Further, for each edge $e = (u, v) \in E$, we introduce two paths, the path $P_e = (f_u, g_e, f_v)$ with source delay δ with unit weight, and the source punctual path $P_e^u = \{f_u\}$ with weight 3. More precisely, the latter weighted path can be replaced by three parallel paths of unit weight. Note that each outgoing edge $(u, v) \in E$ induces one path P_e^u on f_u.

We set $\delta = 1$ and $T = 4$, and ask for a delay policy inducing a total delay of $d = mT - K\delta = 4m - K$. There is a direct correspondence of a delay policy in G' to a cut in G: if f_u waits, $u \in V_1$, otherwise $u \in V_2$. It remains to prove that we have a cut of size at least K if and only if there is a delay policy with at most d total delay. To this end, it is sufficient to analyze the delay caused by the two paths P_e and P_e^u for the different policies. If f_u does not wait, P_e is dropped and P_e^u is on time. So, independent of f_v, these two paths together contribute T to the objective. If both f_u and f_v wait, the paths contribute $4\delta = T$ to the objective, as both paths arrive with a delay. Only if f_u waits and f_v departs as scheduled, the two paths contribute 3δ to the objective. Now, G has a maximum directed cut of size K if and only if $(G, \mathcal{S}, \mathcal{P}, \mathcal{D}, w, T)$ has a delay policy causing $4m - K = d$ delay. Using the described correspondence between a cut in G and a delay policy in G', for every edge e of G there is a contribution of 3 units to the total delay if e crosses the cut, and of 4 units otherwise. □

In contrast to Lemma 2, no source punctual paths are dropped in the above construction. Note that the reduction can be adapted to any $T = k\delta$ by introducing $k - 1$ paths P_e^i per edge e instead of three. The special case $k = 1$ is also feasible, but it is unclear how this should be interpreted. Furthermore, dynamic path choices do not influence the construction of Theorem 5, since the first and the last edge of the paths P_e cannot be changed. This observation allows us to simplify the network topology even further, as stated in the following theorem and proven in [GJPS04].

Corollary 2. *The decision delay management problem with slack times is NP-complete with binary delays, binary slack times, and unweighted passenger paths, even if the network forms a line.*

In general, the proof of Corollary 2 yields paths with an arbitrary number of transfers, as opposed to the proof of Theorem 2. Actually, the delay management problem with slack times is already NP-complete on a series-parallel network where passengers transfer at most twice. For a proof, see [GJPS04].

Corollary 3. *The decision delay management problem with slack times is NP-complete on a series-parallel network if passenger paths transfer twice, with unit path weights, binary delays, and binary slack times.*

4.2 Polynomially Solvable Cases

Although the general setting on the line is NP-hard, some variants of the delay management problem with slack times can be solved efficiently by simple strategies. Below we describe two such variants.

Let G be a graph that forms a line. Contrary to the models analyzed so far, we consider a single train traveling on the line with intermediate stops. This implies that a passenger path does not need to connect to other trains, once it has entered the train. Hence, a path can either be dropped before boarding the train, or it reaches its destination, possibly with some delay.

First, assume that all paths $P \in \mathcal{P}$ end at the terminal station of the considered train. This can be interpreted as passengers traveling to the city center on an urban rail line. We refer to this model as *all passengers to a unique destination on a single train.*

Theorem 6. *The delay management problem with slack times and all passengers to a unique destination on a single train can be solved in polynomial time.*

Proof. This problem can be solved by the following pedal-to-the-metal strategy. The driver a priori fixes a target delay at the terminal stop, exhausts all slack times, and drives at maximum velocity to achieve that target delay. For a more precise analysis, see [GJPS04].

The above delay policy can be extended to the case where single trains operate between the stations, the graph is a rooted in-tree, and all passengers travel to the root of the tree. The passengers must thus connect to a new train at each intermediate station on their trip to the root node. For further details, see [GJPS04].

Acknowledgment. We thank Peter Widmayer for many fruitful discussions.

References

[BS04]   C. Biederbick and L. Suhl. Improving the quality of railway dispatching tasks via agent-based simulation. In *Computers in Railways IX*, pages 785–795. WIT Press, 2004.

[GGJ⁺04]   M. Gatto, B. Glaus, R. Jacob, L. Peeters, and P. Widmayer. Railway delay management: Exploring its algorithmic complexity. In *Algorithm Theory - Proceedings SWAT 2004*, pages 199–211. Springer-Verlag LNCS 3111, 2004.

[GJ79]   M. Garey and D. Johnson. *Computers and Intractability – A Guide to the Theory of NP-Completeness*. Freeman and Company, New York, 1979.

[GJPS04]   M. Gatto, R. Jacob, L. Peeters, and A. Schöbel. The computational complexity of delay management. Technical Report 456, ETH Zurich, 2004.

[GJPW04]   M. Gatto, R. Jacob, L. Peeters, and P. Widmayer. On-line delay management on a single line. In *Proc. Algorithmic Methods and Models for Optimization of Railways (ATMOS)*. Springer-Verlag LNCS, 2004. To appear.

[Hås01]   J. Håstad. Some optimal inapproximability results. *Journal of ACM*, 48:798–859, 2001.

[HK81]   C. Hendrickson and G. Kocur. Schedule delay and departure time decisions in a deterministic model. *Transportation Science*, 15:62–77, 1981.

[Meg04]   C. Megyeri. Bicriterial delay management. Konstanzer Schriften in Mathematik und Informatik 198, University of Konstanz, 2004.

[Pol74]   S. Poljak. A note on the stable sets and coloring of graphs. *Comment. Math. Univ. Carolin.*, 15:307–309, 1974.

[PY91]   C. H. Papadimitriou and M. Yannakakis. Optimization, approximation, and complexity classes. *Journal of Computer and System Sciences*, 43:425–440, 1991.

[Sch01]   A. Schöbel. A model for the delay management problem based on mixed-integer-programming. In Christos Zaroliagis, editor, *Electronic Notes in Theoretical Computer Science*, volume 50. Elsevier, 2001.

[Sch03]   A. Schöbel. *Customer-oriented optimization in public transportation*. Habilitation Thesis, University of Kaiserslautern, 2003. To appear.

Acyclic Choosability of Graphs with Small Maximum Degree

Daniel Gonçalves and Mickaël Montassier

LaBRI UMR CNRS 5800, Université Bordeaux I,
33405 Talence Cedex, France
{goncalve, montassi}@labri.fr

Abstract. A proper vertex coloring of a graph $G = (V, E)$ is acyclic if G contains no bicolored cycle. A graph G is L-list colorable if for a given list assignment $L = \{L(v) : v \in V\}$, there exists a proper coloring c of G such that $c(v) \in L(v)$ for all $v \in V$. If G is L-list colorable for every list assignment with $|L(v)| \geq k$ for all $v \in V$, then G is said k-choosable. A graph is said to be acyclically k-choosable if the coloring obtained is acyclic. In this paper, we study the acyclic choosability of graphs with small maximum degree. In 1979, Burstein proved that every graph with maximum degree 4 admits a proper acyclic coloring using 5 colors [Bur79]. We give a simple proof that (a) every graph with maximum degree $\Delta = 3$ is acyclically 4-choosable and we prove that (b) every graph with maximum degree $\Delta = 4$ is acyclically 5-choosable. The proof of (b) uses a backtracking greedy algorithm and Burstein's theorem.

1 Introduction

Let G be a graph. Let $V(G)$ be its set of vertices and $E(G)$ be its set of edges. A proper vertex coloring of G is an assignment f of integers (or labels) to the vertices of G such that $f(u) \neq f(v)$ if the vertices u and v are adjacent in G. A k-coloring is a proper vertex coloring using k colors. A proper vertex coloring of a graph is *acyclic* if there is no bicolored cycle. The *acyclic chromatic number* of G, $\chi_a(G)$, is the smallest integer k such that G is acyclically k-colorable. Acyclic colorings were introduced by Grünbaum in [Grü73] and studied by Mitchem [Mit74], Albertson, Berman [AB77], and Kostochka [Kos76]. In 1979, Borodin proved Grünbaum's conjecture:

Theorem 1. *[Bor79] Every planar graph is acyclically 5-colorable.*

This bound is best possible: in 1973, Grünbaum gave an example of a 4-regular planar graph [Grü73] which is not acyclically colorable with four colors. Moreover, there exist bipartite 2-degenerate planar graphs which are not acyclically 4-colorable [KM76].

Borodin, Kostochka and Woodall improved this bound for planar graphs with a given girth. We recall that the girth of a graph is the length of its shortest cycle.

Theorem 2. *[BKW99]*

1. *Every planar graph with girth at least 7 is acyclically 3-colorable.*
2. *Every planar graph with girth at least 5 is acyclically 4-colorable.*

D. Kratsch (Ed.): WG 2005, LNCS 3787, pp. 239–248, 2005.

In 1979, Burstein studied graphs with small maximum degree and proved :

Theorem 3. *[Bur79] Every graph with maximum degree 4 is acyclically 5-colorable.*

There are graphs with maximum degree 4 which need 5 colors, for example K_5.

A graph G is L-list colorable if for a given list assignment $L = \{L(v) : v \in V(G)\}$ there is a coloring c of the vertices such that $c(v) \in L(v)$ and $c(v) \neq c(u)$ if u and v are adjacent in G. If G is L-list colorable for every list assignment with $|L(v)| \geq k$ for all $v \in V(G)$, then G is said k-choosable. In [Voi93], Voigt proved that there are planar graphs which are not 4-choosable, and in [Tho94], Thomassen proved that every planar graph is 5-choosable. In this paper we focus on acyclic choosability of graphs. This is, for which value k, any list assignement L, with $|L(v)| \geq k$ for all $v \in V(G)$, allows an acyclic coloring of G. In [BFDFK$^+$02], the following theorem is proved and the next conjecture is given:

Theorem 4. *[BFDFK$^+$02] Every planar graph is acyclically 7-choosable.*

This means that for any given list assignment L, with $|L(v)| \geq 7$ for all $v \in V(G)$, there is an acyclic coloring c of G, such that it is possible to choose for each vertex v a color in $L(v)$. The *acyclic list chromatic number* of G, $\chi_a^l(G)$, is the smallest integer k such that G is acyclically k-choosable.

Conjecture 1. [BFDFK$^+$02] Every planar graph is acyclically 5-choosable.

The acyclic choosability has been studied for other families of graphs. In [MS04], an upper bound on χ_a^l for the graphs with bounded degree is given:

Theorem 5. *[MS04] Let G be a graph with maximum degree Δ, then $\chi_a^l(G)$* $\leq \lceil 50\Delta^{4/3} \rceil$.

In [MOR05], the authors studied the acyclic choosability of graphs with bounded maximum average degree. The maximum average degree, $Mad(G)$, of the graph G is defined as

$$Mad(G) = \max\{2|E(H)|/|V(H)|, H \subseteq G\}$$

Theorem 6. *[MOR05]*

1. *Every graph G with $Mad(G) < \frac{8}{3}$ is acyclically 3-choosable.*
2. *Every graph G with $Mad(G) < \frac{19}{6}$ is acyclically 4-choosable.*
3. *Every graph G with $Mad(G) < \frac{24}{7}$ is acyclically 5-choosable.*

This result implies that every graph with maximum degree 2 (resp. 3) is acyclically 3-choosable (resp. acyclically 4-choosable). The proof of this theorem is based on discharging methods. In this paper, we give a simpler proof of the acyclic 4-choosability of subcubic graphs and we prove the next theorem:

Theorem 7. *Let G be a graph with maximum degree $\Delta \leq 4$, then $\chi_a^l(G) \leq 5$.*

Note that Theorem 7 improves Burstein's result on maximum degree four graphs. In what follows, we call k-vertex a vertex of degree k. The next section is dedicated to the acyclic 4-choosability of subcubic graphs. In Section 3, we prove Theorem 7.

2   Acyclic 4-Choosability of Subcubic Graphs

Let H be a subcubic graph with minimum order which is not acyclically 4-choosable. Let $L = \{L(v) : v \in V(H)\}$ be a list assignment such that there exists no extracted acyclic coloring. Let c be a proper coloring of H, with $c(v) \in L(v)$ for all $v \in V(H)$, such that the number a of bicolored cycles is minimal. There is such coloring, since subcubic graphs are 4-choosable. Let $\mathcal{C}$ be a bicolored cycle. We prove that we can recolor a part of $\mathcal{C}$ such that $\mathcal{C}$ is 3-colored and the total number of bicolored cycle is at most $a - 1$. The coloring obtained contradicts the minimality of a, completing the proof.

Claim. The counterexample H does not contain 1-vertices nor 2-vertices, so H is 3-regular.

Proof.   1. Suppose that H contains a 1-vertex u adjacent to a vertex v. By minimality of H, the graph $H' = H \setminus \{u\}$ is acyclically 4-choosable. Let c be an acyclic coloring of H' such that $c(v) \in L(v)$ for all $v \in V(H')$. We extend this coloring to H by coloring u with any color in $L(u) \setminus \{c(v)\}$. Since u cannot be in a cycle, the coloring obtained is an acyclic coloring of H, contradicting the definition of H.
   2. Suppose that H contains a 2-vertex v adjacent to two other vertices u and w. By minimality of H, the graph $H' = (V(H) \setminus \{v\}, E(H) \setminus \{uv, vw\} \cup \{uw\})$ is acyclically 4-choosable. There is an acyclic coloring c of H' which we can extend to H by coloring v with a color in $L(v) \setminus \{c(u), c(w)\}$. Indeed, v cannot be part of a bicolored cycle since $c(u) \neq c(w)$, u and v being adjacent in H'.

Assume w.l.o.g. that the cycle $\mathcal{C} = x_1 x_2 x_3 \ldots x_k$ with $k \geq 4$ is bicolored using the colors 1 and 2, with $c(x_1) = 1$. Each vertex x_i is adjacent to the vertices x_{i-1}, x_{i+1}, and y_i. Each vertex y_i is adjacent to x_i and to two other vertices z_i, t_i (see Figure 1). The vertices x_i, y_j, z_k, t_l are not necessarily distinct. We consider two cases according to the color of the vertex y_3 : first case, y_3 is colored with a color used in $\mathcal{C}$, so $c(y_3) = 2$; second case, y_3 is not colored 1 or 2, let $c(y_3) = 3$.

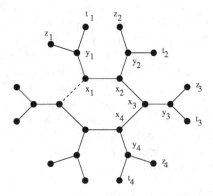

Fig. 1

1. Suppose that $c(y_3) = 2$. We know that $1 \in L(x_3)$. The vertex x_3 cannot be colored 2 because its neighbours are colored 2. There is at most two other problematic colors, say 3 and 4, because they create bicolored cycles passing through y_2, x_2, x_3, y_3, and z_3 and through y_4, x_4, x_3, y_3, and t_3. So we consider that $L(x_3) = \{1, 2, 3, 4\}$, $c(y_2) = c(z_3) = 3$ and $c(y_4) = c(t_3) = 4$. In this case we have to modify $c(y_3)$. We color y_3 with a color in $L(y_3) \backslash \{2, 3, 4\}$. Then we finally color x_3 with 3 or 4.

2. Suppose that $c(y_3) = 3$. We know that $1 \in L(x_3)$. The vertex x_3 cannot be colored 2 or 3 because its neighbours are colored 2 or 3. There is at most one other problematic color, say 4, because it creates a bicolored cycle passing through y_2, x_2, x_3, x_4, and y_4. So we consider that $L(x_3) = \{1, 2, 3, 4\}$ and $c(y_2) = c(y_4) = 4$. If there was a color $b \in L(x_2) \backslash \{1, 2, 3, 4\}$, we could set $c(x_2) = b$ and $c(x_3) = 1$. So we consider that $L(x_2) = L(x_4) = \{1, 2, 3, 4\}$. In this case we can let $c(x_2) = c(x_4) = 3$ and $c(x_3) = 2$.

This completes the proof.

3   Acyclic 5-Choosability of Graphs with Maximum Degree 4

Let H be a counterexample to Theorem 7 with minimum order, and L a list assignment such that there is no extracted acyclic coloring. In the first subsection, we prove some structural properties of H, that will allow us to use an algorithm (presented in the second and in the third subsection) which gives an acyclic coloring of H from L, contradicting the definition of L.

3.1   Structural Properties of H

Claim. The counterexample H is 4-regular.

Proof.   1. H does not contain any 1-vertices nor 2-vertices (see the first claim).

2. H does not contain any 3-vertices. Suppose that H contains a 3-vertex v adjacent to three vertices x, y, z with $d(x) \geq 3, d(y) \geq 3, d(z) \geq 3$. Let x_1, x_2, (x_3 if $d(x) = 4$) be the other neighbours of x (y_1, y_2, y_3 for y and z_1, z_2, z_3 for z). By minimality of H, the graph $H' = (V(H) \backslash \{v\}, E(H) \backslash \{vx, vy, vz\} \cup \{xy\})$ is acyclically 5-choosable. So, there is an acyclic coloring c of H'. If $c(x), c(y), c(z)$ are all distinct, it is easy to extend the coloring c to H by coloring v with a color different from $c(x), c(y), c(z)$. Hence suppose that w.l.o.g. $c(x) = c(z) = 1$ and $c(y) = 2$. Observe that if $L(v) \neq \{1, 2, c(x_1), c(x_2), c(x_3)\}$, we can extend the coloring c to H. So, $L(v) = \{1, 2, c(x_1), c(x_2), c(x_3)\}$ and $\{c(x_1), c(x_2), c(x_3)\} = \{c(z_1), c(z_2), c(z_3)\} = \{3, 4, 5\}$. If $L(x) \neq \{1, 2, 3, 4, 5\}$, we are done : we color x with a color different from 1,2,3,4,5; the colors of x, y, z are all distinct and finally we color v. For the same reason, $L(z) = \{1, 2, 3, 4, 5\}$. In this case, we recolor x and z with 2 and we color v with 1.

Corollary 1. *The counterexample H is bridgeless.*

Claim. The counterexample H does not contain a cut vertex.

Proof. By contradiction, let v be a cut vertex with neighbours x_1, x_2, y_1 and y_2 such that x_1 and x_2 (resp. y_1 and y_2) are in the same connected component of $G\backslash\{v\}$ (by Corollary 1, H is bridgeless). Let G' be the graph obtained from $G\backslash\{v\}$ by adding the edges x_1x_2 and y_1y_2. Since G' is smaller than G, it has an acyclic coloring c where $c(x_1) \neq c(x_2)$ and $c(y_1) \neq c(y_2)$. If we consider this coloring in the graph G, the only uncolored vertex is v and since it is a cut vertex, whatever its assigned color we cannot create a bicolored cycle going through $x_i v y_j$. Furthermore, since $c(x_1) \neq c(x_2)$ (resp. $c(y_1) \neq c(y_2)$) we cannot create a bicolored cycle going through $x_1 v x_2$ (resp. $y_1 v y_2$). So v can be colored with a color in $L(v)\backslash\{c(x_1), c(x_2), c(y_1), c(y_2)\}$.

Claim. The graph H contains an edge uv such that $L(u) \neq L(v)$.

Proof. If $\forall v \in V(H), L(v) = \{1, 2, 3, 4, 5\}$, then by Burstein's Theorem, there exists an acyclic coloring c of H, which contradicts the definition of H. Hence, there exists an edge uv with $L(u) \neq L(v)$.

From now, we suppose that we have an edge uv, which is not a bridge (by Corollary 1), with $L(u) \neq L(v)$ and such that u is not a cut vertex.

Claim. There is an order $x_1, x_2, \ldots, x_n$ on the vertices, such that x_1 and x_n are adjacent, $L(x_1) \neq L(x_n)$, and the vertices x_i, with $i < n$, have a neighbour x_j with $j > i$.

Proof. Since u is not a cut vertex, consider a spanning tree T of $H \setminus \{u\}$ rooted in v. Let $x_1 = u$ and order the others vertices from x_2 to x_n, according to a post order walk on T. Notice that $x_n = v$ and for $i < n$, each x_i has a father in T which is posterior in the order.

In the next subsections, we use this order to acyclically color the vertices of H. We will successively color $x_1, x_2, \ldots, x_n$. During this process, when we color x_i, we may change the color of x_j, for $1 < j < i < n$ (that is why we say that our algorithm is a backtracking greedy algorithm; at each step, we try to color almost greedily and for this we may change the colors of a bounded number of vertices). Note that the color of x_1 remains unchanged until coloring x_n. At the beginning there is no constraints; so, let the color of x_1 be such that $c(x_1) \in L(x_1) \setminus L(x_n)$. In the next subsection we explain how to color the vertices x_i, for $i < n$. In the last subsection, we finally color x_n; that will complete the proof of Theorem 7.

3.2   The Backtracking Greedy Algorithm: The Coloring of x_i, $1 < i < n$

At Step 1, we colored x_1 with a color a with $a \notin L(x_n)$. The following Claim allows us to color all the vertices until x_{n-1}.

Claim. Let c be a partial acyclic coloring of H on the vertices $\{x_1, \ldots, x_{i-1}\}$. Then, there exists a partial acyclic coloring c' of H on the vertices $\{x_1, \ldots, x_i\}, i < n$, which do not modify the color of x_1.

Proof. Let c be a partial acyclic coloring of H on the vertices $\{x_1, \ldots, x_{i-1}\}$. We would like to extend the coloring c to x_i. We know that x_i has at most three colored neighbours by the definition of the order. Let x_j, x_k, x_l be these vertices. We consider two cases following the adjacency of x_i to x_1. However, the analysis are almost the same.

1. The vertex x_i is adjacent to x_1, x_j, x_k. We recall that x_1 is adjacent to x_n which is not colored. Let x_1^1, x_1^2 be the other neighbours of x_1. Let x_j^1, x_j^2, x_j^3 be the other neighbours of x_j (x_k^1, x_k^2, x_k^3 for x_k). We consider the different cases following the coloring of x_j, x_k.

   1.1. The colors of x_j, x_k, x_1 are all distinct. We just let $c'(x_i) \in L(x_i) \setminus \{c(x_j), c(x_k), c(x_1)\}$.

   1.2. A color appears exactly twice on x_j, x_k, x_1.

      1.2.1. The color of x_1 appears twice. W.l.o.g., suppose that $c(x_1) = c(x_j) = a$ and $c(x_k) = 1$, $a \neq 1$. We just let $c'(x_i) \in L(x_i) \setminus \{1, a, c(x_1^1), c(x_1^2)\}$ (we recall that x_1 is adjacent to x_n which is not colored).

      1.2.2. A color different from $c(x_1)$ appears twice. Let $c(x_j) = c(x_k) = 1$ and $c(x_1) = a$, $a \neq 1$. If $L(x_i) \neq \{1, a, c(x_j^1), c(x_j^2), c(x_j^3)\}$, we are done : we could color x_i with $c'(x_i) \in L(x_i) \setminus \{1, a, c(x_j^1), c(x_j^2), c(x_j^3)\}$. Hence, $L(x_i) = \{1, a, c(x_j^1), c(x_j^2), c(x_j^3)\}$ and $\{c(x_j^1), c(x_j^2), c(x_j^3)\} = \{c(x_k^1), c(x_k^2), c(x_k^3)\}$. Set $\{c(x_k^1), c(x_k^2), c(x_k^3)\} = \{2, 3, 4\}$. Now, we re-color x_j with a color different from $1, 2, 3, 4$ and we get case 1.1 or 1.2.1.

   1.3. A color appears three times. So, suppose that $c(x_1) = c(x_j) = c(x_k) = a$. It is easy to see that if we cannot color x_i, this implies that all the neighbours of x_1 (resp. x_j, x_k) have distinct colors. So we recolor x_j with a color different from $a, c(x_j^1), c(x_j^2), c(x_j^3)$ and we get case 1.2.1.

2. The vertex x_i is not adjacent to x_1. Let x_j^1, x_j^2, x_j^3 be the other neighbours of x_j (x_k^1, x_k^2, x_k^3 for x_k and x_l^1, x_l^2, x_l^3 for x_l). Following the coloring of the vertices of x_j, x_k, x_l, we consider the different cases :

   2.1. The colors of x_j, x_k, x_l are all distinct. We just color x_i with $(x_i) \in L(x_i) \setminus \{c(x_j), c(x_k), c(x_l)\}$.

   2.2. A color appears exactly twice on x_j, x_k, x_l. W.l.o.g. We suppose that $c(x_j) = c(x_k) = 1$ and $c(x_l) = 2$. If $L(x_i) \neq \{1, 2, c(x_j^1), c(x_j^2), c(x_j^3)\}$, we are done : let $c'(x_i) \in L(x_i) \setminus \{1, 2, c(x_j^1), c(x_j^2), c(x_j^3)\}$. So, $L(x_i) = \{1, 2, c(x_j^1), c(x_j^2), c(x_j^3)\}$ and $\{c(x_j^1), c(x_j^2), c(x_j^3)\} = \{c(x_k^1), c(x_k^2), c(x_k^3)\}$; say $\{c(x_j^1), c(x_j^2), c(x_j^3)\} = \{3, 4, 5\}$. Now, if $L(x_j) \neq \{1, 2, 3, 4, 5\}$, we recolor x_j such that $c'(x_j) \in L(x_j) \setminus \{1, 2, 3, 4, 5\}$ and let $c'(x_i) \in L(x_i) \setminus \{c'(x_j), c'(x_k), c'(x_l)\}$. Consequently, $L(x_j) = \{1, 2, 3, 4, 5\}$ and for the same reason, $L(x_k) = \{1, 2, 3, 4, 5\}$. In this case, let $c'(x_j) = c'(x_k) = 2$, and $c'(x_i) = 1$.

   2.3. A color appears three times on x_j, x_k, x_l. It is easy to observe that if we cannot color x_i, this implies that at least one vertex of x_j, x_k, x_l has a neighbourhood colored with three distinct colors. Hence we can recolor this vertex with a different color and get case 2.2.

3.3 The Final Step: The Coloring of x_n

At this point, we have a partial acyclic coloring such that $c(x_1) = a$ with $a \notin L(x_n)$. Let x_1, u, v, w be the neighbourhood $N(x_n)$ of x_n. Let u_1, u_2, u_3 be the other neighbours of u (v_1, v_2, v_3 for v, and w_1, w_2, w_3 for w, and x_1^1, x_1^2, x_1^3 for x_1).

We show that we can extend the partial acyclic coloring to x_n by recoloring if necessary one or some vertices of $N(x_n)$. For this, we consider the different cases according to the coloring of $N(x_n)$:

1. The vertices of $N(x_n)$ have all distinct colors. In this case, it is easy to extend the coloring to x_n by coloring x_n with $c(x_n) \in L(x_n) \setminus \{c(u), c(v), c(w)\}$ (recall that $c(x_1) \notin L(x_n)$).
2. Exactly one color appears twice in $N(x_n)$:
   2.1. Suppose that $c(u) = c(v) \neq a$, $c(w) \neq c(u)$, $c(w) \neq a$. If we can color x_n with a color different from $c(u), c(w), c(u_1), c(u_2), c(u_3)$ ($a \notin L(x_n)$), we are done. Hence, $L(x_n) = \{c(u), c(w), c(u_1), c(u_2), c(u_3)\}$; the colors of the u_i are distincts, the colors of the v_i are distinct and $\{c(u_1), c(u_2), c(u_3)\} = \{c(v_1), c(v_2), c(v_3)\}$. Now, we color x_n with $c(u)$ and we recolor u and v with a proper color. The coloring obtained is acyclic.
   2.2. The color of x_1, i.e. a, appears twice. Set $c(u) = c(x_1) = a$, $c(v) = b$, and $c(w) = c$ (a, b, c are distinct). If $L(x_n) \neq \{b, c, c(u_1), c(u_2), c(u_3)\}$, we can color x_n (with a color different from these of v, w, u_1, u_2, u_3) and the coloring obtained is an acyclic coloring. Otherwise, this implies that : the colors of the u_i are distinct ($i = 1, 2, 3$); the colors of the x_1^i are distinct; $S = \{c(u_1), c(u_2), c(u_3)\} = \{c(x_1^1), c(x_1^2), c(x_1^3)\}$, $a \notin S$, and $L(x_n) = \{c(u_1), c(u_2), c(u_3), b, c\}$. Now, we recolor u with a color different from $c(u_1), c(u_2), c(u_3), a$. If this new color is equal to b or c, we have case 2.1, else, we have case 1.
3. Exactly two colors appear twice. W.l.o.g., set $c(u) = c(v) = 1$ and $c(w) = c(x_1) = a$.
   First, we show that $L(x_n)$ contains necessarily the color 1. If $1 \notin L(x_n)$, say $L(x_n) = \{2, 3, 4, 5, 6\}$. If we cannot color x_n, this implies that there exists at least one of the vertices u, v, w, x_1 whose the neighbours have distinct colors; say u. So, we recolor u with a color different from $c(u_1), c(u_2), c(u_3), 1$. If this new color is a, then we can color x_n with a color of the neighbours of u according to the coloring of the neighbours of w and x_1, otherwise, we get case 2.2.
   Hence, we suppose that $1 \in L(x_n)$ and we assume $L(x_n) = \{1, 2, 3, 4, 5\}$. If we cannot color x_n with $2, 3, 4, 5$; this implies that by coloring x_n with one of these colors, we will create a bicolored cycle. So, each of the colors $2, 3, 4, 5$ appears at least twice among the colors $c(u_i), c(v_i), c(w_i), c(x_1^i), i = 1, 2, 3$. We can form four bicolored cycles, we have the following different cases: *Case 1*, we have three bicolored cycles going through $u x_n v$ and one going through $w x_n x_1$. *Case 2*, we have one bicolored cycle going through $u x_n v$ and three going through $w x_n x_1$. *Case 3*, we have two bicolored cycle going through $u x_n v$ and two going through $w x_n x_1$.
   3.1. Suppose that we are in *Case 1*. Three bicolored cycles can be created (using one of the colors of $L(x_n)$), going through u and v; this implies that the colors of the neighbours of u (resp. v) are distinct and $\{c(u_1), c(u_2), c(u_3)\} =$

$\{c(v_1), c(v_2), c(v_3)\}$. assume, w.l.o.g. that $c(u_1) = 2, c(u_2) = 3, c(u_3) = 4$ and $c(w_1) = c(x_1^1) = 5$ (if the color 5 does not appear in the neighbourhood of w and x_1, we can color x_n with 5). Now, if we can recolor u with a color different from $1, 2, 3, 4, a$, we get case 2.2. So $L(u) = \{1, 2, 3, 4, a\}$ and by the same way, $L(v) = \{1, 2, 3, 4, a\}$. We set $c(u) = a$. Now, we try to color x_n with one of the colors $2, 3, 4$: if we still create three bicolored cycles, then this implies that $\{2, 3, 4\} \subset \{c(w_2), c(w_3), c(x_1^2), c(x_1^3)\}$ and consequently the color 1 appears at most one time on u_i, v_i, w_i, x_1^i for $1 \leq i \leq 3$. In this case, we recolor v with the color a and we color x_n with 1. *Case 2* can be dealt analogously.

3.2. Suppose now that we are in *Case 3* and that two bicolored cycles going through u and v and two bicolored cycles going through w and x_1 can be created by choosing a color of x_n in $\{2, 3, 4, 5\}$. So, we have : $c(u) = c(v) = 1, c(w) = c(x_1) = a$ and w.l.o.g $c(u_1) = c(v_1) = 2, c(u_2) = c(v_2) = 3, c(w_1) = c(x_1^1) = 4$ and $c(w_2) = c(x_1^2) = 5$.

Suppose now that there exists a vertex y_3 of $\{u_3, v_3, w_3, x_1^3\}$ such that $c(y_3) \notin \{c(y_1), c(y_2)\}$. Assume w.l.o.g. that $c(u_3) = b$ with $b \neq 2, 3$ (it may happen that $a = b$). If we can recolor u with a color different from $1, 2, 3, a, b$, we obtain case 2.2. Hence, $a \neq b$, $L(u) = \{1, 2, 3, a, b\}$. If $\{c(w_3), c(x_1^3)\} \neq \{2, 3\}$, set $c(u) = a$ and $c(v) = 2$ or 3 following the colors of w_3 and x_1^3. Now, assume that $\{c(w_3), c(x_1^3)\} = \{2, 3\}$ and say that $c(w_3) = 2, c(x_1^3) = 3$. If we can recolor w with a color different from $1, 2, 4, 5, a$ or x_1 with a color different from $1, 3, 4, 5, a$, then we obtain case 2.1. Consequently, $L(w) = \{1, 2, 4, 5, a\}$ and $L(x_1) = \{1, 3, 4, 5, a\}$. We set $c(w) = 1$. If we cannot color x_n then $\{c(u_3), c(v_3)\} = \{4, 5\}$, say $c(u_3) = 4$ and $c(v_3) = 5$. As previously, we can prove that $L(v) = \{1, 2, 3, 5, a\}$. Finally, we set $c(u) = c(v) = c(w) = c(x_1) = a$ and $c(x_n) = 1$. The obtained coloring is acyclic.

Hence, $c(u_3) \in \{2, 3\}, c(v_3) \in \{2, 3\}, c(w_3) \in \{4, 5\}, c(x_1^3) \in \{4, 5\}$. W.l.o.g. set $c(u_3) = 2$. Now, we will recolor u and/or v. If we can recolor u with a color different from $1, 2, 3$, we can color x_n with 2 or 3.

So we must study the coloring of the neighbourhood of u (at distance 2). Let u_1^1, u_1^2, u_1^3 be the other neighbours of u_1 (u_2^1, u_2^2, u_2^3 for u_2, and u_3^1, u_3^2, u_3^3 for u_3). We recall that at least one of u_1^1, u_1^2, u_1^3 or u_3^1, u_3^2, u_3^3 is colored by 1; say u_1^1 (as well, one of u_2^1, u_2^2, u_2^3 is colored by 1; say u_2^1). So, if we can recolor u with a color different from $1, 2, 3, c(u_1^2), c(u_1^3)$, we are done. Assume that $L(u) = \{1, 2, 3, b, c\}$ ($b \neq c, b \notin \{1, 2, 3\}, c \notin \{1, 2, 3\}$), $c(u_1^2) = c(u_3^1) = b$, $c(u_1^3) = c(u_3^2) = c$. Since $c(u_1^1), c(u_1^2), c(u_1^3)$ are distinct, let us recolor u_1. Assign to u_1 a color different from $1, 2, b, c$. If its new color is different from 3, we are done (we can then easily recolor u with a color different from 1). So, suppose that the new color of u_1 is 3. Hence, we cannot recolor u with a color in $\{1, 2, 3, b, c\}$, i.e. with b or c, if and only if $\{c(u_2^2), c(u_2^3)\} = \{b, c\}$. However, if $L(u_2) \neq \{1, 2, 3, b, c\}$, we can recolor u_2, then u. Finally, this implies that we have $L(u) = L(u_1) = L(u_2) = \{1, 2, 3, b, c\}, \{c(u_1^1), c(u_1^2), c(u_1^3)\} = \{c(u_2^1), c(u_2^2), c(u_2^3)\} = \{1, b, c\}$. In this case, we assign the color 2 to u_1 and u_2, the color 3 to u and the color 2 to x_n.

4. Suppose that a color appears exactly three times. It is easy to observe that if we cannot color x_n, this implies that the neighbours of at least one of the vertices u, v, w, x_1 have distinct colors, say u. so we can recolor u and get case 2.2.
5. A color appears four times : there is only one possibility, i.e. $c(u) = c(v) = c(w) = c(x_1) = a$. Since $a \notin L(x_n)$, if we cannot color x_n, this implies that the neighbours of at least one of the vertices u, v, w, x_1 have distinct colors, say u. So, we can recolor u and get case 4.

4   Conclusion

Finally, we propose the following question:

Problem 1. Is it NP-hard or not to decide if a subcubic graph is acyclically 3-choosable?

In 1978, Kostochka proved that it is an NP-complete problem to decide for a given graph G if it is acyclically 3-colorable [Kos78]. Recently, Ochem proved that it is NP-complete to decide if a graph with maximum degree $\Delta \leq 4$ is acyclically 3-colorable [Och05]. However, the complexity of the problem to decide for a given subcubic graph G if it is acyclically 3-colorable is unknown.

Acknowledgement

We would like to thank André Raspaud for fruitful discussions about the problem.

References

[AB77]   M.O. Albertson and D.M. Berman. Every planar graph has an acyclic 7-coloring. *Israel J. Math.*, (28):169–174, 1977.

[BFDFK+02] O.V. Borodin, D.G. Fon-Der Flaass, A.V. Kostochka, A. Raspaud, and E. Sopena. Acyclic list 7-coloring of planar graphs. *J. Graph Theory*, 40(2):83–90, 2002.

[BKW99]   O.V. Borodin, A.V. Kostochka, and D.R. Woodall. Acyclic colourings of planar graphs with large girth. *J. London Math. Soc.*, 2(60):344–352, 1999.

[Bor79]   O.V. Borodin. On acyclic coloring of planar graphs. *Discrete Math.*, 25:211–236, 1979.

[Bur79]   M.I. Burstein. Every 4-valent graph has an acyclic 5-colouring. *Bulletin of the Academy of Sciences of the Georgian SSR*, 93(1):21–24, 1979. In Russian.

[Grü73]   B. Grünbaum. Acyclic colorings of planar graphs. *Israel J. Math.*, 14:390–408, 1973.

[KM76]   A.V. Kostochka and L.S. Mel'nikov. Note to the paper of Grünbaum on acyclic colorings. *Discrete Math.*, 14:403–406, 1976.

[Kos76]   A.V. Kostochka. Acyclic 6-coloring of planar graphs. *Metody Diskret. Anal.*, (28):40–56, 1976. In Russian.

[Kos78]   A.V. Kostochka. *Upper bounds of chromatic functions of graphs*. PhD thesis, Novosibirsk, 1978.

[Mit74]   J. Mitchem. Every planar graph has an acyclic 8-coloring. *Duke Math. J.*, (41):177–181, 1974.

[MOR05]   M. Montassier, P. Ochem, and A. Raspaud. On the acyclic choosability of graphs. *Journal of Graph Theory*, 2005. To appear.

[MS04]   M. Montassier and O. Serra. Acyclic choosability and probabilistic methods. Technical report, 2004. Manuscript.

[Och05]   P. Ochem. Negative results on acyclic improper colorings. Manuscript, 2005.

[Tho94]   C. Thomassen. Every planar graph is 5-choosable. *J. Combin. Theory Ser. B*, 62:180–181, 1994.

[Voi93]   M. Voigt. List colourings of planar graphs. *Discrete Math.*, 120:215–219, 1993.

Generating Colored Trees

(Extended Abstract)

Shin-ichi Nakano[1] and Takeaki Uno[2]

[1] Gunma University, Kiryu-Shi 376-8515, Japan
nakano@cs.gunma-u.ac.jp
[2] National Institute of Informatics, Tokyo 101-8430, Japan
uno@nii.jp

Abstract. A c-tree is a tree such that each vertex has a color $c \in \{c_1, c_2, \cdots, c_m\}$. In this paper we give a simple algorithm to generate all c-trees with at most n vertices and diameter d, without repetition. Our algorithm generates each c-tree in constant time. By using the algorithm for each diameter $2, 3, \cdots, n - 1$, we can generate all c-trees with n vertices.

1 Introduction

It is useful to have the complete list of graphs for a particular class. One can use such a list to search for a counter-example to some conjecture, to find the best graph among all candidate graphs, or to experimentally measure the average performance of an algorithm over all possible input graphs.

Many algorithms to generate a particular class of graphs are already known [B80, LN01, LR99, M98, N02, R78, W86]. Many excellent textbooks have been published on the subject [G93, KS98, W89].

Algorithms to generate all trees with n vertices without repetition are already known. The algorithm [LR99, W86, NU03] generates each tree in $O(1)$ time on average, and the algorithm [NU04] generates each tree in $O(1)$ time.

Let $C = \{c_1 = a, c_2 = b, c_3 = c, \cdots, c_m\}$ be a set of colors. A c-tree is a tree such that each vertex has a color $c \in C$.

In this paper we give a simple algorithm to generate, without repetition, all c-trees with at most n vertices and diameter d. Our algorithm generates each c-tree in constant time. It does not output each c-tree entirely, but outputs the difference from the preceding c-tree. Our algorithm is based on our algorithm in [NU03], and completely different from [W86].

The main idea of our algorithm is first to define a simple relation among the c-trees, that is "a family tree" of c-trees (see Fig. 1), then outputs c-trees by traversing the family tree. *The family tree*, denoted by $T_{n,d,m}$, is the (huge) tree such that the vertices of $T_{n,d,m}$ correspond to the c-trees with at most n vertices and diameter d, and each edge corresponds to some relation between two c-trees. We give a formal definition in Section 4. By traversing the family tree we can generate all c-trees corresponding to the vertices of the family tree without repetition.

D. Kratsch (Ed.): WG 2005, LNCS 3787, pp. 249–260, 2005.

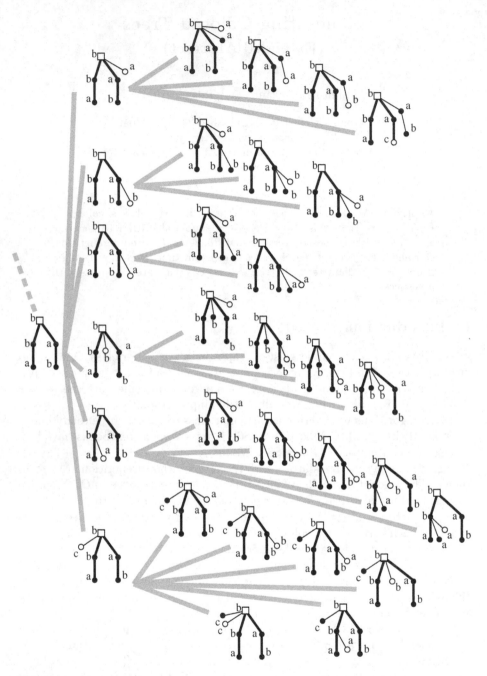

Fig. 1. The family tree $T_{7,4,3}$ sharing c-spine (a, b, b, a, b)

We have designed several generation algorithms based on the family trees [N02, NU03]. In this paper we first extend the method for c-trees.

Our algorithm has an application for a problem in tree mining. Many tree mining algorithms based on systematic enumeration of subtrees are already known. See a survey paper [C05]. Given a huge size of XML data, we wish to discover frequent patterns in the data. The frequent "patterns" are candidates for new "knowledge" [A03]. We can model XML data as a tree, where each data object is represented by a node with a label (=color), and each relationship between data objects by an edge. If we restrict patterns to frequent occurrences of the same colored subtrees, then we can solve the problem by (1) generating every c-tree, (2) then count the occurrences of each c-tree as a subgraph in the given XML tree, (3) then output the frequently occurred c-trees. By using our algorithm to generate every c-tree based on the family tree, we can efficiently prune rarely occurred c-trees, since in the family tree every "child" c-tree contains its "parent" c-tree as a subtree, so if the occurrence of a c-tree T is rare then the occurrence of any "descendant" c-tree of T is also rare. Thus we need not count the occurrences of each descendant rare c-tree of T.

The rest of the paper is organized as follows. Section 2 gives some definitions. Section 3 assigns a unique ordered c-tree H for each c-tree T, by choosing the root of T and the ordering of each child vertices. Section 4 introduces the family tree. Section 5 generates all c-paths, which are colored paths. Section 6 presents our algorithm to generate all c-trees for the even diameter case. In Section 7 we sketch our algorithm for the odd diameter case. Finally Section 8 is a conclusion.

2   Preliminaries

In this section we give some definitions.

Let G be a connected graph with n vertices. An edge connecting vertices x and y is denoted by (x, y). A *path* is a sequence of distinct vertices $(v_0, v_1, \cdots, v_k)$ such that (v_{i-1}, v_i) is an edge for $i = 1, 2, \cdots, k$. The *length* of a path is the number of edges in the path. The *distance* between a pair of vertices u and v is the minimum length of a path between u and v. The *diameter* of G is the maximum distance between two vertices in G.

A *tree* is a connected graph without cycles. A *rooted* tree is a tree with one vertex r chosen as its *root* . A *c-tree* is a tree such that each vertex has a color $c \in \{c_1, c_2, \cdots, c_m\}$. For each vertex v in a rooted tree, let $UP(v)$ be the unique path from v to the root r. If $UP(v)$ has exactly k edges then we say that the *depth* of v is k, and write $dep(v) = k$. The *parent* of $v \neq r$ is its neighbor on $UP(v)$, and the *ancestors* of $v \neq r$ are the vertices on $UP(v)$ except v. The parent of the root r and the ancestors of r are not defined. We say that if v is the parent of u then u is *a child* of v, and if v is an ancestor of u then u is a *descendant* of v. A *leaf* is a vertex that has no child.

An *ordered tree* is a rooted tree with left-to-right ordering specified for the children of each vertex. We denote by $T(v)$ the ordered subtree of an ordered

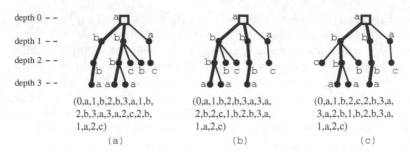

depth 0 – –

depth 1 – –

depth 2 – –

depth 3 – –

(0,a,1,b,2,b,3,a,1,b,   (0,a,1,b,2,b,3,a,3,a,   (0,a,1,b,2,c,2,b,3,a,

2,b,3,a,3,a,2,c,2,b,   2,b,2,c,1,b,2,b,3,a,   3,a,2,b,1,b,2,b,3,a,

1,a,2,c)   1,a,2,c)   1,a,2,c)

(a)   (b)   (c)

Fig. 2. The dc sequences

tree T consisting of a vertex v and all descendants of v with preserving the left-to-right ordering for the children of each vertex.

Let T be an ordered c-tree with n vertices, and $(v_1, v_2, \cdots, v_n)$ be the list of the vertices of T in preorder [A95]. Let $dep(v_i)$ be the depth of v_i and $c(v_i)$ be the color of v_i for $i = 1, 2, \cdots, n$. Then, the sequence $L(T) = (dep(v_1), c(v_1), dep(v_2), c(v_2), \cdots, dep(v_n), c(v_n))$ is called *the dc-sequence* of T. Some examples are shown in Fig. 2. Note that those trees in Fig. 2 are isomorphic as unordered c-trees, but non-isomorphic as ordered c-trees.

Let T_1 and T_2 be two ordered c-trees, and $L(T_1) = (a_1, b_1, a_2, b_2, \cdots, a_n, b_n)$ and $L(T_2) = (x_1, y_1, x_2, y_2, \cdots, x_z, y_z)$ be their dc-sequences. If there is some j such that $a_i = x_i$ and $b_i = y_i$ for each $i = 1, 2, \cdots, j - 1$ (possibly $j = 1$) and either (i) $a_j > x_j$, (ii) $a_j = x_j$ and $b_j > y_j$, or (iii) $n > z = j - 1$, then we say that $L(T_1)$ is *heavier* than $L(T_2)$, and write $L(T_1) > L(T_2)$. For example, in Fig. 2, (a)<(b)<(c).

3   The Left-Heavy Embeddings

In Section 3–6, we only consider the case where the diameter is even.

Let T be a c-tree with diameter $2k$, and $(v_0, v_1, \cdots, v_{2k})$ be a path in T having length $2k$. One can observe that T may have many such paths, but the vertex v_k, called *the center* of T, is unique [W01–p72]. We assign to T the rooted c-tree R derived from T by choosing v_k as the root. Then we assign to R a unique ordered c-tree H as follows.

Given a rooted c-tree R, since we can choose many left-to-right orderings for the children of each vertex, we can observe that R corresponds to many non-isomorphic ordered c-trees. Let H be the ordered c-tree corresponding to R that has the heaviest dc sequence $L(H)$. Then we say that H is the *left-heavy embedding* of R. For example, the ordered c-tree in Fig. 2(c) is the left-heavy embedding of a rooted c-tree, however the ordered c-trees in Fig. 2(a) and (b) are not, since the one in Fig. 2(c) is heavier. We assign the ordered c-tree H to R.

Given a c-tree T, we have assigned to T a unique distinct rooted c-tree R, and then we have assigned to R a unique distinct ordered c-tree H, which is the left-heavy embedding of R. Note that T, R and H have the same diameter $2k$. Let $S_{n,2k,m}$ be the set of all left-heavy embeddings of c-trees with at most n vertices and diameter $2k$. If we generate all ordered c-trees in $S_{n,2k,m}$, then it also means the generation of all c-trees with at most n vertices and diameter $2k$. We are going to generate all ordered c-trees in $S_{n,2k,m}$.

We have the following lemma.

Lemma 1. *An ordered c-tree H is the left-heavy embedding of a rooted c-tree if and only if for every pair of consecutive child vertices v_1 and v_2, they appear in this order in the left-to-right ordering, $L(T(v_1)) \geq L(T(v_2))$ holds.*

Proof. By contradiction.   □

In the rest of the paper the condition "$L(T(v_1)) \geq L(T(v_2))$ for each consecutive child vertices v_1 and v_2", is called *the left-heavy condition*.

4   The Family Tree of c-Trees Sharing a c-Spine

Let H be a left-heavy embedding in $S_{n,2k,m}$ with root r. Let p_k be the first leaf of H at depth k in preorder, and $P_L = (r = p_0, p_1, \cdots, p_k)$ be the path between $r = p_0$ and p_k. We say that P_L is *the left spine* of H. Let H' be the ordered tree derived from H by removing $T(p_1)$, that is the subtree rooted at p_1. We can observe that H' is also a left-heavy embedding. Let q_k be the first leaf in H' at depth k in preorder, and $P_R = (r = q_0, q_1, \cdots, q_k)$ be the path between $r = q_0$ and q_k. We say that P_R is *the right spine* of H. We call $P_L \cup P_R$ *the spine* of H. We can observe that $P_L \cup P_R$ corresponds to a path with $2k$ edges. Since the diameter of H is $2k$, such p_k and q_k always exist.

An left-heavy embedding H in $S_{n,2k,m}$ is *trivial* if it consists of only $P_L \cup P_R$. Observe that any non-trivial $H \in S_{n,2k,m}$ has at least three leaves, so we can choose one leaf except p_k and q_k.

Assume $H \in S_{n,2k,m}$ is non-trivial. The last leaf x of H in preorder except p_k and q_k is called *the removable vertex* of H. Let $P(H)$ be the ordered c-tree derived from H by removing x.

Now we consider whether the left-heavy condition still holds in $P(H)$ or not. We have the following seven cases, depending on the location of x in H. Let $r_1, r_2, \cdots, r_{d(r)}$ be the children of r. Assume that they appear in this order in the left-to-right ordering of them. Also assume that p_k in P_L is a descendant of r_y and q_k in P_R is a descendant of r_z. See Fig. 3.

Case 1: $x \in T(r_i)$ for some $i > z$.

Then the left-heavy condition still holds in $P(H)$, since we remove the rightmost leaf, so a "right" subtree may loose some weight, but it never destroys the left-heavy condition.

Case 2: $x \in T(r_z)$, and x succeeds q_k in preorder.

Then the left-heavy condition still holds in $P(H)$. Similar to Case 1.

Case 3: $x \in T(r_z)$, and x precedes q_k in preorder.

Now there is no leaf x satisfying Case 1 or 2.

Let q_j on P_R be the ancestor of x having maximum depth, and $q_j = q'_j, q'_{j+1}, q'_{j+2}, \cdots, q'_s = x$ be the path between q_j and x. See Fig. 4. Note that by the

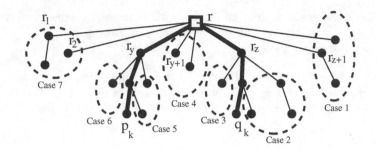

Fig. 3. Illustration for the seven cases

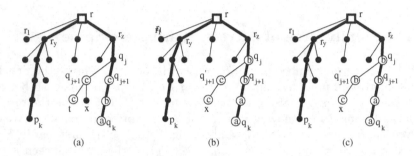

Fig. 4. Illustration for Case 3

definition of P_R, the depth of any descendant of q'_{j+1} is at most $k-1$. (Otherwise, q'_{j+1} has a descendant at depth k, and P_R must pass through q'_{j+1}. Now P_R is the path between r and the leftmost descendant of q'_{j+1} at depth k, a contradiction.)

We have the following two subcases.

Case 3(a): $T(q'_{j+1})$ is not a path.

Then the left-heavy condition still holds in $P(H)$. See Fig. 4(a), where the set of color is $\{c_1 = a, c_2 = b, c_3 = c\}$. Let t be the first leaf of $T(q'_{j+1})$ in preorder. Note that the dc sequence of the path from q'_{j+1} to t is heavier than the dc sequence of the path from q_{j+1} to q_k, since the left-heavy condition holds in H.

Case 3(b): $T(q'_{j+1})$ is a path.

Then we have two subcases.

If $c(q'_{j+1}) = c(q_{j+1}), c(q'_{j+2}) = c(q_{j+2}), \cdots c(q'_{s-1}) = c(q_{s-1})$ holds then the left-heavy condition destroyed in $P(H)$, since $L(T(q_{j+1}))$ is heavier than $L(T(q'_{j+1}))$ in $P(H)$. See Fig. 4(c). In this case, by swapping the order of q'_{j+1} and q_{j+1}, the left-heavy condition again holds. We re-define the resulting ordered c-tree as $P(H)$.

Otherwise the left-heavy condition still holds in $P(H)$. See Fig. 4(b).

Case 4: $x \in T(r_i)$ for some $i, y < i < z$.

Now r_{z-1} is the ancestor of x at depth one, and there is no leaf x satisfying Case 1, 2 or 3.

Case 4(a): $T(r_{z-1})$ is not a path.

Then the left-heavy condition still holds in $P(H)$. (Similar to Case 3(a).)

Case 4(b): $T(r_{z-1})$ is a path.

Similar to Case 3(b). We have two subcases as follows.

Let $q_0' = r, q_1', q_2', \cdots, q_s' = x$ be the path between r and x.

If $c(q_1') = c(q_1), c(q_2') = c(q_2), \cdots, c(q_{s-1}') = c(q_{s-1})$ holds, then the left-heavy condition destroyed in $P(H)$, since $L(T(q_1))$ is heavier than $L(T(q_1'))$ in $P(H)$. In this case, by swapping the order of $q_1' = r_{z-1}$ and $q_1 = r_z$, the left-heavy condition again holds. We re-define the resulting ordered c-tree as $P(H)$.

Case 5: $x \in T(r_y)$, and x succeeds p_k in preorder.

Then the left-heavy condition still holds in $P(H)$. Similar to Case 1 and 2.

Case 6: $x \in T(r_y)$, and x precedes p_k in preorder.

Similar to Case 3.

Case 7: $x \in T(r_i)$ for some $i < y$.

Similar to Case 4.

Since we never remove p_k and q_k, the spine always remains as it was. Note that $P(H)$ is left-heavy unless Case 3(b), 4(b) or 6(b) occurs, and even if Case 3(b), 4(b) or 6(b) occurs, by a possible modification, the resulting $P(H)$ is left-heavy.

Now we have the following lemma.

Lemma 2. *For any non-trivial $H \in S_{n,2k,m}$, $P(H)$ is also in $S_{n,2k,m}$ (after possible modification in Case 3(b), 4(b) or 6(b)).*

Given an ordered c-tree H in $S_{n,2k,m}$, by repeatedly removing the removable vertex, we can have the unique sequence $H, P(H), P(P(H)), \cdots$ of ordered c-trees in $S_{n,2k,m}$, which eventually ends with the trivial ordered c-tree H_1. By merging these sequences we can have *the family tree* of $S_{n,2k,m}$, denoted by $T_{n,2k,m}$, such that the vertices of $T_{n,2k,m}$ correspond to the c-trees in $S_{n,2k,m}$ having the same c-spine, and each edge corresponds to each relation between some H and $P(H)$. For instance, $T_{7,4,3}$ with c-spine (a, b, b, a, b) is shown in Fig. 1.

We say that $P(H)$ is *the parent tree* of H and H is *a child tree* of $P(H)$. We also say $P(H)$ is *a Type i child* of H if Case i occurs to find $P(H)$ from H.

5   Algorithm for c-Paths

A *c-path* is a path such that each vertex has a color $c \in \{c_1, c_2, \cdots, c_m\}$. Given an integer $2k$, one can generate every c-path with length $2k$ in constant time for each on average[RS00]. The detail is not mentioned in [RS00], but we can design a naive recursive algorithm as follows.

Let $S_{2k,m}$ be the set of all c-path with length $2k$. Let $(v_0, v_1, \cdots, v_{2k})$ be a c-path with edge (v_{i-1}, v_i) for $1 \le i \le 2k$. Let $(c(v_0), c(v_1), \cdots, c(v_{2k}))$ be the sequence of colors. Given a c-path, since we can choose the direction of the path we have two such sequences of colors, each one is the reverse of the other. Assume

S_1 and S_2 be the two sequence of colors for a c-path. We say a sequence S_1 is *a forward sequence* if S_1 is lexicographically larger than or equal to S_2.

If we generate all forward sequences with length $2k+1$ over alphabet $\{c_1, c_2, \cdots, c_m\}$, then they correspond to all c-paths in $S_{2k,m}$.

Each forward sequence $(x_0, x_1, \cdots, x_{2k})$ with length $2k + 1$ is one of the following two types.

Type 1: $x_0 > x_{2k}$.

Then subsequence $(x_1, x_2, \cdots, x_{2k-1})$ is any sequence.

Type 2: $x_0 = x_{2k}$.

Then subsequence $(x_1, x_2, \cdots, x_{2k-1})$ is any forward sequence corresponds to a c-path in $S_{2k-1,m}$.

Based on the recursive structure above we can generate every c-path in constant time for each on average.

6   Algorithm for c-Trees

In this section we give an algorithm to construct $T_{n,2k,m}$.

Using the algorithm in [RS00] or Section 5, we can generate every c-path in constant time for each. During the generation above, at the time we generate each c-path P_c, we wish to generate all c-trees in $S_{n,2k,m}$ sharing the c-spine P_c.

All we need to do is, given a c-tree H having the c-spine P_c, to generate all "child" c-trees of H. Then in a recursive manner we can generate all c-trees in $T_{n,2k,m}$ sharing the c-spine P_c. Now we are going to give an algorithm to generate all child c-trees of a given ordered c-tree.

Let H be an ordered c-tree in $S_{n,2k,m}$. We have eight cases depending on the location of the removable vertex x in H as follows.

Again let $r_1, r_2, \cdots, r_{d(r)}$ be the children of the root r. Assume they appear in this order in the left-to-right ordering of them. Let $P_L = (p_0 = r, p_1, \cdots, p_k)$, and $P_R = (q_0 = r, q_1, \cdots, q_k)$. Also assume that p_k in P_L is a descendant of r_y and q_k in P_R is a descendant of r_z. See Fig. 3.

Case 0: H is trivial, that means H has only two leaves p_k and q_k.
Case 1: $x \in T(r_i)$ for some $i > z$.
Case 2: $x \in T(r_z)$, and x succeeds q_k in preorder.
Case 3: $x \in T(r_z)$, and x precedes q_k in preorder.
Case 4: $x \in T(r_i)$ for some $i, y < i < z$.
Case 5: $x \in T(r_y)$, and x succeeds p_k in preorder.
Case 6: $x \in T(r_y)$, and x precedes p_k in preorder.
Case 7: $x \in T(r_i)$ for some $i < y$.

For each case we can generate all child c-trees of H. In this paper we only explain for Case 2 and Case 3, since other cases are similar.

Case 2: $x \in T(c_z)$, and x succeeds q_k in preorder.

If H has a child c-tree H_c with Type 4, 5, 6 or 7, then $P(H_c) \neq H$, a contradiction. Thus H has no child c-tree with Type 4, 5, 6 or 7.

Then consider for child c-trees with Type 1, 2 and 3.

Case 2(1): Child c-trees with Type 1.

Let $H_1[i]$ be the c-tree derived from H by adding the rightmost child leaf of r with color c_i. Assume that r_z has color c_j. The child c-trees of H with Type 1 are $H_1[0], H_1[1], \cdots, H_1[j]$. Note that $H_1[j+1]$ is not left heavy.

Case 2(2): Child c-trees with Type 2.

We need some definitions here.

Let $P = (u_0 = r, u_1, \cdots, u_{dep(x)} = x)$ be the path between $r = u_0$ and x. Let u_y on P_R be the ancestor of x having maximum depth. Thus P and P_R share the subpath $u_0 = q_0, u_1 = q_1, \cdots, u_y = q_y$. Let s_{i+1} be the child vertex of u_i preceding u_{i+1} (if such s_{i+1} exists), for $0 \leq i \leq dep(x)$.

We say that H is *active at depth i* if (i)u_i has two or more child vertices, and (ii)$L(H(u_{i+1}))$ is a prefix of $L(H(s_{i+1}))$. Intuitively, if H is active at depth i, then we are copying subtree $H(u_{i+1})$ from $H(s_{i+1})$. We say the *copy-depth of H* is d if H is active at depth d but not active at any depth in $\{0, 1, \cdots, d-1\}$. If H is not active at any depth, then we say the copy-depth of H is $dep(x)$. Assume that H is active at depth d.

Let $H_2[i,j]$ be the c-tree derived from H by adding the rightmost child leaf s to u_j with color c_i. Thus u_{j+1} precede the new vertex s in $H_2[i,j]$, if $j + 1 \leq dep(x)$. Any child c-tree of H with Type 2 is $H_2[i,j]$ for some i, j, however not all of them are child c-trees of H with Type 2. We need to check each carefully.

For $j = 0, 1, \cdots, d - 1$, if $c(u_{j+1}) \geq c_i$ then $H_2[i,j]$ is a child c-tree of H, and otherwise $H_2[i,j]$ is not a child c-tree of H, since it is not left heavy. The copy-depth of each derived c-tree is j if c_i equal to $c(u_{j+1})$, and is $j+1$ otherwise.

Then consider for $j = d, d + 1, \cdots, dep(x)$. Let n_R be the number of vertices in the subtree $H(u_{j+1})$ rooted at u_{j+1}, and t be the $(n_R + 1)$-th vertex in the subtree $H(s_{j+1})$ rooted at s_{j+1}. Assume t has a color c_ℓ.

If $j > dep(t)$ then $H_2[i,j]$ is not a child c-tree of H, since it is not left heavy. If $j = dep(t)$ but $\ell < i$ then $H_2[i,j]$ is not a child c-tree of H, since it is not left heavy. If $j = dep(t)$ and $\ell = i$ then $H_2[i,j]$ is a child c-tree of H. The copy-depth of the derived c-tree is again d. If $j = dep(t)$ and $\ell > i$ then $H_2[i,j]$ is a child c-tree of H. The copy-depth of each derived c-tree is j if c_i equal to $c(s_{j+1})$, and is $j + 1$ otherwise. If $j < dep(t)$ then $H_2[i,j]$ is a child c-tree of H for any i. The copy-depth of each derived c-tree is j if c_i equal to $c(s_{j+1})$, and is $j + 1$ otherwise.

Case 2(3): Child c-trees with Type 3.

In this case we need to check the reverse of Case 3(b) in Section 4. Thus a c-tree with Type 2 may have a child c-tree with Type 3.

Define $P = (u_0 = r, u_1, \cdots, u_{dep(x)} = x)$, u_y, ℓ as in Case 2(2).

If H has only one leaf succeeding q_k in preorder, $H(u_{y+1})$ is a path, $H(q_{y+1})$ is a path, and $L(H(u_{y+1}))$ is a prefix of $L(H(q_{y+1}))$, then, for each $i > \ell$, $H_2[i, dep(x)]$ is a child c-tree with Type 3, after swapping the order of u_{y+1} and q_{y+1}.

Case 3: $x \in T(c_z)$, and x precedes q_k in preorder.

If H has a child c-tree H_c with Type 4, 5, 6 or 7, then $P(H_c) \neq H$, a contradiction. Thus H has no child c-tree with Type 4, 5, 6 or 7.

Then consider for child c-trees with Type 1, 2 and 3.

Case 3(1): Child c-trees with Type 1.

Omitted. Similar to Case 2(1).

Case 3(2): Child c-trees with Type 2.

Omitted. Similar to Case 2(2).

Case 3(3): Child c-trees with Type 3.

Let $P = (u_0 = r, u_1, \cdots, u_{dep(x)} = x)$ be the path between $r = u_0$ and x. Let u_y on P_R be the ancestor of x having maximum depth. Let s_{i+1} be the child vertex of u_i preceding u_{i+1} (if such s_{i+1} exists), for $0 \leq i \leq dep(x)$.

We say that H is *active at depth* i if (i)u_i has two or more child vertices, and (ii)$L(H(u_{i+1}))$ is a prefix of $L(H(s_{i+1}))$. We say the *copy-depth* of H is d if H is active at depth d but not active at any depth in $\{0, 1, \cdots, d-1\}$. If H is not active at any depth, then we say the copy-depth of H is $dep(x)$. Assume that H is active at depth d.

For $j \geq y$, let $H_3[i, j]$ be the c-tree derived from H by adding the new child leaf s to u_j succeeding u_{j+1} with color c_i.

Any child c-tree of H with Type 3 is $H_3[i, j]$ for some i, j, however not all of them are child c-trees of H with Type 3.

For $j = y$, if $s \leq i < t$, where $c_s = c(u_{j+1})$ and $c_t = c(q_{j+1})$, then $H_2[i, j]$ is a child c-tree of H.

For $j = y + 1, y + 2, \cdots, d - 1$, if $c(u_{j+1}) \geq i$ then $H_3[i, j]$ is a child c-tree of H, and otherwise $H_3[i, j]$ is not a child c-tree of H, since it is not left heavy. The copy-depth of each derived c-tree is j if c_i equal to $c(u_{j+1})$, and is $j + 1$ otherwise.

Then consider for $j = d, d+1, \cdots, dep(x)$. Let n_R be the number of vertices in the subtree $H(u_{j+1})$ rooted at u_{j+1}, and t be the $(n_R + 1)$-th vertex in the subtree $H(s_{j+1})$ rooted at s_{j+1}. Assume t has a color c_ℓ.

If $j > dep(t)$ then $H_3[i, j]$ is not a child c-tree of H, since it is not left heavy. If $j = dep(t)$ but $\ell < i$ then $H_3[i, j]$ is not a child c-tree of H, since it is not left heavy. If $j = dep(t)$ and $\ell = i$ then $H_3[i, j]$ is a child c-tree of H. The copy-depth of the derived c-tree is again d. If $j = dep(t)$ and $\ell > i$ then $H_3[i, j]$ is a child c-tree of H. The copy-depth of each derived c-tree is j if c_i equal to $c(s_{j+1})$, and is $j + 1$ otherwise. If $j < dep(t)$ then $H_3[i, j]$ is a child c-tree of H for any i. The copy-depth of each derived c-tree is j if c_i equal to $c(s_{j+1})$, and is $j + 1$ otherwise.

Based on the case analysis above, we have the following theorem.

Theorem 1. *One can generate all c-trees in $O(f(n))$ time and $O(n)$ space, where $f(n)$ is the number of nonisomorphic c-trees with at most n vertices and diameter $2k$.*

Proof. Since we traverse the family tree $T_{n,2k,m}$ and output each ordered c-tree at each corresponding vertex of $T_{n,2k,m}$, we can generate all c-trees in $S_{n,2k,m}$.

We maintain the last two occurrences of each depth in two arrays of length k. We record the update of the arrays and restore the arrays if return occur. Thus we can find u_i in constant time for each i.

We also maintain the current copy-depth d and the vertex next to be copied.

Other parts of the algorithm need only constant time of computation for each edge of $T_{n,2k,m}$.

Thus the algorithm runs in $O(f(n))$ time. Note that the algorithm does not output each tree entirely, but the difference from the preceding tree.

For each recursive call we need a constant amount of space, and the depth of the recursive call is bounded by n. Thus the algorithm uses $O(n)$ space.   □

7   The Odd Diameter Case

In this section we sketch the case where the diameter is odd.

It is known that a tree with odd diameter $2k+1$ may have many paths of length $2k+1$, but all of them share a unique edge, called *the center* of T [W01–p72].

Intuitively, by treating the edge as the root, we can define the family tree $T_{n,2k+1,m}$ in a similar manner to the even diameter case. The detail is omitted.

8   Conclusion

In this paper we gave a simple algorithm to generate all c-trees with at most n vertices and diameter d. The algorithm generates each c-tree in constant time on average.

By slightly modifying the algorithm as shown below [NU03, NU04] we can improve the worst case running time. Since we traverse at most three edges to generate next c-tree, the algorithm generates each c-tree in constant time.

Procedure find-all-children$(T, depth)$
{ T is the current c-tree, and *depth* is the depth of the recursive call.}
begin
01   **if** *depth* is even
02   **then** Output T { before outputting its child c-trees.}
03   Generate child c-trees $T_1, T_2, \cdots, T_x$ by the method in Section 6 and 7, and
04   recursively call **find-all-children** for each child c-tree.
05   **if** *depth* is odd
06   **then** Output T { after outputting its child c-trees.}
end

References

[A95] A. V. Aho and J. D. Ullman, *Foundations of Computer Science*, Computer Science Press, New York, (1995).

[A03] T. Asai, H. Arimura, T. Uno, and S. Nakano, *Discovering Frequent Substructures in Large Unordered Trees*, The 6th International Conference on Discovery Science (DS'03) LNAI 2843, (2003), pp.47-61.

[B80] T. Beyer and S. M. Hedetniemi, *Constant Time Generation of Rooted Trees*, SIAM J. Comput., 9, (1980), pp.706-712.

[C05] Y. Chi, S. Nijsseny, R. R. Muntz and J. N. Kok, *Frequent Subtree Mining–An Overview*, Fundamenta Informaticae, Special Issue on Graph and Tree Mining, 2005. (to appear)

[G93] L. A. Goldberg, *Efficient Algorithms for Listing Combinatorial Structures*, Cambridge University Press, New York, (1993).

[KS98] D. L. Kreher and D. R. Stinson, *Combinatorial Algorithms*, CRC Press, Boca Raton, (1998).

[LN01] Z. Li and S. Nakano, *Efficient Generation of Plane Triangulations without Repetitions*, Proc. ICALP2001, LNCS 2076, (2001), pp.433–443.

[LR99] G. Li and F. Ruskey, *The Advantage of Forward Thinking in Generating Rooted and Free Trees*, Proc. 10th Annual ACM-SIAM Symp. on Discrete Algorithms, (1999), pp.939–940.

[M98] B. D. McKay, *Isomorph-free Exhaustive Generation*, J. of Algorithms, 26, (1998), pp.306-324.

[N02] S. Nakano, *Efficient Generation of Plane Trees*, Information Processing Letters, 84, (2002), pp.167–172.

[NU03] S. Nakano and T. Uno, *Efficient Generation of Rooted Trees*, NII Technical Report (NII-2003-005E) (2003). (http://research.nii.ac.jp/TechReports/03-005E.html)

[NU04] S. Nakano and T. Uno, *Constant Time Generation of Trees with Specified Diameter*, Proc. WG2004, LNCS, 3353, (2004), pp.33-45.

[RS00] F. Ruskey and J. Sawada, *A Fast Algorithm to Generate Unlabeled Necklaces*, Proc. of SODA (2000), pp.256–262

[R78] R. C. Read, *How to Avoid Isomorphism Search When Cataloguing Combinatorial Configurations*, Annals of Discrete Mathematics, 2, (1978), pp.107–120.

[W01] D. B. West, *Introduction to Graph Theory, 2nd Ed*, Prentice Hall, NJ, (2001).

[W89] H. S. Wilf, *Combinatorial Algorithms : An Update*, SIAM, (1989).

[W86] R. A. Wright, B. Richmond, A. Odlyzko and B. D. McKay, *Constant Time Generation of Free Trees*, SIAM J. Comput., 15, (1986), pp.540-548.

Optimal Hypergraph Tree-Realization

Ephraim Korach and Margarita Razgon[*]

Department of Industrial Engineering and Management,
Ben-Gurion University of the Negev, Beer-Sheva, 84105, Israel
korach@bgu.ac.il
rita19or@hotmail.com

Abstract. Consider a hyperstar H and a function ω assigning a non-negative weight to every unordered pair of vertices of H and satisfying the following restriction: for any three vertices u, v, x such that u and v belong to the same set of hyperedges, $\omega(\{u, x\}) = \omega(\{v, x\})$. We provide an efficient method that finds a tree-realization T of H which has the maximum weight subject to the minimum number of leaves.

We transform the problem to the construction of an optimal degree-constrained spanning arborescence of a non-negatively weighted directed acyclic graph (DAG). The latter problem is a special case of the weighted matroid intersection problem. We propose a faster method based on finding the maximum weighted bipartite matching.

1 Introduction

Consider the following notion. A *realization* of a hypergraph $H = (V, F)$ is an undirected graph $G = (V, E)$ such that every edge of G is contained in some hyperedge of H and the subgraph of G induced by every hyperedge of H is connected. If G is a tree, it is called a *tree-realization* of H.

We state the following optimization problem, which we term the TreeMinLeaves problem. Given a hypergraph $H = (V, F)$ and a function assigning non-negative weights to every unordered pair of vertices of H contained in some hyperedge of H. The task is to find a tree-realization of H that has the minimum number of leaves and, subject to this requirement, has the maximum weight.

We prove that the TreeMinLeaves problem is NP-hard even if H has exactly one hyperedge. Then we solve it polynomially under the following two restrictions.

1. H is a *hyperstar*, that is to say H is a hypergraph where the intersection of all its hyperedges is not empty.
2. For any three vertices u, v, x such that u and v belong to the same set of hyperedges, the weight of $\{u, x\}$ is equal to the weight of $\{v, x\}$.

To solve the problem, we define on vertices of H a special directed acyclic graph (DAG) G having exactly one root. Then we show that there is a bijection

---

[*] This research was partially supported by the Paul Ivanier Center for Robotics Research & Production Management, Ben-Gurion University of the Negev.

D. Kratsch (Ed.): WG 2005, LNCS 3787, pp. 261–270, 2005.

between the tree-realizations of H and the spanning arborescences of G. Moreover, given a spanning arborescence of G, the corresponding tree-realization of H can be obtained by removing the edge directions of the arborescence. Therefore, to find an optimal tree-realization of H with the minimum number of leaves, we have to find an optimal spanning arborescence of G with the minimum number of vertices with degree 1 (we call the latter problem MinDegree1) and then to transform the arborescence into the tree-realization of H.

The MinDegree1 problem is a special case of the weighted matroid intersection problem. We propose a faster method for solving the MinDegree1 problem based on finding the maximum weighted bipartite matching.

The TreeMinLeaves problem is similar to the problem of finding a maximum-weight tree-realization of a hypergraph that was presented in [4], with two differences. On the one hand, we restrict a hypergraph to a hyperstar. And on the other hand, we find a tree-realization of a hypergraph that has the maximum weight subject to the minimum number of leaves.

Several works have investigated problems related to tree-realizations of hypergraphs. A necessary and sufficient condition for a hypergraph to have a tree-realization is described in [5]. Algorithms for checking this condition are presented in [1] and [9]. An algorithm, that for a given hypergraph H recognizes whether there is a tree-realization T such that the subgraph of T induced by every hyperedge of H is a path, was presented in [8].

Our method for solving the TreeMinLeaves problem has applications in the area of communication network design. One possible motivation could be as follows. Given a collection of groups of customers, how do we construct the minimum cost communication tree over a global set of customers, such that each two customers of some group are connected through other customers from this group. The network has to acquire group fault tolerance, group privacy and the minimum number of customers having only one connection (leaves). Group fault tolerance means that customers within a specified group are not sensitive to disconnections of the global set. Group privacy means that communication between two customers of a specified group does not require participation of customers outside this group. Achieving the minimum number of leaves reduces the probability of creating an isolated customer in the case of disconnection.

The rest of the paper is organized as follows. Section 2 provides preliminary definitions. Section 3 describes the transformation of the TreeMinLeaves problem into the MinDegree1 problem. In Section 4 we solve the MinDegree1 problem by the method based on finding the maximum weighted bipartite matching. Section 5 summarizes the paper.

2   Preliminaries

Definition 1. *A* **hypergraph** *is a pair $H = (V, F)$ where V is a finite set and F is a family of subsets of V. The elements of V and F are called vertices and hyperedges, respectively.*

Definition 2. *A* **hyperstar** *is a hypergraph $H = (V, F)$ with $(\bigcap F \neq \emptyset)$.*

Definition 3. *Let $H = (V, F)$ be a hypergraph. A graph G is called a* **realization** *of H, if it satisfies the following two properties.*

1. *Each $e \in E(G)$ is contained in some $s \in F$.*
2. *For any $s \in F$, $G[s]$ is a connected graph.*

Definition 4. *If a realization T of a hypergraph $H = (V, F)$ is a tree, T is called a* **tree-realization** *of H.*

Definition 5. *A* **hypertree** *is a hypergraph that has a tree-realization.*

Remark. A hyperstar is a special case of a hypertree [5].

Definition 6. *Let $H = (V, F)$ be a hypergraph and $v \in V$ be a vertex. The* **incidence set of** v *denoted by $I(v)$ is the set of all hyperedges of H such that v is contained in them.*

Definition 7. *Let $H = (V, F)$ be a hypergraph. Two vertices $u, v \in V$ are* **relatives** *in H, if $I(u) = I(v)$. (For example, vertices a and b are relatives in H in Figure 1.) $S \subseteq V$ is a set of relatives in H, if any two of its elements are relatives.*

Proposition 1. *A binary relation R such that $(u, v) \in R$ if and only if u and v are relatives is an equivalence relation. The maximal sets of relatives are the equivalence classes of R.* ∎

Remark. We denote the equivalence class containing $v \in V$ by $[v]$.

Definition 8. *A directed acyclic graph or a* **DAG** *is a directed graph containing no directed cycles.*

Definition 9. *An* **arborescence** *is a directed tree in which one vertex called the root has indegree 0 and the remaining vertices have indegree 1.*

Let us fix the notations relating to degrees of vertices of a graph. We denote the degree of a vertex v of a graph G by $d_G(v)$. If G is a directed graph then by $d_G^{in}(v)$ and $d_G^{out}(v)$ we denote the indegree and the outdegree of v, respectively.

3   Optimal Tree-Realization of a Hypergraph

In this section we define the TreeMinLeaves problem and describe the transformation of the TreeMinLeaves problem into the MinDegree1 problem.

The TreeMinLeaves Problem. Let $H = (V, F)$ be a hypergraph and let $\omega : \{\{u, v\} \mid \exists s \in F \ (\{u, v\} \subseteq s)\} \to \mathbb{R}^+$ be a function of weights. The task is to find a tree-realization of H that has the minimum number of leaves and, subject to this requirement, has the maximum weight.

The TreeMinLeaves problem is NP-hard even if H has exactly one hyperedge. NP-hardness can be shown by reduction from the Hamiltonian Path problem. We show that the TreeMinLeaves problem can be solved polynomially under the following two restrictions.

1. $H = (V, F)$ is a hyperstar.
2. For every $u, v, x \in V$ such that u and v are relatives, $\omega(\{u, x\}) = \omega(\{v, x\})$. (For example, $\omega(\{a, c\}) = \omega(\{b, c\})$ in H in Figure 1).

To solve the TreeMinLeaves problem under these restrictions, we define the notion of a weighted hierarchical DAG of a hyperstar.

Definition 10. *Let* $H = (V, F)$ *be a hyperstar. Let* $\omega : \{\{u, v\} \mid \exists s \in F (\{u, v\} \subseteq s)\} \to \mathbb{R}^+$ *be a function of weights. A DAG* $G = (V, E)$ *is called a* weighted **hierarchical DAG** *of* H *if it constructed as follows.*

1. *Define an arbitrary linear order on every maximal set of relatives.*
2.

$$\forall u, v \in V : (u, v) \in E \Leftrightarrow \begin{cases} I(v) \subset I(u) \\ \quad or \\ I(v) = I(u) \text{ and } (v < u) \text{ by the linear order for } [v]. \end{cases}$$

3. *Set the weight of every* $(u, v) \in E(G)$ *equal to* $\omega(\{u, v\})$.

An example of the construction of G *see in Figure 1.*

Let us prove two properties of a weighted hierarchical DAG of a hyperstar.

Proposition 2. *A weighted hierarchical DAG of a hyperstar* $H = (V, F)$ *can be constructed in* $O(V^2 * F)$ *time.*

Proof.
$O(V^2)$ pairs of vertices must be checked and every $I(v)$ can be computed in $O(F)$ time. ∎

Proposition 3. *A weighted hierarchical DAG of a hyperstar* $H = (V, F)$ *has exactly one root.*

Proof.
Assume by contradiction that there are two roots r_1 and r_2. Considering that H is a hyperstar, $I(r_1) = I(r_2) = F$. Hence r_1 and r_2 are relatives. According to the linear order defined on $[r_1]$, either $(r_1 < r_2)$ or $(r_2 < r_1)$. Thus indegree of one of these roots is not zero. ∎

The following claims establish a bijection between the tree-realizations of a hyperstar H and the spanning arborescences of a weighted hierarchical DAG of H.

Lemma 1. *Any two weighted hierarchical DAGs of a hyperstar* H *are isomorphic (taking into account the weights of the edges).*

Proof.
Let $R_1, \ldots, R_k, R'_1, \ldots, R'_k$ be two linear orders on k maximal sets of relatives, and let G and G' be two weighted hierarchical DAGs of H constructed under

these linear orders, and let $p_1, \ldots, p_k$ be permutations on $R_1, \ldots R_k$ respectively, such that p_i transforms R_i into R_i'. Consider the function $h = p_1 \cup \ldots \cup p_k$. Then we can show that h is an isomorphism.

Assume that $(u, v) \in E(G)$.

If u and v are not relatives, then $I(v) \subset I(u)$. Observe that $h(u)$ is a relative of u and $h(v)$ is a relative of v; therefore, $(h(u), h(v)) \in E(G')$. According to the restriction on weights, $\omega(u, v) = \omega(h(u), v) = \omega(h(u), h(v))$.

If u and v are relatives, assume that R_i is the order on $[v]$. Then $v < u$ according to R_i and $h(v) < h(u)$ according to R_i'. Therefore, $(h(u), h(v)) \in E(G')$. Equality of the weights can be proved analogously.

In the same manner we can prove that if $(u, v) \in E(G')$ then $(h^{-1}(u), h^{-1}(v)) \in E(G)$ and $\omega(u, v) = \omega(h^{-1}(u), h^{-1}(v))$. ∎

Proposition 4. *Let T be a tree-realization of a hyperstar $H = (V, F)$. For any $u, v \in V$ connected by an edge in T, either $I(u) \subseteq I(v)$ or $I(v) \subseteq I(u)$.*

Proof.
Assume by contradiction that there is $(u, v) \in E(T)$ such that $I(u) \nsubseteq I(v)$ and $I(v) \nsubseteq I(u)$. This means that $\exists x, y \in F \ ((x \in I(u) \wedge x \notin I(v)) \wedge (y \in I(v) \wedge y \notin I(u)))$. Thus $((u \in x \wedge u \notin y) \wedge (v \in y \wedge v \notin x))$. Let $w \in V$ such that $I(w) = F$. In particular, $w \in x$ and $w \in y$. According to Definition 3, w is connected with u in T by a path P_1 that passes through vertices of x only and w is connected with v in T by a path P_2 that passes through vertices of y only. On the one hand, $(u, v) \notin P_1$ and $(u, v) \notin P_2$. On the other hand, by our assumption $(u, v) \in E(T)$. Thus there is a cycle in T in contradiction to Definition 4. ∎

Lemma 2. *Let T be a tree-realization of a hyperstar $H = (V, F)$. Then there is a weighted hierarchical DAG G of H such that T can be transformed into a spanning arborescence of G by assigning directions to its edges.*

Proof.
Let us fix some vertex $r \in V$ such that $I(r) = F$. Let us assign directions to edges of T such that T is transformed into an arborescence T' with root r. We make the following observation: if $(u, v) \in E(T')$ then $I(v) \subseteq I(u)$. Assume by contradiction that the observation does not hold for some $(u, v) \in E(T')$. According to Proposition 4, $I(u) \subset I(v)$. Note that $I(u) \subset I(r) = F$ and that the path P_1 from r to v passes through u in T. Let $s \in I(v)$ be a set such that $s \notin I(u)$. According to Definition 3, r is connected to v in T by a path P_2 that passes through vertices of s only. Considering that $u \notin s$, we get that $P_1 \neq P_2$. Therefore there are two paths between r and v in T in contradiction to Definition 4. Thus the observation holds.

Let $T_1, \ldots, T_k$ be the subgraphs of T induced by maximal sets of relatives, and let $T_1', \ldots T_k'$ be the graphs obtained by transitive closure of $T_1, \ldots T_k$, respectively. Define partial order relations $R_1', \ldots R_k'$ as follows: $(u, v) \in R_i'$ if and only if $(v, u) \in E(T_i')$. Extend the relations to linear orders on maximal set of relatives. Let $R_1, \ldots, R_k$ be the relations obtained as the result of the extension of $R_1', \ldots, R_k'$, respectively. Construct G given $R_1, \ldots R_k$.

It remains to be shown that T' is a spanning arborescence of G. Let us prove that if $(u,v) \in E(T')$ then $(u,v) \in E(G)$. If u and v are not relatives, then $I(v) \subset I(u)$ according to the observation discussed above. If u and v are relatives then $v < u$ by the order we defined on $[v]$. In both cases, according to the construction of G, $(u,v) \in E(G)$.   ∎

Lemma 3. *Let $H = (V, F)$ be a hyperstar. Let T' be a spanning arborescence of a weighted hierarchical DAG G of H. Then T' is transformed into a tree-realization T of H by removing the edge directions.*

Proof.
Let us remove the edge directions of T' and get a tree T. Take $s \in F$ and $u, v \in s$. We have to prove that u and v are connected in T by a path that passes through vertices of s only, according to Definition 4. Let w be the vertex with $d_G^{in}(w) = 0$, which is unique according to Proposition 3. According to Definition 9, there are unique paths from w to u and from w to v in T'. Let $w \to u_1 \to \ldots \to u_k \to u$ and $w \to v_1 \to \ldots \to v_l \to v$ be paths in T' from w to u and from w to v, respectively. According to the construction of G, $s \in I(u) \subseteq I(u_k) \subseteq \ldots \subseteq I(u_1) \subseteq I(w)$ and $s \in I(v) \subseteq I(v_l) \subseteq \ldots \subseteq I(v_1) \subseteq I(w)$. Therefore, $w, u_1, \ldots, u_k, v_1, \ldots, v_l$ belong to s. Hence, the walk $[u, u_k, \ldots, u_1, w, v_1, \ldots, v_l, v]$ contains a path in T between u and v that passes through vertices of s only.   ∎

In order to formulate the central theorem of the section, consider the following problem.

The MinDegree1 Problem. Let $G = (V, E)$ be a DAG with exactly one root r and let $\omega : E(G) \to \mathbb{R}^+$ be a function of weights. The task is to find a spanning arborescence of G that has the minimum number of vertices with degree 1 and, subject to this requirement, has the maximum weight.

Theorem 1. *There is a bijection between the set of solutions of the TreeMinLeaves problem on H and the set of solutions of the MinDegree1 problem on any arbitrary weighted hierarchical DAG of H. Moreover, a solution of TreeMinLeaves can be obtained from a solution of MinDegree1 by removing edge directions.*

Proof.
The proof immediately follows from Lemmas 1, 2 and 3.   ∎

The MinDegree1 problem is a special case of the weighted matroid intersection problem. Let us explain the idea behind the statement. Let $G = (V, E)$, r and ω be a DAG, a root and weights, respectively, as given in the MinDegree1 problem. Consider a set E' of copies of edges of E such that $\forall e = (u,v) \in E : e' = (u,v) \in E'$. Let B be a very large number, for example, $B = (\sum_{e \in E} \omega(e)) * |V|^2$. Consider a function of weights $\omega' : (E \cup E') \to \mathbb{R}^+$ defined as follows:

- $\forall e \in E \; : \; \omega'(e) = \omega(e);$
- $\forall e' \in E' \; : \; \omega'(e') = \omega'(e) + B$ (E' are heavy edges).

Let $M_{in} = (E \cup E', \mathcal{I})$ be a pair such that for every $S \in \mathcal{I}$ and $v \in V(G)$ at most one edge of S enters into v. Let $M_{out} = (E \cup E', \mathcal{I})$ be a pair such that for every $S \in \mathcal{I}$ and $v \in V(G)$ at most two edges of $S \cap E'$ exit from root r and at most one edge of $S \cap E'$ exits from non-root v. $M_{in} = (E \cup E', \mathcal{I})$ and $M_{out} = (E \cup E', \mathcal{I})$ are matroids [6]. We solve the weighted matroid intersection problem for M_{in}, M_{out} and ω' to obtain a maximum-weight set T of $M_{in} \cap M_{out}$. The solution of the MinDegree1 problem is T with the original weights of edges.

According to our best knowledge, the best upper bound of the weighted matroid intersection problem is $O(Ek(logE + k))$ [2], where k is the cardinality of the resulting solution. In our case, $k = O(V)$, hence the upper bound is $O(EV(logE + V))$. In Section 4 we propose a method designed specially for solving the MinDegree1 problem that takes $O(V^2(logV + \triangle G))$ time, where $\triangle G$ is the maximum outdegree of V(G). Clearly, our method is faster for graphs with $|E| > |V|$.

4   Optimal Spanning Arborescence

Let $G = (V, E)$ be a DAG with exactly one root r, and let $\omega : E(G) \rightarrow \mathbb{R}^+$ be a function of weights. We construct a bipartite graph $G' = (V_1, V_2, E')$ as follows.

- $\forall u \in V : V_1$ includes one copy u' of u,
    V_2 includes $d_G^{out}(u)$ copies u: $\langle u'', 1 \rangle, \ldots, \langle u'', d_G^{out}(u) \rangle$.
- $\forall (u, v) \in E$: E' includes $(\langle u'', i \rangle, v')$, $i = 1, \ldots, d_G^{out}(u)$.

Let B be a very large number, for example, $B = (\sum_{e \in E} \omega(e)) * |V(G)|^2$. We define a function $\omega' : E' \rightarrow \mathbb{R}^+$ of weights as follows.
    $\forall (\langle u'', i \rangle, v') \in E' \; :$

$$\omega'(\langle u'', i \rangle, v') = \begin{cases} \omega(u, v) + B, & \text{if } i = 1 \text{ (for } u \neq r) \text{ or } i \leq 2 \text{ (for } u = r) \\ \omega(u, v), & \text{otherwise.} \end{cases}$$

Let us call the graph G' a weighted **bipartite representative** of a DAG G. An example of the construction of G' see in Figure 1.

We say that $\omega(u, v)$ is an **original** weight of an edge $(\langle u'', i \rangle, v')$ of G'. We call edges of G' having weights greater than B **heavy** edges. Also, we refer to a vertex $\langle u'', i \rangle \in V_2$ as a **copy** of $u \in V(G)$. If a copy of $u \in V(G)$ is incident to some heavy edge in G', we call it a **heavy copy** of u.

Theorem 2. *Consider a maximum-weight matching M of a weighted bipartite representative G' of a given DAG G. Then the graph T with $V(T) = V(G)$ such that $\forall (u, v) \in E \; : \; (u, v) \in E(T) \Leftrightarrow \exists i \, (\langle u'', i \rangle, v') \in M$ is a solution of the MinDegree1 problem.*

To prove Theorem 2, we need the two following lemmas.

Lemma 4. T *is a spanning arborescence of* G.

Proof.
We must show that T does not have a vertex with indegree greater than 1 and T connects all vertices of G.

For the former assume by contradiction that there is a vertex $v \in V(T)$ such that $d_T^{in}(v) \geq 2$ and edges $(u, v) \in E(T)$ and $(w, v) \in E(T)$ enter into v. There are i, j such that $(\langle u'', i\rangle, v') \in M$ and $(\langle w'', j\rangle, v') \in M$ in contradiction to the definition of a matching.

For the latter assume by contradiction that T does not connect all vertices of G. This means that there is a vertex $l \in V(G)$ other than the root such that $d_T^{in}(l) = 0$. Therefore, the vertex $l' \in V_1$ is exposed by M. Because of $d_G^{in}(l) \geq 1$, $\exists u \in V(G)$: $(u, l) \in E(G)$. Hence, according to definition of G', $\forall i \in \{1, \ldots, d_G^{out}(u)\}$: $(\langle u'', i\rangle, l') \in E(G')$. Observe that there is $i \in \{1, \ldots, d_G^{out}(u)\}$ such that the vertex $\langle u'', i\rangle \in V_2$ is exposed by M. Actually, if for all $i \in \{1, \ldots, d_G^{out}(u)\}$ vertices $\langle u'', i\rangle$ are matched by M, then $\forall v' \in V_1$: $(u, v) \in E(G) \Rightarrow (\langle u'', i\rangle, v') \in M$ for some i. But we have already seen that for l' this is not so. Therefore, we take such i that the vertex $\langle u'', i\rangle$ is exposed by M and add the edge $(\langle u'', i\rangle, l')$ to M in contradiction to the maximality of M. ∎

Lemma 5. *The maximum-weight matching* M *of* G' *has the maximum possible number of heavy edges.*

Proof.
Let M_1 and M_2 be two matchings of G' containing k_1 and k_2 heavy edges, respectively, such that $k_1 > k_2$. To prove the lemma we must show that $\sum_{e \in M_1} \omega'(e) > \sum_{e \in M_2} \omega'(e)$. Let W_1 and W_2 be sums of original weights of edges of M_1 and M_2, respectively. Then $\sum_{e \in M_1} \omega'(e) = k_1 * B + W_1$ and $\sum_{e \in M_2} \omega'(e) = k_2 * B + W_2$ are the weights of M_1 and M_2, respectively.

According to our selection of B, the sum of all original weights is less then B. Therefore: $k_2 * B + W_2 < k_2 * B + B = (k_2 + 1) * B \leq k_1 * B + W_1$. ∎

Now, we are ready to prove Theorem 2.

Proof of Theorem 2.
First, let us prove that T of G has the minimum number of vertices with degree 1. Let k be the number of heavy edges in M. The three properties below follow from the maximality of M.

1. For every $v \in V(G)$ such that $v \neq r$, if at least one copy of v is matched by M in V_2, then the vertex $\langle v'', 1\rangle$ is matched by M.
2. One copy of r, $\langle r'', 1\rangle$, is matched by M in V_2.
3. If at least two copies of r are matched by M in V_2, then both $\langle r'', 1\rangle$ and $\langle r'', 2\rangle$ are matched by M.

According to these properties, if property 3 holds, then $V(G)$ has exactly $(k-1)$ vertices, whose copies are matched by M in V_2. Otherwise, $V(G)$ has exactly

k such vertices. In both cases, M is transformed into a spanning arborescence T (Lemma 4) with $(k-1)$ vertices, whose degree is at least 2. According to Lemma 5, k is the maximum possible number of heavy edges in M. Thus $(k-1)$ is the maximum possible number of vertices with degree at least 2 in T. Hence $(|V|-k+1)$ is the minimum possible number of vertices with degree 1 in T.

Second, to show that the weight of T is maximum, assume by contradiction that there is a spanning arborescence T' of G having the same number of vertices with degree 1 as T but $\sum_{e \in T'} \omega(e) > \sum_{e \in T} \omega(e)$. For every vertex $v \in V(T)$, let $(v, v_1), \dots, (v, v_m)$ be the edges of T that exit from v. Transform them into the edges $(\langle v'', 1 \rangle, v_1'), \dots, (\langle v'', m \rangle, v_m')$ of G'. It is not difficult to see that T' corresponds to a matching M' of G'. Note that $\sum_{e \in M} \omega'(e) = k * B + \sum_{e \in T} \omega(e)$ and $\sum_{e \in M'} \omega'(e) = k * B + \sum_{e \in T'} \omega(e)$. Therefore, $\sum_{e \in M'} \omega'(e) > \sum_{e \in M} \omega'(e)$ in contradiction to the maximality of M. ∎

We have shown that the MinDegree1 problem can be solved for DAG $G = (V, E)$ by computing the maximum-weight matching of its weighted bipartite representative $G' = (V_1, V_2, E')$. A maximum-weight matching of G' can be computed by the Hungarian method ([7] Chapter 17). However, the *straightforward* application of the Hungarian method to G' is inappropriate. Actually, the execution of the Hungarian method on G' consists of V_1 applications of Dijkstra's algorithm. A rough estimation of the complexity of Dijkstra's algorithm is $O((V_1 + V_2)^2)$. Taking into account that $|V_1| = |V|$, $|V_2| = O(E)$, we get that the whole complexity is $O(V * (V + E)^2)$, which is even worse than the complexity of weighted matroid intersection problem.

The complexity of the matching computation can be considerably reduced if we take into account the structure of G'. According to the definition of the function ω', weights of edges incident to each non-heavy copy of v in V_2 of G' correspond to weights of edges that exit from v in DAG G. Hence, in every iteration of the Hungarian method, non-heavy copies of every vertex $v \in V$, that are exposed in V_2 by the *current* matching, can be contracted into a single vertex v^*. The *new* matching contains at most one edge incident to v^*, thus v^* can be replaced by any non-heavy copy of v in V_2 that was involved in the contraction into v^*. It is easy to obtain the "contracted version" of G' in $O(V)$ time, given DAG G and the current matching. The contracted version of G' has $O(V)$ vertices, thus application of Dijkstra's algorithm to the graph takes $O(V log V + E^*)$ time [3], where E^* is the set of edges of the "contracted version" of G'. Note that $E^* = V * \triangle G$, where $\triangle G$ is the maximum outdegree of $V(G)$. Hence we achieve $O(V log V + V * \triangle G)$ time per application. Thus the whole complexity is $O(V^2 (log V + \triangle G))$.

5   Summary

We presented a technique for the TreeMinLeaves problem that transforms the input into an instance of the MinDegree1 problem. The latter problem is solved by our method based on finding the maximum weighted bipartite matching. An example of execution of the proposed technique is shown in Figure 1.

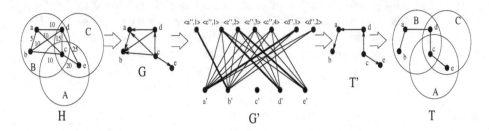

Fig. 1. A process of solving of the TreeMinLeaves problem

A natural extension of the proposed method is a generalization to other types of hypergraphs. It is unlikely that the proposed technique could be applied straightforwardly to hypergraphs which are not a hyperstars, but using it as a part of a more complicated scheme seems quite possible. Finally, note that the complexity of the unweighted version of the TreeMinLeaves problem for hypertrees remains unknown, to our best knowledge.

References

1. C. Beeri, R. Fagin, D. Maier, and M. Yannakakis. On the desirability of acyclic database schemes. *Journal of the ACM*, 30:479–513, 1983.
2. C. Brezovec, G. Cornuejols, and F. Glover. Two algorithms for weighted matroid intersection. *Mathematical Programming, Series A*, 36:39–53, 1986.
3. T. Cormen, C. Leiserson, R. Rivest, and C. Stein. *Introduction to Algorithms*. MIT Press, 2001.
4. E. Korach and M. Stern. The clustering matroid and the optimal clustering tree. *Mathematical Programming, Series B*, 98(1-3):385–414, 2003.
5. T. McKee and F. McMorris. *Topics in Interesection Graph Theory*. Siam Monographs on Discrete Mathematics and Applications, 1999.
6. A. Recski. *Matroid Theory and its Applications in Electric Network Theory and in Statics*. Springer, 1989.
7. A. Schrijver. *Combinatorial Optimization: Polyhedra and Efficiency*. Springer, 2003.
8. R. Swaminathan and D. Wagner. On the consecutive-retrieval problem. *SIAM Journal of Computing*, 23(2):398–414, 1994.
9. R. Tarjan and M. Yannakakis. Simple linear-time algorithms to test chordality of graphs, test acyclicity of hypergraphs, and selectively reduce acyclic hypergraphs. *SIAM Journal of Computing*, 13:566–579, 1984.

Fixed-Parameter Algorithms for Protein Similarity Search Under mRNA Structure Constraints

Guillaume Blin[1], Guillaume Fertin[1],
Danny Hermelin[2], and Stéphane Vialette[3]

[1] Laboratoire d'Informatique de Nantes-Atlantique (LINA),
FRE CNRS 2729 Université de Nantes,
2 rue de la Houssinière, 44322 Nantes Cedex 3, France
{blin, fertin}@lina.univ-nantes.fr
[2] Department of Computer Science, University of Haifa,
Mount Carmel, Haifa, Israel
danny@cri.haifa.ac.il
[3] Laboratoire de Recherche en Informatique (LRI), UMR CNRS 8623
Faculté des Sciences d'Orsay - Université Paris-Sud, 91405 Orsay, France
vialette@lri.fr

Abstract. In the context of protein engineering, we consider the problem of computing an mRNA sequence of maximal codon-wise similarity to a given mRNA (and consequently, to a given protein) that additionally satisfies some secondary structure constraints, the so-called MRSO problem [2]. Since the MRSO problem is known to be **APX**-hard [8], Bongartz proposed in [8] to attack the problem using the concept of parameterized complexity. In this paper we follow this suggested approach by devising fixed-parameter algorithms for several interesting parameters of MRSO. We believe these algorithms to be relevant for practical applications today, as well as for several future applications. Furthermore, our results extend the known tractability borderline of MRSO, and provide new research horizons for further improvements of this sort.

1 Introduction

In [2,3], Backofen *et al.* introduced the problem of computing an mRNA sequence of maximum codon-wise similarity to a given mRNA (and consequently, to a given protein) that additionally satisfies some secondary structure constraints, the so-called MRSO problem.

The initial motivation of MRSO is concerned with selenocysteine insertion, *i.e.* generating new amino acid sequences containing selenocysteine. This rare amino acid was discovered as the 21st amino acid [5], giving another clue to the complexity and flexibility of the mRNA translation mechanism. Selenocysteine is encoded by the UGA codon, which is usually a stop codon encoding the end of translation. It has been shown [5] that in case of selenocysteine, termination of translation is inhibited in the presence of a sequence of nucleotides which

D. Kratsch (Ed.): WG 2005, LNCS 3787, pp. 271–282, 2005.
© Springer-Verlag Berlin Heidelberg 2005

forms a hairpin-like structure in the $3'$-region after the UGA codon. It is argued in [2] that modifying existing proteins by incorporating selenocysteine instead of a catalytic cysteine is an important problem for catalytic activity enhancement and X-ray crystallography.

Selenocysteine insertion is concerned with a restricted type of secondary structure, *i.e.* a secondary structure without pseudo-knots, and hence the linear-time algorithm presented in [2] provides an optimal solution. However, it is reasonable to assume that the discovery of selenocysteine will lead to the discovery of several other amino acids of similar kind, some of which are likely to require more complex secondary structures. Even today, similar problems occur in programmed frameshifts which allow to encode two different amino acid sequences in one mRNA sequence [12,11]. This motivates the investigation of MRSO for more elaborate secondary structures [2,8], and is the starting point of our study.

For the MRSO problem, it has been shown in [2] that there exists a linear-time algorithm if the considered secondary structure corresponds to an outer-planar graph (as it is the case of selenocysteine insertion). In this paper, we refer to this algorithm as $\mathcal{A}_{\text{OP}}$. For the general case, the problem was proved to be **NP**-complete [2], and Bongartz showed recently that the problem is in fact **APX**-hard [8]. An algorithm for approximating MRSO within ratio 2 is given in [2]. A slightly slower but somewhat simpler 4-approximation algorithm is given in [8]. We mention also that an extension of MRSO, where insertions and deletions are allowed in the amino acid sequence, is presented in [1].

Since MRSO for general secondary structures is known to be **APX**-hard [8], Bongartz proposed in [8] to attack the problem using the concept of parameterized complexity [10]. Parameterized complexity is an approach to complexity theory which offers an alternative method of analyzing computational problems in terms of their tractability. For many hard problems, the seemingly unavoidable combinatorial explosion can be restricted to a small part of the input, the *parameter*, so that the problems can be solved in polynomial-time when the parameter is fixed.

In the last decade, parameterized complexity has proven to be useful in several applications within computational biology [7]. In this paper we attempt to follow this line by presenting fixed-parameter algorithms for several interesting parameters of MRSO. We believe these algorithms to be relevant for practical applications today, as well as for several future applications. Furthermore, our results extend the known tractability borderline of MRSO, and provide new research horizons for further improvements of this sort.

The paper is organized as follows. In the next section we briefly discuss basic notations and definitions that we will use throughout. In Section 3, we present a fixed-parameter algorithm for two natural parameters of MRSO, namely the number of degree three vertices, and the number of edge crossings in the given implied structure graph (see Definition 2 in the following section). In Section 4, we give a tighter **NP**-completeness result for MRSO, by showing that the problem is **NP**-complete even if the given implied structure graph has page number two. In Section 5, we consider the cutwidth of the implied structure graph as a parameter, and show that the problem is polynomial-time solvable in case this parameter is fixed. Finally, in Section 6 we prove that a slightly restricted version of MRSO is polynomial-time solvable in case the score of the optimal solution is fixed.

2 Preliminaries

An mRNA is a string over the alphabet $\Sigma = \{A, C, G, U\}$, where Σ represents the four different types of nucleotides in the molecule. The pairs $\{A, U\}$, $\{G, C\}$, and $\{G, U\}$ are known as *complementary nucleotide pairs*. Hydrogen bonds can only be formed between complementary nucleotides in an mRNA folding. A *codon* of an mRNA sequence is a sequence of three consecutive nucleotides, *i.e.* a string in Σ^3. Thus, an mRNA sequence $S = s_1 \cdots s_{3n}$ is a concatenation of n consecutive codons, where the ith codon of S is $s_{3i-2}s_{3i-1}s_{3i}$.

Given a *source* mRNA sequence $S = s_1 \ldots s_{3n}$, we wish to evaluate the codon-wise similarity between S and another *target* mRNA sequence $T = t_1 \ldots t_{3n}$. For this, we are provided with a set of n functions, $\mathcal{F} = f_1, \ldots, f_n$, called *similarity functions* of S, such that for all $1 \leq i \leq n$, each function f_i is of the form $f_i : \Sigma^3 \to \mathbb{Q}$. Thus, f_i assigns a value to the ith codon of T according to its level of similarity in comparison with the ith codon of S. The total level of similarity between S and T is then given by $\sum_{i=1}^{n} f_i(t_{3i-2}t_{3i-1}t_{3i})$. Note that given a set of similarity functions $\mathcal{F} = f_1, \ldots, f_n$ for S, one does not need to know anything else about S in order to compute the similarity score of S and T.

The *structure constraints* $\Gamma \subseteq \{\{i, j\} \mid 1 \leq i < j \leq 3n\}$ for a target mRNA sequence T of length $3n$, are pairings between distinct integers in $\{1, 2, \ldots, 3n\}$. These represent necessary hydrogen bonds in the folding of T. Since we assume that each nucleotide can pair with at most one other nucleotide in any folding, each integer appears in at most one pair in Γ. Furthermore, there are no pairs of the form $\{i, i+1\}$ or $\{i, i+2\}$ in Γ, for all $1 \leq i \leq 3n - 2$.

Given a set of structure constraints $\Gamma \subseteq \{\{i, j\} \mid 1 \leq i < j \leq 3n\}$, and an arbitrary target mRNA sequence $T = t_1 \cdots t_{3n}$, we say that nucleotides t_i and t_j in T are *compatible* with respect to Γ, if either $\{t_i, t_j\}$ is a complementary nucleotide pair or $\{i, j\} \notin \Gamma$. The entire sequence T is compatible with respect to Γ, if all pairs of nucleotides in T are compatible with respect to Γ.

Definition 1 (mRNA Structure Optimization (MRSO) [2]). *Let $\mathcal{F}$ be a set of n similarity functions for a source mRNA sequence of length $3n$, and let $\Gamma \subseteq \{\{i, j\} \mid 1 \leq i < j \leq 3n\}$ be a set of structure constraints. The MRSO problem asks to find a target mRNA sequence which is compatible with respect to Γ, and which achieves the highest possible similarity score with respect to $\mathcal{F}$.*

It is convenient to formalize MRSO in a slightly different manner using graph theoretic concepts. For a graph G, we let $\mathbf{V}(G)$ denote the set of vertices of G, and $\mathbf{E}(G)$ the set of edges of G. A linear graph G is a graph with $\mathbf{V}(G) = \{1, \ldots, |\mathbf{V}(G)|\}$. That is, it is a graph with vertices which have a fixed ordering. Therefore, we now view Γ as a linear graph with $3n$ vertices which has a maximum degree of one. As we are really interested in codon-wise similarity, we use a more suitable representation of Γ.

Definition 2 (Implied structure graph [2]). *Let $\Gamma \subseteq \{\{i, j\} \mid 1 \leq i < j \leq 3n\}$ be a set of structure constraints for a target mRNA sequence of length $3n$. The implied structure graph of Γ, is the linear graph G_Γ with:*

$\mathbf{V}(G_\Gamma) = \{1, 2, \ldots, n\},$ *and*

$\mathbf{E}(G_\Gamma) = \Big\{\{i, j\} \,\Big|\, \exists \{x, y\} \in \Gamma : x \in \{3i-2, 3i-1, 3i\} \wedge y \in \{3j-2, 3j-1, 3j\}\Big\}.$

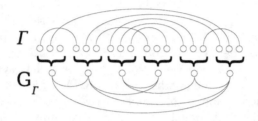

Fig. 1. An example of an implied structure graph obtained from a set of structure constraints. Note that G_Γ is outerplanar since swapping the two middle vertices yields an ordering of the vertices with no edge crossings.

Hence, G_Γ is a subcubic graph (*i.e.* a graph with a maximum degree of three) where vertex i in $\mathbf{V}(G_\Gamma)$ corresponds to the ith codon of a target mRNA sequence, and $i, j \in \mathbf{V}(G_\Gamma)$ are connected in $\mathbf{E}(G_\Gamma)$ if there are any structure constraints in Γ between the ith and jth codons of the sequence. Note that there can be at most three structure constraints between any pair of codons.

Given a subset of vertices $V \subseteq \mathbf{V}(G_\Gamma)$, we let $G_\Gamma[V]$ denote the subgraph of G_Γ *induced* by V, *i.e.* the subgraph with vertex set V and edge set $\mathbf{E}(G_\Gamma) \cap (V \times V)$. Similarly, given a subset of edges $E \subseteq \mathbf{E}(G_\Gamma)$, $G_\Gamma[E]$ denotes the subgraph of G_Γ with vertex set $\{i \mid \{i, j\} \in \mathbf{E}(G_\Gamma)\}$ and edge set E. Furthermore, we let $G_\Gamma[i, j]$ denote the subgraph of G_Γ induced by $\{i, \ldots, j\} \subseteq \mathbf{V}(G_\Gamma)$.

Two edges $\{i, j\}$ and $\{i', j'\}$ *cross* in G_Γ if either $i < i' < j < j'$ or $i' < i < j' < j$. Note that two crossing edges might not cross under a different ordering of $\mathbf{V}(G_\Gamma)$. If there exists an ordering of $\mathbf{V}(G_\Gamma)$ which introduces no edge crossings then G_Γ is *outerplanar*. Recall that in this case, algorithm $\mathcal{A}_{\mathrm{OP}}$ [2] solves MRSO in linear time.

A *codon assignment* for G_Γ is a mappings from some $V \subseteq \mathbf{V}(G_\Gamma)$ to Σ^3. An assignment for a pair of vertices $i, j \in \mathbf{V}(G_\Gamma)$, $i \rightarrow t_{3i-2}t_{3i-1}t_{3i}$ and $j \rightarrow t_{3j-2}t_{3j-1}t_{3j}$, is compatible with respect to G_Γ, if either $\{i, j\} \notin \mathbf{E}(G_\Gamma)$ or $t_{i'}$ and $t_{j'}$ are complementary nucleotides for any $\{i', j'\} \in \Gamma \cap \{3i-2, 3i-1, 3i\} \times \{3j-2, 3j-1, 3j\}$. More generally, an assignment $\phi : V \rightarrow \Sigma^3$ for some $V \subseteq \mathbf{V}(G_\Gamma)$ is compatible with respect to G_Γ, if for any $i, j \in V$, the assignment $i \rightarrow \phi(i)$ and $j \rightarrow \phi(j)$ is compatible with respect to G_Γ. Henceforth, we consider instances for MRSO of the form $(G_\Gamma, \mathcal{F})$. Our goal in this setting is then to find an assignment $\phi : \mathbf{V}(G_\Gamma) \to \Sigma^3$ (*i.e.* a target mRNA sequence $T = \phi(1) \cdots \phi(n)$), which is compatible with G_Γ, and which maximizes $\sum_{i=1}^{n} f_i(\phi(i))$.

3   Two Natural Parameters for MRSO

Our discussion begins by considering two natural parameters for MRSO. These are the number of edge crossings in G_Γ, and the number of degree three vertices in G_Γ. We use χ and δ to denote these two parameters respectfully throughout the section.

Our initial interest in parameters χ and δ arises from the fact that we believe them to be small in many practical applications. Consider parameter χ. It is widely believed that many natural mRNA secondary structures form an outerplanar formation, *i.e.* a formation with no edge crossings. Consequently, exploring this parameter was suggested explicitly in [8]. As for parameter δ, recall that a vertex of degree three in G_Γ represents a codon with three nucleotides, each pairing with complementary nucleotides in three different codons. Although this situation can occur in a folding of an mRNA molecule, it can be expected to be quite rare due to the natural geometric and thermodynamic constraints imposed on any such folding.

It turns out that MRSO is polynomial-time solvable when either χ or δ are fixed. To show this, we will first present an initial algorithm, and later demonstrate how it can be applied for both cases. We will need the following definition:

Definition 3 (Nice edge bipartition). *Let G_Γ be an implied structure graph with n vertices. An edge bipartition $\mathcal{P} = (E_t, E_b)$ of G_Γ is a partitioning of the edges in G_Γ into E_t and E_b, the top and bottom edges of $\mathcal{P}$ respectfully, such that $E_t \cup E_b = \mathbf{E}(G_\Gamma)$, $E_t \cap E_b = \emptyset$ and $E_t \neq \emptyset$. Furthermore, $\mathcal{P}$ is said to be nice if the subgraph $G_\Gamma[E_t]$ is outerplanar.*

Our initial algorithm is called $\mathcal{A}_{\mathrm{NEB}}$. This algorithm will apply only for cases where a nice edge bipartition of G_Γ with a fixed number of bottom edges is given alongside the input. Following the description of $\mathcal{A}_{\mathrm{NEB}}$, we show that when considering either χ or δ to be fixed, one can easily obtain such a bipartition.

The heart of algorithm $\mathcal{A}_{\mathrm{NEB}}$ is the following simple observation. Suppose we want to find the highest scoring compatible mRNA sequence which starts with codon AAA. For this, we can replace the similarity function $f_1 \in \mathcal{F}$ by a different function f', where $f'(AAA) = f_1(AAA)$ and $f'(C) = -\infty$ for all codons $C \neq AAA$. Solving MRSO for the instance $(G_\Gamma, \mathcal{F}')$, where $\mathcal{F}' = f', f_2, \ldots, f_n$, will then give us our desired mRNA. The following definition generalizes this example.

Definition 4 (Corresponding similarity functions). *Let $(G_\Gamma, \mathcal{F})$ be an instance of MRSO with $\mathcal{F} = f_1, \ldots, f_n$. Also, let $\phi : V \to \Sigma^3$ be a codon assignment for some $V \subseteq \mathbf{V}(G_\Gamma)$. The corresponding set of similarity functions of assignment ϕ, denoted $\mathcal{F}_\phi = f_1^\phi, \ldots, f_n^\phi$, is defined as follows:*

- For all $i \in V : f_i^\phi(\phi(i)) = f_i(\phi(i))$, and $f_i^\phi(C) = -\infty$ for any $C \neq \phi(i)$.
- For all $j \in \mathbf{V}(G_\Gamma) - V : f_j^\phi = f_j$.

Algorithm $\mathcal{A}_{\text{NEB}}$ uses $\mathcal{A}_{\text{OP}}$, the algorithm given in [2] for outerplanar implied structure graphs, as a subprocedure in its computation. At its core, $\mathcal{A}_{\text{NEB}}$ is basically an exhaustive search procedure that searches through all possible codon assignments for vertices which are incident to edges in E_b. For each such assignment, $\mathcal{A}_{\text{NEB}}$ first checks if the assignment is compatible with respect to $G_\Gamma[E_b]$, and if so, it invokes $\mathcal{A}_{\text{OP}}$ with the set of similarity functions corresponding to this assignment. Finally, $\mathcal{A}_{\text{NEB}}$ outputs the maximum solution over all target mRNAs returned by $\mathcal{A}_{\text{OP}}$. A schematic description of $\mathcal{A}_{\text{NEB}}$ is given in Figure 2.

Algorithm $\mathcal{A}_{\text{NEB}}(G_\Gamma, \mathcal{F}, \mathcal{P})$

Data   : An implied structure graph G_Γ of order n, a set of similarity functions $\mathcal{F} = f_1, \ldots, f_n$ and a nice edge bipartition $\mathcal{P} = (E_t, E_b)$.

Result : An optimal target mRNA sequence T which is compatible with respect to G_Γ.

begin

   foreach *possible codon assignment ϕ to vertices incident to edges in E_b* **do**

      if *ϕ is compatible with respect to $G_\Gamma[\mathcal{E}_b]$* **then**

         (a) Construct $\mathcal{F}_\phi$, the similarity functions corresponding to ϕ.

         (b) Invoke $\mathcal{A}_{\text{OP}}(G_\Gamma[E_t], \mathcal{F}_\phi)$.

      end

   end

   return *the target mRNA sequence found in Step (b) with the highest similarity score.*

end

Fig. 2. Algorithm $\mathcal{A}_{\text{NEB}}$

Lemma 1. *Given an instance $(G_\Gamma, \mathcal{F})$ for MRSO accompanied by a nice edge bipartition $\mathcal{P} = (E_t, E_b)$ of G_Γ, $\mathcal{A}_{\text{NEB}}$ computes an optimal target mRNA sequence for this instance in $\mathcal{O}(64^{2\epsilon}n)$ time, where $n = |\mathbf{V}(G_\Gamma)|$ and $\epsilon = |E_b|$.*

Proof. Consider the schematic description of $\mathcal{A}_{\text{NEB}}$ in Figure 2 and let $V_b = \{i \mid \{i, j\} \in E_b\}$ be the subset of vertices incident to E_b. Any assignment $\phi : V_b \to \Sigma^3$ enumerated in the algorithm is verified for compatibility with respect to $G_\Gamma[E_b]$. Hence, by the correctness of $\mathcal{A}_{\text{OP}}$, any target mRNA outputted by $\mathcal{A}_{\text{NEB}}$ with a similarity score higher than $-\infty$ is compatible with respect to G_Γ. Furthermore, by the optimality of $\mathcal{A}_{\text{OP}}$, and since all possible codon assignments to V_b are considered by $\mathcal{A}_{\text{NEB}}$, this target mRNA is optimal with respect to $\mathcal{F}$.

For the time complexity bound, note that the number of codon assignments enumerated by the algorithm is $|\Sigma^3|^{|V_b|} \leq 64^{2\epsilon}$. Furthermore, constructing any such assignment and checking it for compatibility with respect to $G_\Gamma[E_b]$ can be done in $\mathcal{O}(n)$ time. Therefore, since each call to $\mathcal{A}_{\text{OP}}$ requires $\mathcal{O}(n)$ time [2], the overall time complexity of $\mathcal{A}_{\text{NEB}}$ is bounded by $\mathcal{O}(64^{2\epsilon}n)$. $\qquad\square$

We now return to our two parameters χ and δ, starting with χ. Recall that if $\chi = 0$ then G_Γ is outerplanar. Hence, a nice edge bipartition with χ bottom edges is available by definition. To see this, consider an edge bipartition with one bottom edge for each pair of edge crossings in G_Γ. Such an edge bipartition is nice, has at most χ bottom edges, and can be constructed in linear time. We therefore obtain the following proposition.

Proposition 1. MRSO *is polynomial-time solvable in case* $\chi = \mathcal{O}(\lg n)$.

Proof. According to the above discussion, G_Γ has a nice edge bipartition with at most χ bottom edges and this partitioning can be constructed in $\mathcal{O}(n)$ time. Thus, by Lemma 1, algorithm $\mathcal{A}_{\text{NEB}}$ can be applied to solve MRSO in $\mathcal{O}(64^{2\chi}n)$ time, and so the proposition follows.   □

Next consider parameter δ. Constructing a nice edge bipartition with δ bottom edges is immediate when considering the following easy lemma.

Lemma 2. *If G is a graph with maximum degree 2, then G is outerplanar.*

Proof. If G is a graph with maximum degree 2, then every connected component in G is either a path or a cycle. Since paths and cycles are outerplanar, the lemma immediately follows.   □

Consider an edge bipartition of G_Γ such that for each degree three vertex $i \in \mathbf{V}(G_\Gamma)$, exactly one edge incident to i is a bottom edge. Clearly, such a bipartition has at most δ bottom edges and can be constructed in linear time. Let $\mathcal{P} = (E_t, E_b)$ be an edge bipartition obtained in this fashion. Since G_Γ is subcubic, every vertex is incident to at most two top edges in $\mathcal{P}$. Thus, by Lemma 2, $G[E_t]$ is outerplanar and $\mathcal{P}$ is nice.

Proposition 2. MRSO *is polynomial-time solvable in case* $\delta = \mathcal{O}(\lg n)$.

Proof. Replace δ with χ in the proof of Proposition 1.   □

4   Page-Number Characterization of G_Γ

In light of algorithm $\mathcal{A}_{\text{NEB}}$ and Lemma 1, a natural question to ask is whether MRSO is polynomial-time solvable in case we are provided an edge bipartition in which both parts induce no edge crossing under the same vertex ordering. Alternatively, since the problem is polynomial-time solvable in case G_Γ is outerplanar, one might inquire if MRSO is still tractable when the implied structure graph is planar. In this section we provide a negative answer to both these question by proving that MRSO remains **NP**-hard even for a restrictive class of implied structure graphs.

Given a graph G, the *page-number* of G is the smallest partitioning of $\mathbf{E}(G)$ possible, such that each subset of edges in the partition induces no edge crossings under the same vertex ordering. Clearly the page-number of an outerplanar graph is one. Also, it is known that four pages are necessary and sufficient for planar graphs [17]. We show that MRSO is **NP**-complete even if the implied structure graph has page number two.

Proposition 3. MRSO *is* **NP**-*complete even when restricted to implied structure graphs with page-number two.*

Proof. We describe a reduction from the MAXIMUM INDEPENDENT SET problem, which is known to be **NP**-complete even when restricted to cubic planar bridegeless connected graphs [4]. The proof is a direct extension of the **APX**-completeness proof for MRSO given in [8].

Let an instance of the MAXIMUM INDEPENDENT SET problem be given by a cubic planar bridgeless connected graphs G of order n. According to [14], there exists a linear-time algorithm for finding a 2-page embedding of a cubic planar bridgeless graph, and hence there is no loss of generality in assuming that G is given in the form of a linear graph with page-number two. We now turn to defining the corresponding instance of MRSO. The implied structure graph G_Γ is merely the input graph G and the set of similarity functions $f_i : \Sigma^3 \to \mathbb{Q}$, $1 \le i \le n$, is defined as follows:

$$\forall i, \ 1 \le i \le n, \quad f_i(t_{3i-2}t_{3i-1}t_{3i}) = \begin{cases} 1 & \text{if } t_{3i-2}t_{3i-1}t_{3i} = AAA \\ 0 & \text{otherwise} \end{cases}$$

Quoting [8], the idea of the reduction is simply to identify the set of vertices which are assigned to AAA in a solution for the corresponding instance of the MRSO problem, with an independent set in G. Correctness of the proof now follows directly from [8], Theorem 3.   □

Corollary 1. MRSO *is* **NP**-*complete even when restricted to planar implied structure graphs.*

5   The Cutwidth of G_Γ

Let $(G_\Gamma, \mathcal{F})$ be an instance of MRSO with $\mathbf{V}(G_\Gamma) = \{1, \ldots, n\}$. For $p \in \{1, \ldots, n-1\}$, the p-*cutwidth* of G_Γ is defined as the number of edges connecting vertices in $\{1, \ldots, p\}$ to vertices in $\{p+1, \ldots, n\}$. The *cutwidth* of G_Γ is defined as the maximum p-cutwidth over all $p \in \{1, \ldots, n-1\}$. In the following we consider the cutwidth of G_Γ as a parameter for MRSO. We begin by showing that the problem is polynomial-time solvable in case G_Γ has bounded cutwidth. Following this, we show this result implies that MRSO is polynomial-time solvable for several other interesting cases. We let ψ denote the cutwidth of G_Γ throughout the section.

For obtaining our initial result, we present an algorithm which we call $\mathcal{A}_{\text{CUT}}$. This algorithm works by recursively partitioning G_Γ into two subgraphs $G_\Gamma[1, p]$ and $G_\Gamma[p+1, n]$, and then concatenating two optimal target mRNA sequences $T' = C_1, \ldots, C_p$ and $T'' = C_{p+1}, \ldots, C_n$ which are compatible with respect to these two subgraphs. To ensure that the concatenated solution $T = T'T''$ is compatible with respect to G_Γ, $\mathcal{A}_{\text{CUT}}$ enumerates all codon assignments between connected vertices of the two subgraphs.

In order to prevent unnecessary assignments from being enumerated, we distinguish in $\mathcal{A}_{\text{CUT}}$ between vertices which were assigned a codon in a previous recursive step, and those which have not yet been assigned one. We enforce two invariants. First, all assigned vertices are compatible throughout the entire execution of the algorithm. Second, once a vertex is assigned at some recursive step of the algorithm, no assignments are enumerated for this vertex in any subsequent step.

As in $\mathcal{A}_{\text{NEB}}$, algorithm $\mathcal{A}_{\text{CUT}}$ uses corresponding similarity functions (Definition 4) to apply codon assignments. A similarity function f is *degenerate*, if there is some codon C such that $f(C) > -\infty$, and $f(C') = -\infty$ for any other codon $C' \in \Sigma^3$, $C' \neq C$. In $\mathcal{A}_{\text{CUT}}$, we use degenerate similarity functions both to recognize the assigned vertices along the recursion, and also to propagate their corresponding codon assignment. A schematic description of $\mathcal{A}_{\text{CUT}}$ is given in Figure 3.

Algorithm $\mathcal{A}_{\text{CUT}}(G_\Gamma, \mathcal{F})$

Data   : An implied structure graph G_Γ with $\mathbf{V}(G_\Gamma) = \{1, \ldots, n\}$, and a set of similarity functions $\mathcal{F} = f_1, \ldots, f_n$.

Result : An optimal target mRNA sequence T which is compatible with respect to G_Γ.

begin

    1. if $\mathbf{E}(G_\Gamma) = \emptyset$ **then return** T that maximizes $\mathcal{F}$.

    2. Select $p \in \{1, \ldots, n-1\}$ with maximum p-cutwidth.

    3. Set $E_p = \{\{i, j\} \in \mathbf{E}(G_\Gamma) \mid 1 \leq i \leq p, \; p+1 \leq j \leq n\}$.

    4. Set $V_p = \{i \mid \{i, j\} \in E_p\}$ to be the vertices incident to E_p.

    5. Let $A_p = \{i \in V_p \mid f_i \text{ is degenerate}\}$ be the assigned vertices in V_p.

    6. Define $\phi^{A_p} : A_p \to \Sigma^3$ such that $\phi^{A_p}(i) = C \Leftrightarrow f_i(C) > -\infty$.

    7. foreach *possible codon assignment* $\phi^{V_p - A_p} : V_p - A_p \to \Sigma^3$ **do**

        if $\phi = \phi^{A_p} \cup \phi^{V_p - A_p}$ *is compatible with respect to* $G_\Gamma[E_p]$ **then**

            (a) $T' \leftarrow \mathcal{A}_{\text{CUT}}(G_\Gamma[1, p], f_1^\phi, \ldots, f_p^\phi)$.

            (b) $T'' \leftarrow \mathcal{A}_{\text{CUT}}(G_\Gamma[p+1, n], f_{p+1}^\phi, \ldots, f_n^\phi)$.

        end

    end

    return *the highest similarity scoring target mRNA sequence* $T = T'T''$ *found in step 7.*

end

Fig. 3. Algorithm $\mathcal{A}_{\text{CUT}}$

Lemma 3. *Given an instance* $(G_\Gamma, \mathcal{F})$ *for MRSO, algorithm* $\mathcal{A}_{CUT}$ *computes an optimal target mRNA sequence for this instance in* $\mathcal{O}(64^{2\psi} n)$ *time, where* $n = |\mathbf{V}(G_\Gamma)|$ *and* ψ *is the cutwidth of* G_Γ.

Proof. Consider the schematic description of $\mathcal{A}_{\text{CUT}}$ in Figure 3. We prove the correctness and optimality of the algorithm by induction on its recursion. At the recursive basis, the solution returned is optimal and compatible by construction. For the inductive step, assume T' and T'' are the two target mRNAs computed at steps (a) and (b) respectfully. Then T' and T'' are compatible with respect to $G_\Gamma[1, p]$ and $G_\Gamma[p+1, n]$ respectfully. Hence, since by construction $T'T''$ is compatible with respect to $G_\Gamma[E_p]$, it is also compatible with respect to G_Γ. Furthermore, since the algorithm considers all assignments to vertices in V_p with score higher than $-\infty$, the target mRNA returned at this step is optimal.

For the time complexity bound of $\mathcal{A}_{\text{CUT}}$, note that the number of codon assignments enumerated by the algorithm in each recursive step is $|\Sigma^3|^{|V_p|} \le 64^{2\psi}$. Since the number of recursive steps is bounded by $\mathcal{O}(n)$, the overall time complexity of $\mathcal{A}_{\text{CUT}}$ is bounded by $\mathcal{O}(64^{2\psi}n)$.   □

Corollary 2. MRSO *is polynomial-time solvable in case* $\psi = \mathcal{O}(\lg n)$.

We next consider the implications of Corollary 2. The treewidth [15] of a graph is a graph property that has been studied extensively in the literature. Informally, it measures in some sense the degree of tree-likeness of the graph. In [13] (via [9]), the authors showed that for a graph with n vertices, constant maximum degree, and constant treewidth, one can obtain an ordering of the vertices such that the linear graph under this ordering has cutwidth bounded by $\mathcal{O}(\lg n)$.

Corollary 3. MRSO *is polynomial-time solvable in case* G_Γ *has constant treewidth.*

Note that the tree width of any outerplanar graph is bounded by two [16], and so the algorithm above generalizes $\mathcal{A}_{\text{OP}}$, although the time complexity bound of $\mathcal{A}_{\text{OP}}$ is better. In [6], Bodlaender gives a list of several other interesting graph classes which are subclasses of constant treewidth graphs. Among many others, we state only a few in the following corollary.

Corollary 4. MRSO *is polynomial-time solvable in case* G_Γ *is either a chordal graph, an interval graph, a circular arc graph, or a* k-*outerplanar graph where* k *is any constant.*

Hence, Corollary 1 and the last case in the corollary above give a fine borderline between tractable and intractable instances of MRSO.

6   Parameterizing by the Similarity Score

We next turn to consider the score of the optimum solution as a parameter for MRSO. For this, we suggest a relaxation on the similarity functions of an MRSO instance. More specifically, we consider instances with similarity functions of the form $f_i : \Sigma^3 \to \mathbb{N}$. We call similarity functions of this sort *natural similarity functions*, and denote MRSO$_\mathbb{N}$ the MRSO problem restricted to instances with this type of similarity functions. Most of the interest in restrictive similarity functions stems from the following proposition.

Proposition 4. MRSO$_\mathbb{N}$ *is polynomial-time solvable in case the similarity score of the optimal solution is fixed.*

Proof. Let $(G_\Gamma, \mathcal{F})$ be an instance of MRSO$_\mathbb{N}$ and let κ denote the similarity score of the optimal target mRNA of this instance. Set $n = |\mathbf{V}(G_\Gamma)|$. We may assume without loss of generality that for all $1 \leq i \leq n$, $f_i(C) > 0$ for some codon $C \in \Sigma^3$. Otherwise, if there exists any function $f_i \in \mathcal{F}$ which fails to meet this requirement, we solve the sub-instance $(G'_\Gamma, \mathcal{F}')$ obtained by deleting i from G_Γ and f_i from $\mathcal{F}$. Any feasible solution for $(G'_\Gamma, \mathcal{F}')$ can then be extended to a feasible solution of the same score for the original instance since Γ has maximum degree one. We present an algorithm which searches for a target mRNA string T, by focusing on finding κ pairwise compatible codons with respect to G_Γ. The proof is divided into two separate parts depending on $\alpha(G_\Gamma)$, the cardinality of a maximum independent set in G_Γ.

Suppose $\kappa \leq \alpha(G_\Gamma)$. Let $V \subseteq \mathbf{V}(G_\Gamma)$ be an independent set of size κ in G_Γ. Since G_Γ is at most cubic, such a subset V can be found in $\mathcal{O}(4^\kappa n)$ time using the bounded search tree technique [10]. We define a string T of length $3n$ as follows. For each $i \in V$, assign codon $C_i \in \Sigma^3$ such that $f_i(C_i) \geq 1$. This is always possible since V is an independent set in G_Γ, and since for all $1 \leq i \leq n$, $f_i(C) > 0$ for some $C \in \Sigma^3$. For each $j \in \mathbf{V}(G_\Gamma) - V$, assign codon C_j which is compatible with all codons assigned to vertices in V with respect to G_Γ. Again this is always possible since Γ has maximum degree one. We check at once that $T = C_1 C_2 \ldots C_n$ is compatible with respect to G_Γ and $\sum_{i=1}^n f_i(C_i) \geq |V| = \kappa$.

Now suppose $\kappa > \alpha(G_\Gamma)$. Since G_Γ is at most cubic, we have $\alpha(G_\Gamma) \geq \frac{n}{4}$, and hence $\kappa > \frac{n}{4}$. Here, the algorithm is by direct enumeration. More precisely, the algorithm tries in turn to obtain a solution mRNA string T by finding ℓ pairwise compatible codons, where ℓ ranges from 1 to κ. So, let $\ell \in \{1, 2, \ldots, \kappa\}$. We search through all ℓ-subsets of $\mathbf{V}(G_\Gamma)$ for an ℓ-subset with an assignment which is compatible with respect to G_Γ. Such an exhaustive search can be executed in $\mathcal{O}(\binom{n}{\ell} 64^\ell)$ time. Summing-up over ℓ and neglecting the time to check $\kappa > \alpha(G_\Gamma)$, i.e., $\mathcal{O}(4^\kappa)$, we obtain $\mathcal{O}(\sum_{\ell=1}^\kappa \binom{n}{\ell} 64^\ell)$, which is $\mathcal{O}(2^{\mathcal{O}(\kappa)} \kappa^{\kappa+1})$ since G_Γ is at most cubic and $\kappa > \alpha(G_\Gamma) \geq \frac{n}{4}$.

Hence, MRSO$_\mathbb{N}$ can be solved in $\mathcal{O}(2^{\mathcal{O}(\kappa)} \kappa^{\kappa+1} + 4^\kappa n)$ time, and the proposition above follows.   □

Note that all hardness results obtained for MRSO still hold for MRSO under natural similarity functions. Nevertheless, using a simple combinatorial argument, we can easily obtain an optimal algorithm if we consider the score of the optimal solution for MRSO$_\mathbb{N}$ to be fixed. Even so, it is a challenging problem to investigate the parameterized complexity of the MRSO problem for more general similarity functions. We do believe that it might be worth considering similarity functions of the form $f_i : \Sigma^3 \to \mathbb{N} \cup \{-\infty\}$ since these capture most of the information necessary in most practical applications. Here, the $-\infty$ value can be used in case a certain codon (*e.g.* a stop codon) is not acceptable in a certain position of T.

Acknowledgments

The authors would like to thank Gad Landau for his support and valuable advice.

References

1. R. Backofen and A. Busch. Computational design of new and recombinant se-lenoproteins. In *Proc. of the 15th Annual Symposium on Combinatorial Pattern Matching (CPM)*, volume 3109 of *LNCS*, pages 270–284, 2004.
2. R. Backofen, N.S. Narayanaswamy, and F. Swidan. On the complexity of protein similarity search under mRNA structure constraints. In *Proc. of the 19th Sympo-sium on Theoretical Aspects of Computer Science (STACS)*, volume 2285 of *LNCS*, pages 274–286, 2002.
3. R. Backofen, N.S. Narayanaswamy, and F. Swidan. Protein similarity search under mRNA structural constraints: application to targeted selenocystein insertion. *In Silico Biology*, 2(3):275–290, 2002.
4. T.C. Biedl, G. Kant, and M. Kaufmann. On triangulating planar graphs under the four-connectivity constraints. *Algorithmica*, 19:427–446, 1997.
5. A. Böch, K. Forchhammer, J. Heider, and C. Baron. Selenoprotein synthesis: a review. *Trends in Biochemical Sciences*, 16(2):463–467, 1991.
6. H.L. Bodlaender. Classes of graphs with bounded tree-width. Research Report RUU-CS-86-22, Utrecht University, Padualaan 14, Utrecht, The Netherlands, De-cember 1986.
7. H.L. Bodlaender, R.G. Downey, M.R. Fellows, M.T. Hallett, and H.T. Wareham. Parameterized complexity analysis in computational biology. *Computer Applica-tions in the Biosciences*, 11:49–57, 1995.
8. D. Bongartz. Some notes on the complexity of protein similarity search under mRNA structure constraints. In *Proc. of the 30th Conference on Current Trends in Theory and Practice of Computer Science (SOFSEM)*, volume 2932 of *LNCS*, pages 174–183, 2004.
9. F.R. Chung and P.D. Seymour. Graphs with small bandwidth and cutwidth. *Dis-crete Math.*, 75(1-3):113–119, 1989.
10. R. Downey and M. Fellows. *Parameterized Complexity*. Springer-Verlag, 1999.
11. T. Jacks, M. Power F. Masiarz, P. Luciw, P. Barr, and H. Varmus. Characterization of ribosomal frameshifting in HIV-1 gag-pol expression. *Nature*, 331:280–283, 1988.
12. T. Jacks and H. Varmus. Expression of the Rous sarcoma virus pol gene by ribo-somal frameshifting. *Science*, 230:1237–1242, 1985.
13. E. Korach and N. Solel. Tree-width, path-width, and cutwidth. *Discrete Applied Mathematics*, 43(1):97–101, 1993.
14. G. Lin, Z-Z. Chen, T. Jiang, and J. Wen. The longest common subsequence prob-lem for sequences with nested arc annotations. *Journal of Computer and System Sciences*, 65(3):465–480, 2002. Special issue on computational biology.
15. N. Robertson and P.D. Seymour. Graph minors II: algorithmic aspects of tree-width. *Journal of Algorithms*, 7:309–322, 1986.
16. J.A. Wald and C.J. Colbourn. Stiener trees, partial 2-trees, and minimum ifi networks. *Networks*, 13:159–167, 1983.
17. M. Yannakakis. Embedding planar graphs in four pages. *Journal of Computer and System Sciences*, 38:36–67, 1986.

On the Fixed-Parameter Enumerability
of Cluster Editing

Peter Damaschke

School of Computer Science and Engineering,
Chalmers University, 41296 Göteborg, Sweden
ptr@cs.chalmers.se

Abstract. CLUSTER EDITING is the problem of changing a graph G by
at most k edge insertions or deletions into a disjoint union of cliques.
The problem is motivated from computational biology and known to
be FPT. We study the enumeration of all solutions with a minimal set
of edge changes. Enumerations can support efficient decisions between
ambiguous solutions. We prove that all minimal solutions differ only on a
so-called a full kernel of at most $k^2/4 + O(k)$ vertices. This bound is tight.
For ambiguous edges we get an optimal bound up to a constant factor.
Finally we give an algorithm that outputs a compressed enumeration in
$O^*(2.4^k)$ time.

1 Introduction

The CLUSTER EDITING problem requires to transform a graph $G = (V, E)$ with
n vertices by at most k edge changes into a *cluster graph*, that is, a disjoint
union of complete graphs (G, k are given). A *change* is an edge insertion or
deletion, whereas V remains fixed. A graph is a cluster graph iff it is free of
induced P_3 (path of three vertices). In CLUSTER DELETION, only edge deletions
are allowed. These problems from [2,1,14] have applications in computational
biology, such as phylogeny reconstruction [2], and classification of gene expression
data [15,16], where vertices represent genes, and edges join co-regulated genes
belonging to the same functional groups. We imagine that there is a hidden
clustering, but the observed graph G is up to k changes away from a cluster
graph, due to experimental errors, noisy data, vertices belonging to different
groups, and incomplete clusters due to non-transitive similarity relations. If $k \ll
n$, we may look for possible underlying clusterings close to G.

Known complexity results include the NP-hardness of computing the smallest
possible k, given G [2,1,14]. Efficient FPT algorithms with parameter k have
been devised in [9]. Main results are a problem kernel for CLUSTER EDITING
with $O(k^2)$ vertices and $O(k^3)$ edges, computable in $O(n^3)$ time, and algorithms that solve CLUSTER EDITING and CLUSTER DELETION in $O(2.27^k + n^3)$
and $O(1.77^k + n^3)$ time, respectively. (For an introduction to fixed-parameter
tractability see e.g. [5].)

In the present paper we follow a line of research initiated in [8,4]. We want
an *enumeration* of all possible solutions for given G and k. (Actually, we have to

D. Kratsch (Ed.): WG 2005, LNCS 3787, pp. 283–294, 2005.

modify this goal slightly.) A particular smallest set of changes may not always yield the true hidden clustering, especially if several optimal solutions exist. It is not even clear that the smallest possible number k_0 of changes is the correct explanation of data. It is more cautious to consider all solutions within some parameter value $k \geq k_0$, and to judge them afterwards by further criteria specific to the application. Pairs of items may belong to the same cluster with different prior probabilities, which allows discrimination of solutions, made by a Bayesian inference algorithm or ad-hoc by an expert, e.g., a biologist who examines gene expression data and has additional knowledge about gene functions. In the language of [12] the enumeration is a version space, i.e., the set of hypotheses (clusterings) consistent with the data and the assumptions (here: at most k changes appeared). The version space is used a a basis for inference. A concise description of it is crucial to efficiency of any inference procedure. However, the scope of the present paper is on the complexity of the enumeration itself. Anyway, our first main result for CLUSTER EDITING is that all solutions agree on most pairs of vertices, except vertices in a small kernel and in small clusters, i.e., with a size depending on k only. For any two vertices outside these small sets we can safely conclude whether they belong to the same cluster or not, provided that parameter value k is valid.

Preliminaries. Given an instance G, k of CLUSTER EDITING, let G', G'' be two solutions, i.e., two cluster graphs obtained from G by at most k changes. We say that G'' *contains* G' if the changes leading from G to G' are a subset of the changes leading from G to G''. (This should not be confused with containment of the graphs.) A solution not containing any other solution is *minimal*. The enumeration version of CLUSTER EDITING can be split in a nontrivial and a trivial part. The former part is to enumerate *all minimal* solutions G'. Once we know them, it is pretty easy to characterize *all* solutions reachable from G by at most k changes: One can just divide or merge clusters, using the remaining number of allowed changes. Note that only the smallest clusters in G' are subject to further changes. In particular, if k is close to k_0, further changes are very limited. From now on we consider the nontrivial part only.

Vertices u, v form an *ambiguous* pair if uv is an edge in some minimal solution but a non-edge in some other minimal solution. A vertex is called ambiguous if it belongs to an ambiguous pair. A *full kernel* is a set of vertices containing all ambiguous vertices. This notion, introduced in [4], is a counterpart of problem kernels of FPT optimization problems. The actual enumeration process can be restricted to a full kernel.

Finally we clarify some graph-theoretic notation. A *clique* is any complete subgraph of $G = (V, E)$, not necessarily maximal. For any $X \subseteq V$, $G - X$ is graph G with X and all incident edges removed. For an induced subgraph H of G, and w a vertex not in H, we denote by $H + w$ the subgraph induced by the vertices of H and w. The open neighborhood $N(X)$ of a vertex set X is the set of vertices being not in X but incident to some vertex of X. We call two subsets of vertices *pair-disjoint* if they share at most one vertex. A *module* is a set $M \subseteq V$ such that every vertex outside M is adjacent to either no vertex or all vertices in M. The number of vertices and edges of G is n and m, respectively. Symbols P_n, C_n, K_n denote a chordless path, a chordless cycle, and a clique, respectively, of n vertices. The star graph $K_{1,s}$ has one central vertex adjacent to s other vertices, and no further edges. The $O^*()$ complexity notation suppresses polynomial factors in exponential bounds.

Our Results. In Section 2 we show that a full kernel with $O(k^2)$ vertices for CLUSTER DELETION is efficiently computable. Our main tool is a certain decomposition of graphs. (We remark that novel graph decompositions like "crowns" recently proved very powerful in FPT algorithms, see e.g. [3,7] and several papers in [6]. In particular, [13] consider P_3 packings, which is loosely related to our subject.) We also find the optimal constant factor $1/4$. For the number of ambiguous pairs we get the optimal bound $\Theta(k^4)$, however with coarse estimates of the constant factor.

We think that it is worth the effort to figure out a matching bound for the exact size of a full kernel, *including the constant factor*. (The bound in [9] is 8 times as high.) The full kernel is interesting for its own, for it includes all deviations between any two minimal solutions, whereas problem kernels in traditional FPT theory only serve as computational tools for finding one optimal solution.

However, another benefit of a full kernel is that it can be used to enumerate all minimal solutions in $O(c^k)$ time for some constant c, whereas polynomial time in n is needed for kernel construction only. Section 3 is devoted to the enumeration of all minimal solutions. A compressed description can be easily computed in $O^*(2.562^k)$ time. In the rest of Section 3 we further reduce the base to 2.4 by deeper analysis of the P_3 structure of graphs.

2 The Number of Ambiguities

2.1 The Cluster Decomposition

The following decomposition is the basis for our estimate of the full kernel size.

Definition 1. *A* cluster decomposition *of a connected graph* $G = (V, E)$ *is a partition of* V *in disjoint sets, namely a* head Q *and* pre-clusters, *with the following properties:*

(i) Every pre-cluster is a clique.
(ii) There are no edges between any two pre-clusters.
(iii) Every pre-cluster C is a module.
(iv) For any two pre-clusters C and D, sets $N(C), N(D)$, called the tags *of C and D in Q, are disjoint.*

Lemma 1. *If a G graph is at most k edit steps away from a cluster graph, then a cluster decomposition of G with $|Q| \le 3k$ can be computed in $O(k^2 n + m)$ time.*

Proof. Consider a maximal set $\mathcal{P}$ of mutually pair-disjoint induced P_3 in G. Let Q be the vertex set spanned by $\mathcal{P}$. Since $\mathcal{P}$ is maximal, $G - Q$ does not contain another P_3, hence $G - Q$ induces a cluster graph. Defining Q to be the head and the connected components of $G - Q$ to be the pre-clusters, we see that (i),(ii) are fulfilled. Furthermore (iii) is true, since otherwise there would exist $u, v \in C$ and $w \in Q$ that induce a P_3 $u - v - w$, contradicting the maximality of $\mathcal{P}$. Finally, if some C, D violate (iv), then some vertices $u \in C$, $v \in Q$, $w \in D$ would form a P_3, again contradicting the maximality of $\mathcal{P}$. In conclusion, a set $\mathcal{P}$ as specified yields a cluster decomposition. We also remark that Q contains at most $3k$ vertices, hence no more than $3k$ pre-clusters are connected to Q.

In order to construct $\mathcal{P}$ efficiently, we start from the subgraph H of G with empty vertex set and from an empty $\mathcal{P}$. Then we insert vertex by vertex in H and maintain a cluster decomposition of the current H, until $H = G$. For any new vertex w added to H we form several new P_3, each consisting of w and a pair of vertices u, v in H. We call u, v an eligible pair if u, v, w actually form a P_3, and u, v are not already together in some P_3 in $\mathcal{P}$. Then, we take disjoint eligible pairs and add the resulting new P_3 to $\mathcal{P}$ in a greedy fashion, until $\mathcal{P}$ cannot be extended further in the current H. Correctness of this procedure is straightforward. Next we sketch the time analysis.

We have to find a greedy set of eligible pairs u, v efficiently in every step. First of all, u, v must form an induced P_3 with w. Since Q has size $O(k)$, only $O(k)$ of the pre-clusters are connected with Q, and the pre-clusters are modules in H, subgraph $H + w$ can be easily partitioned in $O(k)$ modules, using the cluster decomposition of H. Now, it suffices to choose $O(k)$ representative vertices, one from each module of $H + w$, and to check $O(k^2)$ pairs u, v whether they form P_3 with w, since the results carry over to all vertices u', v' in the same modules. Once we know the pairs of candidate modules, our greedy extension procedure can be restricted to vertices from these pairs of modules. Vertices that became members of any new P_3 in $\mathcal{P}$ will move, of course, from the pre-clusters to Q.

As we have seen above, for every w we need $O(k^2)$ preprocessing time for identifying the new members of $\mathcal{P}$ from a partitioning of $H + w$ into modules. This gives the $O(k^2 n)$ term. The time for all other operations can be limited globally, i.e. for the whole algorithm: Since the final $\mathcal{P}$ has still at most k P_3, inserting them in $\mathcal{P}$ costs only $O(k)$ time. Edges incident to every new w are inserted in the growing subgraph H in $O(m)$ time. Thus, we also need only $O(m)$ time in total, in order to compute modules of $H + w$ from the cluster decomposition of H, and for all updates of the cluster decomposition. $\square$

Next we define a cleaning procedure which transforms an instance G, k of CLUSTER EDITING into another instance while preserving the set of minimal solutions. The purpose is to do some forced changes immediately and to remove parts of G being irrelevant for the problem. Recall that every $v \in Q$ is in the tag of at most one pre-cluster C. To avoid case distinctions, we introduce a dummy clique of size 0 and define its tag as the set of all vertices in Q that are not in the tag of any (real) pre-cluster.

Cleaning Procedure:

(1) For every pre-cluster C with more than k vertices, insert an edge uv between any two $u, v \in N(C)$ that are not yet adjacent.

(2) For any two pre-clusters C and D with c and d vertices, respectively, where $c + d > k$, delete every edge uv with $u \in N(C)$ and $v \in N(D)$. (In particular, D may be the dummy pre-cluster and $d = 0$.)

(3) Remove every clique which is disconnected from the rest of the graph.

Apply these rules as long as possible.

Lemma 2. *The cleaning procedure does not alter the set of minimal solutions and can be implemented to run in $O(k^2 + m)$ time if a cluster decomposition as in Lemma 1 is already given.*

Proof. In any solution, every clique with $k+2$ vertices in G must entirely belong to one cluster, otherwise we had to disconnect the clique, which is impossible with k deletions. To see the correctness of (1), note that both $C \cup \{u\}$ and $C \cup \{v\}$ are cliques of size at least $k + 2$, hence u, v must be in the same cluster. As for (2), observe that if we keep edge uv, every vertex in $C \cup D$ must be incident to an inserted or deleted edge incident one of u, v, but this amounts to more than k changes, a contradiction. Correctness of (3) is trivial. Extra edges that connect an isolated clique to other vertices cannot be part of any minimal solution.

Note that (1) and (2) only add edges inside (or delete edges between) tags in a set of $O(k)$ vertices, hence these rules are applicable at most once to each of $O(k^2)$ vertex pairs. Rule (3) simply removes isolated cliques. Hence the time is linear in the size of G. □

Corollary 1. *After the cleaning procedure, every pre-cluster C has at most k vertices.*

Proof. Assume that a larger C exists. Since (1) does not apply, $C \cup N(C)$ is a clique. Since (2) does not apply, no edge connects $N(C)$ and vertices outside $C \cup N(C)$. This gives an isolated clique, and (3) applies, a contradiction. □

Corollary 2. *A full kernel with at most $3k^2 + 3k$ vertices can be computed in $O(k^2 n + m)$ time.*

Proof. Observe that the remaining graph is a full kernel, $|Q| \leq 3k$, all tags are disjoint, and the pre-clusters are bounded due to Corollary 1. □

2.2 Ambiguous Vertices

Corollary 2 establishes an $O(k^2)$ bound for the full kernel. Next we considerably reduce the constant factor by a tighter analysis. In the following we *fix the cluster decomposition and a certain minimal solution.* With respect to this minimal solution, we distinguish several cases of pre-clusters C and "charge" them for changes of edges that touch vertices in C. All pre-clusters will be charged. Since at most k changes are allowed in total, this will eventually limit the full-kernel size. Before we can present our sophisticated charging scheme, we need:

Lemma 3. *If the graph (full kernel) G' remaining after the cleaning procedure is disconnected, then a minimal solution never adds edges between vertices from different connected components of G'.*

Proof. We merely need the fact that the cleaning procedure performs only enforced changes, i.e., changes that must be done in any minimal solution. Any cluster with vertices from different connected components of G' can be split in smaller clusters, each containing vertices from one component. This finer clustering used a proper subset of further changes (starting from G'), hence the given clustering was not a minimal solution. □

Theorem 1. *At most $k^2/4+7k/2+1/4$ vertices are ambiguous, and a full kernel of that size can be computed in $O(k^2n+m)$ time.*

Proof. Consider any connected component H of the full kernel after the cleaning procedure. Let $c_1 \geq \ldots \geq c_r$ be the vertex numbers of all pre-clusters in H, in descending order. If the tags of all pre-clusters in H are cliques, we define the weight of H as $c_1 + c_r$. Otherwise, the weight of H is defined to be only c_1. Finally, w is the maximum weight of a connected component in our full kernel. Note that every tag is non-empty, otherwise the pre-cluster would be an isolated clique and would have been removed by the cleaning procedure.

We say that a pre-cluster is *inert* with respect to a fixed minimal solution, if the vertex set of this pre-cluster and its tag forms exactly one cluster there. Now let C, with c vertices, be a largest pre-cluster in a connected component H of weight w. *In the following we suppose that C is not inert in some minimal solution, and we fix such a minimal solution.* (Only later we discuss the case that C is inert in all minimal solutions.)

Phase 1:

First assume that all other pre-clusters in H are inert. Since, by Lemma 3, changes after the cleaning procedure occur only inside connected components, it follows as the only remaining possibility that $C \cup N(C)$ is split in at least two clusters. This splitting requires at least c deletions of edges incident to vertices of C, this is easily seen from $N(C) \neq \emptyset$ and the fact that C is both a clique and a module. Moreover, since we are considering a minimal solution, $N(C)$ was not a clique. Hence, by the definition of w, we have $c = w$.

The other case is that some other pre-cluster D in H, say with d vertices, is not inert either. We claim that at least $c + d$ changes of edges that involve vertices of $C \cup D$ must be done. It is easy to check the few different cases that $C \cup N(C)$ (or $D \cup N(D)$) is cut in different clusters or stays in one cluster that gets at least one more vertex. Trivially, we also have $c + d \geq w$.

Thus we can already charge one or two non-inert pre-clusters for at least w changes, in both cases.

Phase 2:

Next we will also charge *all* the remaining pre-clusters in *all* components. In the following, consider any connected component with, say, r pre-clusters. By connectivity, at least $r - 1$ edges tie their tags together. Two pre-clusters are said to be neighbors if there exists an edge between their tags.

If the considered component has at least one non-inert pre-cluster C, or an inert pre-cluster C whose tag $N(C)$ is not a clique, we can successively charge all r pre-clusters, each for one change, as follows. First charge all pre-clusters C of the two mentioned types: If C is not inert, there is a change involving a vertex of C. If C is inert but $N(C)$ not a clique, an edge insertion must be done in $N(C)$. Next, pick a yet uncharged pre-cluster D (inert, with a clique as tag) which has already a charged neighbor C. Since D is inert, an edge between $N(C)$ and $N(D)$ is deleted in the solution, and we charge D for this deletion. Since C was already charged, we do not count this edge twice. By connectivity, this procedure never gets stuck, until all pre-clusters are charged. If one of the pre-clusters in the component is dummy, we also consider it as "charged" in the beginning and proceed as above. In either case, every non-empty pre-cluster in the component is now charged for a different change. Also note that every pre-cluster contains at most w vertices.

It remains to discuss components where all pre-clusters are inert and have cliques as tags. Obviously, at least $r - 1$ edges must be deleted. We charge the largest and the smallest pre-cluster together for one deletion, and the other $r - 2$ pre-clusters together for $r - 2$ deletions. By definition of w, at most w vertices from pre-clusters are now charged for each of the $r - 1$ edge deletions.

Putting things together:
Recall that we have at most k changes in total. In Phase 1 we charged, for $y \geq w$ of them, no more than y vertices from one or two pre-clusters. In Phase 2, at most $w \leq y$ vertices from nonempty pre-clusters have been charged for each of the, at most, $k - y$ other changes. Since all pre-clusters are charged, the pre-clusters contain in total no more than $(1 + k - y)y$ vertices. This expression is maximized if $y = (k+1)/2$, hence the pre-clusters contain at most $k^2/4 + k/2 + 1/4$ vertices. Adding the at most $3k$ vertices from Q yields the result.

Finally we give the time complexity. Let C again be a largest pre-cluster in a connected component of maximum weight. (Hence C is easy to find in the cluster decomposition.) Our construction and analysis shows: If C is not inert in some minimal solution, the union of pre-clusters has already a size of at most $k^2/4 + k/2 + 1/4$, and we can stop. By contraposition, if this union is larger, we know that C is inert in all minimal solutions. Hence, removing $C \cup N(C)$ leaves us with a smaller full kernel. Moreover, we still have a cluster decomposition of this smaller graph with $|Q| \leq 3k$, since the part outside Q is still a cluster graph. Thus we can simply iterate the procedure. A cluster decomposition with $|Q| \leq 3k$ must be computed only once in the beginning. The only thing to recompute after every removal is the connected components of Q. Now the time bound follows from the previous results.   □

The size bound in Theorem 1 is asymptotically tight (even for the more restrictive CLUSTER DELETION problem), as the following example shows. For simplicity let k be even.

Proposition 1. *There exist graphs with $k^2/4 + 3k/2 + 2$ ambiguous vertices.*

Proof. Take $k/2 + 1$ disjoint cliques, each with $k/2 + 1$ vertices, and attach to every clique another vertex that is adjacent to one vertex in the clique. We put any one of the extra vertices x and its only neighbor y in one cluster, the other $k/2$ edges incident to y are deleted. The other $k/2$ extra edges are also deleted. Every such solution is minimal, hence all vertices are ambiguous.   □

There remains a gap in the linear term only. However note that for the graphs in the previous proof, the minimum number of changes to reach a cluster graph is only $k_0 = k/2 + 1$. It arises the question whether the full kernel size is even smaller than $k^2/4$ for $k = k_0$, or already for some $k < 2k_0$.

2.3   Ambiguous Pairs

The worst case number of ambiguous pairs which can be $\Theta(k^4)$, due to:

Theorem 2. *The number of ambiguous pairs is bounded by $k^4/32 + 7k^3/8 + 97k^2/16 - 7k/8 - 3/32$. On the other hand, there exist graphs with $k^4/800$ ambiguous pairs.*

Proof. The first statement follows readily from Theorem 1. The worst case would be that all pairs in a full kernel of maximum size are ambiguous.

To construct an example for the lower bound, take ak disjoint cliques, each with bk vertices, and another special vertex z. Constants a, b are specified later. For simplicity assume that all numbers that denote cardinalities are in fact integer. Join one vertex from every clique by an edge to z. In the following, K_v denotes the clique with vertex v connected to z. For any K_u and K_v and $w \in K_v$ ($w \neq v$) we specify a minimal solution as follows. We keep edges uz, zv, vw, and all edges in K_u. This implies that K_u and z, v, w belong to the same cluster. In order to put z, v, w in K_u we have to insert $3bk$ edges. The other $ak - 2$ edges indicent to z are deleted, as well as the $2bk - 4$ edges between v and w, respectively, and the rest of K_v. Obviously, these $(a + 5b)k - 6$ changes yield a cluster graph. This solution is minimal, by the following argument. In any solution using a subset of changes we have to keep uz, zv, vw and the edges in K_u, too, hence K_u, z, v, w are in the same cluster. Furthermore we have to keep $K_v \setminus \{v, w\}$ and all other cliques. Since merging two $\Theta(k)$ cliques would require $\Theta(k^2)$ insertions, we must also delete the same edges as above. Note that all $a^2b^2k^4/2$ edges (lower-order terms neglected) are ambiguous. The constant factor is maximized under constraint $a + 5b \leq 1$ if $a = 1/2$ and $b = 1/10$.   □

Closing the huge gap in the constant factor is left as an open problem. Similar remarks as above apply to the graphs used in this proof: Note that $k_0 = ak = k/2$. We conjecture that the number of ambiguous pairs is much smaller for k close to k_0. Another open question arises from Theorem 1. Our construction guarantees a bound in terms of k, but not the smallest full kernel for every graph. We conjecture that, in order to compute the exact set of ambiguous pairs, one must actually compute all minimal solutions, so that exponential time in k is required. This leads us to the next subject: We can compute the smallest full kernel, via a concise description of all minimal solutions, by a search tree smaller than the obvious $O^*(3^k)$ bound.

3   Enumerating the Minimal Solutions

We suppose that the reader is familiar with the search tree technique which is standard in FPT algorithmics, and with the notion of branching number or, synonymously, splitting number, we refer to [10]. The branching number is the positive root of the characteristic polynomial of the recurrence that bounds the size of the search tree. A trivial algorithm for enumerating all minimal sets of at most k changes that turn a given graph G into a cluster graph works as follows: At every node of the search tree it chooses some induced P_3 in G and deletes one of the two edges or inserts the missing edge, then it branches on these three cases, and so on. The complexity is $O^*(3^k)$. It is also a trivial fact that this time is needed in the worst case to enumerate all minimal solutions. For instance, the disjoint union of k copies of P_3 has that many solutions. On the other hand, this solution space has a simple structure and can be described implicitly. Due to an analoguous situtation for the VERTEX COVER problem, we introduced in [4] the notion of a concise description of the solution space. This gives the possibility to develop more clever branching rules and to describe all solutions by a smaller search tree.

Next we give such an algorithm for CLUSTER EDITING which is still fairly simple. Our branching rule is based on a lemma that can be esily proved by case inspection:

Lemma 4. *If two P_3 in G share exactly one vertex, there also exist two P_3 that share exactly two (adjacent) vertices. If G is connected and all P_3 in G are disjoint, then G is a complete graph or $G = P_3$.*

Consequently, a graph is either a disjoint union of cliques and P_3, or it contains a 4-vertex subgraph with two different P_3. In the former case, the set of minimal solutions is simply the cartesian product of sets of size 3 (the vertex pairs in each P_3 component), as stated above. In the latter case we branch on such a 4-vertex subgraph.

Theorem 3. *A description of all minimal solutions (more precisely, a list of single solutions and cartesian products of certain sets of vertex pairs) can be computed in $O^*(2.562^k)$ time.*

Proof. Every tree node represents a graph and a parameter value. The root represents the given G and k. Every tree node is processed as follows. If the graph is a disjoint union of cliques and P_3 then stop at this node and output this part of the list. Otherwise take four vertices that form two different P_3, say with vertex set $\{w_1, u, v\}$ and $\{w_2, u, v\}$, respectively, note that the common pair uv can be an edge or not. Clearly, we have to change either the status of uv, or the status of one of w_iu, w_iv, for $i = 1, 2$. One option reduces k by 1, four options reduce k by 2. Thus, the branching number is the root of equation $x^2 = x + 4$, which yields $x = (1 + \sqrt{17})/2 < 2.562$. $\square$

The idea of this theorem can be immediately generalized to an arbitrary but fixed p: Always choose a pair of vertices u, v that belongs to more than p

different P_3. Then, either change the status of uv or of one pair in each of these P_3. As long such u, v exist, the branching number is given by the recurrence $x^{p+1} = x^p + 2^{p+1}$. In particular, $p = 2$ yields $x^3 = x^2 + 8$, hence $x < 2.4$. Larger p would further improve the branching number. However, it remains to describe the minimal solutions in the leaves of the resulting search tree, where the branching rule is nolonger applicable. Therefore, we first have to characterize graphs G where every vertex pair belongs to at most p P_3. Clearly, it is enough to study connected graphs of this type. In the following we manage the case $p = 2$. The plan is as follows. We completely characterize our graphs G of interest, showing that G is either a "trivial" graph, i.e. a large path, cycle or clique, or one of a handful of small exceptional graphs. Hence the leaves of our search trees represent only "small" or "trivial" graphs for which the solutions to CLUSTER EDITING with the residual parameter are easy to enumerate. Thus we end up with a search tree algorithm with branching number 2.4.

Theorem 4. *A connected graph G where every vertex pair belongs to at most two P_3 is either a P_n, C_n or K_n or one of the 6-vertex graphs (a)-(g) shown below, or an induced subgraph thereof.*

Proof. Predicate $P(., .)$ with two vertices as arguments says that this pair is in at most two P_3. If every vertex of G is simplicial (its neighborhood is a clique), then $G = K_n$. Otherwise we choose some vertex r such that $N(r)$ contains two non-adjacent vertices u, v. Let r be of maximum degree among all non-simplicial vertices. By $P(r, v)$, vertex u has at most one neighbor $x \in N(r)$ which is not adjacent to v. Similarly, v has at most one neighbor $y \in N(r)$ which is not adjacent to u. Due to $P(u, v)$, vertices u, v have at most one common neighbor $z \neq r$. Thus $N(r)$ consists of u, v and perhaps x, y, z. (Each of x, y, z may or may not exist.) $P(r, t)$ gives the following *Claim*: Any $t \in N(r)$ has at most two non-neighbors in $N(r)$, and if t has exactly two, t cannot possess neighbors outside $N(r)$. Now we examine the cases for $N(r)$.

x and y exist. Then each of u, v has two non-neighbors in $N(r)$ and hence no further neighbors. If z exists too, z cannot have a further neighbor $z' \notin N(r)$ either: Since z' is not adjacent to u and v, edge zz' would violate $P(z, z')$. Next observe that x must be adjacent to at least one of y, z, and y to at least one of x, z, due to the claim. From the resulting cases, only one (and its symmetric counterpart) is impossible: If xy, xz are edges but yz is not, $P(y, z)$ does not hold. The other cases give the graphs (a),(b),(c). In (a) and (b), vertices x and y have no further neighbors, due to the claim. Hence (a) and (b) are maximal with the desired property. We show the same for (c), in a different way: Assuming that x has a further neighbor x', edge $P(x, x')$ is violated, and similarly with y.

x and y exist, but not z. Then similar arguments apply. If x, y are not adjacent, none of u, v, x, y can have further neighbors due to the claim, and we get an induced subgraph of (a). If xy is an edge then $N(r) = P_4$. Both x and y can have at most one further neighbor. It must be a common neighbor, or we get violations of predicate P. This vertex w in turn cannot have further neighbors, by $P(x, w)$. Thus we get graph (d).

Only x and z exist. (Case y and z is symmetric). If x, z are not adjacent then $N(r) = P_4$ as before. Hence suppose that xz is an edge. By the claim, the maximum number of further neighbors of u, v, x, z is $1, 0, 1, 2$, respectively. But a neighbor of z would contradict $P(v, z)$. The neighbors of u and x must be identical, and this new vertex w cannot have further neighbors, by $P(u, w)$. This case yields graph (e).

Only z exists. Since r is the non-simplicial vertex with maximum degree, z has no further neighbors. Each of u, v may have one further neighbor, and they are different, which gives the maximal graph (f).

Only x exists. (Case y is symmetric.) Each of u and x can have one further neighbor. If these neighbors are different, we obtain graph (g) which is maximal by similar arguments as before. If u, x have a common neighbor w, we can append another edge to w, and the maximal graph is isomorphic to (f).

None of x, y, z exists. Since r is a non-simplicial vertex with maximum degree 2, each of u and v has at most one further neighbor. Applying this argument inductively to the new vertices, we easily find that G is either C_n or P_n, depending on whether the last neighbors are equal, different and adjacent, or different and non-adjacent.   □

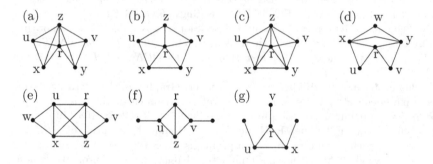

Coming back to the motivation, we can get a description of all minimal solutions to CLUSTER EDITING as follows: Branch on pairs that belong to three or more P_3, as long as possible. Every leaf of the resulting search tree is labeled by either a disjoint union of cliques (that can be ignored, as discussed earlier), chordless paths and cycles, and other connected components of constantly bounded size. Given the "residual" k for the leaf, these paths and cycles must be of length $O(k)$, otherwise there is no solution for this leaf (dead end). Characterizing all minimal solutions in chordless paths and cycles is trivial, due to their highly regular structure. For the other connected components we may precompute the (constantly many) minimal solutions, this part does not even depend on parameter k. Finally we have to combine solutions from the connected component so that the total number of changes is at most the current k. Details are straightforward but tedious. We summarize our findings in:

Theorem 5. *A description of all minimal solutions is computable in $O^*(2.4^k)$ time.*

Acknowledgement

Support has been received from the Swedish Research Council (Veten-skapsrådet), project title "Algorithms for searching and inference in genetics", file no. 621-2002-4574. I am grateful to Tomas Nolander and to an anonymous referee who pointed out some inaccuracies in earlier drafts.

References

1. N. Bansal, S. Chawla, A. Blum. Correlation clustering, *43rd IEEE FOCS'2002*, 238-247
2. Z.Z. Chen, T. Jiang, G. Lin. Computing phylogenetic roots with bounded degrees and errors, *SIAM J. Comp.* 32 (2003), 864-879
3. B. Chor, M.R. Fellows, D. Juedes. Linear kernels in linear time, or how to save k colors in $O(n^2)$ steps, *30th WG'2004, LNCS*
4. P. Damaschke. Parameterized enumeration, transversals, and imperfect phylogeny reconstruction. In [6], 1-12, journal version accepted for a special issue of *Theoretical Computer Science*
5. R.G. Downey, M.R. Fellows. *Parameterized Complexity*, Springer, 1999
6. R.G. Downey, M.R. Fellows, F. Dehne (eds.). *Parameterized and Exact Computation*, 1st Int. Workshop IWPEC'2004, Proceedings, *LNCS* 3162
7. M.R. Fellows, P. Heggernes, F.Rosamond, C. Sloper, J.A. Telle. Exact algorithms for finding k disjoint triangles in an arbitrary graph, *30th WG'2004, LNCS*
8. H. Fernau. On parameterized enumeration, *8th COCOON'2002, LNCS* 2387, 564-573
9. J. Gramm, J. Guo, F. Hüffner, R. Niedermeier. Graph-modeled data clustering: Fixed-parameter algorithms for clique generation, to appear in *Theory of Computing Systems*, preliminary version in *5th CIAC'2003, LNCS* 2653, 108-119
10. J. Gramm, J. Guo, F. Hüffner, R. Niedermeier. Automated generation of search tree algorithms for hard graph-modification problems, *Algorithmica* 39 (2004), 321-347
11. R.E. Greenwood, A.M. Gleason. Combinatorial relations and chromatic graphs, *Canad J. Math.* 7 (1955), 1-7
12. T.M. Mitchell. *Machine Learning*, McGraw-Hill 1997
13. E. Prieto, C. Sloper. Looking at the stars. In [6], 138-148
14. R. Shamir, R. Sharan, D. Tsur. Cluster graph modification problems, *Discrete Applied Math.* 144 (2004), 173-182, preliminary version in: *28th WG'2002, LNCS* 2573, 379-390
15. R. Sharan, A. Maron-Katz, R. Shamir. CLICK and EXPANDER: A system for clustering and visualizing gene expression data, *Bioinformatics* 19 (2003), 1787-1799
16. R. Sharan, R. Shamir. Algorithmic approaches to clustering gene expression data, in: *Current Topics in Computational Molecular Biology*, MIT Press, 2002, 269-300

Locally Consistent Constraint Satisfaction Problems with Binary Constraints

Manuel Bodirsky[1] and Daniel Král'[2,3,*]

[1] Institut für Informatik, Abteilung Algorithmen und Komplexität I,
Humboldt-Universität zu Berlin,
Unter den Linden 6, 10099 Berlin, Germany
bodirsky@informatik.hu-berlin.de
[2] Institute for Mathematics, Technical University Berlin[**],
Strasse des 17. Juni 136, 10623 Berlin, Germany
[3] Department of Applied Mathematics,
Faculty of Mathematics and Physics, Charles University,
Malostranské náměstí 25, 118 00 Praha 1, Czech Republic
kral@kam.mff.cuni.cz

Abstract. We study constraint satisfaction problems (CSPs) that are *k-consistent* in the sense that any k input constraints can be simultaneously satisfied. In this setting, we focus on constraint languages with a single binary constraint type. Such a constraint satisfaction problem is equivalent to the question whether there is a homomorphism from an input digraph G to a fixed target digraph H. The instance corresponding to G is k-consistent if every subgraph of G of size at most k is homomorphic to H. Let $\rho_k(H)$ be the largest ρ such that every k-consistent G contains a subgraph G' of size at least $\rho||E(G)||$ that is homomorphic to H. The ratio $\rho_k(H)$ reflects the fraction of constraints of a k-consistent instance that can be always satisfied. We determine $\rho_k(H)$ for all digraphs H that are not acyclic and show that $\lim_{k\to\infty} \rho_k(H) = 1$ if and only if H has tree duality. In addition, for graphs H with tree duality, we design an algorithm that computes in linear time for a given input graph G either a homomorphism from almost the entire graph G to H, or a subgraph of G of bounded size that is not homomorphic to H.

1 Introduction

Constraint satisfaction problems (CSPs) form an important model for problems arising in many areas of computer science. This is witnessed by the interest in the computational complexity of various variants of CSPs [1,2,8,9,10,11,23]. However, sometimes not all the constraints need to be satisfied, but it suffices to satisfy a large fraction of them. In order to maximize this fraction, the input can

---

[*] Supported by Institute for Theoretical Computer Science (ITI), project 1M0021620808 of Czech Ministry of Education.
[**] The author was a postdoctoral fellow at TU Berlin within the framework of the European training network COMBSTRU from October 2004 to July 2005.

D. Kratsch (Ed.): WG 2005, LNCS 3787, pp. 295–306, 2005.

be pruned in the beginning by removing small sets of contradictory constraints so that the input becomes "locally" consistent. Formally, an instance of CSP is said to be *k-consistent* if any k constraints can be simultaneously satisfied.

A similar notion of local consistency can be defined in terms of variables: an instance is k-consistent if the values of any k variables can be chosen so that any constraint on only these variables is satisfied. Our results extend to this setting.

Both these notions of local consistency differ fundamentally from the notion of *k-consistency* of Freuder [11] (and also the notion of *relational k-consistency* of Dechter and van Beek [4]), where a CSP instance is *k-consistent* if every solution for the constraints on $k - 1$ variables (constraints) can be extended to another variable (constraint).

1.1 History of Locally Consistent CSPs

The notion of local consistency considered in this paper can be traced back to the early 1980's. Lieberherr and Specker [19,20] studied the problem for CNF formulas: they require that any k clauses of a given formula can be satisfied and asked what fraction of all the clauses can be satisfied. They settled the case $k = 1, 2, 3$. A simpler proof of their results was found by Yannakakis [24]. The case $k = 4$ was settled in [17] (exploring a connection to Usiskin's numbers [21]). Locally consistent CNF formulas can also be found in Chapter 20 of [16].

Huang et al. [14] and Trevisan [22] resolved the asymptotic behavior of locally consistent CNF formulas as k approaches infinity. Trevisan [22] was the first to define the notion of local consistency for CSPs with constraints that are Boolean predicates. For a set Π of Boolean constraints, $\rho_k(\Pi)$ is the maximum ρ such that a fraction of at least ρ constraints can be satisfied in any k-consistent input. Note that we now allow negations in the arguments of the constraints (the domain is the Boolean field). If Π is the set of all the predicates of arity ℓ, then $\lim_{k\to\infty} \rho_k(\Pi) = 2^{1-\ell}$ [22]. The ratios $\rho_k(\Pi)$, $k \geq 1$, for a set Π consisting of a single predicate of arity at most three were determined by Dvořák et al. [6]. The asymptotic behavior for sets Π of predicates was studied in [18], where $\lim_{k\to\infty} \rho_k(\Pi)$ was expressed as the minimum of a certain functional on a convex set of polynomials derived from Π. Efficient algorithms for locally consistent CSPs with constraints that are Boolean predicates were also designed [6,7,18].

1.2 Our Contribution

We initiate the study of locally consistent CSPs on larger finite domains and focus on the case where all the constraints are of the same binary relation. The relation can be described by a digraph H whose vertices correspond to the elements of the domain. Two vertices are joined by an arc if the ordered pair of the corresponding elements is contained in the relation. Similarly, the input can be described by a digraph G: the vertices of G correspond to the variables and the arcs to the given constraints. There is a satisfying assignment for the input if and only if G is *homomorphic* to H, i.e., there exists a mapping $h : V(G) \to V(H)$ such that $h(u)h(v) \in E(H)$ for every $uv \in E(G)$.

The notion of local consistency translates to digraphs as follows: G corresponds to an k-consistent input if every subgraph of G of size at most k, i.e., with at most k edges, is homomorphic to H. The ratio $\rho_k(H)$ denotes the largest ρ such that for any k-consistent G there is a mapping $h : V(G) \rightarrow V(H)$ preserving at least $\rho\|G\|$ arcs of G. The version defined in terms of variables also translates to digraphs: for that we require that each subgraph G' of order at most k is homomorphic to H. The corresponding ratio is denoted by $\rho_k^v(H)$.

We can restrict our attention to digraphs H that are cores. H is a *core* if it does not have a homomorphism to a proper subgraph. Every digraph H contains a unique (up to isomorphism) subgraph H' such that H is homomorphic to H' and H' is a core. Obviously, $\rho_k(H) = \rho_k(H')$ and $\rho_k^v(H) = \rho_k^v(H')$.

We show that if H contains a directed cycle (or a loop), then $\rho_k(H)$ and $\rho_k^v(H)$ coincide and are equal to the fractional relative density $\delta'_{\mathrm{rel}}(H)$ of H as defined in Section 2. For such digraphs H, we also design a simple linear time algorithm that finds a mapping $h : V(G) \rightarrow V(H)$ preserving at least $\delta'_{\mathrm{rel}}(H) \cdot \|G\|$ arcs of G.

Then, we focus on the asymptotic behavior of $\rho_k(H)$. We find a close relation to the notion of tree duality from [13] by showing that $\lim_{k \rightarrow \infty} \rho_k(H) = 1$ if and only if H has tree duality. In particular, the limit is equal to one for orientations of paths or acyclic tournaments. Finally, in Section 6, we sketch possible generalizations of our results for CSPs with larger constraint languages.

2   Target Graphs with Cycles or Loops

We first focus on the binary relations (constraints) whose corresponding target graph H such that H contains a directed cycle. This includes the case when the relation is symmetric.

For a digraph H, we define the *fractional relative density* of H as follows.

$$\delta'_{\mathrm{rel}}(H) = \max_{\substack{x:V(H)\rightarrow\langle 0,1\rangle | \\ \sum_{v\in V(H)} x(v)=1}} \sum_{uv\in E(H)} x(u) \cdot x(v)$$

The maximum is taken over all functions $x : V(H) \rightarrow \langle 0,1 \rangle$ such that the sum of $x(v)$ is equal to one. In particular, if H contains a loop, then $\delta'_{\mathrm{rel}}(H) = 1$. This notion of density is similar to that of *relative density* as used e.g. in [15]:

$$\delta_{\mathrm{rel}}(H) = \max_{\emptyset \neq H' \subseteq H} \frac{\|H'\|}{|H'|^2}.$$

The two notions are in general different. Consider the digraph H depicted in Figure 1 that is obtained from the complete graph of order five by replacing each edge by a bigon and removing two non-incident arcs. The relative density $\delta_{\mathrm{rel}}(H)$ is $18/25 = 0.720$ but the fractional relative density $\delta'_{\mathrm{rel}}(H)$ is $88/121 \approx 0.727$ (set $x(v) = 3/11$ for the vertex incident with 8 arcs and $x(v) = 2/11$ for the other vertices). However, the two notions coincide when H corresponds to a symmetric binary relation [5]. In this case, both the target and the input graph can be viewed as an undirected digraph (replace each bigon by a single undirected edge).

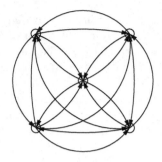

Fig. 1. A digraph with different relative density and fractional relative density

Let $\omega(H)$ denote the size of the largest subset $A \subseteq V(H)$ such that any two distinct vertices of A are joined by an arc.

Proposition 1. *If a digraph H corresponds to a symmetric binary relation, then $\delta'_{\mathrm{rel}}(H) = \delta_{\mathrm{rel}}(H)$. Moreover, $\delta_{\mathrm{rel}}(H)$ equals 1, if H contains a loop, and $\delta_{\mathrm{rel}}(H) = 1 - 1/\omega(H)$ otherwise.*

Proof. If H contains a loop, then $\delta_{\mathrm{rel}}(H) = \delta'_{\mathrm{rel}}(H) = 1$. In the rest, we assume that H has no loops. Let $x : V(H) \to \langle 0, 1 \rangle$ be the function such that $\delta'_{\mathrm{rel}}(H) = \sum_{uv \in E(H)} x(u) \cdot x(v)$, $\sum_{v \in V(H)} x(v) = 1$ and the support of x is minimal. We show that $uv \in E(H)$ for any two vertices u and v contained in the support.

Assume the opposite and let u and v be two non-adjacent vertices such that $x(u) > 0$ and $x(v) > 0$. Let $X_u = 2\sum_{uw \in E(H)} x(w)$ and $X_v = \sum_{uw \in E(H)} x(w)$. By symmetry, we can assume that $X_u \le X_v$. Consider the following function x':

$$x'(w) = \begin{cases} x(u) - \min\{x(u), x(v)\} & \text{if } w = u, \\ x(v) + \min\{x(u), x(v)\} & \text{if } w = v, \text{ and} \\ x(w) & \text{otherwise.} \end{cases}$$

Since the vertices u and v are non-adjacent in H, the following holds:

$$\sum_{uv \in E(H)} x'(u) \cdot x'(v) - \sum_{uv \in E(H)} x(u) \cdot x(v) = 2(X_v - X_u)\min\{x(u), x(v)\}.$$

The choice of x implies that $X_u = X_v$ and $\delta'_{\mathrm{rel}}(H) = \sum_{uv \in E(H)} x'(u) \cdot x'(v)$. Since $X_u = X_v$, the configuration is again symmetric with respect to u and v and we may assume that $x(u) < x(v)$. Consequently, $x'(u) = 0$ and the support of x is not minimal. We conclude that the support of x induces a complete graph.

It is an easy exercise in calculus to show that if $\delta'_{\mathrm{rel}}(H) = \sum_{uv \in E(H)} x(u) \cdot x(v)$, then $x(u) = 1/k$ where k is the size of the support of x. Hence, $\delta'_{\mathrm{rel}}(H) = 1 - 1/\omega(H)$. Since $\delta_{\mathrm{rel}}(H) \le \delta'_{\mathrm{rel}}(H)$, we have $\delta_{\mathrm{rel}}(H) = \delta'_{\mathrm{rel}}(H)$.

The density $\delta'_{\text{rel}}(H)$ is always a lower bound on $\rho_k(H)$ (even if H is acyclic):

Lemma 1. *Let H be a digraph. The following holds for every $k \geq 1$:*

$$\rho_k(H) \geq \delta'_{\text{rel}}(H) .$$

Moreover, there exists a deterministic algorithm that for any digraph G finds a mapping $h : V(G) \to V(H)$ that preserves at least $\delta'_{\text{rel}}(H) \cdot \|G\|$ arcs of G. The running time of the algorithm is linear in the size of G (if H is fixed).

Proof. Let $x : V(H) \to \langle 0,1 \rangle$ be the function such that $\sum_{v \in V(H)} x(v) = 1$ and $\delta'_{\text{rel}}(H) = \sum_{uv \in E(H)} x(u) \cdot x(v)$. Consider a mapping $h : V(G) \to V(H)$ that maps each vertex of G to a vertex $v \in V(H)$ with probability $x(v)$. The probability that an arc of G is mapped to an arc of H is exactly $\sum_{uv \in E(H)} x(u) \cdot x(v) = \delta'_{\text{rel}}(H)$. Hence, the expected number of arcs preserved by h is $\delta'_{\text{rel}}(H) \cdot \|G\|$. The mapping h can be found deterministically in linear time using the derandomization method based on conditional expectations as described in [24]. $\square$

Let us now recall Markov's inequality and Chernoff's inequality:

Proposition 2. *If X is a non-negative random variable with the expected value E, then the following holds for every $\alpha \geq 1$:*

$$\text{Prob}(X \geq \alpha) \leq \frac{E}{\alpha} .$$

Proposition 3. *If X is the sum of N independent random zero-one variables, each equal to one with probability p, then the following holds for $0 < \delta \leq 1$:*

$$\text{Prob}(X \geq (1+\delta)pN) \leq e^{-\frac{\delta^2 pN}{3}} \quad and \quad \text{Prob}(X \leq (1-\delta)pN) \leq e^{-\frac{\delta^2 pN}{2}}.$$

We now prove the converse inequality of Lemma 1:

Theorem 1. *If H is a non-acyclic digraph, then $\rho_k(H) = \delta'_{\text{rel}}(H)$ for all $k \geq 1$.*

Proof. Fix $k \geq 1$ and ε, $0 < \varepsilon \leq 1/2$. Let n be a sufficiently large integer. We find a digraph G such that every subgraph of G of size at most k is homomorphic to H, but every mapping $h : V(G) \to V(H)$ preserves at most $(\delta'_{\text{rel}}(H) + \varepsilon)\|G\|$ arcs.

We first consider a random digraph G_0 and we later prune it to obey all the requirements. G_0 is a random graph of order n in which the arc from u to v, $u \neq v$, is included with probability $n^{-1+1/2k}$ independently of the other arcs. G_0 contains no loops. Since the expected number of arcs of G_0 is $n(n-1)n^{-1+1/2k} = n^{1/2k}(n-1)$, Proposition 3 implies that the probability that the number of arcs is smaller than $(1 - \varepsilon/4)n^{1+1/2k}$ does not exceed $1/4$ for a sufficiently large n.

Next, we estimate the number of (not necessarily directed) cycles of G_0. The expected number of bigons is $\binom{n}{2}n^{-2+2/2k} \leq n^{1/k}$ and that of cycles of length $\ell = 3, \ldots, k$ is at most $n^\ell 2^\ell (n^{-1+1/2k})^\ell \leq 2^\ell n^{1/2}$. By Proposition 2, the number of such bigons and cycles does not exceed $4 \cdot k 2^k n^{1/2}$ with probability at least $1/4$. Hence, if n is sufficiently large (and k is fixed), the number of arcs contained in such bigons and cycles is bounded by $\varepsilon n/4$ with probability at least $3/4$.

Fix a mapping $h : V(G_0) \to V(H)$. Set $x(v) := |h^{-1}(v)|/n$ for $v \in V(H)$. By the definition of $\delta'_{\mathrm{rel}}(H)$, the following holds:

$$\sum_{uv \in E(H)} x(u)x(v) \leq \delta'_{\mathrm{rel}}(H) \tag{1}$$

The expected number of arcs of G_0 preserved by h can be estimated using (1):

$$\sum_{uv \in E(H)} |h^{-1}(u)| \cdot |h^{-1}(v)| n^{-1+1/2k} = \sum_{uv \in E(H)} x(u)x(v)n^{1+1/2k} \leq \delta'_{\mathrm{rel}}(H)n^{1+1/2k}.$$

By Proposition 3, the probability that the number of arcs preserved by h exceeds $(1+\varepsilon/4)\delta'_{\mathrm{rel}}(H)n^{1+1/2k}$ is at most $e^{-\frac{\varepsilon^2 \delta'_{\mathrm{rel}}(H)n^{1+1/2k}}{48}}$. Since there are $|V(H)|^n$ possible choices of h, and since the target graph H and the numbers k and ε are fixed, the probability that there exists $h : V(G_0) \to V(H)$ preserving more than $(1+\varepsilon/4)\delta'_{\mathrm{rel}}(H)n^{1+1/2k}$ arcs of G_0 is at most $1/4$ if n is sufficiently large.

We conclude that the following holds with positive probability:

1. G_0 contains at least $(1-\varepsilon/4)n^{1+1/2k}$ arcs,
2. the size of the set E of the arcs contained in bigons or cycles of length at most k in G_0 does not exceed $\varepsilon n/4$, and
3. every $h : V(G_0) \to V(H)$ preserves at most $(1+\varepsilon/4)\,\delta'_{\mathrm{rel}}(H)n^{1+1/2k}$ arcs.

Therefore, there exists a graph G_0 with the above three properties. The final graph G is obtained from G_0 by removing the arcs contained in the set E.

The size of G is at least $(1-\varepsilon/2)n^{1+1/2k}$. Since every mapping $h : V(G_0) \to V(H)$ preserves at most $(1+\varepsilon/4)\delta'_{\mathrm{rel}}(H)n^{1+1/2k}$ arcs and G is a subgraph of G_0, every $h : V(G) \to V(H)$ also preserves at most this number of arcs. We now infer the bound on the fraction of arcs preserved by h (recall that $\varepsilon \leq 1/2$):

$$\frac{(1+\varepsilon/4)\delta'_{\mathrm{rel}}(H)n^{1+1/2k}}{\|G\|} \leq \frac{1+\varepsilon/4}{1-\varepsilon/2}\delta'_{\mathrm{rel}}(H) \leq (1+\varepsilon)\delta'_{\mathrm{rel}}(H) .$$

Next, we show that any subgraph of G of size at most k is homomorphic to H. Let G' be such a subgraph. Since the size of G' is at most k, G' is an orientation of a forest. Hence, there is a homomorphism from G' to any directed cycle. In particular, there is a homomorphism from G' to H.

Since for every $\varepsilon > 0$, there is a digraph G corresponding to a k-consistent input and the fraction of arcs preserved by any $h : V(G) \to V(H)$ is at most $\delta'_{\mathrm{rel}}(H)+\varepsilon$, we have $\rho_k(H) \leq \delta'_{\mathrm{rel}}(H)$. The opposite inequality holds by Lemma 1.

Proposition 1 and Theorem 1 imply the following for symmetric relations:

Corollary 1. *Let H be a digraph corresponding to a symmetric binary relation $\mathcal{R}$. The following holds for every $k \geq 1$:*

$$\rho_k(H) = \begin{cases} 1 & \text{if there exists an element } a \text{ such that } [a,a] \in \mathcal{R} \\ 1 - 1/\omega(H) & \text{otherwise} \end{cases}$$

Since every subgraph of order at most k of G from the proof of Theorem 1 is an orientation of a forest and thus homomorphic to H, the following also holds:

Corollary 2. *If a digraph H contains a directed cycle, then $\rho_k^v(H) = \delta'_{\mathrm{rel}}(H)$ for all $k \geq 1$.*

3   Graph Homomorphisms and Tree Duality

In the next two sections, we discover a close connection between the limit $\lim_{k \to \infty} \rho_k(H)$ and the concept of tree duality. A digraph H has *tree duality* if G is homomorphic to H whenever every (directed) tree homomorphic to G is also homomorphic to H. E.g., every orientation of a simple path or every acyclic tournament has tree duality. Feder and Vardi [10] and Hell, Nešetřil and Zhu [13] observed that if H has tree duality, then the H-coloring problem (the decision problem whether a given graph is homomorphic to H) can be solved in polynomial time [12] by the so-called *consistency check algorithm*, which is also called *arc-consistency procedure* in artificial intelligence.

An equivalent definition of having tree duality uses the notion of set graphs. For a digraph H, the *set graph* 2^H is the graph whose vertices are non-empty subsets of $V(H)$ and two subsets U and V are joined by an arc if the following holds: for every vertex $u \in U$, there exists a vertex $v \in V$ such that uv is an arc of H, and for every vertex $v \in V$, there exists a vertex $u \in U$ such that uv is an arc of H (see Figure 2). H has tree duality if and only if 2^H is homomorphic to H [3,10]. Note that this criterion can be used to decide algorithmically whether a given digraph H has tree duality.

We now describe the *consistency check algorithm* studied already in [11]. At the beginning, each vertex v of an input graph G is assigned the set $\ell_0(v)$ of all the vertices of the target graph H and the set assigned to v after i steps of the algorithm is denoted by $\ell_i(v)$. At the i-th step, a vertex $w \in V(H)$ is removed from the set of v if G contains an arc vv' such that H does not contain an arc ww' for any $w' \in \ell_{i-1}(v')$ or G contains an arc $v'v$ such that H does not contain

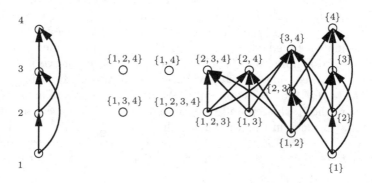

Fig. 2. An example of a digraph H with tree duality and its set graph

an arc $w'w$ for any $w' \in \ell_{i-1}(v')$. We say that such an arc vv' is *violated* at the i-th step. The procedure terminates when there are no violated arcs. The number of steps never exceeds $|G| \cdot |H|$. The running time of the procedure is linear in $|G| + ||G||$ when H is fixed (and when the assignments ℓ_i at each step are implicitly represented).

If there is a vertex v whose final set is empty, then G is not homomorphic to H. Otherwise, the mapping $h : V(G) \to V(2^H)$ that maps each v to its final set is a homomorphism from G to 2^H. If H has tree duality, then 2^H is homomorphic to H and thus G is homomorphic to H.

4   Target Graphs with Tree Duality

In this section, we show that if H has tree duality, then $\lim_{k\to\infty} \rho_k(H) = 1$. We also design an algorithm that either finds a good mapping from G to H or detects a subgraph of G of bounded size not homomorphic to H (under the assumption that H has tree duality). Note that even if H has tree duality, the problem to maximize the number of satisfied constraints can be hard. For instance the problem whether G is homomorphic to the digraph consisting of a single arc can be solved in polynomial time, but the maximization problem is hard: for an undirected graph G_0, let G be the digraph obtained by replacing each edge with a bigon. The maximum number of arcs that can be preserved by a mapping from G to the arc is equal to the size of the maximum cut of G_0.

Theorem 2. *If H is a digraph that has tree duality, then the following holds:*

$$\lim_{k\to\infty} \rho_k(H) = 1.$$

Moreover, there is an algorithm that for an input graph G and $\varepsilon > 0$ either finds a mapping $h : V(G) \to V(H)$ preserving at least $(1 - \varepsilon) \cdot ||G||$ arcs or finds a subgraph of G of size at most $|H|^{\lceil 2|H|/\varepsilon\rceil}$ not homomorphic to H. The running time of the algorithm is linear in $|G| + ||G||$ if the target graph H is fixed.

Proof. We first describe the algorithm. The algorithm runs the consistency check algorithm for the first $\lceil 2|H|/\varepsilon\rceil$ steps and constructs the corresponding assignments ℓ_i. It then distinguishes two cases. The first case is that there exists a vertex v with $\ell_i(v) = \emptyset$. Let i be the smallest index with this property. For every $w \in V(H)$, there exists a step of the algorithm when w was removed from the set of v because an incident edge was violated. For $w \in V(H)$, consider such an edge $v_w v$ and the corresponding step i_w. Note that $i_w < i$. Now, for every $w' \in V(H)$ missing in $\ell_{i_w}(v_w)$, consider the step when w' was removed from the sets of v_w. Note that $\ell_{i_w}(v_w) \neq \emptyset$ by the choice of i. We obtain new sets of arcs that were violated before the i_w-th step and that caused vertices w' to be removed from the set of v_w. Continue in this way unless the sets assigned to the vertices of G are equal to $V(H)$. This terminates because the numbers i_w of steps decrease. Since $i \leq \lceil 2|H|/\varepsilon\rceil$, the number of considered violated arcs does not exceed:

$$|H| + |H|(|H| - 1) + |H|(|H| - 1)^2 + \cdots + |H|(|H| - 1)^{\lceil 2|H|/\varepsilon\rceil - 1} \leq |H|^{\lceil 2|H|/\varepsilon\rceil}.$$

This set of arcs contains a subgraph of G that is not homomorphic to H.

The other case is that $\ell_i(v) \neq \emptyset$ for all $v \in V(G)$ and i. Let E_i be the set of the arcs violated at the i-th step. Since each edge is violated at most $2|H|$ times (at each step when the edge is violated, the size of the set of one of its end-vertices decreases), the sum $|E_1| + \cdots + |E_{\lceil 2|H|/\varepsilon \rceil}|$ is bounded by $2|H| \cdot ||G||$. Hence, there exists $i = 1, \ldots, \lceil 2|H|/\varepsilon \rceil$ such that $|E_i| \leq \varepsilon ||G||$. Consider a mapping $h' : V(G) \to V(2^H)$ defined as $h'(v) := \ell_{i-1}(v)$; h' preserves all arcs possibly except for those of E_i. Since 2^H is homomorphic to H, there is a homomorphism $h : V(G) \to V(H)$ that preserves at least $(1 - \varepsilon) ||G||$ arcs. The bound on the running time of our algorithm follows from the discussions in Section 3.

Since the algorithm finds for a $|H|^{\lceil 2|H|/\varepsilon \rceil}$-consistent G a mapping preserving at least $(1 - \varepsilon) ||G||$ arcs, $\rho_{|H|^{\lceil 2|H|/\varepsilon \rceil}}(H) \geq 1 - \varepsilon$ and $\lim_{k \to \infty} \rho_k(H) = 1$.

Since $\rho_{2k}^v(H) \geq \rho_k(H)$ and $\rho(H)$ is non-decreasing in k, Theorem 2 implies:

Corollary 3. *If H is a digraph that has tree duality, then the following holds:*

$$\lim_{k \to \infty} \rho_k^v(H) = 1.$$

Moreover, there is an algorithm that for an input graph G and $\varepsilon > 0$ either finds a mapping $h : V(G) \to V(H)$ preserving at least $(1 - \varepsilon) \cdot ||G||$ arcs or finds a subgraph of G of order at most $1 + |H|^{\lceil 2|H|/\varepsilon \rceil}$ not homomorphic to H. The running time of the algorithm is linear in $|G| + ||G||$ if H is fixed.

5   Target Graphs Without Tree Duality

In this section, we show that if a digraph H does not have tree duality, then $\rho_k(H)$ is bounded away from 1 by a constant (that depends on the digraph H):

Theorem 3. *Let H be a digraph of order n without tree duality. The following holds for every k:*
$$\rho_k(H) \leq 1 - 2^{-2n} .$$

Proof. Fix $k \geq 1$. Let m be the number of arcs of 2^H. We show that for every $\varepsilon > 0$, there exists a digraph G such that every subgraph of G of size at most k can be mapped to H but no mapping $h : V(G) \to V(H)$ preserves more than $(1 - m^{-1} + \varepsilon) ||G||$ arcs of G. Since $m \leq 2^{2n}$, the statement of the theorem would then follow.

The proof proceeds similarly to that of Theorem 1. Let N be a (sufficiently) large integer. Associate with every vertex v_i, $1 \leq i \leq 2^n - 1$, of 2^H a set V_i of N distinct vertices. We construct a random graph G_0 with the vertex set $\cup_i V_i$. Two vertices $u \in V_i$ and $v \in V_j$ are joined by an arc in G_0 with probability $N^{-1+1/2k}$ if $v_i v_j$ is an arc of H, and they are not joined by an arc, otherwise. The expected number of arcs of G_0 is $m N^{1+1/2k}$. In particular, by Proposition 3, the probability that G_0 has less than $(1 - \varepsilon/4) m N^{1+1/2k}$ arcs does not exceed $1/4$ (for a sufficiently large N).

The expected number of (not necessarily directed) cycles of G_0 of length $\ell = 1, \ldots, k$ does not exceed $2^{n\ell} N^\ell N^{\ell(-1+1/2k)} = O(N^{1/2})$ (recall that k and H

are fixed). Hence, by Proposition 2, the number of arcs of G_0 contained in such cycles does not exceed εN with probability at least $3/4$.

Fix now a mapping $h : V(G_0) \to V(H)$. Consider the following random mapping $h_0 : V(2^H) \to V(H)$. A vertex $v_i \in V(2^H)$ is mapped to a vertex $w \in V(H)$ with probability $|h^{-1}(w) \cap V_i|/|V_i|$. Let m_0 be the expected number of arcs of 2^H (the expectation is taken over mappings h_0) that are mapped to arcs of H. Since 2^H cannot be homomorphically mapped to H, $m_0 \le m - 1$. Observe that the expected number of arcs of G_0 (the expectation is now taken over the choices of G_0) mapped to arcs of H is $m_0 N^{1+1/2k}$. Similarly, as in the proof of Theorem 1, it can be shown that the number of arcs of G_0 mapped to arcs of H by h exceeds $(1 + \varepsilon/4)m_0 N^{1+1/2k}$ with probability $2^{-\Omega(\varepsilon^2 N^{1+1/2k})}$. Since there are $n^{nN} = 2^{O(N)}$ mappings h, G_0 has the following three properties with positive probability:

1. G_0 contains at least $(1 - \varepsilon/4)mN^{1+1/2k}$ arcs,
2. the number of arcs contained in cycles of length at most k is at most $\varepsilon N/4$, and
3. no mapping $h : V(G_0) \to V(H)$ preserves more than $(1+\varepsilon/4)(m-1)N^{1+1/2k}$ arcs of G_0.

The desired graph G is obtained from G_0 by removing the arcs contained in cycles of length at most k. A calculation analogous to that in the proof of Theorem 1 yields that every mapping $h : V(G) \to V(H)$ preserves at most $(1-m^{-1}+\varepsilon)\|G\|$ arcs of G.

Theorem 3 immediately yields:

Corollary 4. *The limit $\lim_{k\to\infty} \rho_k(H)$ for a digraph H is equal to 1 if and only if H has tree duality.*

6   Directions for Future Research

The main interest in locally consistent CSPs comes from the question how much it helps if the input is locally consistent. This is reflected by the behavior of $\rho_k(H)$ as a function of k. In case that H does not have tree duality, in particular, if H contains a loop or a directed cycle, the assumption on local consistency does not help. On the other hand, if H has tree duality, this assumption helps a lot.

In [6,7,18], the authors also addressed the weighted versions of the problems. Let us mention that all our results, in particular Theorems 1, 2 and 3, Corollaries 1, 2, 3 and 4, hold for the weighted versions of the problems, too. The reader is welcomed to check him/her/itself that the proofs translate to this setting.

The ultimate goal is to find an expression for the limit $\lim_{k\to\infty} \rho_k$ for CSPs with more types of constraints and with constraints of arbitrary arity. The characterization of CSPs with $\lim_{k\to\infty} \rho_k = 1$ that is based on tree duality applies to all constraint languages that admit a set function, even if the constraint language contains several constraint types. Note that this class of computational problems contains many previously known tractable families of problems including Horn, constant, and ACI problems [3]. At the moment, we are trying to extend our arguments and to find an formula that expresses the limit for CSPs without tree duality.

Acknowledgement

The authors are indebted to Zdeněk Dvořák for discussions on relative density of graphs and to Pavol Hell for his comments on tree duality.

References

1. A. Bulatov, A. Krokhin, P. Jeavons: The Complexity of Maximal Constraint Languages. In: Proc. 33rd Symp. on Theory of Computation, STOC (2001) 667–674.
2. S. Cook, D. Mitchell: Finding Hard Instances of the Satisfiability Problem: A Survey. In: Satisfiability Problem: Theory and Applications. DIMACS Series in DMTCS Vol. 35 AMS (1997).
3. V. Dalmau, J. Pearson: Closure Functions and Width 1 Problems. In: Proc. 5th International Conferences on Principles and Practice of Constraint Programming, CP, LNCS 1713, Springer Berlin (1999) 159–173.
4. R. Dechter, P. van Beek: Local and Global Relational Consistency. Theor. Comput. Sci. 173 (1997) 283–308.
5. Z. Dvořák: personal communication.
6. Z. Dvořák, D. Král', O. Pangrác: Locally Consistent Constraint Satisfaction Problems. In: Proc. 31st International Colloquium on Automata, Languages and Programming, ICALP, LNCS 3142, Springer-Verlag Berlin (2004) 469–480.
7. Z. Dvořák, D. Král', O. Pangrác: Locally Consistent Constraint Satisfaction Problems, to appear in Theor. Comput. Sci.
8. D. Eppstein: Improved Algorithms for 3-coloring, 3-edge-coloring and Constraint Satisfaction. In: Proc. 12th ACM-SIAM Symposium on Discrete Algorithms, SODA (2001) 329–337.
9. T. Feder, R. Motwani: Worst-case Time Bounds for Coloring and Satisfiability Problems. J. Algorithms 45(2) (2002) 192-201.
10. T. Feder, M. Vardi: Monotone monadic SNP and constraint satisfaction. In: Proc. 25th Symposium on the Theory of Computation, STOC (1993) 612–622.
11. E. C. Freuder: A sufficient condition for backtrack-free search. J. ACM 29 (1982) 24–32.
12. P. Hell, J. Nešetřil: Graphs and homomorphisms. Oxford University Press (2004).
13. P. Hell, J. Nešetřil, X. Zhu: Duality and polynomial testing of tree homomorphisms. Trans. Amer. Math. Soc. 348(4) (1996) 1281–1297.
14. M. A. Huang, K. Lieberherr: Implications of Forbidden Structures for Extremal Algorithmic Problems. Theor. Comput. Sci. 40 (1985) 195–210.
15. S. Janson, T. Łuczak, A. Ruciński: Random Graphs. Wiley, New York (2000).
16. S. Jukna: Extremal Combinatorics with Applications in Computer Science. Springer, Heidelberg (2001).
17. D. Král': Locally Satisfiable Formulas. In: Proc. 15th Annual ACM-SIAM Symposium on Discrete Algorithms (SODA). SIAM (2004) 323-332.

18. D. Král', O. Pangrác: An Asymptotically Optimal Linear-Time Algorithm for Locally Consistent Constraint Satisfaction Problems, submitted.
19. K. Lieberherr, E. Specker: Complexity of Partial Satisfaction. J. of the ACM, 28(2) (1981) 411–422.
20. K. Lieberherr, E. Specker: Complexity of Partial Satisfaction II. Technical Report 293, Dept. of EECS, Princeton University (1982).
21. Z. Usiskin: Max-min Probabilities in the Voting Paradox. Ann. Math. Stat. 35 (1963) 857–862.
22. L. Trevisan: On Local versus Global Satisfiability, SIAM J. Discrete Math., 17(4) (2004), 541–547. A preliminary version available as ECCC report TR97-12.
23. G. J. Woeginger: Exact Algorithms for NP-hard Problems: A Survey. In: Proc. Worksh. Comb. Opt.—Eureka, You Shrink. LNCS 2570, Springer (2003) 185–207.
24. M. Yannakakis: On the Approximation of Maximum Satisfiability. J. Algorithms 17 (1994) 475–502.

On Randomized Broadcasting in Star Graphs*

Robert Elsässer and Thomas Sauerwald

University of Paderborn,
Fürstenallee 11, 33102 Paderborn
{elsa, sauerwal}@upb.de

Abstract. One of the most frequently studied problems in the context of information dissemination in communication networks is the broadcasting problem. In this paper, we study the following robust, simple, and scalable randomized broadcasting protocol: At some time t an information is placed at one of the nodes of a graph G, and in the succeeding steps, each informed node choses one of its neighbors in G uniformly at random, and sends the information to this neighbor. We show that this algorithm spreads an information to all nodes in a Star graph S_n of dimension n within $O(\log(N))$ steps, with high probability, where N denotes the number of nodes in S_n. In our proofs, we apply some methods which may be of independent interest, and extend the results of [10] concerning randomized broadcasting in hypercubic graphs.

1 Introduction

Broadcasting algorithms have been extensively studied in the context of information dissemination in communication networks. These algorithms are designed to solve the problem of distributing a particular message from a distinguished node called *source* to all other nodes in the network. Several efficient (deterministic and randomized) broadcasting schemes have been developed for different graph classes and communication models.

In this paper we study the following randomized algorithm (also known as "rumor spreading"): A vertex of a graph G initially has an information that has to be transmitted to all nodes of G. In each succeeding round, any informed node chooses one of its neighbors uniformly at random, and transmits the information to this neighbor. The goal is to determine the number of steps needed to spread the information to all nodes of G.

The algorithm described before has several advantages such as simplicity, scalability, and robustness [10]. It can be applied in standard point to point communication networks, described by connected, undirected graphs in which the vertices represent the processors and the edges represent bidirectional communication channels between the nodes.

* This work was partially supported by the German Research Foundation (DFG) within the SFB 376 "Massive Parallelität: Algorithmen, Entwurfsmethoden, Anwendungen" and by the EU within the 6th Framework Programme under contract 001907 "Dynamically Evolving, Large Scale Information Systems" (DELIS).

D. Kratsch (Ed.): WG 2005, LNCS 3787, pp. 307–318, 2005.

Another application comes from the maintenance of replicated databases. There are updates injected at various nodes, and these updates must propagate to all the nodes in the network. At each step, a processor and its neighbor check whether their copies of the database agree, and if not then they make the necessary reconciliation. The goal is that all copies of the database converge to the same contents. See [8] for details.

The performance of the algorithm we consider has been analyzed in several graph classes in the past. Pittel proved that in a complete graph every node is informed by the algorithm described before within $\log(N) + \ln(N) + O(1)$ steps, with high probability[1], where N denotes the number of vertices in the graph. Frieze and Molloy [12] showed that in a random graph a lower bound of $\Theta(\log(N)/N)$ is required on the edge density in order to deterministically broadcast information in $\lceil \log(N) \rceil$ to all nodes in the graph. This result has been improved by Chen in [5]. Feige et al. [10] analyzed the performance of the aforementioned broadcasting algorithm in random graphs, bounded degree graphs, and hypercubes. Moreover they prooved an upper bound of $O(N \log(N))$ which holds for any graph. Karp et al. [16] showed that, in a similar model, the number of messages can be bounded by $O(N \log(\log(N)))$, w.h.p.

As described above, we are interested in analyzing the run time of the aforementioned broadcasting algorithm in Star graphs. These graphs have been introduced by Akers et al. in [2,1]. The n dimensional Star graph S_n has $N = n!$ vertices corresponding to the $n!$ permutations of $(1, 2, \ldots, n)$. There is an edge from one permutation $(x_1, \ldots, x_n)$ to some other one $(y_1, \ldots, y_n)$ iff an index $i \in \{2, \ldots, n\}$ exists such that $x_1 = y_i$, $x_i = y_1$, and $x_j = y_j$ for any $j \neq 1, i$. It is easy to see that S_n is $n - 1$ regular, and as shown in [2], it is a bipartite edge-transitive Cayley graph.

The Star graph has several very attractive properties. Like the Hypercube, the Star graph is strongly hierarchical, maximally fault tolerant, and strongly resilient. However, the Star graph has significantly less connections and smaller distances than the Hypercube with comparible number of vertices [1]. Several communication algorithms like routing and broadcasting have also been analyzed in Star graphs as well [1,3,4,9,7,11,15,19]. However, these papers mostly consider the deterministic case.

In this paper we analyze the performance of the simple randomized broadcasting algorithm described above in S_n. In Section 2, we state new results concerning the expansion properties of small subsets of vertices in a Star graph. These results are needed in Section 3 to prove that, with high probability, the algorithm described above broadcasts an information to every node of S_n within $O(\log(N))$ steps. We conclude the paper by summarizing our results and by pointing to some open problems.

---

[1] When we write "with high probability" or "w.h.p." in this paper, we mean with probability at least $1 - O(1/N)$.

2   Auxiliary Combinatorial Results

In this section we derive some new results on the expansion properties of small subsets of vertices in $S_n = (V_n, E_n)$. Let $E(m) := \min_{|X|=m} \frac{1}{m} |E(X, \overline{X})|$ be called the *expansion for m* in the graph S_n, where $X \subset V_n$ and $|E(X, \overline{X})|$ denotes the number of edges connecting X with its complement $\overline{X}$. We know that for the Hypercube the edge isoperimetric problem is solved [14], and it holds that $E(m) \geq n - \lceil \log(m) \rceil$ for any integer m with $1 \leq m \leq N/2$. However, a similar result for the Star graph like $E(m) \geq n + 1 - \lceil \Gamma^{-1}(|m|) \rceil$, where $\Gamma(x) = \int_0^\infty t^{x-1} e^{-t} dt$ for real-valued $x \neq 0$, is not known yet. In fact, there are two major difficulties in analyzing the expansion on this graph. First, the Star graph falls into $n - 1$ subgraphs, and is clearly *not* the Cartesian Product of these subgraphs. On the other hand, the n-Cube is simply the product of only two $n - 1$-dimensional subcubes.

The following simple propositions gives an upper bound on the expansion for some m. We omit the proofs due to space limitations.

Proposition 1. *For any Star graph S_n it holds that $E((n - c)!) \leq c$, $c \in \{0, \ldots, n\}$, and $E((\delta n)!) \leq (1 - \delta)n$, where $(\delta n)! \in I\!\!N$.*

Proposition 2. *If $m \in [(n - c)!, (n - c + 1)!]$, then for any Star graph S_n we have*

$$E(m) \leq \frac{(c + 1)(n - c - 2) + n - 1}{n - c - 1},$$

where $c \in \{1, \ldots, n - 2\}$.

Combining these upper bounds, it follows that if we assume in every round the worst-case expansion, we are only able to achieve a runtime of $O(n(\log n)^2)$.

There is another way to derive upper bounds which based on the hierachical structure. Denote by $E_{S_n}(m)$ the expansion for an integer m in the n-Star Graph.

Lemma 1. *It holds for any integer n and integer $0 \leq m \leq n!$ that $E_{S_n}(m) \leq c \Rightarrow E_{S_{n+1}}(m) \leq c + 1$.*

In order to compute an adequate lower bound, we will identify a proper subgraph in S_n which enables us to examine the expansion of small subsets of V_n (cf. the embedding of well studied graphs in the Star graph [15,17,18]). First consider the following definition (cf. [7]).

Definition 1. *Let the tree $B_n(v) \subseteq S_n$ with root $v \in V_n$ defined in the following way. $B_n(v)$ consists of n levels, where level $B_n^i(v)$ contains some vertices at distance i from v in $B_n(v)$, $i \in \{0, \ldots n - 1\}$. We begin by $B_n^0(v) := \{v\}$ and $B_n^1(v) := \{v_i \mid v_i = (1\,i) \circ v, i \in \{2, \ldots, n\}\}$, where $(1\,i)$ denotes the transposition which interchanges the first and ith entries in the permutation represented by v and $\circ$ denotes the composition of both permutations. Similarly, $B_n^2(v) = \{v_{ij} \mid v_{ij} = (1\,j) \circ (1\,i) \circ v, j \in \{2, \ldots, i-1, i+1, \ldots, n\}, i \in \{2, \ldots, n\}\}$,* and $B_n^k(v) = \{v_{i_1 \ldots i_k} \mid v_{i_1 \ldots i_k} = (1\,i_k) \circ \cdots \circ (1\,i_1) \circ v,\, i_k \in \{2, \ldots, n\},\, i_q \neq i_r \text{ for any } q \neq r\}$.

Lemma 2. B_n *is indeed a tree and a subgraph of* S_n, *that is no additional edges between two nodes in* B_n *exists.*

Proof. Since S_n is vertex-transitive, we define id to be the root in B_n. The construction of B_n implies that once a number has been put to any component, this component remains unchanged during the remaining process. As a consequence, every node except for the root contains exact one non-trivial cycle. It also follows that there exist only one node in B_n, namely id, with the first component 1. Now consider any node $u \neq id \in B_n$ in level l, where we may assume $l \geq 1$, since all neighbours of the root are in B_n by definition. The node u can be represented as a cylce of length l: $u = (c_1 \ldots c_l)$, where $c_1 = 1$, since id is the root of our tree. Applying the transposition $(1\,c_l) \circ (1 \ldots c_l) = (1 \ldots c_{l-1})$ leads obviously to the predecessor of u. If we apply the transposition $(1\,c_2) \circ (1 \ldots c_l)$, we reach the node $(c_2 \ldots c_l)$ which first component must be 1. Any transposition like $(1\,c_i)$, where $2 \leq i \leq l-1$, leads to the node $(1\,c_i) \circ (1 \ldots c_l) = (1 \ldots c_{i-1})(c_i \ldots c_l)$ which contains two disjoint cycles and can therefore not be in B_n. On the other hand, if we apply $(1\,j)$, where $j \neq c_i$ for all $2 \leq i \leq l$, we reach the node node $(1 \ldots c_l c_{l+1})$ with $c_{l+1} = j$ which is one of the successors of u in B_n.

Now we use this tree to compute a lower bound on the expansion of small subsets in S_n.

Theorem 1. *Let $c \geq 1$ be an arbitrary but fixed constant. Then, for sufficiently large n it holds that $E(m) \geq \frac{c-1}{c}(n-1)$, where $m \leq n^{n/((2+\epsilon)c)}$, where $\epsilon > 0$ is a constant.*

Proof. For our proof we consider the following algorithm:

FIND GOOD EXPANSION SET

Input: $I \subseteq V_n, |I| \leq n^{n/((2+\epsilon)c)}$
Output: $\emptyset \neq X \subseteq I, |E(X, \bar{I})| \geq \frac{2c-1}{2c}(n-1)|X|$

Select an arbitrary $v \in I$
$X := \{v\}$
while $\frac{|E(X, \bar{I} \cap B_n(v))|}{|X|} < \frac{2c-1}{2c}(n-1)$
    $X = \text{succ}(X) \cap I$
end while
return X

In this proof, the set I will be called *the set of informed nodes*. Let t be the final step of the algorithm and let $\frac{2c-1}{2c}$ be denoted by d. Before step i will be executed, all nodes in X are at level $i-1$ and have therefore $n-1-i+1 = n-i$ successors. If the condition of the loop is satisfied, then we have $|E(X, \bar{I} \cap B_n(v))| < d(n-1)|X|$, and it follows that $|E(X, I \cap B_n(v))| \geq ((1-d)$

$(n-1)-i+1)|X|$. Consequently, the number of nodes in X after the last iteration t is at least:

$$\prod_{i=1}^{t}\left((1-d)(n-1)-i+1\right) \geq \left(\frac{n}{c'}\right)^t,$$

c' a proper constant, if the smallest factor is at least 1. Now we will show that a t exists such that $n-t-1 \geq d(n-1)$ and $\left(\frac{n}{c'}\right)^t \geq n^{\frac{n}{(2+\epsilon)c}}$. The first inequality ensures that the number of successors of the nodes in $B_n(v)$ at level t is still large enough and the second one implies that within at most t iterations all informed nodes have to be used. If we set $t = \frac{n-1}{2c}$, then both inequalities hold. The first inequality means that, after t iterations, every informed node on level t has at least $d(n-1)$ successors. This implies that $E(X, \overline{I} \cap B_n) \geq d(n-1)|X|$ after the termination of our algorithm. Now, since $X \subseteq I$, it follows that

$$|E(I, \overline{I})| = |E(X, \overline{I})| + |E(I \backslash X, \overline{I})|$$
$$= \underbrace{|E(X, \overline{I})|}_{\geq d(n-1)|X|} - \underbrace{|E(I \backslash X, X)|}_{\leq (1-d)(n-1)|X|} + |E(I \backslash X, \overline{I} \cup X)|$$
$$\geq \frac{c-1}{c}(n-1)|X| + |E(I \backslash X, \overline{I} \cup X)|,$$

We can now apply our algorithm again with the input $I \backslash X$, and by induction we get

$$\frac{|E(I, \overline{I})|}{|I|} \geq \frac{c-1}{c}(n-1),$$

and the statement is proved. □

Combining Theorem 1 with Propositions 1,2 it is easy to see that for subsets of size $(\delta n)!$ with $\delta < 1/3$, the bound of Theorem 1 is tight up to a small constant factor.

In addition, we can state the following corollary

Corollary 1. *Let $X \subseteq V_n$, where $|X| = (n-c)!$, $c \in \{0, \ldots, n\}$ and $x \in X$ arbitrary. Then it holds*

$$\exists Y \subseteq X : x \in Y \text{ and } \frac{|E(Y, \overline{X})|}{|Y|} \geq (c-1).$$

Recall that $E((n-c)!) \leq c$ due Proposition 1, so the last Corollary is already near to the optimum, but there is still a gap since our last result holds only for a subset of X.

3   Runtime Analysis

In this section we use the result of Theorem 1 to show that the randomized broadcasting algorithm[2] spreads an information to all nodes of a Star graph

---

[2] When we write the broadcasting algorithm, we always refer to the algorithm defined at the beginning of this section.

within $O(n \log(n))$ steps, w.h.p. Since this algorithm always requires at least $\Omega(\log(N))$ steps to inform all nodes in a graph [10], and since a node in the Star graph exists, which is never contacted within $O(n \log(n))$ steps with probability $(1/n)^{O(n)}$, we can conclude that our result is asymptotically optimal.

Let P be an arbitrary but fixed partition of S_n in $\frac{n}{4}$-dimensional substars. In order to obtain such a partition, we first partition S_n into n disjoint components S'_i, $i \in \{1, \dots, n\}$, where $S'_i = \{(x_1, i, x_3, \dots, x_n) \in V_n\}$ Then, we partition each S'_i into $n-1$ further components among the third entry in the permutations, and so on... Consequently, the partition P consists of sets $S'_{i_1, \dots, i_{3n/4}} = \{(x_1, i_1, \dots, i_{3n/4}, x_{3n/4+2}, \dots, x_n) \in V_n\}$. Note, that $\frac{n}{4}$ must not be an integer, but for simplicity we will assume this in the sequel. It is easy to see, that $\frac{n}{4}$ can be replaced by $\lfloor \frac{n}{4} \rfloor$ in the following proofs.

Our first objective is to show that after $O(n \log(n))$ steps, in every substar of this partition P at least one node has been informed with a probability of $1 - (1/n)^{O(n \log(n))}$ by the broadcasting algorithm.

Lemma 3. *After $120n \log(n)$ steps at least n^{20} nodes of S_n will be informed by the randomized broadcasting algorithm, with probability $1 - (1/n)^{\Omega(n \log(n))}$.*

Proof. We consider the tree $B_n(v)$ defined in Definition 1, but here we are only interested in nodes belonging to the levels $0, \dots, 40$. First, we show that the root informs more than $\sqrt{n}$ successors within $3n \log(n)$ steps with probability $1 - O(1/n^{n \log(n)})$. The probability for informing at most $\sqrt{n}$ successors of the root within this time is

$$P_{\sqrt{n}}(v) \le \binom{n-1}{\sqrt{n}} \left(\frac{\sqrt{n}}{n-1} \right)^{3n \log(n)} \le \left(\frac{1}{n} \right)^{n \log(n)}.$$

The conditional probability $P'_{\sqrt{n}}(v')$ that a successor v' of v informs at most $\sqrt{n}$ successors, given that v' has been informed by v, is

$$P_{\sqrt{n}}(v') \le \binom{n-1}{\sqrt{n}+1} \left(\frac{\sqrt{n}+1}{n-1} \right)^{3n \log(n)} \le \left(\frac{1}{n} \right)^{n \log(n)}.$$

Therefore, the probability for having more than n informed nodes in level $B_n^2(v)$ after $6n \log(n)$ steps is higher than $(1 - (1/n)^{n \log(n)}) \cdot ((1 - (1/n)^{n \log(n)})^{\sqrt{n}}$. Generally, the probability that an informed node v'' in some level $B_n^i(v)$, where $i \le 39$, informs at most $\sqrt{n}$ successors is

$$P_{\sqrt{n}}(v'') \le \binom{n-1}{\sqrt{n}+39} \left(\frac{\sqrt{n}+39}{n-1} \right)^{3n \log(n)} \le \left(\frac{1}{n} \right)^{n \log(n)}.$$

This implies that after $120n \log(n)$ steps, in level $B_n^{40}(v)$ are n^{20} informed nodes with probability $P > (1 - (1/n)^{n \log(n)})^{n^{21}} = 1 - (1/n)^{\Omega(n \log(n))}$, and the lemma follows. □

Lemma 4. *Let $Y \subseteq V_n$ and $d \in \{1, \dots, \lfloor \frac{3}{2}(n-1) \rfloor \}$. Then, there exists $X \subseteq Y$ such that $|X| \ge |Y|/(n-1)^d$ and $\forall u, v \in X : \text{dist}(u, v) \ge d$.*

We omit the proof due to space limitations.

Lemma 5. *Assume that any fixed S_d, where $d \in \{\frac{n}{4}, \ldots, n\}$, contains n^{20} informed nodes. Then, with probability $1 - O(\exp(-n^{11}))$, there exists after 2 additional steps in every $S_{d-1} \subset S_d$ at least n^6 informed nodes.*

Proof. It is easy to see, that every substar $S_{d-1} \subset S_d$ can be reached from any node in S_d within at most 2 steps. Using Lemma 4, we can find a set of nodes $X \subset I$ of size n^{15}, in which any two nodes are at distance at least 5 from each other. Now we can divide X into $d \leq n$ disjoint subsets so that every subset contains at least n^{14} nodes. We assign each substar S_{d-1} one of these subsets, and consider the propagation of the information into each substar from its assigned subset only. Since these nodes are at distance at least 5 from each other, the propagations caused by the nodes of X within the next two steps are independent from each other. Therefore, we can model the problem of informing n^6 nodes in a fixed substar $S_{d-1} \subset S_d$ by n^{14} Bernoulli-distributed random variables X_i, $i \in \{1, \ldots, n^{14}\}$, where $\Pr[X_i = 1] \geq \frac{1}{(n-1)^2} \geq \frac{1}{n^2}$, and $\Pr[X_i = 0] = 1 - \Pr[X_i = 1]$ for any i. Let $X := \sum_{i=1}^{n^{14}} X_i$. Then $\mu := \mathbb{E}[X] \geq n^{14}/(n^2) = n^{12}$. If we apply the Chernoff bound [6,13], we obtain by setting $\delta = 1 - (1/n^6)$ that $\Pr[X \leq n^6] \leq \exp(-(n^{12}(1 - \frac{1}{n^6})^2)/2)$. Since we assigned each substar S_{d-1} a different subset of nodes, and since all the events of informing the substars are independent from each other, the probability for informing at least n^6 nodes in each substar S_{d-1} equals $(1 - \Pr[X \leq n^6])^d$, and the lemma follows. □

Lemma 6. *Let a substar $S_d = (V_d, E_d)$ of S_n contain n^6 informed nodes, where $d \in \{\frac{n}{4}, \ldots, n\}$. Then, at least n^{20} nodes will be informed in S_d, after $O(\log(n))$ additional rounds, with probability $1 - O(\exp(-n^5))$.*

Proof. Let $I(t)$ denote the set of informed nodes in S_d at time t. We show that if $n^6 \leq |I(t)| \leq n^{20}$, then a constant $\tau > 1$ exists such that $|I(t+1)| \geq \tau |I(t)|$. Since $|I(t)| \leq n^{20}$ here, Theorem 1 implies that $|E(I(t), \overline{I(t)} \cap V_d)| \geq c(n-1)|I(t)|$, where $c < 1$ is a proper constant value. Now we have to take into consideration that a node can be possibly informed by different nodes in one round. However, we will try to construct a process with mutually independent random variables. Let x denote the fraction of nodes in $I(t) \cap V_d$ with at least $\frac{c}{2}(n-1)$ neighbours in $\overline{I(t)} \cap V_d$. By simple calculation it follows that x is at least $2c/(2-c)$.

Let $X^0 = \{v \in I(t) \mid N(v) \cap \overline{I(t)} \cap V_d \geq \frac{c}{2}(n-1)\}$, where $N(v)$ denotes the neighbours of v in S_n. Due to the algorithm, each of these nodes select within one step a neighbor, and transmit the information to this neighbor. For the analysis, we divide this one step into substeps, in which only one selected node is allowed to transmit. After every substep i, the set X^{i-1} is updated, leading to some new set X^i as described in the sequel. In every substep i, we select a node $v \in X^{i-1}$ which maximizes $|E(v, \overline{I(t)} \cap V_d)| = |N(v) \cap \overline{I(t)} \cap V_d|$. After substep i in which v informs its chosen neighbor w, $X^i = X^{i-1} \setminus \{v\}$, and if

$w \in \overline{I(t)} \cap V_d$, then $I(t)$ is extended by w. Obviously, after every substep i (in step $t + 1$), $|E(X^i, \overline{I(t)} \cap V_d)| > |E(X^{i-1} \setminus \{v\}, \overline{I(t)} \cap V_d)| - (n-1)$. After y substeps, the number of edges between the set X^y and $\overline{I(t)} \cap V_d$ is more than $c(|X^0| - y)(n-1)/2 - y(n-1)$. If $y \leq m|X^0|$, where $m = c/(4+c)$, then there are at least $(|X^0| - y) \cdot c(n-1)/4$ edges between the set X^y and $\overline{I(t)} \cap V_d$. Therefore, a node v in the set X^y must exist which has at least $\frac{c}{4}(n-1)$ edges to $\overline{I(t)} \cap V_d$, and contacts in the next substep an uninformed node in V_d with probability $p = c/4$. This implies that in step $t+1$, we can model the number of newly infected nodes by $m|X^0|$ mutually independent Bernoulli-distributed random variables X_i with success probabilities $p_i \geq p \, \forall i \in \{1, \ldots, m|X^0|\}$. Again, let $X = \sum_{i=1}^{m} X_i$, and $\mu := \mathbb{E}[X] \geq m|X^0|p$. Using the Chernoff bounds as before, we obtain

$$\Pr\left[X \leq \frac{m|X^0|p}{4}\right] \leq e^{-\frac{m|X^0|p\frac{9}{16}}{2}}.$$

This implies that with probability $1 - O(\exp(-n^6))$, for any $I(t)$ with $n^6 \leq |I(t)| \leq n^{20}$, it holds that $|I(t+1)| \geq \tau|I(t)|$, where $\tau > 1$ is a constant, and the lemma follows. $\square$

Let us now summarize the results of Lemmas 3, 5, and 6 in the following theorem.

Theorem 2. *There exists a constant α such that after $\alpha n \log(n)$ steps at least n^{20} nodes are informed in all $\frac{n}{4}$-dimensional substars of the partition P with probability $1 - 1/(n^{\Omega(n \log(n))})$.*

Proof. Due to Lemma 3, with a probability of $1 - (1/n)^{\Omega(n \log(n))}$ at least n^{20} nodes will be informed after $O(n \log(n))$ rounds. Recall the description of P by an balanced tree. Clearly, P has at most $\frac{n!}{(n/4)!} \leq n^{\frac{3}{4}n}$ edges and the nodes on the last level represent the $\frac{n}{4}$-dimensional substars of our partition P. Using the Lemmas 5 and 6, on every edge the information is successfully transmitted with a probability of at least

$$1 - \frac{1}{e^{n^5}} = \left(1 - \frac{1}{e^{n^5}}\right)^{\frac{e^{n^5}}{e^{n^5}}} \geq \left(\frac{1}{4}\right)^{\frac{1}{e^{n^5}}}$$

This probability is such large that the transmission is successful along all considered edges with a probability of at least

$$\left(\left(\frac{1}{4}\right)^{\frac{1}{e^{n^5}}}\right)^{n^{\frac{3}{4}n}} \geq \left(\frac{1}{4}\right)^{\frac{1}{e^{n^3}}} = 1 - (1/n)^{\Omega(n \log(n))},$$

because all events are independent. Recall, that the lowest probability appeared in our proof was $1 - (1/n)^{\Omega(n \log(n))}$ and so the theorem follows. $\square$

We should mention here that the technique applied for the hypercube seems to be inappropriate here on the Star Graph. With our new method of analyzing the stochastic process from behind, we are able to rely on the techniques which lead to our completely informed partition. Looking at the last theorem, we might ask for a conversion, i.e., for every node $w \in V_n$ there exist at least one node v in each $\frac{n}{4}$-dimensional substar, which contacts w within the next $\gamma n \log(n)$ rounds, where γ is a large but fixed constant. In order to show this, we need the following definition.

Definition 2. *A node $u \in V_n$ contacts another node $v \in V_n$ within the time-interval $[a, b]$, if there exists a path $(u_1 := u, u_2, \ldots, u_{m-1}, u_m := v)$ in S_n with the following properties:*

$$\exists t_1 < t_2 < \cdots < t_{m-1} \in [a, b] : u_i \text{ contacts } u_{i+1} \text{ in round } t_i, i \in \{1, \ldots, m-1\}$$

That is, if u is informed at the beginning of round a, v will be informed after round b.

Similar to the definition of informed nodes, a node u contacts a subset of nodes V', if there exists at least one node in V', which is contacted by u.

In the sequel, we fix one arbitrary node $w \in V_n$. Similar to the definitions of I and $I(t)$, we define $K(t)$ to be the set of nodes in round t, which will contact the fixed node w within the time interval $[t, fn \log(n)]$, where f is a large but fixed constant. In contrast to $I(t)$, $K(t)$ increases while t decreases. The round $fn \log(n)$ can be viewed as a time step in the future from which we analyze the propagation backwards. Our aim is to show that in round $fn \log(n)$ all nodes have already been informed. To prove this, we consider first the following lemma.

Lemma 7. $|K(fn \log(n) - 120n \ln(n))| \geq n^{20}$ *with probability* $1 - O(1/n^{2n})$.

Proof. Again, we consider the tree $B_n(w)$ up to level 40. We show that w is contacted by at least $\sqrt{n}$ out of his successors within $3n \ln(n)$ rounds, with probability $1 - O(1/n^{2n})$. The probability that an arbitrary direct successor of w fails to contact w during this time is less than $(1 - \frac{1}{n-1})^{3n \ln(n)} \leq (\frac{1}{n})^3$. Therefore, the probability that less than $\sqrt{n}$ successors contact w is

$$P_1' \leq \sum_{i=n-\sqrt{n}}^{n-1} \binom{n-1}{i} \left(\frac{1}{n^3}\right)^i \left(1 - \frac{1}{n^3}\right)^{n-1-i}$$

$$\leq \left(\frac{(1/n^3)}{(n-\sqrt{n})/(n-1)}\right)^{n-\sqrt{n}} \cdot \left(\frac{(n^3-1)/n^3}{(n-\sqrt{n})/(n-1)}\right)^{\sqrt{n}-1} < o\left(\frac{1}{n^{2n}}\right).$$

Similarly, a node w' at level $j-1$ in $B_n(w)$ is contacted by less than $\sqrt{n}$ successors with probability

$$P_j' \leq \sum_{i=n-j+1-\sqrt{n}}^{n-j} \binom{n-j}{i} \left(\frac{1}{n^3}\right)^i \left(1 - \frac{1}{n^3}\right)^{n-j-i}$$

$$\leq \left(\frac{(1/n^3)}{(n-j+1-\sqrt{n})/(n-j)}\right)^{n-j+1-\sqrt{n}} \cdot \left(\frac{(n^3-1)/n^3}{(n-j+1-\sqrt{n})/(n-j)}\right)^{\sqrt{n}-1}$$

Since we have less than n^{41} nodes in these 40 levels, w is contacted by at least n^{20} nodes within $120n \ln(n)$ steps with probability $1 - O(1/n^{2n})$. $\square$

For the following lemmas, we can apply the methods already used in this section.

Lemma 8. *Let $d \in \{\frac{n}{4}, \ldots, n\}$ and $|K(t) \cap V_d| \geq n^{20}$. Then, with probability $1 - O(\exp(-n^{11}))$, there exist in every substar $S_{d-1} \subset S_d$ at least n^6 nodes, which contact $K(t)$ within the time-interval $[t - 2, t]$.*

Proof. Analogous to Lemma 5, since for every node we consider a fixed path. $\square$

Lemma 9. *Let $d \in \{\frac{n}{4}, \ldots, n\}$ and $|K(t) \cap V_d| = n^6$. Then, with probability $1 - O(\exp(-n^5))$ it holds that $|K(t - \rho\log(n))) \cap V_d| \geq n^{20}$, where ρ is a large but fixed constant.*

Proof. For simplicity, we denote by K the set $K(t) \cap V_d$. In contrast to lemma 6, the problem that one node can be informed by different nodes in one round does not appear. On the other hand, we allow the possibility that some nodes in K can be contacted by more than one node.

We apply theorem 1 which implies that at least $c(n - 1)|K|$ edges, where c is a suitable constant value, are connecting K and $\overline{K} \cap V_d$. Let $l := |N(K) \cap V_d|$ be the number of neighbours in V_d of nodes in K. For every node in $N(K) \cap V_d$ we define:

$$\forall i \in \{1, \ldots, l\} \, X_i := \begin{cases} 1 & \text{if the } i\text{-th node from } \overline{K} \cap V_d \text{ contacts } K \\ 0 & \text{else} \end{cases}$$

So the random variable $X := \sum_{i=1}^{l} X_i$ describes the number of nodes in the set $K(t-1) \backslash K(t)$. Every X_i is bernoulli-distributed and it holds that $\mu := \mathbb{E}[X] \geq cn^6$. Since the X_i are mutually independent, we can use the Chernoff-Bound again, we obtain $\Pr[X \leq cn^6/4] \leq \exp(-(c(9/16)n^6)/2)$, where $\delta = \frac{3}{4}$. The more t decreases, the more the last probability decreases and so we are able to raise the power of $O(\log(n))$ to the inverse probability, which finishes our proof.
     $\square$

We summarize now the results of the previous lemmas in the following theorem.

Theorem 3. *With probability $1 - O(1/n^{2n})$ it holds that*

$$K\left(fn\log(n) - \gamma n\log(n)\right) \cap S_{\frac{n}{4}} \neq \emptyset$$

for all $\frac{n}{4}$-dimensional substars of P, where γ is a large but fixed constant.

The proof is very similar to the proof of Theorem 2, and we omit it here due to space limitations.

Obviously, the goal is now to show that $I(t) \cap K(t) \neq \emptyset$ in a proper round t, with high probability, which means in fact that the node w will be informed in round $fn\log(n)$.

Theorem 4. *Suppose that there exists a round t such that for every $\frac{n}{4}$-dimensional substar $S_{\frac{n}{4}}$ of P the following statements hold:*

$$I(t) \cap S_{\frac{n}{4}} \neq \emptyset \text{ and } K\left(t + \frac{3}{8}n\right) \cap S_{\frac{n}{4}} \neq \emptyset.$$

Then, with probability $1 - \exp(-n^{\frac{2}{8}n})$ it holds that

$$I\left(t + \frac{3}{8}n\right) \cap K\left(t + \frac{3}{8}n\right) \neq \emptyset.$$

Proof. Since the diameter of an n-dimensional Star graph is $\lfloor 3(n-1)/2 \rfloor \leq 3n/2$, the diameter of every $S_{\frac{n}{4}}$ is at most $3n/8$. Consequently, the informed node in every substar $S_{\frac{n}{4}}$ contacts a node in $K(t + 3n/8) \cap S_{\frac{n}{4}}$ within $3n/8$ steps with a probability of at least $(1/n)^{\frac{3}{8}n}$. Since the events in these substars are mutually independent, and there are $n!/(n/4)! \geq (n/4)^{\frac{3n}{4}} \geq n^{\frac{5n}{8}}$ substars, at least one informed node succeeds with probability

$$1 - \left(1 - \frac{1}{n^{\frac{3}{8}n}}\right)^{n^{\frac{5}{8}n}} \geq 1 - \left(\frac{1}{e}\right)^{n^{\frac{2}{8}n}}.$$

$\square$

We are now able to state our main result.

Theorem 5. *Given one informed node at the beginning, then after $O(n \log(n))$ rounds the randomized broadcasting algorithm informs all nodes in a Star graph S_n with probability $1 - O(1/n^n)$.*

The main result of the paper can easily be generalized to other simple single port randomized broadcasting algorithms such as the pull model. See [8] for details concerning the pull model.

4   Conclusion

Let us now summarize the results of the paper. In Section 2, we stated new results concerning the expansion properties of small subsets of vertices in a Star graph. We used these results in Section 3 to prove that, with high probability, the algorithm described above broadcasts an information to every node of S_n within $O(\log(N))$ steps. Since this algorithm requires in every graph $\Omega(\log(N) + D)$ steps, where D is the diameter of the graph, the result is asymtotically optimal. We also considered Cayley graphs on wich the randomized algorithm requires $\omega(\log(N) + D)$ steps, however we omit the details here due to space limitations.

Acknowledgments

The authors would like to thank Ulf Lorenz for some helpful discussions. Moreover, we are thankful to Sven-Ake Wegner and the referees for finding some flaws.

References

1. S. Akers, D. Harel, and B. Krishnamurthy. The star graph: An attractive alternative to the n-cube. In *Proc. of the International Conference on Parallel Processing, (ICPP)*, pages 393–400, 1987.
2. S. Akers and B. Krishnamurthy. A group-theoretic model for symmetric innterconnection networks. In *Proc. of the International Conference on Parallel Processing, (ICCP)*, pages 555–565, 1986.
3. C. Chen and J. Chen. Vertex-disjoint routings in Star graphs. In *Proc. of the first IEEE Int'l Conf. Algorithms and Architectures in Parallel Processing*, pages 460–464, 1995.
4. C. Chen and J. Chen. Nearly optimal one-to-many parallel routing in Star networks. *IEEE Transactions on Parallel and Distributed Systems*, 8:1196–1202, 1997.
5. H. Chen. Threshold of broadcast in random graphs. *DIMACS Technical Report 97-12*, 1997.
6. H. Chernoff. A measure of asymptotic efficiency for tests of a hypothesis based on the sum of observations. *Ann. Math. Stat.*, 23:493–507, 1952.
7. K. Day and A. Tripathi. A comparative study of topological properties of hypercubes and star graphs. *IEEE Transactions on Parallel and Distributed Systems*, 5, 1994.
8. A. Demers, D. Greene, C. Hauser, W. Irish, J. Larson, S. Shenker, H. Sturgis, D. Swinehart, and D. Terry. Epidemic algorithms for replicated database maintenance. In *Proc. of the 6th ACM Symposium on Principles of Distributed Computing*, pages 1–12, 1987.
9. M. Dietzfelbinger, S. Madhavapeddy, and I. Sudborough. Three disjoint path paradigms in Star networks. In *Proc. of the 3rd IEEE Symp. Parallel and Distributed Processing*, pages 400–406, 1991.
10. U. Feige, D. Peleg, P. Raghavan, and E. Upfal. Randomized broadcast in networks. *Random Structures and Algorithm*, I(4):447–460, 1990.
11. P. Fragopoulou and S. Akl. Edge-disjoint spanning trees on the Star network with applications to fault tolerance. *IEEE Trans. Computers*, 45:174–185, 1996.
12. A. Frieze and M. Molloy. Broadcasting in random graphs. *Discrete Applied Mathematics 54*, pages 77–79, 1994.
13. T. Hagerup and C. Rüb. A guided tour of chernoff bounds. *Information Processing Letters*, 36(6):305–308, 1990.
14. L. Harper. Optimal assignment of numbers to vertices. *J. Soc. Ind. Appl. Math.*, 12:131–135, 1964.
15. J. Jwo, S. Lakshmivarahan, and S. Dhall. Characterization of node disjoint (parallel) path in Star graphs. In *Proc. of the 5th Intl. Parallel Processing Symp.*, pages 404–409, 1991.
16. R. Karp, C. Schindelhauer, S. Shenker, and B. Vöcking. Randomized rumor spreading. *Proc. of FOCS'00*, pages 565–574, 2000.
17. M. Nigam, S. Sahni, and N. Yeh. Embedding hamiltonians and hypercubes in Star interconnection graphs. In *Proc. of the Intl. Conf. on Parallel Processing*, volume III, pages 340–342, 1990.
18. S. Ranka, J.-C. Wang, and N. Yeh. Embedding meshes on the Star graph. *Journal of Parallel and Distributed Computing*, 19:131–135, 1993.
19. J. Sheu, C. Wu, and T. Chen. An optimal broadcasting algorithm without message redundancy in Star graphs. *IEEE Trans. Computers*, 45:174–185, 1996.

Finding Disjoint Paths on Directed Acyclic Graphs

Torsten Tholey

Institut für Informatik, Universität Augsburg,
D-86135 Augsburg, Germany
tholey@informatik.uni-augsburg.de

Abstract. Given $k+1$ pairs of vertices $(s_1, s_2), (u_1, v_1), \ldots, (u_k, v_k)$ of a directed acyclic graph, we show that a modified version of a data structure of Suurballe and Tarjan can output, for each pair (u_l, v_l) with $1 \le l \le k$, a tuple (s_1, t_1, s_2, t_2) with $\{t_1, t_2\} = \{u_l, v_l\}$ in constant time such that there are two disjoint paths p_1, from s_1 to t_1, and p_2, from s_2 to t_2, if such a tuple exists. Disjoint can mean vertex- as well as edge-disjoint. As an application we show that the presented data structure can be used to improve the previous best known running time $O(mn)$ for the so called 2-disjoint paths problem on directed acyclic graphs to $O(m(\log_{2+m/n} n) + n \log^3 n)$. In this problem, given a tuple (s_1, s_2, t_1, t_2) of four vertices, we want to construct two disjoint paths p_1, from s_1 to t_1, and p_2, from s_2 to t_2, if such paths exist.

1 Introduction

The problem of finding disjoint paths is one of the fundamental problems in graph theory with many applications concerning network reliability, routing problems, VLSI-design, ... Such problems have been studied extensively and a variety of efficient algorithm are known for undirected graphs (cf. [1] and [2]), whereas much less is known about finding disjoint paths on directed graphs.

Previous results. Given $2k$ vertices $s_1, \ldots, s_k, t_1, \ldots, t_k$, one simple path finding problem consists of determining k disjoint paths p_i ($i \in \{1, \ldots, k\}$) between the vertices $\{s_1, \ldots, s_k\}$ and $\{t_1, \ldots, t_k\}$ with p_i leading from s_i to $t_{\pi(i)}$ such that π is a permutation of the numbers $1, \ldots, k$. This problem can be solved with standard network flow techniques for directed as well as for undirected graphs and for both, vertex- and edge-disjoint paths. For fixed $k \in \mathbb{N}$, this leads to a running time of $O(m + n)$, where here and in the following m will denote the number of edges and n the number of vertices of the graph under consideration.

For undirected graphs and $k \in \{2, 3\}$, Di Battista, Tamassia, and Vismara [1] have shown that allowing a preprocessing time of $O(m + n)$ (if $k = 2$) or $O(n^2)$ (if $k = 3$) one can construct a data structure that can test the existence of k vertex-disjoint paths between each pair of two vertices in constant time and output k such paths, if they exist, in a time linear in the number of the edges visited by these paths. Di Battista, Tamassia, and Vismara also gave an overview

D. Kratsch (Ed.): WG 2005, LNCS 3787, pp. 319–330, 2005.

over other data structures supporting the above queries for $k \geq 4$. For results concerning edge-disjoint paths between pairs of vertices, we refer the reader to the paper of Dinitz and Westbrook [2].

For a directed graph $G = (V, E)$ and a fixed vertex $s \in V$, Suurballe and Tarjan [13] presented a data structure with a preprocessing time of $O(n + m \log_{2+m/n} n)$ which, for each $t \in V$, can test in constant time whether there are two disjoint paths from s to t, and, if so, can output such paths in linear time. The result holds for both, vertex- and edge-disjoint paths.

Another interesting paths finding problem is the k-disjoint paths problem. In this problem we are given a tuple $(s_1, t_1, \ldots, s_k, t_k)$ of $2k$ vertices and we want to construct k disjoint paths p_i $(1 \leq i \leq k)$, from s_i to t_i. For short, we will refer to this problem as the k-DPP or, more precisely, as k-VDPP, if disjoint means vertex-disjoint, and as k-EDPP, if disjoint means edge disjoint.

The first polynomial time algorithms for the k-VDPP on undirected graphs where given by Ohtsuki [6], Seymour [11], Shiloach [12], and Thomassen [16], for $k = 2$, and by Robertson and Seymour [9] for general but fixed k. With the line-graph reduction described by Perl and Shiloach in [8] the k-EDPP can also be solved in polynomial time. If we let α be the inverse Ackerman function as defined in [14], the currently best known time bounds for the k-DPP on undirected graphs, are $O(m\alpha(m, n) + n)$ for the 2-VDPP, $O(m\alpha(m, n) + n \log n)$ time for the 2-EDPP as shown by the author of this paper in [15], and $O(mn^2)$ time for the k-VDPP with fixed $k > 2$, and $O(m^2 n^2)$ for the k-EDPP with fixed $k > 2$ as shown by Perković and Reed in [7].[1]

For directed graphs, the decision versions of the k-EDPP and the k-VDPP are $\mathcal{NP}$-complete, even for $k = 2$, as shown by Fortune, Hopcroft, and Wyllie [3]. However, in [8] Perl and Shiloach presented an $O(mn)$-time algorithm for solving the 2-VDPP and the 2-EDPP on dags (directed acyclic graphs). Fortune, Hopcroft, and Wyllie [3] generalized this result of Perl and Shiloach to an $O(mn^{k-1})$-time algorithm for the k-VDPP on dags for all $k \geq 2$. Lucchesi and Giglio [5] described a linear time reduction from the decision version of the 2-VDPP on dags to the decision version of the 2-VDPP on undirected graphs, such that there is always a solution of the 2-VDPP on the undirected graph after the reduction, if this graph is non-planar. Since Perl and Shiloach [8] have shown that the 2-VDPP on undirected planar graphs is solvable in linear time, the decision version of the 2-VDPP on dags is also solvable in linear time. Finally, applying the reduction from the 2-EDPP on dags to the 2-VDPP on dags given in [15] there is an $O(n + m \log_{2+m/n} n)$ time algorithm for solving the decision version of the 2-EDPP on dags. As an application of the k-EDPP on dags, Schrijver [10] described an airplane routing problem that can be solved with an algorithm for the k-EDPP on dags.

New results. In some scenarios, given a tuple (s_1, s_2, t_1, t_2) of vertices, apart from testing whether there are two disjoint paths leading from the vertices

---

[1] For the last two results we also use the line graph reduction from the k-EDPP to the k-VDPP as well as a reduction from the decision version to the general version of the k-DPP that increases the running time by factor m.

in $\{s_1, s_2\}$ to the vertices in $\{t_1, t_2\}$ we might also be interested in knowing whether the path starting in s_1 leads to t_1 or t_2 without constructing such paths. Given $k+1$ pairs of vertices $(s_1, s_2), (u_1, v_1), \ldots, (u_k, v_k)$ of a directed graph, we present in Section 3 a modified version of a data structure of Suurballe and Tarjan which can output, for each pair (u_l, v_l) with $1 \le l \le k$, a tuple (s_1, t_1, s_2, t_2) with $\{t_1, t_2\} = \{u_l, v_l\}$ in constant time such that there are two vertex- or, alternatively, edge-disjoint paths p_1, from s_1 to t_1, and p_2, from s_2 to t_2, if such a tuple exists. This data structure can be constructed in $O((m+k)(\log_{2+(m+k)/(n+k)} n) + n \log^2 n)$ time.

As an application of this data structure and main result of this paper, extending some ideas of Lucchesi and Giglio [5] concerning a reduction for the decision version of the 2-VDPP, we show that it can be used to improve the running time for the 2-VDPP on dags from $O(mn)$ to $O(m(\log_{2+m/n} n) + n \log^3 n)$ time. Applying the reduction from the 2-EDPP to the 2-VDPP given in [15] results in an $O(m(\log_{2+m/n} n) + n \log^3 n)$ time algorithm for the 2-EDPP on dags.

2   Preliminaries

Paths referred to in this paper are always simple paths, i.e. paths on which no vertex appears more often than once. If a vertex v or an edge e is visited by a path p, we write $v \in p$ or $e \in p$. For a path p and vertices $a, b \in p$, we let $p[a, b]$ be the sub-path of p from a to b. $p(a, b], p[a, b)$, and $p(a, b)$ will denote the sub-paths of $p[a, b]$ starting in the vertex visited immediately after a, or ending in the vertex visited immediately before b, or both, respectively. The *length* of a path p is the number of edges visited by p and denoted by $|p|$. Finally, for two paths p_1 and p_2, $p_1 \circ p_2$ is the concatenation of the two paths.

As for paths, given a tree $T = (V, E)$ and a vertex v or an edge e, we write $v \in T$ if $v \in V$ and $e \in T$ if $e \in E$. $f_T(v)$ denotes the father of v in T.

A *topological numbering* τ of the vertices of a dag $G = (V, E)$ is an injective mapping from V to $\{1, \ldots, n\}$ such that for each pair (v, w) of vertices for which there is a path from v to w, $\tau(v) < \tau(w)$ holds. It is well known that for each dag G a topological numbering can be computed in linear time.

3   Finding Disjoint Paths Between Pairs of Vertices

Suurballe and Tarjan presented in [13] a data structure which, given a directed graph $G = (V, E)$ and a fixed vertex $s \in V$, for each vertex v, can test the existence of two disjoint paths from s to v in constant time. This data structure can be constructed in $O(n + m \log_{2+m/n} n)$ time. It consists of a shortest-path tree T with source node s and stores with each vertex $v \in V$ two vertices $p(v)$ and $q(v)$ which on dags have the following properties:

1. If τ is a topological numbering of the vertices of V, then, for each $v \in V$ with two edge-disjoint paths from s to v, $\tau(q(v)) < \tau(v)$ and $(p(v), v) \in E$.

2. If there are two edge-disjoint paths from s to v, then there are also two edge-disjoint paths from s to $q(v)$.

3. Two edge-disjoint paths p_1 and p_2 from s to v, if they exist, can be constructed in $O(|p_1| + |p_2|)$ time as follows:

In a first round, mark v and, beginning in v with each marked vertex x, also mark $q(x)$ until reaching s. This process must stop because of property 1. In a second round p_1 is constructed in reverse direction starting in v and, when reaching a vertex x, following edge $(p(x), x)$ in reverse direction if x is marked, or, if it is not, following edge $(f_T(x), x)$ in reverse direction. Moreover, when visiting a marked vertex x, un-mark x. In a third round p_2 is constructed in the same way as p_1 following $(p(x), x)$ and un-marking x, if x is marked, and following $(f_T(x), x)$, if x is not marked.

Suurballe and Tarjan also observed that the construction of p_1 and p_2 un-marks all vertices marked in the first round of the construction. This guarantees that prior to the construction of a further pair of disjoint paths no vertex in our graph is marked. Note that we do not claim that property 3 follows immediately from the first two other properties. We only claim that the values $p(v)$ and $q(v)$ computed by the data structure of Suurballe and Tarjan have the above three properties.

Let $G' = (V', E')$ be the graph obtained from a dag G by replacing each vertex $v \in V$ with two vertices v_1 and v_2 and each edge (u, v) with an edge (u_2, v_1) and by adding new edges (v_1, v_2) for every $v \in V$. Then, there are two internally vertex-disjoint paths from a vertex $s \in V$ to a vertex $t \in V$ in G, if, and only if, there are two edge-disjoint paths from s_2 to t_1 in G'. Hence, the data structure of Suurballe and Tarjan can be also used to test the existence of two vertex-disjoint paths of a dag $G = (V, E)$ and to construct such paths p_1 and p_2 in $O(|p_1| + |p_2|)$ time.

In this section we want to show:

Lemma 1. *Let $G = (V, E)$ be a dag. Then, given $k + 1$ pairs of vertices (s_1, s_2), $(u_1, v_1), \ldots, (u_k, v_k)$ it is possible to construct in $O((m+k)(\log_{2+(m+k)/(n+k)} n) + n \log^2 n)$ time a data structure that can output, for each pair (u_l, v_l) with $1 \le l \le k$ a tuple (s_1, t_1, s_2, t_2) with $\{t_1, t_2\} = \{u_l, v_l\}$ in constant time such that there are two disjoint paths p_1, from s_1 to t_1, and p_2, from s_2 to t_2, if such a tuple exist. The paths themselves can be output in $O(|p_1| + |p_2|)$ time.*

In our proof of Lemma 1 disjoint means edge-disjoint, but with the previous reduction it also holds for vertex-disjoint paths.

Proof. Let $G' = (V', E')$ be the graph obtained by adding vertices $s, w_1, \ldots, w_k$ and edges $(s, s_1), (s, s_2), (u_1, w_1), (v_1, w_1), \ldots, (u_k, w_k), (v_k, w_k)$ to G. Then our problem reduces to the problem of determining a data structure able to output, for each $i \in \{1, \ldots, k\}$, a tuple (s_1, y_1, s_2, y_2) such that there are disjoint paths p_1 and p_2 from s to w_i with p_j $(j \in \{1, 2\})$ using (s, s_j) as first and (y_j, w_i) as last edge, if such a tuple exists.

We start with constructing in $O(n + (m+k) \log_{2+(m+k)/(n+k)} n)$ time the data structure of Suurballe and Tarjan for graph G' with s as fixed source node and we define $T, p,$ and q to be the shortest-path tree and the mappings constructed by this data structure with the properties described at the beginning of this section. Moreover, we determine in $O(n)$ time a tree T' consisting of all vertices $v \in V'$ for which there are two disjoint paths from s to v, s being the root of T', and $f_{T'}(v) = q(v)$ for all $v \in T'$.

In the following, for each $v \in T'$, let $p_1(v)$ and $p_2(v)$ be the two disjoint paths from s to v which would be constructed by Suurballe's and Tarjan's data structure or, more precisely, $p_1(v)$ should be the path visiting $(p(v), v)$ as last edge, and $p_2(v)$ should be the path visiting $(f_T(v), v)$ as last edge. Moreover, for $i \in \{1, 2\}$, we define $r_i(v)$ to be first vertex visited after s on $p_i(v)$.

We now try to determine a lookup table containing the vertices $r_1(v)$ for all $v \in V$ (hence, $r_2(v)$ is the vertex $w \in \{s_1, s_2\}$ with $w \neq r_1(v)$). We start with a depth-first-search in T' and when visiting a vertex y, we colour the vertices of T such that all vertices $x \neq s$ on the tree path from s to y in T' are coloured black in T if $p_1(x)$ starts with edge (s, s_1), whereas, if $p_1(x)$ starts with edge (s, s_2), x is coloured red. All other vertices of T should be coloured white. In other words, if x is coloured black, we have $r_1(x) = s_1$, whereas, if x is coloured red, we have $r_1(x) = s_2$. Note that the red or black coloured vertices are exactly the vertices marked before the construction of $p_1(y)$ and $p_2(y)$.

Suppose our depth-first-search reaches a child y of s in T'. For constructing two disjoint paths from s to y with the data structure of Suurballe and Tarjan in the first round of the construction process only y and s are to be marked. Hence, it follows from the construction process described above that $r_1(y)$ is equal to y if $p(y) = s$, and equal to the first vertex $z \neq s$ on the tree path from s to $p(y)$ in T, if $p(y) \neq s$ (z can be determined in constant time if in a preprocessing step taking $O(m)$ time we determine for each $v \in T$ the first vertex $\neq s$ on the tree path from s to v in T). Hence, we know how to colour y correctly.

When reaching a vertex y not equal to a child of s in T', we will determine the last red or black coloured vertex $x \neq s$ before $p(y)$ on the tree path from s to $p(y)$ in T. Note that the ancestors of y in T' are exactly the vertices that would be marked by the data structure of Suurballe and Tarjan in the first round of constructing two disjoint paths from s to y and that all these nodes have already been coloured black or red by the depth-first-search in T'. If no red or black coloured vertex exists on the tree path from s to $p(y)$ in T, we know from the construction process of path $p_1(y)$, that $p_1(y)$ between s and $p(y)$ follows the tree path from s to $p(y)$ in T. Hence, y should be coloured black if (s, s_1) is the first edge on the path from s to $p(y)$ in T, and, if (s, s_2) is the first edge on this path, y should be coloured red. If x exists, from the properties of the data structure of Suurballe and Tarjan given at the beginning of this section it follows that $p_1(y)[s, y] = p_1(x)[s, x] \circ T[x, p(y)] \circ (p(y), y)$, where $T[x, p(y)]$ denotes the tree path from x to $p(y)$ in T (note that, if τ is a topological numbering of the vertices in G', the vertices that would be marked before the construction of two disjoint paths from s to x by the data structure of Suurballe and Tarjan are exactly the vertices v with $\tau(v) \leq \tau(x)$ that would by marked before the construction of disjoint paths from s to y and that $p_1(x)$ visits only vertices v with $\tau(v) \leq \tau(x)$). Hence, if by induction we have already shown that all ancestors of y in T' are coloured correctly, then y is also coloured correctly by colouring it in the same colour as x.

For the computation of the last coloured vertex x on the tree path from s to a vertex y in T, we maintain two copies T_1 and T_2 of our shortest-path tree T. We delete all black and red coloured vertices from T_1, as well as all black coloured vertices from T_2. Let y' be the vertex that appears in the middle of the tree path from s to y in T (with an appropriate encoding of the vertices of T, y' can be computed in constant time). We then ask whether y is reachable from y' in T_1. If so, x does not exist or lie on the tree path from s to y' in T. Otherwise, our search can be reduced to the tree path from y' to y in T. In other words, x can be determined by a binary search. We can also identify the colour of x by testing whether y is reachable from $f_T(x)$ in T_2. We use the dynamic data structure of Holm, de Lichtenberg, and Tarjan [4] for updating T_1 and T_2 and for answering our connectivity queries. This data structure allows us to delete a vertex with r adjacent edges or to reinsert such a vertex in $O(r \log^2 n)$ amortized time and to decide whether two vertices are connected in $O(\log n / \log \log n)$ worst case time.

Since our algorithm consists of $O(n)$ deletions of vertices and (adjacent) edges, $O(n)$ reinsertions, and only $O(n \log n)$ queries for determining the vertices $r_1(v)$ for all $v \in V$, the construction time of our data structure is bounded by $O((m+k)(\log_{2+(m+k)/(n+k)} n) + n \log^2 n)$. For each $v \in V$, $p_1(v)$ and $p_2(v)$ can be output with the data structure of Suurballe and Tarjan in $O(|p_1(v)| + |p_2(v)|)$ time.   □

4   Solving the 2-VDPP on Dags

In this section we present an $O(m(\log_{2+m/n} n) + n \log^3 n)$-time algorithm for solving the 2-VDPP on dags. In the following, disjoint means always vertex-disjoint.

Let us call an instance $I = (G, s_1, s_2, t_1, t_2)$ of the 2-VDPP on a dag $G = (V, E)$ to be *irreducible* if the in-degree of each vertex $v \in V - \{s_1, s_2\}$ and the out-degree of each vertex $v \in V - \{t_1, t_2\}$ is at least two, and if t_1, t_2 have no outgoing and s_1, s_2 no incoming edges. On irreducible instances the following lemma holds:

Lemma 2. *Let (G, s_1, s_2, t_1, t_2) be an irreducible instance of the 2-VDPP on a dag $G = (V, E)$. Then, for each pair $v, w \in V$ with $v \neq w$, there are two disjoint paths p_1 and p_2 such that p_i ($1 \leq i \leq 2$) leads from a vertex in $\{v, w\}$ to a vertex in $\{t_1, t_2\}$ as well as two disjoint paths leading from $\{s_1, s_2\}$ to $\{v, w\}$.*

Corollary 3 (Thomassen [17]). *If (G, s_1, s_2, t_1, t_2) is an irreducible instance of the 2-VDPP on a dag $G = (V, E)$, then, for each vertex $v \in V - \{s_1, s_2, t_1, t_2\}$, there exist four paths p_1 from s_1 to v, p_2 from s_2 to v, p_3 from v to t_1, and p_4 from v to t_2 such that the only vertex visited by more than one of the paths is v.*

As observed by Thomassen [17], given an algorithm for solving the 2-VDPP on irreducible instances in $T(m, n)$ time, the 2-VDPP on dags can be solved in

$O(T(m,n)+m+n)$ time. Hence, in the following, we let $I = (G, s_1, s_2, t_1, t_2)$ be an irreducible instance of the 2-VDPP on a dag $G = (V, E)$. Moreover, we define $U(G)$ to be the undirected graph obtained from G by replacing each directed edge (u, v) of G with an undirected edge $\{u, v\}$.

Let us first describe how the original algorithm of Lucchesi and Giglio finds two disjoint paths solving the 2-VDPP. In a first step it determines two disjoint paths p_1, from s_1 to t_1, and p_2, from s_2 to t_2, in $U(G)$. Like Lucchesi and Giglio, for two consecutive edges (u, v) and (v, w) on p_1 or p_2, let us refer to v as a *switch* if either both edges (u, v) and (w, v) are part of E (i.e. $(v, u), (v, w) \notin E$, since G is a dag), or $(v, u), (v, w) \in E$. Lucchesi and Giglio [5] proved that there is a choice of four vertices u, u', v, and v' in the following also called *boundary vertices* and of four paths r_1, r_2, q_1, q_2 in G such that p_1 and p_2 depending on the positions of u, u', v, v' can be replaced by one of the four pairs of paths given in the left column of Table 1 such that the resulting paths are disjoint (ignore the other columns of this table). Moreover, in Lucchesi's and Giglio's algorithm u can be chosen as the switch with the smallest and v as the switch with the largest topological number among all switches on p_1 and p_2. This guarantees that in each replacement of Table 1 replacing sub-paths of p_1 and p_2 by sub-paths of q_1 and q_2 the vertex u is not a switch of the new paths p_1^* and p_2^* and in all other cases v is not a switch of p_1^* and p_2^*. u' and v' are chosen in such a way that they are not switches of p_1 and p_2 neither before nor after the replacement. Being paths in G the paths q_1, q_2, r_1, and r_2 cannot contain any switch. Consequently, the set of switches of p_1^* and p_2^* is a proper subset of the switches of p_1 and p_2 before the replacement. Therefore, after $O(n)$ replacements as shown in Table 1 the resulting paths can no longer contain any switch and they solve the 2-VDPP. Since the running time for identifying the vertices u, u', v, and v' and the construction of the paths r_1, r_2, q_1 and q_2 is bounded by $O(m)$, Lucchesi's and Giglio's algorithm runs in $O(mn)$ time.

The main idea of the algorithm of this paper is to choose the vertices u, u', v, and v' much more carefully such that after each replacement at least a constant fraction of the switches of p_1 and p_2 are removed. This would reduce the number of replacements from $O(n)$ to $O(\log n)$. Unfortunately, this approach will not always be successful. In some sub-cases we will not be able to reduce the number of switches by a constant fraction. However, in all these cases using the data structure presented in Section 3 we will be able to guess the boundary vertices of the next sub-rounds without constructing the paths r_1, r_2, q_1, q_2. This will reduce the running time between two replacements which remove a constant fraction of switches to $O(m \log^2 n)$ time.

Let us now describe our new algorithm for the 2-VDPP on dags. Like Lucchesi and Giglio we start with the construction of two disjoint paths p_1, from s_1 to t_1, and p_2, from s_2 to t_2, in $U(G)$. This can be done in $O(m\alpha(m, n))$ time (cf. [15]). The remaining part of the algorithm is divided into several rounds.

Let us describe what is done in each round. For $i \in \{1, 2\}$, let n_i be the number of switches on p_i at the beginning of the round, let c_i be the vertex on p_i visited immediately after the $\lfloor \frac{1}{4} n_i \rfloor$-th switch of p_i and let d_i be the vertex

Table 1. The path replacements in the different sub-cases

Sub-case	Description: For i,j with $\{1,2\}=\{i,j\}$	Replacements
1a	$v\in p_i, v'\in r_j$	$p_i^* := p_i[s_i,v]\circ r_i[v,t_i]$
2a	$v\in p_i, v'\in r_j$	$p_j^* := p_j[s_j,v']\circ r_j[v',t_j]$
1b.α	$u\in p_i[c_i,t_i], u'\in q_j$	$p_i^* := q_i[s_i,u]\circ p_i[u,t_i]$
1b.β	$u\in p_i[s_i,c_i), u'\in q_j$	$p_j := q_j[s_j,u']\circ p_j[u',t_j]$
2b	$u\in p_i, u'\in q_j$	
1c.α	$u,v\in p_i, u'\in q_i, v'\in r_i, u\notin p_i(d_i,t_i]$	$p_i^* := q_i[s_i,u']\circ p_j[u',v']\circ r_i[v',t_i]$
1c.β	$u,v\in p_i, u'\in q_i, v'\in r_i, u\in p_i(d_i,t_i]$	$p_j^* := q_j[s_j,u]\circ p_i[u,v]\circ r_j[v,t_j]$
2c	$u,v\in p_i, u'\in q_i, v'\in r_i$	
1d	$u\in p_i, v\in p_j, u'\in q_i, v'\in r_j$	$p_i^* := q_i[s_i,u']\circ p_j[u',v]\circ r_i[v,t_i]$
2d	$u\in p_i, v\in p_j, u'\in q_i, v'\in r_j$	$p_j^* := q_j[s_j,u]\circ p_i[u,v']\circ r_j[v',t_j]$

on p_i visited immediately before $(n_i - \lfloor\frac{1}{4}n_i\rfloor)$-th switch of p_i. If $\lfloor\frac{1}{4}n_i\rfloor = 0$, let $c_i := s_i, d_i := t_i$.

Let τ be a topological numbering of the vertices of G. Like Lucchesi and Giglio in [5] we define v to be the switch with largest topological number on p_1 and p_2, but unlike Lucchesi and Giglio we let v' be the first vertex x with $\tau(x) > \tau(v)$ on the path p_1 or p_2 not visiting v. We distinguish between Case 1, where $v \in p_1[s_1,c_1)$ or $v \in p_2[s_2,c_2)$, and Case 2, where $v \in p_1[c_1,t_1]$ or $v \in p_2[c_2,t_2]$. In Case 1, we define u to be the switch with the lowest topological number on p_1 or p_2, whereas in Case 2, unlike Lucchesi and Giglio, we let u be the switch on $p_1[c_1,t_1]$ or $p_2[c_2,t_2]$ with the smallest topological number. In both cases we let u' be the last vertex x with $\tau(x) < \tau(u)$ on the path p_1 or p_2 not visiting u. We define q_1 and q_2 to be disjoint paths from s_1 and s_2 to u and u' such that q_i starts in s_i $(1 \leq i \leq 2)$, and, similarly, we let r_1 and r_2 be disjoint paths from v and v' to t_1 and t_2 such that r_i ends in t_i $(1 \leq i \leq 2)$. These paths exist because of Lemma 2.

We consider different sub-cases and replace p_1 and p_2 with two paths p_1^* and p_2^* as shown in Table 1. For $i \in \{1,2\}$, sub-cases with prefix number i should be sub-cases of Case i. The new paths are disjoint:

Lemma 4. p_1^* and p_2^* are disjoint.

Proof. p_1^* and p_2^* are disjoint: For the Cases 1a, 2a, 1b.α, 1b.β, and 2.b this follows from the fact that the remaining sub-paths of p_1 and p_2 used for the construction of p_1^* and p_2^* apart from u' and v' visit only vertices x with $\tau(x) \leq \tau(v)$ (Cases 1a, 2a) or only vertices x with $\tau(x) \geq \tau(u)$ (Cases 1b.α, 1b.β, 2b). Let p_1' and p_2' be the sub-paths of p_1 and p_2, respectively, that were used for the construction of p_1^* and p_2^* in one of the remaining cases. Then the disjointness from p_1^* and p_2^* in the remaining cases follows if we can show that $\tau(u) \leq \tau(x) \leq \tau(v)$ holds for all $x \in p_1'$ and all $x \in p_2'$ with $x \notin \{u',v'\}$. It is easy to see that this holds if, for each ordered pair of vertices $(x,y) \in \{(u,v'),(u',v),(u',v')\}$ with $x,y \in p_i'$ for an $i \in \{1,2\}$, x appears before y on p_i'. The latter statement is true since v' must appear after the last switch on p_1 or p_2, whereas u', in Case 1, must appear before the first switch on p_1 or p_2, and, in Case 2, must appear before the first switch on $p_1[c_1, t_1]$ or $p_2[c_2, t_2]$, and, therefore, before v or v' on p_1 or p_2.   □

After the path replacements of Table 1, u and v as vertices with the smallest or largest topological number can no longer be switches of p_1 or p_2. Unfortunately, u and v may be the only switches deleted from p_1 and p_2 in the Cases 1b.β, 1c.β, or 2a. Therefore, in these cases the idea is to consider not only one round but a series of k rounds such that in the first $k-1$ rounds we are in one of the Cases 1b.β, 1c.β, or 2a, and in the last round we are in one of the other cases.

We will from now on consider the k rounds as exactly one round sometimes also called *super-round* and the k rounds as *sub-rounds* of this round. For a simpler implementation we will not update the vertices c_i and c_j, after each of the first $k-1$ sub-rounds. There is one exception: In a sub-round corresponding to Case 1c.β we replace c_i with d_i and d_i with c_i (since p_i after the replacement visits the vertices between c_i and d_i in reverse direction).[2]

The k-th sub-round then guarantees that enough switches are being removed from p_1 and p_2 in each super-round. More precisely, from Table 1 we can conclude that after each round (super-round in Case 1b.β, 1c.β, or 2a) at least $1 + \min\{\lfloor \frac{1}{4}n_1 \rfloor, \lfloor \frac{1}{4}n_2 \rfloor\}$ switches (or $1 + \max\{\lfloor \frac{1}{4}n_1 \rfloor, \lfloor \frac{1}{4}n_2 \rfloor\}$ switches if $n_1 = 0$ or $n_2 = 0$) are removed from p_1 and p_2: Apart from u and v, in the Cases 1a, 1c.α, and 1d at least all switches of $p_1(d_1, t_1]$ or $p_2(d_2, t_2]$ and in the Cases 1b.α, 2b, 2c, and 2d at least all switches of $p_1[s_1, c_1)$ or $p_2[s_2, c_2)$ are removed (for the Cases 1d and 2d note that, as shown in the proof of Lemma 4, u appears before v' on p_i and u' before v on p_j). Thus, our algorithm terminates after $O(\log n)$ rounds with two disjoint paths p_1, from s_1 to t_1, and p_2, from s_2 to t_2. Each round can be implemented efficiently:

Lemma 5. *Each round has a running time of $O(m(\log_{2+m/n} n) + n \log^2 n)$.*

Proof. For each round $n_1, n_2, c_1, c_2, d_1, d_2$ and therefore the boundary vertices u, u', v, v' (of the first sub-round in the case of a super-round) can be computed in $O(n)$ time. With standard network flow techniques two disjoint paths from s_1 and s_2 to u and u' as well as two disjoint paths from v and v' to t_1 and t_2 can be computed in $O(m)$ time. Given these paths, it is easy to decide in which case we are and to implement the path replacements for the Cases 1a, 1b.α, 1c.α, 1d, 2b, 2c, and 2d, again in $O(m)$ time.

We now consider the time complexity of the Cases 1b.β, 1c.β, and 2a. When talking about p_1 and p_2 at the beginning of the l-th sub-round or the boundary vertices in the l-th sub-round we denote them by $p_1^l, p_2^l, u^l, u'^l, v^l$, or v'^l, respectively. If we mean the paths after the last sub-round we write p_1^{k+1} and p_2^{k+1}.

Let us define the *original part* of p_1^l and p_2^l to be the part of p_1^l and p_2^l that is equal to the corresponding part of p_1^1 or p_2^1. More precisely, if before the first sub-round we mark all edges of p_1^1 and p_2^1 and in the j-th sub-round when replacing

---

[2] More precisely, if one of the vertices c_1, c_2, d_1, d_2 does no longer exist on these paths, we know that the replacement by which it was removed resulted in the deletion of at least a constant fraction of all switches from the paths given in the first sub-round and the corresponding sub-round can be defined as the last sub-round.

p_1^j and p_2^j with p_1^{j+1} and p_2^{j+1} we un-mark all edges not lying on the sub-paths of p_1^j and p_2^j used for the construction of p_1^{j+1} and p_2^{j+1}, then the original part of p_1^l (p_2^l) is the sub-path of p_1^l (p_2^l) consisting of the marked edges.

We next want to show that the boundary vertices of each sub-round must lie on the original parts of the paths given in this sub-round. For u^l and v^l ($1 \leq l \leq k$), this is true since all switches of p_1^l and p_2^l lie on the original part.

For $l \in \{1, \ldots, k\}$, let us define numbers a_l and b_l such that the a_l-th sub-round is the last sub-round before the l-th sub-round corresponding to Case 1b.β or 1c.β and the b_l-th sub-round is the last sub-round before the l-th sub-round corresponding to Case 2a or 1c.β (a_l or b_l should be 0 if no such sub-round exists). Then by induction one can show that the endpoints of the original parts of p_1^l and p_2^l consist of the vertices $u^{a_l}, u'^{a_l}, v^{b_l}$, and v'^{b_l}, where we define $u^0 = s_1, u'^0 = s_2, v^0 = t_1$, and $v'^0 = t_2$. Moreover, again by induction one can show that $\tau(u^{a_l}) \leq \tau(x) \leq \tau(v^{b_l})$ holds for all vertices $x \notin \{u'^{a_l}, v'^{b_l}\}$ on the original parts of p_1^l and p_2^l. Now, from $\tau(u'^{a_l}) < \tau(u^{a_l}) \leq \tau(u^l) \leq \tau(v^l) \leq \tau(v^{b_l}) < \tau(v'^{b_l})$ we can conclude that the vertices u^l and v^l must appear after a vertex $x \in \{u^{a_l}, u'^{a_l}\}$ on p_1^l or p_2^l or be equal to x and they must appear before a vertex $y \in \{v^{b_l}, v'^{b_l}\}$ on p_1^l or p_2^l or be equal to y. Therefore, u^l and v^l lie on the original part of p_1^l or p_2^l. We can use the knowledge that the boundary vertices always lie on the original part of p_1 or p_2 which always is a sub-path of p_1^1 or p_2^1 for an efficient computation of the boundary vertices:

Knowing the original parts of p_1 and p_2 for each sub-round, we can easily compute v^l for all $l \in \{1, \ldots, k\}$ if, before the first sub-round, we construct in $O(n)$ time a list of all switches on p_1 and p_2 sorted by their topological numbers. We then repeatedly delete the vertex with the largest topological number from this list until we find a vertex x lying on the original part of p_1 or p_2. We always start the search with the last vertex deleted in the previous sub-round. Hence, the time needed to compute the boundary vertex v taken over all sub-rounds is bounded by $O(n)$, and, similarly, this also holds for the boundary vertex u.

We now describe the computation of u'^l and v'^l for $1 \leq l \leq k$: For each $i \in \{1, 2\}$, let us number the vertices visited by the paths p_i^1 in the order in which they appear on p_i^1 and let us construct a list of all switches of p_i^1 sorted by these numbers. Using these lists we can identify the last switch of p_i^l by a binary search in $O(\log n)$ time. If $v^l \in p_j^l$ holds for $j \in \{1, 2\}$ with $j \neq i$, then v'^l is the first vertex x with a topological number larger than that of v^l on the part of p_i^l between the last switch of p_i^l and the endpoint of the original part of p_i^l appearing after the last switch. Since the vertices on this part of p_i^l are sorted by their topological numbers, v'^l can be determined by a further binary search again in $O(\log n)$ time. Hence, the time needed for the construction of the boundary vertices v'^l for all sub-rounds can be bounded by $O(n \log n)$ time and, similarly, this also holds for the computation of the boundary vertices u'^l.

Therefore, if we know for each sub-round which case is applicable, i.e. if we know the original parts of p_1^l and p_2^l of the following sub-round, we can efficiently compute u^l, u'^l, v^l, and v'^l for each sub-round. In order to determine the relevant case, the super-round is split into two phases. In the first phase, if in a sub-round

u and v lie on p_1 and p_2 in such a way that we might be in Case 1b.β, 1c.β, or 2a, we assume that we are in this case and, under this assumption, we compute the boundary vertices of the next sub-round. For example, if $v \in p_i[s_i, c_i)$ and $u \in p_j[s_j, c_j)$ with $i, j \in \{1, 2\}$ we assume that we are in Case 1b.β (note that we will never encounter more than one of the Cases 1b.β, 1c.β, and 2a).

After the first phase we construct in maximal $O((m + n)(\log_{2+(m+n)/2n} n) + n^2 \log n) = O(m(\log_{2+m/n} n) + n^2 \log n)$ time the data structure described in Lemma 1 with $(u_1, v_1), \ldots, (u_k, v_k)$ being equal to the pairs of boundary vertices (u, u') of each sub-round considered in the first phase of our super-round.

In the second phase, starting again with the first sub-round we use this data structure to determine for each pair (u^l, u'^l) a tuple (s_1, w_1, s_2, w_2) with $\{w_1, w_2\} = \{u^l, u'^l\}$ such that there are two disjoint paths q_1, from s_1 to w_1, and q_2, from s_2 to w_2, and in a similar way again using the data structure of Lemma 1 we can construct a tuple (x_1, t_1, x_2, t_2) with $\{x_1, x_2\} = \{v^l, v'^l\}$ such that there are two disjoint paths r_1, from x_1 to t_1, and r_2, from x_2 to t_2. We finally test whether we are in one of the Cases 1b.β, 1c.β, or 2a and, therefore, have correctly computed the boundary vertices of the next sub-round. If we are in one of the other cases we stop the computation of boundary vertices since we must be in the last sub-round of the super-round.

Concerning the paths p_1^{k+1} and p_2^{k+1} resulting from the last sub-round of our super-round, if in the last sub-round we are in one of the c- or d-Cases of Table 1, they consist of three pairs of disjoint paths: q_1 and q_2, from s_1 and s_2 to u^k and u'^k, r_1 and r_2, from v^k and v'^k to t_1 and t_2, and two sub-paths of the original parts of p_1^k and p_2^k. We can determine these paths from the data structure of Lemma 1 and from the paths p_1^1 and p_2^1 in $O(n)$ time. Even if in last sub-round we are in an a- or b-Case, we can construct p_1^{k+1} and p_2^{k+1} in $O(n)$ time. For details see the full version of this paper.   □

Theorem 6. *On dags the 2-VDPP is solvable in $O(m(\log_{2+m/n} n) + n \log^3 n)$ time.*

Proof. In a first step we reduce the problem to a dag G with $O(n)$ edges:

Lucchesi and Giglio [5] have shown that two disjoint paths from s_1 to t_1 and from s_2 to t_2 on a dag $G = (V, E)$ can be constructed from two disjoint paths p_1 and p_2 in $U(G)$ by replacing sub-paths of p_1 and p_2 by sub-paths of a set S of paths. If we add extra vertices x and y as well as four extra edges $(x, s_1), (x, s_2), (t_1, y)$, and (t_2, y) to G, S can be chosen arbitrarily as long as S consists of two disjoint paths from x to v as well as of two disjoint paths from v to y for every $v \in V$. Such paths must exist because of Corollary 3.

If we choose as paths from x to vertices $v \in V$ the paths that would be constructed by the data structure of Suurballe and Tarjan [13], these paths visit only edges of the shortest-path tree T and edges of the form $(p(w), w)$ with T and p being defined as in the beginning of Section 3. Consequently, the graph containing these $O(n)$ edges plus $O(n)$ edges for the construction of disjoint paths from vertices $v \in V$ to y, as well as the edges of p_1 and p_2 is a subgraph of G on which the 2-VDPP is solvable, but which consists of only $O(n)$ edges.

The running time for the reduction of our problem to a sparse graph with only $O(n)$ edges is dominated by the construction time of $O(m(\log_{2+m/n} n)+n\log^2 n)$ for the data structure of Suurballe and Tarjan. After the reduction two disjoint paths on $U(G)$ can be computed in $O(n\alpha(n,n))$ time [15]. The following $O(\log n)$ rounds run in $O(n\log^2 n)$ time (Lemma 5).   □

References

1. G. Di Battista, R. Tamassia, and L. Vismara, Output-sensitive reporting of disjoint paths, *Algorithmica* **23** (1999), pp. 302–340.
2. Y. Dinitz and J. Westbrook, Maintaining the classes of 4-edge-connectivity in a graph on-line, *Algorithmica* **20** (1998), pp. 242–276.
3. S. Fortune, J. Hopcroft, and J. Wyllie, The directed subgraph homeomorphism problem, *Theoret. Comput. Sci.* **10** (1980), pp. 111–121.
4. J. Holm, K. de Lichtenberg, and M. Thorup, Poly-logarithmic deterministic fully-dynamic algorithms for connectivity, minimum spanning tree, 2-edge, and biconnectivity, *J. ACM* **48** (2001), pp. 723–760.
5. C. L. Lucchesi and M. C. M. T. Giglio, On the irrelevance of edge orientations on the acyclic directed two disjoint paths problem, IC Technical Report DCC-92-03, Universidade Estadual de Campinas, Instituto de Computação, 1992.
6. T. Ohtsuki, The two disjoint path problem and wire routing design, Proc. Symposium on Graph Theory and Algorithms, Lecture Notes in Computer Science, Vol. 108, Springer, Berlin, 1981, pp. 207–216.
7. L. Perković and B. Reed, An improved algorithm for finding tree decompositions of small width, *International Journal of Foundations of Computer Science (IJFCS)* **11** (2000), pp. 365–371.
8. Y. Perl and Y. Shiloach, Finding two disjoint paths between two pairs of vertices in a graph, *J. ACM* **25** (1978), pp. 1–9.
9. N. Robertson and P. D. Seymour, Graph minors. XIII. The disjoint paths problem, *J. Comb. Theory, Ser. B*, **63** (1995), pp. 65–110.
10. A. Schrijver, A group-theoretical approach to disjoint paths in directed graphs, *CWI Quarterly* **6** (1993), pp. 257–266.
11. P. D. Seymour, Disjoint paths in graphs, *Discrete Math.* **29** (1980), pp. 293–309.
12. Y. Shiloach, A polynomial solution to the undirected two paths problem, *J. ACM* **27** (1980), pp. 445–456.
13. J. W. Suurballe and R. E. Tarjan, A quick method for finding shortest pairs of disjoint paths. *Networks* **14** (1984), pp. 325-336.
14. R. E. Tarjan and J. van Leeuwen, Worst-case analysis of set union algorithms, *J. ACM* **31** (1984), pp. 245–281.
15. T. Tholey, Solving the 2-disjoint paths problem in nearly linear time. Proc. 21st Annual Symposium on Theoretical Aspects of Computer Science (STACS 2004), Lecture Notes in Computer Science Vol. 2996, Springer-Verlag, Berlin, 2004, pp. 350–361.
16. C. Thomassen, 2-linked graphs, *Europ. J. Combinatorics* **1** (1980), pp. 371–378.
17. C. Thomassen, The 2-linkage problem for acyclic digraphs, *Discrete Math.* **55** (1985), pp. 73–87.

Approximation Algorithms for the Bi-criteria Weighted MAX-CUT Problem

Eric Angel, Evripidis Bampis, and Laurent Gourvès

LaMI, CNRS UMR 8042, Université d'Évry Val d'Essonne, France
{angel, bampis, lgourves}@lami.univ-evry.fr

Abstract. We consider a generalization of the classical MAX-CUT problem where two objective functions are simultaneously considered. We derive some theorems on the existence and the non-existence of feasible cuts that are at the same time near optimal for both criteria. Furthermore, two approximation algorithms with performance guarantee are presented. The first one is deterministic while the second one is randomized.

1 Introduction

Given an undirected graph $G = (V, E)$ with non-negative edge weights w_{ij}, the objective of the Maximum Cut problem (MAX-CUT) is to find a partition of the vertex set into two subsets S and $\overline{S}$, such that the sum of the weights of the edges having endpoints in different subsets is maximum. Formally, the weight of the cut $(S, \overline{S})$ to be maximized is given by

$$W(S, \overline{S}) = \sum_{i \in S, j \in \overline{S}} w_{ij}.$$

This well known combinatorial problem was shown to be **NP**-complete by Karp [10]. It has applications in many fields including VLSI circuit design and statistical Physics [5].

In this article, we study a *bi-criteria* version of the MAX-CUT problem. Formally, we are given an undirected graph $G = (V, E)$ and two distinct weighting functions. Each feasible cut is then evaluated with respect to these two criteria.

In general no feasible solution can meet optimality simultaneously for both criteria. However, a set of solutions which *dominate*[1] all the others (the so-called *Pareto curve*) always exists. Because of the complexity of the classical (mono-criterion) MAX-CUT problem, determining this Pareto curve is computationally problematic. Indeed, the bi-criteria MAX-CUT problem generalizes MAX-CUT. Moreover, the size of the Pareto curve, i.e. the number of non-dominated solutions, may be exponential.

Concerning *multi-criteria optimization* (see [6] for a recent book on the topic), three different approaches are often followed: the *budget approach*, the *Pareto*

---

[1] A solution x dominates another solution y if x is at least as good as y for all criteria and strictly better for at least one criterion.

D. Kratsch (Ed.): WG 2005, LNCS 3787, pp. 331–340, 2005.

curve approach and the *simultaneous approach*. In this article we follow the third one.

By taking as a reference an *ideal solution*, namely a (not necessarily feasible) cut which simultaneously maximizes all objective functions, one tries to compute a feasible cut which approximates this ideal solution with a performance guarantee on each criterion.

In this direction, Stein and Wein [13] considered a scheduling problem with two well studied criteria, namely the *makespan* and the *average weighted completion time*. They derived existence and non-existence theorems on schedules that are simultaneously near-optimal with respect to both objective functions. A series of recent papers follow this approach [12,2,1,3,4].

In this article, we follow the same approach for the bi-criteria MAX-CUT problem. The paper is organized as follows: A formal presentation of the problem is given in Section 2. Sections 3 and 4 are respectively devoted to a deterministic and a randomized bi-criteria approximation algorithm with performance guarantee. Finally, some outlooks and concluding remarks are given in Section 5.

2   Formalization and Notation

We are given an undirected graph $G = (V, E)$ where each edge $e \in E$ has a non-negative weight w_e and a non-negative length l_e. A solution $(S, \overline{S})$ is feasible if it constitutes a partition of V. An edge e belongs to a cut $(S, \overline{S})$, denoted by $e \in (S, \overline{S})$, if e links a vertex in S and a vertex in $\overline{S}$. The following objective functions, namely the total weight and the total length, are considered:

$$W(S, \overline{S}) = \sum_{e \in (S, \overline{S})} w_e \text{ and } L(S, \overline{S}) = \sum_{e \in (S, \overline{S})} l_e.$$

Let $(O, \overline{O})$ (resp. $(P, \overline{P})$) be a feasible cut which maximizes the total weight (resp. length). Let $(I, \overline{I})$ be an *ideal* (not necessarily feasible) cut such that:

$$W(I, \overline{I}) = W(O, \overline{O}) = OPTW \text{ and } L(I, \overline{I}) = L(P, \overline{P}) = OPTL.$$

The bi-criteria weighted MAX-CUT problem is then to find a feasible cut $(A, \overline{A})$ such that:

$$W(A, \overline{A}) \geq \alpha \, OPTW \text{ and } L(A, \overline{A}) \geq \beta \, OPTL$$

where $0 < \alpha \leq 1$ and $0 < \beta \leq 1$. An (α, β)-approximation algorithm outputs a solution which is simultaneously α-approximate on the first criterion (the total weight) and β-approximate on the second criterion (the total length).

3   A Deterministic Approximation Algorithm

Given a deterministic α-approximation algorithm **Al** for the mono-criterion weighted MAX-CUT problem, one can build an $(\alpha/2, \alpha/2)$-approximation algorithm for the bi-criteria weighted MAX-CUT problem. The algorithm called **Bi-Approx** follows:

Bi-Approx	
Input:	G and **Al**
Step 1:	Find $(S_1, \overline{S_1})$ with **Al** s.t. $W(S_1, \overline{S_1}) \geq \alpha.OPTW$
Step 2:	Find $(S_2, \overline{S_2})$ with **Al** s.t. $L(S_2, \overline{S_2}) \geq \alpha.OPTL$
Step 3:	Build $(S_3, \overline{S_3})$ s.t. $S_3 = (S_1 \cap S_2) \cup (\overline{S_1} \cap \overline{S_2})$
Step 4:	*If* $L(S_1, \overline{S_1}) \geq 0.5\, L(S_2, \overline{S_2})$
	Then Return $(S_1, \overline{S_1})$
	Else If $W(S_2, \overline{S_2}) \geq 0.5\, W(S_1, \overline{S_1})$
	Then Return $(S_2, \overline{S_2})$
	Else Return $(S_3, \overline{S_3})$

Theorem 1. Bi-Approx *is a deterministic* $(\alpha/2, \alpha/2)$-*approximation algorithm for the bi-criteria weighted* MAX-CUT *problem if* **Al** *is a deterministic* α-*approxima-tion algorithm for the mono-criterion weighted* MAX-CUT *problem.*

Proof. Clearly, if **Bi-Approx** returns $(S_1, \overline{S_1})$ or $(S_2, \overline{S_2})$ then the solution returned is either $(\alpha, \alpha/2)$ or $(\alpha/2, \alpha)$-approximate, and hence $(\alpha/2, \alpha/2)$-approximate. In the following, we suppose that $(S_3, \overline{S_3})$ is returned by **Bi-Approx** and we prove that it is an $(\alpha/2, \alpha/2)$-approximate cut.

We partition V into four subsets X, Y, Z and T such that $(S_1, \overline{S_1}) = (X \cup Y, Z \cup T)$ and $(S_2, \overline{S_2}) = (X \cup Z, Y \cup T)$. Vertices of each subset are shrunk into *super-nodes* denoted by v_X, v_Y, v_Z and v_T. More precisely, all nodes $v \in X$ fall into v_X, all nodes $v \in Y$ fall into v_Y etc. Edges between two super-nodes are also shrunk into one *super-edge* such that:

$$w_{v_A v_B} = \sum_{v \in A, v' \in B} w_{v v'} \text{ and } l_{v_A v_B} = \sum_{v \in A, v' \in B} l_{v v'}$$

where $A \in \{X, Y, Z, T\}$, $B \in \{X, Y, Z, T\}$ and $A \neq B$. Finally, we get a new graph K_4 as depicted in Figure 2.

Now observe that if $l_{v_X v_T} + l_{v_Y v_Z} \geq l_{v_X v_Y} + l_{v_Z v_T}$ is true then we get a contradiction since instead of $(S_3, \overline{S_3})$, $(S_1, \overline{S_1})$ would have been returned:

$$l_{v_X v_T} + l_{v_Y v_Z} \geq l_{v_X v_Y} + l_{v_Z v_T}$$
$$l_{v_X v_T} + l_{v_Y v_Z} \geq (l_{v_X v_Y} + l_{v_Z v_T} + l_{v_X v_T} + l_{v_Y v_Z})/2$$
$$L(S_1, \overline{S_1}) \geq L(S_2, \overline{S_2})/2$$

Symmetrically, if $w_{v_X v_T} + w_{v_Y v_Z} \geq w_{v_X v_Z} + w_{v_Y v_T}$ is true then we get a contradiction since instead of $(S_3, \overline{S_3})$, $(S_2, \overline{S_2})$ would have been returned:

$$w_{v_X v_T} + w_{v_Y v_Z} \geq w_{v_X v_Z} + w_{v_Y v_T}$$
$$w_{v_X v_T} + w_{v_Y v_Z} \geq (w_{v_X v_Z} + w_{v_Y v_T} + w_{v_X v_T} + w_{v_Y v_Z})/2$$
$$W(S_2, \overline{S_2}) \geq W(S_1, \overline{S_1})/2$$

Thus we have:

$$l_{v_X v_T} + l_{v_Y v_Z} < l_{v_X v_Y} + l_{v_Z v_T} \text{ and} \tag{1}$$

$$w_{v_X v_T} + w_{v_Y v_Z} < w_{v_X v_Z} + w_{v_Y v_T}. \tag{2}$$

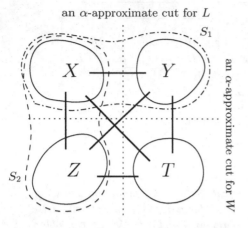

an α-approximate cut for L

Fig. 1. Vertices of G are partitioned into four subsets X, Y, Z and T. This partition depends on $(S_1, \overline{S_1})$ and $(S_2, \overline{S_2})$.

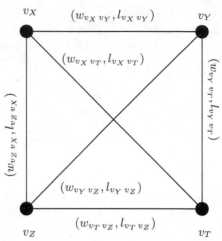

Fig. 2. Vertices and edges of G are shrunk to get a complete graph with four nodes

From inequality (1) we get:

$$(l_{v_X v_Y} + l_{v_Z v_T})/2 > (l_{v_X v_T} + l_{v_Y v_Z})/2$$
$$l_{v_X v_Z} + l_{v_Y v_T} + (l_{v_X v_Y} + l_{v_Z v_T})/2 > (l_{v_X v_T} + l_{v_Y v_Z})/2$$
$$l_{v_X v_Z} + l_{v_Y v_T} + l_{v_X v_Y} + l_{v_Z v_T} > (l_{v_X v_T} + l_{v_Y v_Z} +$$
$$+ l_{v_X v_Y} + l_{v_Z v_T})/2$$
$$L(S_3, \overline{S_3}) > 0.5 L(S_2, \overline{S_2})$$
$$L(S_3, \overline{S_3}) \geq \frac{\alpha}{2} OPTL$$

From inequality (2) we get:

$$(w_{v_X v_Z} + w_{v_Y v_T})/2 > (w_{v_X v_T} + w_{v_Y v_Z})/2$$
$$w_{v_X v_Y} + w_{v_Z v_T} + (w_{v_X v_Z} + w_{v_Y v_T})/2 > (w_{v_X v_T} + w_{v_Y v_Z})/2$$
$$w_{v_X v_Y} + w_{v_Z v_T} + w_{v_X v_Z} + w_{v_Y v_T} > (w_{v_X v_T} + w_{v_Y v_Z} +$$
$$+ w_{v_X v_Z} + w_{v_Y v_T})/2$$
$$W(S_3, \overline{S_3}) > 0.5 W(S_1, \overline{S_1})$$
$$W(S_3, \overline{S_3}) > \frac{\alpha}{2} OPTW \qquad \square$$

The analysis of **Bi-Approx** is tight. To see it, consider the instance given in Figure 3 where K is a large integer. The ideal point has a total weight and a total length equal to 1 while $(S_1, \overline{S_1})$ achieves the values $(\alpha, \alpha \frac{K-1}{2K})$ and $(S_2, \overline{S_2})$ achieves the values $(\alpha \frac{K-1}{2K}, \alpha)$. The algorithm returns a solution $(S_3, \overline{S_3})$ such that $S_3 = \{v_1, v_3, v_5\}$ and its total weight and total length are both equal to $\alpha \frac{K+1}{2K}$. When K tends to infinity, the solution returned tends to be $(\alpha/2, \alpha/2)$-approximate.

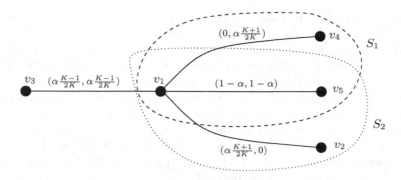

Fig. 3. Instance for which **Bi-Approx** returns an $(\alpha/2, \alpha/2)$-approximate solution

Corollary 1. *There exists a deterministic* $(0.43928, 0.43928)$*-approximate algorithm for the bi-criteria weighted* MAX-CUT *problem.*

Proof. Replace **Al** in **Bi-Approx** by the derandomized algorithm of Goemans and Williamson [7,8] which is a 0.87856-approximate algorithm and the result follows. □

Interestingly, an existence result can be derived from the algorithm **Bi-Approx**.

Corollary 2. *For all instances of the bi-criteria weighted* MAX-CUT *problem, there always exists a feasible solution which approximates the ideal point within a ratio* $1/2$ *on the two criteria.*

Proof. Suppose that **Al** in **Bi-Approx** is an optimal (1-approximate) algorithm for the mono-criterion weighted MAX-CUT problem and the result follows. □

The question whether the above theorem can be improved arises but the following theorem brings a negative answer.

Theorem 2. *No* (α, β)*-approximation algorithm such that* $\alpha > \beta \geq 1/2$ *(or* $\beta > \alpha \geq 1/2$*) exists for the bi-criteria* MAX-CUT *problem.*

Proof. Consider the complete graph K_3 whose edges e, e' and e'' are such that $w_e = l_{e'} = 0$ and $l_e = w_{e'} = w_{e''} = l_{e''} = 1$. The ideal solution $(I, \overline{I})$ has a total weight and a total length both equal to 2 while no feasible cut has a total weight and a total length simultaneously strictly superior to 1. □

4   A Randomized Approximation Algorithm

As usual, we consider that a randomized algorithm for a mono-criterion maximization problem is an α-expected approximate algorithm if the expected value (denoted by $E[X]$) of the solution returned is at least α times the value (denoted by OPT) of an optimal solution: $E[X] \geq \alpha OPT$.

When randomization is considered, the bi-criteria weighted MAX-CUT problem is then to find a feasible cut $(A, \overline{A})$ such that $E[W(A, \overline{A})] \geq \alpha OPTW$ and $E[L(A, \overline{A})] \geq \beta OPTL$ where $0 < \alpha \leq 1$ and $0 < \beta \leq 1$.

There is no hope to get an (α, β)-expected approximate algorithm for the bi-criteria weighted MAX-CUT problem with $\alpha = \beta$ and $\alpha > 2/3$. To see it, consider the example given in Figure 4 where the ideal cut $(I, \overline{I})$ achieves the values $(1,1)$. Four cuts $(S_1, \overline{S_1})$, $(S_2, \overline{S_2})$, $(S_3, \overline{S_3})$ and $(S_4, \overline{S_4})$ are feasible with values respectively $(0,0)$, $(2/3, 2/3)$, $(1/3, 1)$, and $(1, 1/3)$. Let **Ran Al** be a randomized algorithm which outputs $(S_i, \overline{S_i})$ with a probability p_i. Obviously, one has $p_1 + p_2 + p_3 + p_4 = 1$. The expected value of the cut $(S, \overline{S})$ output by **Ran Al** is:

$$E[W(S, \overline{S})] = \frac{2p_2}{3} + \frac{p_3}{3} + p_4 \text{ and } E[L(S, \overline{S})] = \frac{2p_2}{3} + p_3 + \frac{p_4}{3}.$$

The problem is then to find p_1, p_2, p_3 and p_4 such that $E[W(S, \overline{S})] \geq \alpha$, $E[L(S, \overline{S})] \geq \alpha$ and α is maximized. When $p_1 = p_3 = p_4 = 0$ and $p_2 = 1$, α reaches $2/3$ which is the best possible value. As a consequence, no randomized algorithm can be (α, α)-expected approximate with $\alpha > 2/3$.

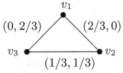

$$v_1$$

$(0, 2/3)$　　　　$(2/3, 0)$

v_3 　　　　　　 v_2

$(1/3, 1/3)$

Fig. 4. The ideal cut $(I, \overline{I})$ has a total weight and a total length both equal to 1

This statement has a consequence in the approximability of the weighted bi-criteria MAX-CUT problem. Indeed, there is no hope to design a deterministic (α, β)-approximate algorithm such that $\alpha + \beta > 4/3$. To see it, suppose that we have such an algorithm[2]. One can build two solutions $(S_1, \overline{S_1})$ and $(S_2, \overline{S_2})$ such that $W(S_1, \overline{S_1}) \geq \alpha OPTW$, $L(S_1, \overline{S_1}) \geq \beta OPTL$, $W(S_2, \overline{S_2}) \geq \beta OPTW$ and $L(S_2, \overline{S_2}) \geq \alpha OPTL$. Now consider the randomized algorithm which consists in returning $(S_1, \overline{S_1})$ with a probability $1/2$ and $(S_2, \overline{S_2})$ with a probability $1/2$. We would get an $(\frac{\alpha+\beta}{2}, \frac{\alpha+\beta}{2})$-expected approximate solution $(S, \overline{S})$ and $\frac{\alpha+\beta}{2} > 2/3$.

The algorithm (called **Ransam** in [9]) which consists in building a cut $(S, \overline{S})$ by putting equiprobably a vertex $v \in V$ to either S or $\overline{S}$ is $1/2$-expected approximate for the mono-criterion weighted MAX-CUT problem. One can remark that it achieves the same performance guarantee for a multi-criteria weighted MAX-CUT problem. However, a better randomized algorithm can be built for the bi-criteria MAX-CUT problem. We propose an algorithm called **Ran Bi-Approx** which uses a mono-criterion α-approximation algorithm (called **Al** in the following).

---

[2] Because of the symmetry of the problem, an (α, β)-approximate algorithm is also (β, α)-approximate.

Ran Bi-Approx
Input: G and **Al**
Step 1: Find $(S_1, \overline{S_1})$ with **Al** s.t. $W(S_1, \overline{S_1}) \geq \alpha\, OPTW$
Step 2: Find $(S_2, \overline{S_2})$ with **Al** s.t. $L(S_2, \overline{S_2}) \geq \alpha\, OPTL$
Step 3: Build $(S_3, \overline{S_3})$ s.t. $S_3 = (S_1 \cap S_2) \cup (\overline{S_1} \cap \overline{S_2})$
Step 4: Let $\gamma = (3 - \sqrt{5})/2$
Step 5: If $W(S_2, \overline{S_2}) \geq \gamma W(S_1, \overline{S_1})$
Then If $L(S_1, \overline{S_1}) \geq \gamma L(S_2, \overline{S_2})$
Then Return $(S_1, \overline{S_1})$ with a probability 0.5
and $(S_2, \overline{S_2})$ with a probability 0.5
Else Return $(S_1, \overline{S_1})$ with a probability γ
and $(S_2, \overline{S_2})$ with a probability $1 - \gamma$
Else If $L(S_1, \overline{S_1}) \geq \gamma L(S_2, \overline{S_2})$
Then Return $(S_1, \overline{S_1})$ with a probability $1 - \gamma$
and $(S_2, \overline{S_2})$ with a probability γ
Else Return $(S_3, \overline{S_3})$

Theorem 3. Ran Bi-Approx *is a randomized* $(\frac{\sqrt{5}-1}{2}\alpha, \frac{\sqrt{5}-1}{2}\alpha)$-*expected approximation algorithm for the bi-criteria weighted* MAX-CUT *problem if* **Al** *is an* α-*approximation algorithm.*

Proof. The algorithm considers four cases. For the first case, we suppose that:

$$W(S_2, \overline{S_2}) \geq \gamma W(S_1, \overline{S_1}) \text{ and } L(S_1, \overline{S_1}) \geq \gamma L(S_2, \overline{S_2}).$$

So, we have:

$$W(S_2, \overline{S_2}) \geq \gamma\alpha OPTW \text{ and } L(S_1, \overline{S_1}) \geq \gamma\alpha OPTL.$$

Since the solution returned in this case is $(S_1, \overline{S_1})$ with a probability 0.5 and $(S_2, \overline{S_2})$ with a probability 0.5, the expected value on each criterion of the solution returned is at least $\frac{\alpha(1+\gamma)}{2}$ times the optimum.

For the second case, we suppose that:

$$W(S_2, \overline{S_2}) \geq \gamma W(S_1, \overline{S_1}) \text{ and } L(S_1, \overline{S_1}) \geq 0.$$

So, we have $W(S_2, \overline{S_2}) \geq \gamma\alpha OPTW$. Since the solution returned in this case is $(S_1, \overline{S_1})$ with a probability $\gamma = \frac{1-\gamma}{2-\gamma}$ and $(S_2, \overline{S_2})$ with a probability $1 - \gamma = \frac{1}{2-\gamma}$, the expected value on each criterion of the solution returned is at least $\frac{\alpha}{2-\gamma}$ times the optimum.

The third case is symmetric to the second case, the expected value on each criterion of the solution returned is at least $\frac{\alpha}{2-\gamma}$ times the optimum.

For the fourth case, we suppose that:

$$W(S_2, \overline{S_2}) < \gamma W(S_1, \overline{S_1}) \text{ and } L(S_1, \overline{S_1}) < \gamma L(S_2, \overline{S_2}).$$

As it was done before, we consider that the set of vertices is partitioned into four subsets (see Figure 1) and the proof is done on a simple K_4 graph (see Figure 2). So, we have:

$$w_{v_X v_Y} + w_{v_Z v_T} + w_{v_X v_T} + w_{v_Y v_Z} < \gamma\big(w_{v_X v_Z} + w_{v_Y v_T} +$$
$$+ w_{v_X v_T} + w_{v_Y v_Z}\big) \qquad (3)$$
$$l_{v_X v_Z} + l_{v_Y v_T} + l_{v_X v_T} + l_{v_Y v_Z} < \gamma\big(l_{v_X v_Y} + l_{v_Z v_T} +$$
$$+ l_{v_X v_T} + l_{v_Y v_Z}\big). \qquad (4)$$

From inequality (3), we get:

$$w_{v_X v_T} + w_{v_Y v_Z} < \gamma\big(w_{v_X v_Z} + w_{v_Y v_T} +$$
$$+ w_{v_X v_T} + w_{v_Y v_Z}\big)$$
$$(1 - \gamma)\big(w_{v_X v_T} + w_{v_Y v_Z}\big) < \gamma\big(w_{v_X v_Z} + w_{v_Y v_T}\big)$$
$$\frac{(1 - \gamma)}{\gamma}\big(w_{v_X v_T} + w_{v_Y v_Z}\big) < w_{v_X v_Z} + w_{v_Y v_T}$$
$$\frac{(1 - \gamma)}{\gamma}\big(w_{v_X v_T} + w_{v_Y v_Z} + w_{v_X v_Z} + w_{v_Y v_T}\big) < \frac{1}{\gamma}\big(w_{v_X v_Z} + w_{v_Y v_T}\big)$$
$$(1 - \gamma)\big(w_{v_X v_T} + w_{v_Y v_Z} + w_{v_X v_Z} + w_{v_Y v_T}\big) < w_{v_X v_Z} + w_{v_Y v_T} +$$
$$+ w_{v_X v_Y} + w_{v_Z v_T}$$
$$(1 - \gamma)W(S_1, \overline{S_1}) < W(S_3, \overline{S_3})$$

Symmetrically, from inequality (4) we get:

$$(1 - \gamma)L(S_2, \overline{S_2}) < L(S_3, \overline{S_3})$$

In this case, $(S_3, \overline{S_3})$ is returned and its value on each criterion is at least $(1-\gamma)\alpha$ times the optimum.

Let $f(\gamma) = \min\{1 - \gamma, \frac{1}{2-\gamma}, \frac{1+\gamma}{2}\}$ for $0 \leq \gamma \leq 1$. This function finds its maximum when $\gamma = \frac{3-\sqrt{5}}{2}$. As a consequence, the solution returned by **Ran Bi-Approx** has an expected value on each criterion which is at least $\frac{\sqrt{5}-1}{2}\alpha$ times the optimum. $\qquad \square$

Corollary 3. *There exists a randomized* $(0.54297, 0.54297)$-*expected approximate algorithm for the bi-criteria weighted* MAX-CUT *problem.*

Proof. Replace **Al** by the algorithm of Goemans and Williamson [7,8] in **Ran Bi-Approx** and the result follows. $\qquad \square$

5 Concluding Remarks

Since we considered a bi-criteria MAX-CUT problem and provided approximation algorithms, the question whether it is possible to get similar results with more than two criteria arises. Unfortunately, the example given in Figure 5 shows that it is not possible to build a deterministic algorithm which approximates the ideal point with a performance guarantee when three criteria are considered. As a consequence, there is no hope to find an approximation algorithm with performance guarantee for the k-criteria weighted MAX-CUT problem where $k > 2$. However, the algorithm which consists in building a cut $(S, \overline{S})$ by putting equiprobably a vertex $v \in V$ to either S or $\overline{S}$ remains a $1/2$-expected approximation algorithm for any k-criteria weighted MAX-CUT problem.

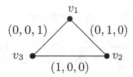

Fig. 5. The ideal cut $(I, \overline{I})$ achieves the values $(1, 1, 1)$ while any feasible cut achieves 0 on at least one coordinate. Thus, no approximation factor can be guaranteed.

Note that approximation results for the k-criteria weighted MAX-CUT problem can be found if another approach is considered. Indeed, if we restrict ourselves to feasible solutions then rarely a solution will dominate all the others (i.e. will be better than the others on every criterion) but a set of solutions which dominate all the others always exists. This set of solutions is called the *Pareto curve* and Papadimitriou and Yannakakis [11] proved that an approximation with performance guarantee of this curve (an *ε-approximate Pareto curve*) always exists.

The algorithms proposed in this article achieve the same ratios for both criteria. Indeed, **Bi-Approx** is a $(\alpha/2, \alpha/2)$-approximation algorithm while **Ran Bi-Approx** is a $(\frac{\sqrt{5}-1}{2}, \frac{\sqrt{5}-1}{2})$-expected approximation algorithm. As a consequence, it would be interesting to obtain results when the ratios are different.

Acknowledgement. We thank Martin Skutella for giving us the example of Figure 3.

References

1. E. Angel and E. Bampis and A.V. Fishkin, A note on scheduling to meet two min-sum objectives, *Proceeding 9th International Workshop on Project Management and Scheduling (PMS'2004)*, 143–146, 2004.
2. J. Aslam and A. Rasala and C. Stein and N. Young, Improved bicriteria existence theorems for scheduling, *Proceedings 10th Annual ACM-SIAM Symposium on Discrete Algorithms*, 846–847, 1999.
3. F. Baille and E. Bampis and C. Laforest, A Note on Bicriteria Schedules with Optimal Approximations Ratios, *Parallel Processing Letters*, 315–323, 14, 2004.
4. E. Bampis and A. Kononov, Bicriteria approximation algorithms for scheduling problems with communications delays, *Annals of Operations Research*, 2005, to appear.

5. F. Barahona, M. Grötschel, M. Jünger and G. Reinelt. An Application of Combinatorial Optimization to Statistical Physics and Circuit Layout Design. *Operations Research*, 36:493–513, 1998.
6. M. Ehrgott. Multicriteria optimization, Lecture Notes in Economics and Mathematical Systems N 491. Springer, 2000.
7. M.X. Goemans and D.P. Williamson. .879-approximation algorithms for MAX CUT and MAX 2SAT, in *Proceedings of STOC*, pp 422–431, 1994.
8. M.X. Goemans and D.P. Williamson. Improved Approximation Algorithms for Maximum Cut and Satisfiability Problems Using Semidefinite Programming, *Journal of the ACM*, 42(6), 1115–1145, 1995.
9. J. Hromkovič. Algorithmics for Hard Problems: Introduction to Combinatorial Optimization, Randomization, Approximation, and Heuristics. Springer, 2001.
10. R.M. Karp. Reducibility Among Combinatorial Problems. In R. Miller and J. Thatcher (editors), *Complexity of Computer Computations*, pp 85–103, Plenum Press, NY, 1972.
11. C.H. Papadimitriou and M. Yannakakis. On the approximability of trade-offs and optimal access of web sources, In *Proceedings of th 41th Annual IEEE Symposium on Foundations of Computer Science*, 86–92, 2000.
12. A. Rasala, C. Stein, E. Torng and P. Uthaisombut. Existence theorems, lower bounds and algorithms for scheduling to meet two objectives, In *Proceedings of the 13th Annual ACM-SIAM Symposium on Discrete Algorithms*, 723–731, 2002.
13. C. Stein and J. Wein. On the existence of schedules that are near-optimal for both makespan and total weighted completion time, *Operations Research Letters*, 21, 115–122, 1997.

Approximation Algorithms for the Weighted Independent Set Problem

Akihisa Kako[1], Takao Ono[1], Tomio Hirata[1], and Magnús M. Halldórsson[2]

[1] Graduate School of Information Science, Nagoya University,
Furou, Chikusa, Nagoya, 464-8603, Japan
Tel: 052-789-3440,   Fax: 052-789-3089
{kako, takao, hirata}@hirata.nuee.nagoya-u.ac.jp
[2] Department of Computer Science, University of Iceland,
IS-107 Reykjavik, Iceland
mmh@hi.is

Abstract. In the unweighted case, approximation ratio for the independent set problem has been analyzed in terms of the graph parameters such as the number of vertices, maximum degree, and average degree. In the weighted case, no corresponding results are possible for average degree, since inserting the vertices with small weight decreases the average degree arbitrarily without significantly changing the approximation ratio. In this paper, we introduce weighted measures, namely "weighted" average degree and "weighted" inductiveness, and analyze algorithms for the weighted independent set problem in terms of these parameters.

1   Introduction

An independent set in a graph is a set of vertices in which no two vertices are adjacent. The (weighted) independent set problem is that of finding a maximum (weight) independent set. Numerous approximation algorithms has been analyzed for this problem. In the unweighted case, an algorithm with approximation ratio $\Delta/6 + O(1)$ was proposed by Halldórsson and Radhakrishnan [6] for the graphs with the maximum degree Δ. Vishwanathan proposed an SDP-based algorithm whose approximation ratio is $O(\Delta \log \log \Delta / \log \Delta)$ [3]. For graphs with the average degree $\bar{d}$, Hochbaum [7] proved that an LP-based algorithm has approximation ratio $(\bar{d} + 1)/2$. Halldórsson and Radhakrishnan [5] improved this approximation ratio to $(2\bar{d} + 3)/5$. Moreover, an algorithm with approximation ratio $O(\bar{d} \log \log \bar{d} / \log \bar{d})$ was proposed by Halldórsson [2]. In the weighted case, Halldórsson and Lau [4] gave an algorithm with approximation ratio $(\Delta + 2)/3$. For δ-inductive graphs approximation ratio $(\delta + 1)/2$ is known due to Hochbaum [7], and Halldórsson [2] proposed an algorithm with approximation ratio $O(\delta \log \log \delta / \log \delta)$. Note that $\delta \leq \Delta$ for any graph.

In this paper, we extend the approximation algorithms of [2,7] to the weighted case. Since inserting the vertices with small weight decreases $\bar{d}$ arbitrarily without significantly changing approximation ratio, we introduce the *weighted average degree* $\bar{d}_w$ and analyze the approximation ratio. For weighted graphs, there exist

D. Kratsch (Ed.): WG 2005, LNCS 3787, pp. 341–350, 2005.

approximation algorithms whose approximation ratio is analyzed in terms of inductiveness. We extend inductiveness to weighted version and introduce the *weighted inductiveness* δ_w.

The rest of this paper is organized as follows. In Section 2 we define the weighted average degree and the weighted inductiveness. We also show the relationship between various degrees. In Section 3 we propose a greedy algorithm whose lower bound is $\max(W/(\overline{d}_w+1), W/(\delta_w+1))$, where W is the total weight. We also prove that this algorithm has approximation ratio $\max(\delta_w, 1)$. In Section 4 we prove that the approximation ratio of $\min((\overline{d}_w + 1)/2, (\delta_w + 1)/2)$ can be achieved by an LP-based algorithm. Finally we will prove that the approximation ratios of $O(\overline{d}_w \log \log \overline{d}_w / \log \overline{d}_w)$ and $O(\delta_w \log \log \delta_w / \log \delta_w)$ can be achieved by an SDP-based algorithm in Section 5. We will assume that the input graphs have no isolated vertices.

2   Preliminaries

2.1   Definitions

Let G be an undirected graph where each vertex v has positive weight w_v. Let $V(G)$ and $E(G)$ denote the vertex set and the edge set of G, respectively, as usual. Let $W(G)$ be the sum of the weights of all vertices. $n(G)$ is the number of vertices in G. Let $\Delta(G)$ and $\overline{d}(G)$ denote the maximum and the average degree of G, respectively. $d(v, G)$ is the degree of vertex v in G. The inductiveness $\delta(G)$ of a graph G is given by

$$\delta(G) = \max_{H \subseteq G} \min_{v \in V(H)} d(v, H), \tag{1}$$

where $H \subseteq G$ denotes that H is a subgraph of G. Let π be an ordering of vertices in V, that is, a one to one map $V \to \{1, 2, \ldots, n\}$ ($n = |V|$). We define the right degree of a vertex v in G with respect to π as follows:

$$d^\pi(v, G) = |\{u \in V | (u, v) \in E, \pi(u) > \pi(v)\}|. \tag{2}$$

The right degree of a vertex v is the number of adjacent vertices to the right when we arrange vertices from left to right according to π. If there exists π such that $m \geq \max_v d^\pi(v, G)$, we call G an m-inductive graph.

For a vertex set X, let $w(X)$ denote the sum of the weights of the vertices in X. Let $N_G(v)$ denote the set of vertices adjacent to vertex v in G. For a vertex v, we define the *weighted degree* $d_w(v, G)$ in G as follows:

$$d_w(v, G) = \frac{w(N_G(v))}{w_v}. \tag{3}$$

$\Delta_w(G) = \max_v d_w(v, G)$ is the maximum weighted degree of G. We will omit G if it is clear from the context. We define the *weighted average degree* $\overline{d}_w(G)$ of graph G as follows:

$$\overline{d}_w(G) = \frac{\sum_{v \in V} w_v d(v)}{W}. \tag{4}$$

In fact, we can represent the weighted average degree in the following form:

$$\bar{d}_w(G) = \frac{\sum_{v \in V} w(N(v))}{W} \tag{5}$$

$$= \frac{\sum_{v \in V} w_v d_w(v)}{W}. \tag{6}$$

The *weighted inductiveness* $\delta_w(G)$ of a graph G is given by

$$\delta_w(G) = \max_{H \subseteq G} \min_{v \in V(H)} d_w(v, H). \tag{7}$$

We define the right weighted degree of a vertex v for an ordering π in G as follows:

$$d_w^\pi(v, G) = \frac{w(\{u \in V | (u, v) \in E, \pi(u) > \pi(v)\})}{w_v}.$$

If there exists π such that $m \geq \max_v d_w^\pi(v, G)$, we call G a weighted m-inductive graph.

We denote by $\alpha_w(G)$ the weight of the optimal solution of the weighted independent set problem on G. For an algorithm A, $A(G)$ denotes the weight of the independent set obtained by A on G. Then the approximation ratio of A is defined by

$$\sup_G \frac{\alpha_w(G)}{A(G)}.$$

We will consider unweighted graphs as weighted ones where each vertex has unit weight. $\alpha(G)$ denotes the size of a maximum independent set on G.

2.2 Weighted Inductiveness

Let π be an ordering of the vertices of G and v_i be a vertex with $\pi(v_i) = i$. We define $V_i^\pi = \{v_j | j \geq i\}$. Let G_i^π be the induced subgraph of G by V_i^π. Smallest-first ordering π is an ordering such that the weighted degree of v_i is minimum in G_i^π for all i ($1 \leq i \leq n$). We can find a smallest-first ordering in polynomial time by a greedy algorithm. We can prove the following theorem in the same manner as in the case of unweighted inductiveness [8].

Theorem 1. *For any ordering π, the inequality*

$$\delta_w(G) \leq \max_v d_w^\pi(v, G)$$

holds. Moreover, if π is a smallest-first ordering, then the equality

$$\delta_w(G) = \max_v d_w^\pi(v, G)$$

holds.

Corollary 1. *A smallest-first ordering π minimizes $\max_v d_w^\pi(v, G)$.*

2.3   Relationship Between Weighted and Unweighted Degrees

Theorem 2. *The following relationships hold for all graphs G:*

$$\delta \leq \Delta_w \tag{8}$$
$$\delta_w \leq \Delta \tag{9}$$
$$\overline{d} \leq \Delta_w \tag{10}$$
$$\overline{d}_w \leq \Delta. \tag{11}$$

Proof. We obtain inequalities (8) and (9) by considering the non-decreasing order and the non-increasing order of weight, respectively. (11) follows immediately from the definition of measures. Finally, we prove inequality (10). We can get the following inequalities:

$$\sum_{v \in V} d_w(v) = \sum_{v \in V} \sum_{u:(u,v) \in E} \frac{w_u}{w_v} = \sum_{(u,v) \in E} \left[\frac{w_u}{w_v} + \frac{w_v}{w_u} \right] \geq 2|E| = n\overline{d}.$$

Thus,

$$\Delta_w = \max_{v \in V} d_w(v) \geq \frac{1}{n} \sum_{v \in V} d_w(v) \geq \overline{d}.$$

Hence, this theorem holds. ☐

There exist graphs where δ_w and $\overline{d}_w$ are arbitrarily smaller than δ: Consider the complete bipartite graph $G = K_{n/2,n/2}$, where vertices have weight 1 on one side and w on the other side. Then, $\delta(G) = n/2$, while $\delta_w(G) = n/(2w)$. For $\overline{d}_w$, we consider an n-clique of $\{v_0, v_1, \ldots, v_{n-1}\}$ plus v_n connected to only v_{n-1}. The weight w_i of v_i is given by $w_i = 1$ for $0 \leq i \leq n-1$ and $w_n = w$. In the graph, $\delta = n - 1$ and

$$\overline{d}_w = \frac{w + (n-1)^2 + n}{w + n} = 1 + O\left(\frac{n^2}{w}\right).$$

3   Greedy Algorithm

3.1   Previous Results

For unweighted graphs, the greedy algorithm can be written as follows. We select a minimum degree vertex as a vertex in the independent set I, and delete this vertex and all of its neighbors from the graph. We repeat this process for the remaining subgraph until the subgraph becomes empty. This algorithm attains the Turán bound [5,7];

$$|I| \geq \frac{n}{\overline{d} + 1}. \tag{12}$$

For weighted graphs, the minimum degree greedy algorithm attains the following lower bound [2,8]

$$w(I) \geq \frac{W}{\delta + 1}. \tag{13}$$

The greedy algorithm for the weighted graphs is almost the same as the unweighted greedy algorithm. The difference is that, instead of selecting a minimum degree vertex, our algorithm selects a vertex of minimum weighted degree. We call this algorithm WG.

Sakai, Togasaki, and Yamazaki proposed an algorithm which is essentially the same as WG and proved the following theorem [10].

Theorem 3 ([10]). WG *finds an independent set with the following lower bound:*

$$\mathsf{WG}(G) \geq \sum_{v \in V} \frac{w_v^2}{w(N(v)) + w_v}.$$

3.2   Lower Bound

We use the following proposition.

Proposition 1. *Assume that $a_i > 0$, $b_i > 0$ for all $1 \leq i \leq n$. Then the inequality*

$$\sum_i \frac{b_i^2}{a_i} \geq \frac{\left(\sum_i b_i\right)^2}{\sum_i a_i}$$

holds.

Proof. The inequality is equivalent to

$$\sum_i a_i \sum_i \frac{b_i^2}{a_i} \geq \left(\sum_i b_i\right)^2.$$

This inequality comes from the Cauchy-Schwarz inequality $\left(\sum_i x_i^2\right)\left(\sum_i y_i^2\right) \geq \left(\sum_i x_i y_i\right)^2$, by assigning $x_i = \sqrt{a_i}$ and $y_i = b_i/\sqrt{a_i}$.   □

Let I be the independent set obtained by WG. Let v_i be the i-th vertex selected into the independent set I. Let G_i be the subgraph induced by the remaining vertices at the beginning of the i-th iteration.

Theorem 4. WG *produces an independent set satisfying the inequality*

$$\mathsf{WG}(G) \geq \frac{W}{\bar{d}_w + 1}.$$

Proof. We obtain the lower bound of $\bar{d}_w W$:

$$\bar{d}_w W = \sum_{v \in V(G)} w_v d_w(v, G) \geq \sum_i \sum_{v \in N_{G_i}(v_i) \cup \{v_i\}} w_v d_w(v, G_i)$$

$$\geq \sum_i \sum_{v \in N_{G_i}(v_i) \cup \{v_i\}} w_v d_w(v_i, G_i) = \sum_i \left(w(N_{G_i}(v_i)) + w_{v_i}\right) d_w(v_i, G_i).$$

Adding $W = \sum_i \left(w(N_{G_i}(v_i)) + w_{v_i} \right)$, we can deduce the inequality

$$\left(\bar{d}_w + 1 \right) W \geq \sum_i \frac{\left(w(N_{G_i}(v_i)) + w_{v_i} \right)^2}{w_{v_i}}.$$

Finally we apply Proposition 1 with $a_i = w_{v_i}$, $b_i = w(N_{G_i}(v_i)) + w_{v_i}$. The inequality

$$\left(\bar{d}_w + 1 \right) W \geq \frac{W^2}{\mathsf{WG}(G)}$$

holds, which implies the theorem.   □

Theorem 3 also leads to Theorem 4.

Theorem 5. WG *produces the independent set satisfying the inequality*

$$\mathsf{WG}(G) \geq \frac{W}{\delta_w + 1}.$$

Proof. Because $\delta_w \geq d_w(v_i, G_i)$ for all i and $W = \sum_i \left(w(N_{G_i}(v_i)) + w_{v_i} \right)$, the inequality

$$W \delta_w \geq \sum_i \left(w(N_{G_i}(v_i)) + w_{v_i} \right) d_w(v_i, G_i)$$

holds. With this inequality, we can prove this theorem in the same way as Theorem 4.   □

Proposition 2. *The lower bounds of Theorems 4 and 5 are tight.*

Proof. We illustrate the tight example for both theorems. Let G be a star graph with n vertices. We assign weight 1 to the central vertex and $1/\sqrt{n-1}$ to the other vertices. In this graph, $\bar{d}_w = \delta_w = \sqrt{n-1}$, $W = \sqrt{n-1} + 1$. WG may output the singleton with the central vertex. In this case, $\mathsf{WG}(G) = 1$ and thus the inequalities in Theorems 4 and 5 hold with equality.   □

3.3   Approximation Ratio

Theorem 6. WG *attains approximation ratio* $\max(\delta_w, 1)$.

Proof. Let $V_i = N_{G_i}(v_i) \cup \{v_i\}$, and H_i be the subgraph of G induced by V_i. In the case $\delta_w \leq 1$, it is easy to see that $\alpha_w(H_i) = w_{v_i}$ and thus $\alpha_w(G) \leq \sum_i \alpha_w(H_i) = \sum_i w_{v_i} = \mathsf{WG}(G)$. Otherwise, by the property of WG and the definition of inductiveness, $\alpha_w(H_i) \leq \max(w_{v_i}, w(N_{H_i}(v_i))) = w_{v_i} \cdot \max(1, d_w(v_i, H_i)) \leq w_{v_i} \cdot \max(1, \delta_w(G)) = w_{v_i} \cdot \delta_w(G)$. The inequalities

$$\alpha_w(G) \leq \sum_i \alpha_w(H_i) \leq \sum_i w_{v_i} \cdot \delta_w(G) = \mathsf{WG}(G) \cdot \delta_w(G)$$

are immediate.   □

This theorem immediately implies that this problem is polynomial time solvable for the graphs with $\delta_w \leq 1$; We will ignore this case hereafter.

The graph in Proposition 2 is the tight example for WG:

Proposition 3. *The approximation ratio* δ_w *of* WG *is tight.*

4   LP-Based Algorithms

4.1   Unweighted Results

We will consider the combination of linear programming and the greedy algorithm. With the lower bound (12), Hochbaum [7] proved that this combination achieves the approximation ratio $(\bar{d}+1)/2$. In this section we extend Hochbaum's analysis to the weighted case and prove that the proposed algorithm has the approximation ratios $(\bar{d}_w + 1)/2$ and $(\delta_w + 1)/2$.

4.2   LP Relaxation for the Weighted Independent Set Problem

The weighted independent set problem can be formulated in the integer programming as follows:

$$\text{maximize } \sum_{i \in V} w_i x_i, \tag{14}$$
$$\text{subject to } x_i + x_j \leq 1 \quad \text{for all } (i,j) \in E,$$
$$x_i \in \{0,1\} \quad \text{for all } i \in V.$$

Relaxing the integral constraint, we can deduce the following linear programming:

$$\text{maximize } \sum_{i \in V} w_i x_i, \tag{15}$$
$$\text{subject to } x_i + x_j \leq 1 \quad \text{for all } (i,j) \in E,$$
$$0 \leq x_i \leq 1 \quad \text{for all } i \in V.$$

We can obtain the optimal solution to this LP each of whose elements is 0, $1/2$, or 1 [12]. Note that this LP can be solved with a combinatorial algorithm [9,11]. We classify the vertices into three sets according to the value of x_i, that is, $S_1 = \{i \in V | x_i = 1\}$, $S_{1/2} = \{i \in V | x_i = 1/2\}$, $S_0 = \{i \in V | x_i = 0\}$. Note that S_1 is an independent set of G and no vertex in $S_{1/2}$ has a neighbor in S_1. We also note that $S_{1/2}$ induces a subgraph with no isolated vertices.

4.3   Algorithm

We first solve the LP relaxation to divide the vertex set V into three subsets S_1, $S_{1/2}$, and S_0 as above. We then apply WG to the subgraph H induced by $S_{1/2}$ to obtain an independent set I_H of H. Finally, we output the independent set $I = S_1 \cup I_H$. We call this algorithm WGL.

4.4   Approximation Ratio

From Theorem 4, we can prove the following theorem in the same manner as the Hochbaum's proof [7] of the approximation ratio $(\bar{d}+1)/2$ for unweighted graphs.

Theorem 7. *Approximation ratio of* WGL *is* $(\bar{d}_w + 1)/2$.

We prove the approximation ratio in terms of the weighted inductiveness.

Theorem 8. *Approximation ratio of* WGL *is* $(\delta_w + 1)/2$.

Proof. From Theorem 5,

$$\frac{\alpha_w(G)}{\mathsf{WGL}(G)} \leq \frac{w(S_1) + \frac{1}{2}w(S_{\frac{1}{2}})}{w(S_1) + \frac{w(S_{\frac{1}{2}})}{\delta_w(H)+1}} \leq \frac{\delta_w(H) + 1}{2} \leq \frac{\delta_w + 1}{2}. \qquad \square$$

Proposition 4. *The approximation ratio of Theorems 7 and 8 is tight.*

Proof. We consider the split graph $G = (V, E)$, where $V = \{u_1, u_2, \ldots, u_t, v_1, v_2, \ldots, v_{2t-1}\}$ and $E = \{(u_i, v_j) | 1 \leq i \leq t, 1 \leq j \leq 2t - 1\} \cup \{(u_i, u_j) | 1 \leq i < j \leq t\}$. The induced subgraph by $\{u_i\}$ is a clique and the set $\{v_i\}$ is an independent set. We give each vertex u_i weight $w/t + \epsilon$, each vertex v_i weight $w/(2t-1)$, where ϵ is a small positive constant. In the optimal solution for LP (15), each value of x_i is $1/2$. Thus, $S_{1/2} = V(G)$. In this graph, $\mathsf{WGL}(G) = w/t + \epsilon$ and $\alpha_w = w$. So, the following equations hold:

$$\overline{d}_w = 2t - 1 + \frac{3t^2 - 2t}{2w}\epsilon, \qquad \frac{\alpha_w(G)}{\mathsf{WGL}(G)} = \frac{\overline{d}_w + 1}{2} - \left(\frac{t^2}{w + \epsilon t} - \frac{3t^2 - 2t}{4w}\right)\epsilon,$$

$$\delta_w = 2t - 1 - \frac{t^2}{w + \epsilon t}\epsilon, \qquad \frac{\alpha_w(G)}{\mathsf{WGL}(G)} = \frac{\delta_w + 1}{2} - \frac{t^2}{2(w + \epsilon t)}\epsilon.$$

Hence, Theorems 7 and 8 are tight. $\qquad \square$

5   SDP-Based Algorithms

5.1   Previous Result

The following theorem was proved in [2]:

Theorem 9 ([2]). *For any fixed real k such that $\vartheta_w(G) \geq 2W/k$, we can construct an independent set in G whose weight is $\Omega(W/(k\delta^{1-1/(2k)}))$.*

The function $\vartheta_w(G)$, defined in [1], is the weighted version of Lovász's ϑ-function. This function can be computed using semi-definite programming (SDP) in polynomial time, and has the property $\alpha_w(G) \leq \vartheta_w(G)$.

For the unweighted graphs, the combination of this theorem and the greedy algorithm yields the approximation ratio $O(\overline{d} \log \log \overline{d}/ \log \overline{d})$.

5.2   Approximation Ratio for the Weighted Graphs

We will prove the following result for the weighted version of the algorithm with the approximation ratio $O(\overline{d} \log \log \overline{d}/ \log \overline{d})$.

Theorem 10. *For any fixed real t such that $t \geq W(G)/\alpha_w(G)$, we can approximate the weighted independent set problem within $O(t^2 \overline{d}_w^{1-1/(8t)})$.*

Proof. Assume that $t \geq W(G)/\alpha_w(G)$ is fixed. Let K be the subgraph induced by the vertices whose degrees in G are less than $2t\bar{d}_w$. Then we can estimate the value $\bar{d}_w W(G)$ as follows:

$$\bar{d}_w W(G) = \sum_{v \in V(G)} w_v d(v) \geq \sum_{v \in V(G)\backslash V(K)} w_v d(v) \geq 2t\bar{d}_w \sum_{v \in V(G)\backslash V(K)} w_v.$$

Thus, the inequality $\sum_{v \in V(G)\backslash V(K)} w_v \leq W(G)/(2t)$ holds. From this inequality, we can prove the theorem along with [2]. $\qquad\square$

Theorem 11. *For any fixed real t such that $t \geq W(G)/\alpha_w(G)$, we can approximate the weighted independent set problem within $O(t^2 \delta_w^{1-1/(8t)})$.*

Proof. Let π be an ordering of vertices in G with which the value of $\max_v d_w^\pi(v)$ is equal to δ_w. Let π' be the reverse ordering of π. Assume that $t \geq W(G)/\alpha_w(G)$ is fixed. Let K be the subgraph induced by the vertices whose right degrees $d^{\pi'}(v, G)$ are less than $2t\delta_w$. Thus K is a $2t\delta_w$-inductive graph. Then the following inequalities hold:

$$W\delta_w \geq \sum_{v \in V(G)} w_v d_w^\pi(v) = \sum_{v \in V(G)} w_v d^{\pi'}(v)$$

$$\geq \sum_{v \in V(G)\backslash V(K)} w_v d^{\pi'}(v) \geq 2t\delta_w \sum_{v \in V(G)\backslash V(K)} w_v.$$

Thus, we can prove this theorem just like [2]. $\qquad\square$

5.3   Algorithm

In this section we propose two algorithms: WGSA, whose approximation ratio is a function of $\bar{d}_w$, and WGSI, whose approximation ratio is a function of δ_w.

WGSA is the following algorithm. We get an independent set by applying WG. Independently, we apply the algorithm given by Theorem 10 to obtain another independent set. We output the one with larger weight.

Theorem 12. *WGSA achieves approximation ratio $O(\bar{d}_w \log\log \bar{d}_w / \log \bar{d}_w)$ for the weighted independent set problem.*

Proof. From Theorems 4 and 10, we can prove this theorem in the same manner as [2]. $\qquad\square$

WGSI is the following algorithm. We get an independent set by applying WG. Independently, we apply the algorithm given by Theorem 11 to obtain another independent set. We output the one with larger weight.

Theorem 13. *WGSI achieves approximation ratio $O(\delta_w \log\log \delta_w / \log \delta_w)$ for the weighted independent set problem.*

Proof. From Theorems 5 and 11, we can prove this theorem in the same way as [2]. $\qquad\square$

6   Conclusion

In this paper, we defined the weighted average degree $\bar{d}_w$ and the weighted inductiveness δ_w, and proved the lower bound of the weight of the independent set obtained by the weighted greedy algorithm. We also proved that this algorithm has approximation ratio δ_w. Combining with LP, we obtained the approximation ratio $\min((\bar{d}_w+1)/2, (\delta_w+1)/2)$. Also combining with SDP, we proved that approximation ratio can attain $O(\bar{d}_w \log\log \bar{d}_w / \log \bar{d}_w)$ and $O(\delta_w \log\log \delta_w / \log \delta_w)$.

Acknowledgments

We thank Toshihiro Fujito for his fruitful comments.

References

1. M. Grötschel, L. Lovász, A. Schrijver. *Geometric algorithms and combinatorial optimization, 2nd ed.* Springer-Verlag, 1993.
2. M.M. Halldórsson. Approximations of weighted independent set and hereditary subset problems. *Journal of Graphs Algorithms and Applications*, 4(1):1–16, 2000.
3. M.M. Halldórsson. Approximations of independent sets in graphs. In K. Jansen, J. Rolim, editors, *The First International Workshop on Approximation Algorithms for Combinatorial Optimization Problems (APPROX)*, 1–14, 1998.
4. M.M. Halldórsson and H.C. Lau. Low-degree graph partitioning via local search with applications to constraint satisfaction, max cut, and 3-coloring. *Journal of Graph Algorithms and Applications*, 1(3):1–13, 1997.
5. M.M. Halldórsson and J. Radhakrishnan. Greed is good: Approximating independent sets in sparse and bounded-degree graphs. *Algorithmica*, 18:145–163, 1997.
6. M.M. Halldórsson and J. Radhakrishnan. Improved approximations of independent sets in bounded-degree graphs via subgraph removal. *Nordic Journal of Computing*, 1(4):475–482, 1994.
7. D.S. Hochbaum. Efficient bounds for the stable set, vertex cover and set packing problems. *Discrete Applied Mathematics*, 6:243–254, 1983.
8. D.W. Matula and L.L. Beck. Smallest-last ordering and clustering and graph coloring algorithms. *Journal of the Association for Computing Machinery*, 30(2):417–427, 1983.
9. G.L. Nemhauser and L.E. Trotter. Vertex packing: Structural properties and algorithms. *Mathematical Programming*, 8:232–248, 1975.
10. S. Sakai, M. Togasaki, K. Yamazaki. A note on greedy algorithms for maximum weighted independent set problem. *Discrete Applied Mathematics*, 126:313–322, 2003.
11. A. Schrijver. *Combinatorial optimization: polyhedra and efficiency, vol. A.* Springer-Verlag, 2003.
12. V.V. Vazirani. *Approximation algorithms.* Springer-Verlag, 2001.

Approximation Algorithms
for Unit Disk Graphs*

Erik Jan van Leeuwen

CWI, Kruislaan 413, 1098 SJ Amsterdam, The Netherlands
erikjan@cwi.nl

Abstract. We consider several graph theoretic problems on unit disk graphs (Maximum Independent Set, Minimum Vertex Cover, and Minimum (Connected) Dominating Set) relevant to mobile ad hoc networks. We propose two new notions: thickness and density. If the thickness of a unit disk graph is bounded, then the mentioned problems can be solved in polynomial time. For unit disk graphs of bounded density, we present a new asymptotic fully-polynomial approximation scheme for the considered problems. The scheme for Minimum Connected Dominating Set is the first Baker-like asymptotic FPTAS for this problem. By adapting the proof, it implies e.g. an asymptotic FPTAS for Minimum Connected Dominating Set on planar graphs.

1 Introduction

Mobile ad hoc networks and wireless sensor networks have attracted widespread attention in the last few years. Due to their flexibility, they are very interesting for consumer, military, and scientific markets. To solve combinatorial problems on such networks, the networks are often modeled as (unit) disk graphs. A graph $G = (V, E)$ is a *disk graph* if and only if there exists a set of disks $D = \{D_i \mid 1 \leq i \leq |V|\}$, such that each vertex corresponds to a disk D_i and two vertices are connected by an edge if and only if the two corresponding disks intersect. Tangent disks are assumed to intersect as well. In a *unit disk graph*, the radii of all disks are equal. We call D a *disk representation* of G. Observe that (unit) disk graphs are a good model for wireless communication networks.

As (unit) disk graphs have a nice geometric interpretation, classical graph theoretic problems relevant to wireless communication networks seem easier to solve or approximate than they are on general graphs. By making some realistic assumptions about the geometric interpretation, we find various new properties and algorithms. The main contribution of this paper is an asymptotic FPTAS for Maximum Independent Set, Minimum Vertex Cover, and Minimum (Connected) Dominating Set on unit disk graphs of bounded density, improving existing results on such graphs. The proof used for Minimum Connected Dominating Set

---

* This research was supported by the Netherlands Organisation for Scientific Research NWO (project *Treewidth and Combinatorial Optimisation*) and by the Bsik project BRICKS.

D. Kratsch (Ed.): WG 2005, LNCS 3787, pp. 351–361, 2005.

substantially improves previous analyses for this problem and can also be applied to these algorithms (see e.g. [8,10]). The notion of thickness, introduced in Sec. 4, is of interest by itself as a modeling assumption in wireless communication networks. We develop polynomial time algorithms for the considered problems on unit disk graphs of bounded thickness.

2   Preliminaries

An *independent set* $S \subseteq V$ contains only non-adjacent vertices (i.e. $u, v \in S \Rightarrow (u, v) \notin E$). In a mobile ad hoc network, an independent set of nodes is capable of transmitting simultaneously without signal interferences. To maximize communication capabilities, a maximum size independent set (maximum independent set) is sought. A set $S \subseteq V$ is a *vertex cover* if and only if $u \in S$ or $v \in S$ for each $(u, v) \in E$.

A set $S \subseteq V$ is a *dominating set* for $U \subseteq V$ if and only if each $v \in U$ is in S or is adjacent to a vertex in S. The nodes in a dominating set could function as emergency transmitters capable of reaching every node in the network, or as central nodes in node clusters. For a *connected dominating set* S, the subgraph of G induced by S $(G[S] = (S, (S \times S) \cap E))$ must also be connected. It can be used as a backbone to simplify and improve message routing. We seek a minimum size (connected) dominating set (minimum (connected) dominating set).

Several approximation scheme types are used. For each instance x of a maximization (minimization) problem and each $\epsilon > 0$, a *polynomial time approximation scheme (PTAS)* delivers in time polynomial in $|x|$ a feasible solution of value within a factor $1 - \epsilon$ (resp. $1 + \epsilon$) of the optimum. A *fully polynomial time approximation scheme (FPTAS)* delivers such a solution in time polynomial in $|x|$ and $\frac{1}{\epsilon}$. An *asymptotic fully polynomial time approximation scheme (FPTAS$^\omega$)* gives a feasible solution in time polynomial in $|x|$ and $\frac{1}{\epsilon}$ and attains the approximation factor if $|x| > c$, for some constant c only dependent on ϵ. We also use the notion of fixed-parameter tractability from Downey and Fellows [11].

3   Previous Work

Clark, Colbourn, and Johnson [9] proved all problems mentioned above NP-hard for (unit) disk graphs. Marathe et al. [15] give constant factor polynomial time approximation algorithms. Wan et al. [23] present a distributed constant factor approximation algorithm for Minimum Connected Dominating Set.

Matsui [16] and Hunt et al. [14] give different PTASs for Maximum Independent Set on unit disk graphs. Hunt et al. also have PTASs for Minimum Vertex Cover and Minimum Dominating Set. Cheng et al. [8] propose a PTAS for Minimum Connected Dominating Set on unit disk graphs. Nieberg et al. [17,18] give a robust PTAS for Maximum Independent Set and Minimum Dominating Set if no disk representation is known. Erlebach et al. [12] propose a PTAS for Maximum Independent Set and Minimum Vertex Cover on general disk graphs. Chan [7] gives a PTAS for Maximum Independent Set on the intersection graph of a set of fat objects. Under the used definition, a set of disks is fat.

The possibility of creating PTASs and FPTAS$^\omega$s exploiting fixed-parameter tractability has been used before, notably by Baker [4] for planar graphs. Demaine and Hajiaghayi [10] consider minor-closed graphs of locally bounded treewidth. Although bounded density unit disk graphs have locally bounded treewidth, they are not minor-closed [22]. Hunt et al. [14] consider λ-precision unit disk graphs where the distance between any two disk centers is at least λ. Any λ-precision unit disk graph has density $\Theta(\frac{1}{\lambda^2})$. The reverse is not necessarily true. Hence our results generalize the schemes of Hunt et al. Furthermore, we show that path decompositions suffice instead of the more complex tree decompositions used by Hunt et al. Finally, Alber and Fiala [1] describe subexponential exact algorithms for Maximum Independent Set on λ-precision unit disk graphs and show the problem is fixed-parameter tractable on such graphs.

4 Thickness

Assume we are given a unit disk graph $G = (V, E)$ with $n = |V|$ and a known disk representation $D = \{D_i = (c_i, r_i) \mid i = 1, \ldots, n\}$ for G, where $c_i \in \mathbb{R}^2$ is the center of disk i and $r_i = \frac{1}{2}$ its radius.

The thickness of a unit disk graph is determined by a slab decomposition of a disk representation of that graph. Given an angle α ($0 < \alpha < \pi$) and a disk center c_s, partition the plane using an infinite set of parallel lines, such that the distance between each two neighboring lines is 1, each line intersects the x-axis at angle α, and one line goes through c_s. The area between two neighboring lines (slab boundaries) is called a slab. A disk is said to be in a slab if its center is either between two slab boundaries defining the slab, or on the left boundary of the slab. This partition induces a slab decomposition $s = \langle \alpha, c_s \rangle$ of D. It also induces a decomposition of V into mutually disjoint, but collectively exhaustive subsets $Y_1, Y_2, \ldots, Y_b$ ($b \leq n$), such that Y_j contains the vertices corresponding to the disk centers in the j-th non-empty slab of s. We assume there also exist three empty 'dummy' slabs Y_{-1}, Y_0, and Y_{b+1}.

Given a slab decomposition s, the thickness $t(s)$ is the maximum number of disk centers of D (or, equivalently, the maximum number of vertices) in any slab of s, i.e. $t(s) = \max_{1 \leq i \leq b} |Y_i|$. We define the thickness t of D as the minimum thickness over all slab decompositions of D. Both $t(s)$ and t can be computed in polynomial time [21].

4.1 Relation to Pathwidth

A path decomposition of a graph $G = (V, E)$ is a sequence $(X_1, X_2, \ldots, X_p)$ of subsets of V (called bags) such that 1) $\bigcup_{1 \leq i \leq p} X_i = V$, 2) for all $(v, w) \in E$, there is an i ($1 \leq i \leq p$) such that $v, w \in X_i$, and 3) $X_i \cap X_k \subseteq X_j$ for all i, j, k with $1 \leq i < j < k \leq p$. The width of a path decomposition $(X_1, X_2, \ldots, X_p)$ is $\max_{1 \leq i \leq p} |X_i| - 1$. The pathwidth of a graph $G = (V, E)$ is the minimum width of any path decomposition of G [19].

Theorem 1. *Given a slab decomposition s of D, there exists a path decomposition of G of width at most $2t(s) - 1$ and consisting of at most $n - 1$ bags.*

Proof. Construct a sequence of bags $(X_1, X_2, \ldots, X_{b-1})$ with $X_j = Y_j \cup Y_{j+1}$ $(1 \le j \le b - 1)$. As the length of each edge is at most 1, vertices in Y_j can only have edges to vertices in Y_{j-1}, Y_j, and Y_{j+1}. Then clearly $(X_1, \ldots, X_{b-1})$ satisfies all requirements for a path decomposition. The width of the decomposition is $\max_{1 \le j \le b-1} |X_j| - 1 = \max_{1 \le j \le b-1} |Y_j \cup Y_{j+1}| - 1 \le 2 \max_{1 \le j \le b} |Y_j| - 1 = 2t(s) - 1$. Since $b \le n$, the decomposition consists of at most $n - 1$ bags. □

A corollary is that $2t - 1$ is an upper bound on the pathwidth of G.

A path decomposition can be used to solve many optimization problems. This includes the problems focused on in this paper. Given the pathwidth bound of $2t - 1$, Maximum Independent Set and Minimum Vertex Cover can be solved in $O(2^{2t}n)$ time, Minimum Dominating Set in $O(3^{2t}n)$ time, and Minimum Connected Dominating set in $O(2t^{2t}n)$ time [2,6,10,20]. However, a slab decomposition can also be used to solve the optimization problems directly, i.e. without creating a path decomposition, as will be shown below.

4.2　Direct Dynamic Programming Algorithms

Maximum Independent Set. For any j $(1 \le j \le b+1)$ and given $W_j \subseteq Y_j$, in any independent set $S \cup W_j$, with $S \subseteq Y_0 \cup \cdots \cup Y_{j-1}$, of maximum size, S must be an independent set, independent of W_j, of maximum size. Equivalently, as disks have radius $\frac{1}{2}$, $W_{j-1} = S \cap Y_{j-1}$ must be independent of W_j. In a slab-wise dynamic programming algorithm, we can therefore exhaustively enumerate all possible W_j and W_{j-1}, for each slab j.

Theorem 2. *A maximum independent set of a unit disk graph with thickness t can be computed in $O(t^2 2^{2t}n)$ time.*

It follows straightforwardly that a minimum vertex cover of a unit disk graph with thickness t can also be computed in $O(t^2 2^{2t}n)$ time.

Minimum Dominating Set. For some set $W \subseteq V$, we denote the vertices in Y_j dominated by W as $D^j(W)$. Furthermore, given a dominating set DS, we denote DS restricted to slab j $(DS \cap Y_j)$ by A_j, the vertices in slab j dominated by A_j or A_{j-1} $(D^j(A_j) \cup D^j(A_{j-1}))$ by B_j, and the vertices dominated by A_{j+1} $(D^j(A_{j+1}))$ by C_j. In the algorithm, we will always ensure that A_j, B_j, and C_j are mutually exclusive, but collectively exhaust Y_j.

We observe that for any j $(1 \le j \le b+1)$ and given $A_j, B_j \subseteq Y_j$ $(A_j \cap B_j = \emptyset)$, in any dominating set $S \cup A_j$ for $Y_{-1} \cup \cdots \cup Y_{j-1} \cup B_j$, with $S \subseteq Y_{-1} \cup \cdots \cup Y_{j-1}$, of minimum size, S must be a dominating set for $(Y_{-1} \cup \cdots \cup Y_{j-2}) \cup (Y_{j-1} - A_{j-1} - D^{j-1}(A_j))$, with $D^j(A_{j-1}) \supseteq B_j - D^j(A_j)$, of minimum size. Because $D^j(A_{j-1})$ is allowed to be a superset of $B_j - D^j(A_j)$, we can use exhaustive enumeration on just A_j and A_{j-1} and compute B_j as $D^j(A_j) \cup D^j(A_{j-1}) - A_j$. Since this considers only maximal B_j, we need a post-processing step with exhaustive enumeration on A_j and B_j to fix the table. This is shown in Alg. 1.

1.   Set $size_0(\emptyset, \emptyset) = 0$
2.   **for** $j \leftarrow 1$ **to** $b+1$
3.   **do**   Set $size_j(A_j, B_j) = \infty$ for each $A_j \subseteq Y_j,\ B_j \subseteq Y_j - A_j$
4.   **for each** $A_j \subseteq Y_j$
5.   **do**   **for each** $A_{j-1} \subseteq Y_{j-1}$
6.   **do**   Let $B_j = D^j(A_j) \cup D^j(A_{j-1}) - A_j$
                  $C_{j-1} = D^{j-1}(A_j) - A_{j-1}$, and
                  $B_{j-1} = Y_{j-1} - A_{j-1} - C_{j-1}$
7.   **if** $size_{j-1}(A_{j-1}, B_{j-1}) \neq \infty$ **and**
                       $|A_j| + size_{j-1}(A_{j-1}, B_{j-1}) < size_j(A_j, B_j)$
8.   **then** $size_j(A_j, B_j) = |A_j| + size_{j-1}(A_{j-1}, B_{j-1})$
9.   **od**   **od**
10.   **for each** $A_j \subseteq Y_j$
11.   **do**   **for each** $B_j \subseteq Y_j - A_j$ (in order of descending $|B_j|$)
12.   **do**   **for each** $v \in B_j$
13.   **do**   **if** $size_j(A_j, B_j) < size_j(A_j, B_j \backslash \{v\})$
14.   **then** $size_j(A_j, B_j \backslash \{v\}) \leftarrow size_j(A_j, B_j)$
15. **od**   **od**   **od**   **od**
16. **return** $size_{b+1}(\emptyset, \emptyset)$

Alg. 1. Solving Minimum Dominating Set, given a slab decomposition $Y_1, \ldots, Y_b$

Theorem 3. *Algorithm 1 computes a minimum dominating set of a unit disk graph with thickness t in $O(t^2 2^{2t} n)$ time.*

Minimum Connected Dominating Set. The subset of a minimum connected dominating set on slabs 1 to j is not necessarily connected. Therefore we consider *partial* connected dominating sets. For each j $(1 \leq j \leq b+1)$, a set $S_j \subseteq Y_{-1} \cup \cdots \cup Y_j$ is a *partial connected dominating set* of $Y_{-1} \cup \cdots \cup Y_{j-1} \cup B_j$ with $B_j \subseteq Y_j$ if and only if S_j is a dominating set for $Y_{-1} \cup \cdots \cup Y_{j-1} \cup B_j$ and either $C \cap Y_j \neq \emptyset$ for each connected component C of S_j, or $j \geq b$ and S_j is connected. Note that this definition enforces that S_{b+1} is a connected dominating set.

We solve the *minimum partial connected dominating set problem* by building on the solution for Minimum Dominating Set. A major problem is to remember the connected components of each considered partial connected dominating set. For each connected component C_i $(1 \leq i \leq K_j)$ of a partial connected dominating set, denote $C_i \cap Y_j$ by A_j^i, and thus $A_j = A_j^1 \cup \cdots \cup A_j^{K_j}$. The sets $A_j^1, \ldots, A_j^{K_j}$ are mutually exclusive and not connected to each other. Hence each A_j^i must be the union of one or more connected components of A_j. In fact, $A_j^1, \ldots, A_j^{K_j}$ forms a partition of the connected components of A_j. We call $A_j^1, \ldots, A_j^{K_j}$ the *front* of the partial connected dominating set. It can be shown that a front can represent the connectivity information of a partial connected dominating set.

Now we adapt the algorithm for Minimum Dominating Set. We use the same enumeration strategy, but also look at all possible fronts. This can be done by considering all non-crossing partitions of the connected components of A_j. Given a total ordering $\prec$ on the elements of a finite set S, $S_1, \ldots, S_p$ is a *non-crossing partition* of S if and only if $S_i \cap S_j = \emptyset$ for each $1 \le i < j \le p$, $\bigcup_{i=1}^p S_i = S$, and $a \prec b \prec c \prec d$ is false for all i, j $(1 \le i, j \le p, i \ne j)$, any $a, c \in S_i$, and any $b, d \in S_j$. Here, for two connected components X, Y of A_j, we define $\prec$ such that $X \prec Y$ if and only if $c_u^y < c_v^y$ for all vertices $u \in X, v \in Y$, i.e. Y lies above X. Note that either $X \prec Y$ or $Y \prec X$ must hold.

Lemma 1. *Given the above total ordering $\prec$, each front is a non-crossing partition of the connected components of A_j and A_{j-1}.*

The number of non-crossing partitions of an m-element set is C_m [5], the m-th Catalan number. It is well known that C_m is $O(\frac{4^m}{m\sqrt{m}})$. Demaine and Hajiaghayi [10] used Catalan structures similarly for Minimum Connected Dominating Set on planar graphs. In fact, this approach can be used for any graph having an embedding where the induced subgraph of the endpoints of any two crossing edges has at least three edges.

Finally, if A_{j-1} and A_j are in the same partial connected dominating set, we can show that a non-crossing partition of the connected components of A_{j-1} induces a unique non-crossing partition of the connected components of A_j. As the table maintained by the algorithm stores all relevant non-crossing partitions of A_{j-1}, we can derive all needed non-crossing partitions from the table.

Theorem 4. *A minimum connected dominating set of a unit disk graph with thickness t can be computed in $O(t^2 2^{4t} n)$ time.*

The following is a corollary of the theorems in this section.

Corollary 1. *Maximum Independent Set, Minimum Vertex Cover, and Minimum (Connected) Dominating Set are fixed-parameter tractable (in t) for unit disk graphs with a known disk representation.*

Note that this also implies that polynomial time algorithms for the problems exist if $t = t(n) \le c \log n$, for some constant $c > 0$.

5   Density

If the thickness of a unit disk graph is large, we have to resort to approximation algorithms to obtain a polynomial time algorithm. For this purpose, we propose the new notion of *density*. We first define grid decompositions, which determine the density of a unit disk graph. Given an angle α $(0 \le \alpha < \frac{1}{2}\pi)$ and two disk centers c_v and c_h, the plane can be partitioned into a grid of 1×1 squares, such that the vertical lines defining the grid intersect the x-axis at angle α, one vertical line intersects c_v, and one horizontal line intersects c_h. A disk is considered to be *in* a grid square if its center is between the two horizontal and between the two vertical grid boundaries defining the square. If a center is on a vertical (horizontal) grid boundary, then the disk is considered to be in the grid square to the right of (below) the boundary. This partition of the plane induces a *grid decomposition* $g = \langle \alpha, c_v, c_h \rangle$ of D.

Given a grid decomposition g, the *density* $d(g)$ is the maximum number of disk centers of D (or, equivalently, the maximum number of vertices) in any grid square of g. The *density* d of D is defined as the minimum density over all grid decompositions of D. The density of a given grid decomposition and the minimum density can be computed in polynomial time [21].

Unfortunately, the considered graph problems are still NP-hard for unit disk graphs of bounded density [21]. Because the problems are polynomially bounded optimization problems, we know from Ausiello et al. [3] that no FPTAS exists for Maximum Independent Set, Minimum Vertex Cover, or Minimum (Connected) Dominating Set on unit disk graphs of bounded density, unless P=NP. An FPTAS$^\omega$ can however be found, as will be shown below.

5.1 Shifting Strip Decompositions

Let $G = (V, E)$ be a unit disk graph with known disk representation D, containing disks of radius $\frac{1}{2}$. We assume a grid decomposition of D with density d and angle 0 is given. Partition the plane into horizontal *strips* using a set of horizontal lines (*strip boundaries*). Strip boundaries are of the form $y = j$, where $j \in \mathbb{Z}$, and coincide with horizontal boundaries of the grid decomposition. The *height* of a strip is equal to the distance between its two strip boundaries. A disk is *in* a strip defined by two horizontal lines $y = j$ and $y = l$ ($j < l$ and $j, l \in \mathbb{Z}$) if and only if its center is on or below $y = l$ and above $y = j$. The decomposition of the disk centers induced by the strips is called a *strip decomposition*. A strip decomposition also induces a decomposition of the graph, such that a vertex is in a strip if and only if the corresponding disk is in that strip.

A strip decomposition can be used as a decomposition for the well known *shifting technique* [4,13]. By choosing the height of the strips appropriately, we can bound the thickness of each strip and apply the algorithms of the previous section to each strip. Repeating this for several placements of the strip boundaries gives an approximation of the optimum. We show how to apply these ideas to construct an FPTAS$^\omega$ for Minimum Connected Dominating Set.

Minimum Connected Dominating Set. Assume G is connected. Let $f(n)$ be some function ($4d \le f(n) \le n$) and k an integer ($3 < k \le \frac{f(n)}{d}$). Construct a strip decomposition such that each strip has thickness smaller than $f(n) + d$ and height at least $\frac{f(n)}{d}$, and such that there are at most $\frac{n}{f(n)} + 1$ strips. Because the density is d, this can indeed be done, and in $O(n \log n)$ time [21]. Denote such a decomposition, shifted down by a, by D_a ($0 \le a \le k - 1$). We denote the set of disks in the b-th strip of D_a by D_a^b ($b \in \mathbb{Z}$).

Next, for each strip, we add all disks within distance 2 of its boundaries to the strip. So if $y = j$ and $y = l$ ($j, l \in \mathbb{Z}, j < l$) are the two strip boundaries defining strip D_a^b, then a disk will be in $\bar{D}_a^b$ if and only if its center is on or below $y = l+2$ and above $y = j-2$. We call the area between $y = l+1$ and $y = j-1$

the *interior* of the strip. The areas between $y = l + 2$ and $y = l + 1$ and between $y = j - 1$ and $y = j - 2$ are jointly referred to as the *exterior*.

The thickness of any strip in D_0 is at most $f(n) + 5d$. Hence the thickness of any strip in D_a is at most $f(n) + (a + 5)d \leq f(n) + (k + 4)d \leq f(n) + (\frac{f(n)}{d} + 4)d \leq 3f(n)$. We compute a minimum connected dominating set for each connected component in the interior of each strip in D_a, possibly using vertices in the exterior of the strip. By adapting the algorithm proposed in Theorem 4 and using that the thickness of any strip in D_a is at most $3f(n)$, this takes $O(f(n)^2 2^{12f(n)} n)$ time. Denote the union of the minimum connected dominating sets of each connected component in the interior of the b-th strip of D_a by CDS_a^b ($b \in \mathbb{Z}$) and let $CDS_a = \bigcup_{b \in \mathbb{Z}} CDS_a^b$.

Lemma 2. CDS_a ($0 \leq a \leq k - 1$) *is a connected dominating set for G.*

Proof. Because the interiors of the strips overlap, CDS_a is trivially a dominating set for G. Now suppose CDS_a is not connected. Consider two distinct connected components X, Y of CDS_a and the shortest path P between X and Y. If $|P| \leq 2$, then $X = Y$. If $|P| > 2$, consider the first three consecutive vertices v, p_0, p_1 of P with $v \in X$. These vertices must be in the interior of the same strip and belong to the same connected component in this strip. As CDS_a contains a minimum connected dominating set for each connected component in the interior of each strip, there exists a path in CDS_a from v to a vertex $r \in X$ dominating p_1. Then a path between X and Y exists which is shorter than P. This contradicts the definition of P. Hence CDS_a is connected. □

Next we link the size of CDS_a^b to the size of a minimum connected dominating set. We first need the following proposition.

Proposition 1. *Let $G = (V, E)$ be a connected graph and S an arbitrary dominating set of G. If S has n_{cc} connected components, then there exists a connected dominating set for G of size at most $|S| + 2n_{cc} - 2$.*

Proof. In any disconnected dominating set there are two connected components that can be connected by adding at most two vertices to the dominating set. □

Lemma 3. *Let OPT be a minimum connected dominating set for G. Then for each a ($0 \leq a \leq k - 1$) and for each strip b, $|CDS_a^b| \leq |OPT \cap D_a^b| + 2|OPT \cap \text{exterior}(D_a^b)|$.*

Proof. Let $OPT_a^b = OPT \cap D_a^b$. Clearly, OPT_a^b is a dominating set for the interior of D_a^b. However, OPT_a^b may consist of several connected components. We observe that each such connected component must intersect the exterior of D_a^b, or OPT would not be connected. Hence the number of connected components of OPT_a^b is at most $|OPT \cap \text{exterior}(D_a^b)|$. Then, following Proposition 1, there exists a connected dominating set of size at most $|OPT_a^b| + 2|OPT \cap \text{exterior}(D_a^b)|$ for each connected component in the interior of D_a^b, which possibly uses vertices in the exterior of D_a^b. Because CDS_a^b is a *minimum* connected dominating set for each connected component in the interior of D_a^b, possibly using vertices in the exterior of D_a^b, $|CDS_a^b| \leq |OPT \cap D_a^b| + 2|OPT \cap \text{exterior}(D_a^b)|$. □

Now let $CDS_{\min}$ be such that $|CDS_{\min}| = \min_{0 \leq a \leq k-1} |CDS_a|$.

Lemma 4. $CDS_{\min}$ *is at least a* $(1 + \frac{8}{k})$-*approximation of a minimum connected dominating set of* G.

Proof. Let OPT be a minimum connected dominating set of G. Then

$$
\begin{aligned}
k \, |CDS_{\min}| &\leq \sum_{a=0}^{k-1} |CDS_a| \\
&\leq \sum_{a=0}^{k-1} \sum_b |CDS_a^b| \\
&\leq \sum_{a=0}^{k-1} \sum_b |OPT \cap D_a^b| + 2|OPT \cap \text{exterior}(D_a^b)| \\
&= \left(\sum_{a=0}^{k-1} \sum_b |OPT \cap D_a^b| \right) + 2 \left(\sum_{a=0}^{k-1} \sum_b |OPT \cap \text{exterior}(D_a^b)| \right)
\end{aligned}
$$

Observe that no disk can be within distance 2 of a strip boundary for more than four values of a. Hence $\sum_{a=0}^{k-1} \sum_b |OPT \cap D_a^b| \leq (k+4)|OPT|$. Similarly, no disk can be in the exterior of a strip for more than two values of a. Therefore $\sum_{a=0}^{k-1} \sum_b |OPT \cap \text{exterior}(D_a^b)| \leq 2|OPT|$. Then $k \, |CDS_{\min}| \leq (k+4)|OPT| + 2(2|OPT|) = (k+8)|OPT|$. Hence $|CDS_{\min}| \leq (1 + \frac{8}{k})|OPT|$. □

Theorem 5. *There exists an FPTAS$^\omega$ for Minimum Connected Dominating Set on unit disk graphs of bounded density, i.e. of density* $d = d(n) = o(\log n)$.

Proof. For each a $(0 \leq a \leq k-1)$, D_a can be computed in $O(n \log n)$ time [21]. Computing CDS_a^b takes $O(f(n)^2 \, 2^{12f(n)} \, n)$ time. Since there are $\frac{n}{f(n)}$ strips, CDS_a can be computed in $O(f(n) \, 2^{12f(n)} \, n^2)$ time. Then $CDS_{\min}$ can be computed in $O(kn^2 f(n) \, 2^{12f(n)} + kn \log n)$ time. If we choose $f(n) = \frac{1}{12} \log n$, the running time is $O(kn^3 \log n)$. Choosing $k = \lceil \frac{8}{\epsilon} \rceil$, we obtain an $(1+\epsilon)$ approximation of the minimum connected dominating set in time polynomial in n and $\frac{1}{\epsilon}$. Because k can be at most $\frac{f(n)}{d}$, $\frac{96d}{\log n}$ is the lowest possible value of ϵ for a given n. If however $d = d(n) = o(\log n)$, then $\frac{96d}{\log n}$ becomes smaller than any desired value for n large enough. Hence the described scheme is an FPTAS$^\omega$. □

Now consider again the results for Minimum Connected Dominating Set by Cheng et al. [8] and Demaine and Hajiaghayi [10]. Our main improvement is adding vertices within distance 2, where previously this distance would have been $\Theta(\log n)$. By adapting Lemmas 2 and 3, we can improve the running times of the Cheng et al. and Demaine and Hajiaghayi algorithms. Then the almost-PTAS for minor-closed graphs of locally bounded treewidth and PTAS for planar graphs by Demaine and Hajiaghayi improve to an FPTAS$^\omega$. Although Demaine and Hajiaghayi [10] already claim an FPTAS$^\omega$ for planar graphs using their generic algorithms for contraction-bidimensional problems, the new analysis extends Baker's [4] original approach for planar graphs to Minimum Connected Dominating Set.

We mention here that schemes for other problems than Minimum Connected Dominating Set can be constructed in similar ways [22].

Theorem 6. *There exists an FPTAS$^\omega$ for Maximum Independent Set, Minimum Vertex Cover, and Minimum Dominating Set on unit disk graphs of bounded density, i.e. of density* $d = d(n) = o(\log n)$.

For lack of space, we omit a formal proof of this theorem.

Acknowledgements. The author would like to thank Hans Bodlaender and Lex Schrijver for many helpful suggestions and discussions.

References

1. Alber, J., Fiala, J., "Geometric Separation and Exact Solutions for the Parameterized Independent Set Problem on Disk Graphs", *J. Algorithms* **52** 2 (Aug. 2003), pp. 134–151.
2. Alber, J., Niedermeier, R., "Improved Tree Decomposition Based Algorithms for Domination-like Problems" in *LATIN 2002*, LNCS **2286**, Springer-Verlag, Berlin, 2002, pp. 613–628.
3. Ausiello, G., Creszenzi, P., Gambosi, G., Kann, V., Marchetti-Spaccamela, A., Protasi, M., *Complexity and Approximation - Combinatorial Optimization Problems and Their Approximability*, Springer-Verlag, Berlin, 1999.
4. Baker, B.S., "Approximation Algorithms for NP-Complete Problems on Planar Graphs", *JACM* **41** 1 (1994), pp. 153-180.
5. Becker, H.W. "Planar Rhyme Schemes", *Bull. Am. Math. Soc.* **58** (1952), pp. 39.
6. Bodlaender, H.L., "A Tourist Guide through Treewidth", *Acta Cybernetica* **11** 1–2 (1993), pp. 1–22.
7. Chan, T.M., "Polynomial-time Approximation Schemes for Packing and Piercing Fat Objects", *J. Algorithms* **46** 2 (Feb. 2003), pp. 178–189.
8. Cheng, X., Huang, X., Li, D., Wu, W., Du, D.-Z., "A Polynomial-Time Approximation Scheme for the Minimum Connected Dominating Set in Ad Hoc Wireless Networks", *Networks* **42** 4 (Dec. 2003), pp. 202–208.
9. Clark, B.N., Colbourn, C.J., Johnson, D.S., "Unit Disk Graphs", *Discr. Math.* **86** 1–3 (1990), pp. 165–177.
10. Demaine, E.D., Hajiaghayi, M., "Bidimensionality: New Connections between FPT Algorithms and PTASs" in *SODA 2005*, SIAM, 2005, pp. 590–601.
11. Downey, R.G., Fellows, M.R., *Parameterized Complexity*, Springer-Verlag, New York, 1999.
12. Erlebach, T., Jansen, K., Seidel, E., "Polynomial-time Approximation Schemes for Geometric Graphs" in *SODA 2001*, SIAM, 2001, pp. 671–679.
13. Hochbaum, D.S., Maass, W., "Approximation Schemes for Covering and Packing Problems in Image Processing and VLSI", *JACM* **32** 1 (1985), pp. 130–136.
14. Hunt III, D.B., Marathe, M.V., Radhakrishnan, V., Ravi, S.S., Rosenkrantz, D.J., Stearns, R.E., "NC-Approximation Schemes for NP- and PSPACE-Hard Problems for Geometric Graphs", *J. Algorithms* **26** 2 (1998), pp. 238–274.
15. Marathe, M.V., Breu, H., Hunt III, H.B., Ravi, S.S., Rosenkrantz, D.J., "Simple Heuristics for Unit Disk Graphs", *Networks* **25** (1995), pp. 59–68.
16. Matsui, T., "Approximation Algorithms for Maximum Independent Set Problems and Fractional Coloring Problems on Unit Disk Graphs", in *JCDCG*, LNCS **1763**, Springer-Verlag, Berlin, 1998, pp. 194–200.
17. Nieberg, T., Hurink, J.L., Kern, W., "A Robust PTAS for Maximum Weight Independent Sets in Unit Disk Graphs" in *WG 2004*, LNCS **3353**, Springer-Verlag, Berlin, 2004, pp. 214–221.
18. Nieberg, T., Hurink, J.L., *A PTAS for the Minimum Dominating Set Problem in Unit Disk Graphs*, Memorandum No. 1732, Dept. of Appl. Math., Univ. Twente, Enschede, 2004.

19. Robertson, N., Seymour, P.D., "Graph Minors. I. Excluding a Forest", *J. Comb. Th. B* **35** (1983), pp. 39–61.
20. Telle, J.A., Proskurowski, A., "Algorithms for Vertex Partitioning Problems on Partial k-Trees", *SIAM J. Disc. Math.* **10** 4 (1997), pp. 529–550.
21. van Leeuwen, E.J., *Optimization Problems on Mobile Ad Hoc Networks – Algorithms for Disk Graphs*, Master's Thesis INF/SCR-04-32, Inst. of Information and Computing Sciences, Utrecht Univ., 2004.
22. van Leeuwen, E.J., *Approximation Algorithms for Unit Disk Graphs*, Technical Report UU-CS-2004-066, Inst. of Information and Computing Sciences, Utrecht Univ., 2004.
23. Wan, P-J., Alzoubi, K.M., Frieder, O., "Distributed Construction of Connected Dominating Set in Wireless Ad Hoc Networks", *IEEE Infocom 2002*, Volume 3, 2002, pp. 1597–1604.

Computation of Chromatic Polynomials Using Triangulations and Clique Trees

Pascal Berthomé[1], Sylvain Lebresne[2], and Kim Nguyễn[1]

[1] Laboratoire de Recherche en Informatique (LRI), CNRS UMR 8623,
Université Paris-Sud, 91405, Orsay-Cedex, France
{Pascal.Berthome, Kim.Nguyen}@lri.fr
[2] Preuves, Programmes et Systèmes (PPS), CNRS UMR 7126, Université Paris 7,
Case 7014, 2 Place Jussieu, 75251 PARIS Cedex 05, France, and
Projet Logical, LIX, École Polytechnique, 91128 Palaiseau Cedex, France
Sylvain.Lebresne@pps.jussieu.fr

Abstract. In this paper, we present a new algorithm for computing the chromatic polynomial of a general graph G. Our method is based on the addition of edges and contraction of non-edges of G, the base case of the recursion being chordal graphs. The set of edges to be considered is taken from a triangulation of G. To achieve our goal, we use the properties of triangulations and clique-trees with respect to the previous operations, and guide our algorithm to efficiently divide the original problem.

Furthermore, we give some lower bounds of the general complexity of our method, and provide experimental results for several families of graphs.

Keywords: Chromatic polynomial, chordal graphs, minimal triangulation, clique tree.

1 Introduction

Introduced by Birkhoff and Lewis in 1946 [6], the chromatic polynomial of a graph G counts the number of ways of properly coloring G. This polynomial also captures many combinatorial information about a graph, describing acyclic orientations, the all-terminal reliability, and the spanning trees. More surprisingly, it is closely related in physics with the zero-temperature partition function of the q-state Potts antiferromagnet, motivating the computation this polynomial for some class of graphs by physicists (e.g., see [8]). Created in order to give some proof of the famous 4-color theorem, the chromatic polynomial has been studied for itself. Studies include among others the search of the real/complex roots of the polynomial [11,8], and the search of graphs uniquely defined by their chromatic polynomials [9].

Once we have the chromatic polynomial of any graph, its chromatic number is simply the first integral non-zero of the polynomial, thus, it can be computed in polynomial time. As shown in [14], the chromatic polynomial includes

D. Kratsch (Ed.): WG 2005, LNCS 3787, pp. 362–373, 2005.

many other notions than the chromatic number, thus its computation reveals to be quite complex even when the chromatic number is known. Recently, it was shown that computing the coefficients of this polynomial in general graphs is #P-hard [13].

Several papers deal with the effective computation of this polynomial. Most of them use the paradigm of edge contraction/deletion. The main strategies developed in the literature stop the recursion to graphs for which the polynomial is easy to compute. We can note the work of Haggard [10], in which this latter paradigm is used at the extreme: if the considered graph is small enough it checks if it is isomorphic to a graph which polynomial has already been computed. Other papers, such as [14], use as base case chordal graphs for which the chromatic polynomial is easy to determine [7].

In this paper, we also use the chordal graphs as base cases in the recursion tree. However, we improve the global computation by exploiting the clique-tree associated to the considered chordal graphs. Another important difference with most of the literature is that we prefer an edge addition/contraction method rather than the classical edge contraction/deletion method.

In the remaining of the paper, we recall in Section 2 known results on the chromatic polynomials, the chordal graphs and the triangulation of a graph. In Section 3, we provide results that lead to our algorithm for computing the chromatic polynomial. The general algorithm as well as lower bounds on its time complexity is given in Section 4. In Section 5, we provide some experimental results. Finally, we conclude the paper in Section 6. A detailed version of this extended abstract is available at [5].

2   Preliminaries

2.1   Chromatic Polynomials

In the following, we use classical graph theory on undirected graphs. Notions may be found for example in [2]. Let $G = (V, E)$ be an undirected loop-less graph, with $n = |V|$ and $m = |E|$. A proper coloring of G with k colors is simply a function ϕ from V into $I_k = \{1, \ldots, k\}$ ($I_0 = \emptyset$) such that two neighbors in the graph have different colors. For a given integer $\lambda \geq 0$, the *chromatic polynomial* $P(G, \lambda)$ is the number of distinct proper colorings of G using at most λ colors. In [14], the different forms of this polynomial are reviewed. According to the chosen basis, the coefficients reflect different properties of the graph. As example, let give some well-known chromatic polynomials, leading to the most popular basis for writing these polynomials.

Empty graph: λ^n;

Complete graph: $\lambda^{(n)} = \prod\limits_{i=0}^{n-1} (\lambda - i)$;

Trees: $\lambda(\lambda - 1)^{n-1}$.

Here are some simple results that are the base of the computation of the chromatic polynomial. If e is an edge of G, we note $G - e$ the graph obtained by removing e from E, and G/e the graph obtained by contracting e. If e is not an edge of G, we denote $G + e$ the graph obtained by adding e to G. In this case, we can define G/e by $(G+e)/e$. Using this notation, the two following relations can be easily established.

$$P(G, \lambda) = P(G + e, \lambda) + P(G/e, \lambda), \text{ if } e \notin E \tag{1}$$

$$P(G, \lambda) = P(G - e, \lambda) - P(G/e, \lambda), \text{ if } e \in E; \tag{2}$$

Both formulations can be applied to compute recursively the chromatic polynomial. Base case in Formulation 1 is the complete graph, since we add edges to the initial graph, and to any contracted graph, whereas the base cases with the other formulation are the trees or the empty graphs. However, the straightforward use of these formulations clearly lead to an exponential exploration and is not practical in many simple cases. Note that these two equations simply verify that the object we are talking about is a polynomial, since it is obtained by a finite combination (addition/subtraction) of elementary polynomials.

We also introduce the following folklore lemma:

Lemma 1 ([11]). *Let G be a graph and G_1 and G_2 be subgraphs of G such that $G = G_1 \cup G_2$ and $G_1 \cap G_2 \sim K_r$, then we have:*

$$P(G, \lambda) = \frac{P(G_1, \lambda) P(G_2, \lambda)}{\lambda^{(r)}} \tag{3}$$

Using this lemma, we can derive the chromatic polynomial of the trees, by eliminating all the leaves, one by one. Another application is the computation of the chromatic polynomial of the chordal graphs as shown below.

In this paper, we will use Lemma 1 to break down the effective complexity of computation of the chromatic polynomial. The goal in the recursions steps of our algorithm is first to complete clique separators of the graph in order to split the computation of the chromatic polynomial of a large graph into several computations for smaller ones. The edges that should be added in this process should belong to some (minimal) triangulation of the input graph.

2.2   Triangulation and Clique Trees

A *chordal graph* G is a graph in which there are no induced cycles of length > 3. Many problems that are NP-complete in general graphs are polynomially solvable in chordal graphs, such as the colorability problem. One useful representation of chordal graphs is the *clique-tree* initially described in [3]. The following definition is derived from [1]. A **clique-tree** of a given chordal graph $G = (V, E)$ is a tree $T = (\mathcal{V}, \mathcal{E})$ such that:

- $\mathcal{V} = \{C_1, C_2, \ldots, C_j\}$ is the set of the maximal cliques of G;
- for every vertex v in G, the set of maximal cliques containing v induces a subtree of T.

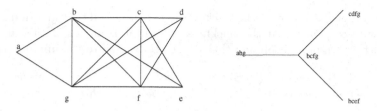

Fig. 1. A chordal graph and its clique tree

A simple example is given in Figure 1. Note that any edge in the clique tree represents a minimal clique separator of G. This provides a simple way to compute the chromatic polynomial for this family of graphs as shown in Lemma 2 below.

For a general graph $G = (V, E)$, a *triangulation* of G is a set of edges F such that $G' = (V, E \cup F)$ is chordal. Finding a triangulation with the minimum number of edges is known as the *minimal fill-in* problem and is NP-complete [15]. A triangulation is minimal if no proper subset of it is a triangulation. Many efficient algorithms have been designed for computing a minimal triangulation. For example, the minimum degree heuristic and its variations [4] appear to be very efficient in practise for the minimum fill-in problem, however it does not constitute an approximation algorithm. In [12], a $O(k)$ approximation polynomial algorithm is given, where k is the minimum fill-in value.

Definition 1. *Let $G = (V, E)$ be a graph and F a minimal triangulation of G. Let $T = (\mathcal{V}, \mathcal{E})$ a clique tree of $G' = (V, E \cup F)$. The **augmented clique tree** of G for F is $T = (\mathcal{V}, \mathcal{E}, \phi)$, where ϕ is a labeling of $\mathcal{E}$ by subsets of F defined by:*

$$\phi(V_i, V_j) = F \cap E(G'[V_i \cap V_j]),$$

where $G'[U]$ denotes the subgraph of G' induced by U a subset of $V(G')$.

Note that the edges of T constitute the minimal clique separators of G', the triangulation of G. Consequently, the labeling defines for any edge of T the subset of F that has to be added in G to complete this separator. In order to measure the quality of a triangulation, we define the **thickness of a triangulation** as the maximal size of a label of the corresponding augmented clique tree.

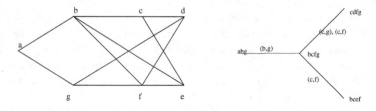

Fig. 2. A graph and its augmented clique tree for the triangulation $F = \{(b, f), (b, g), (c, f)\}$

For example, in Figure 2, the thickness of the proposed triangulation is 2. We will show in Section 6 that the thickness of a graph is an original parameter of the triangulation of a graph and that it is disconnected to the minimum fill-in.

Lemma 2 ([7]). *Let $G = (V, E)$ be a chordal graph. Let $T = (\mathcal{V}, \mathcal{E})$ be a clique-tree representation of G, $\mathcal{V} = \{V_1, \ldots, V_k\}$. Then, the chromatic polynomial of G is:*

$$P(G, \lambda) = \prod_{i=1}^{k} \lambda^{|V_i|} / \prod_{(V_i, V_j) \in \mathcal{E}} \lambda^{|V_i \cap V_j|} \tag{4}$$

3   Augmented Clique Tree, Edge Contraction and Separation

In this section, we exhibit some properties of the augmented clique tree under the edge manipulations and graph separation operations.

Using Equation 1, we perform both operations of edge addition and contraction of a non-edge of the graph. In this section, we show how an augmented clique tree evolves under these two elementary operations. At each step of the computation of the chromatic polynomial, we can associate three elements:

(1) the graph G_1 whose chromatic polynomial has to be computed;
(2) a triangulation of G_1, say F_1;
(3) a clique tree T_1 of $G'_1 = G_1 + F_1 = (V(G_1), E(G_1) \cup F_1)$.

Theorem 1. *Let G be a graph, F a triangulation of G, and T a clique tree of $G + F$. Let e be an element of F. Let $G_1 = G + e$ and $G_2 = G/e$. Then, there exists F_1, F_2, T_1 and T_2 such that:*

1. F_1 is a triangulation of G_1 and T_1 is a clique tree of $G_1 + F_1$;
2. F_2 is a triangulation of G_2, and T_2 is a clique tree of $G_2 + F_2$.

These sets can be computed in quadratic time $(O(|V(G)|^2))$ from G, F and T.

Proof. This proof is divided into two parts.

The **first** one considers the graph $G_1 = G + e$. From the previous remarks, it is easy to see that $F_1 = F - e$ and $T_1 = T$ is a valid choice. These two sets can be clearly obtained in linear time.

The **second** part of the proof concerns $G_2 = G/e$. We will just sketch the proof here, for the sake of clarity. First, let us remark that if a graph is chordal, then it remains chordal after an edge contraction. In our case, $G + F$ is chordal, and consequently, so is $(G + F)/e$. Let denote a and b the extremities of e. The operation, $/e$ on a set or a graph simply consists in identifying both extremities of e and remove redundant elements.

Then F_2 can be defined as $E((G + F)/e) - E(G/e)$. It simply corresponds to the elements of F/e that are not edges of G/e. This occurs when an edge of type (b, x) in F collapse into an edge (a, x) in $E(G)$. Then, the clique tree corresponding to $G_2 + F_2$ can be obtained from the description of this graph. However, there exists a clever way to do this, simply based on the description of T. This method is described in Algorithm 1 which is proven in details in [5]. This algorithm and the contraction algorithms can be performed in $O(n^2)$ operations, where n is the number of vertices of G. $\qquad\square$

Algorithm 1: Clique-Tree-Contraction(T,e)
     ▷ T *is a clique-tree of a given chordal graph* G *and* $e = (i, j)$ *is an edge of* G
1   **begin**
2   Within any clique set, replace the elements j by i
3   **while** there exists two neighbor cliques c_1 and c_2 such that $c_1 \subseteq c_2$
4   **if** $\phi(c_1, c_2)$ is contained in $\cup_{(i,j)\neq(1,2)}\phi(e_i, e_j)$ **then**
5   Contract the edge c_1, c_2 in T
6   **return** T', the resulting tree after renaming and contractions
7   **end**

We illustrate our method on a simple example in Figure 3. Let consider the cycle with 7 vertices and one of its minimal triangulation in Figure 3(a). Figure 3(c) shows the result on C_7 of the contraction of the edge $(1, 5)$. Figure 3(b) shows how the (augmented) clique tree evolves using Algorithm 1.

Note that the augmented clique tree structure can be regarded as another description of the graph itself. Thus, during the overall computation, we can only maintain this structure.

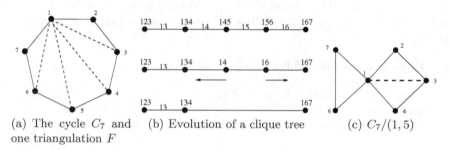

(a) The cycle C_7 and one triangulation F (b) Evolution of a clique tree   (c) $C_7/(1, 5)$

Fig. 3. The clique-tree contraction on the cycle. The triangulation sets are shown in dashed lines.

Let consider now the impact of Lemma 1 on the augmented clique tree.

Theorem 2. *Let* G *be a graph and* G_1 *and* G_2 *be subgraphs of* G *such that* $G = G_1 \cup G_2$ *and* $G_1 \cap G_2 \sim K_r$. *Let* F *be a triangulation of* G *and* T *be the clique tree of* $G + F$. *Then, the triangulation set* F *can be divided into two disjoint sets* F_1 *and* F_2 *such that* F_1 *(resp.* F_2*) is a triangulation of* G_1 *(resp.* G_2*). Furthermore, the clique trees of the triangulated graphs can be obtained from* T *by removing only one edge.*

Proof. First, since $G_1 \cap G_2 \sim K_r$, there is no edge of F that can be added to $G_1 \cap G_2$. Consequently any edge of F has to be added either on $G\backslash G_2$ or $G\backslash G_1$.

Then let consider F_1 (resp. F_2) the subset of F for which both extremities are in G_1 (resp. F_2). Using the above remark, these two sets form a partition of F.

Considering a clique tree of $G + F$, since $G_1 \cap G_2$ is isomorphic to K_r, it corresponds to a clique separator of G, consequently for $G + F$. Thus, there exists an edge e in this tree that represents this separator. Consequently, removing this edge from the tree creates two clique trees T_1 and T_2. It is straightforward to see that T_i is an augmented clique tree of $G_i + F_i$, $i = 1, 2$. □

These two operations performed on an initial triangulation of the input graph lead to our general algorithm presented in the following section.

4   General Algorithm

4.1   Algorithm

Using the results of the previous section, we give an algorithm for computing the chromatic polynomial of a generic graph G as well as some bounds on its complexity. The idea of the algorithm is quite simple. First, using Theorem 1 we pre-compute a triangulation for a given graph G and use it to add (and contract) edges, computing until we reach a chordal graph. The second optimization is to direct the choice of the edge in the addition/contraction algorithm in order to arise to a clique separator. For this choice, we use the augmented clique tree. Actually, if a label of the augmented clique tree is empty, the corresponding edge is a clique separator of the associated graph. We discuss below the choice of the edge to add/contract at each step has a direct impact on the efficiency of the algorithm.

Algorithm 2: Chromatic-Polynomial(G, T)
▷ *T is an augmented clique-tree of the graph G*
▷ *Returns the Chromatic polynomial of G*
1 **begin**
2   **if** G is triangulated **then return** Chromatic-Polynomial(G,T) using Lemma 2

3   **if** $\exists e \in T$ such that $\phi(e) = \emptyset$
4   Decompose G using Theorem 2:
5   Let G_1, T_1, G_2, T_2 and K_r the resulting elements
6              $P_1 \leftarrow$ Chromatic-Polynomial(G_1, T_1)
7              $P_2 \leftarrow$ Chromatic-Polynomial(G_2, T_2)
8   **return** $P_1 \times P_2 / P(K_r)$
9   Let $e = $ ChoiceFunction(G,T)
10   Using Theorem 1, we compute
11         $G_1 = G + e$, and the resulting $T_1 = T$
12         $G_2 = G/e$, and the resulting T_2
13   **return** Chromatic-Polynomial(G_1, T_1) + Chromatic-Polynomial(G_2, T_2)
14 **end**

The correctness of this algorithm is straightforward given Equation 1 and Theorems 1 and 2.

The choice function used in Step 9 has a great impact on the overall efficiency of the algorithm. Let us remind that the main idea of our method is to add all the edges of $\phi(e)$ for a given edge e of the augmented clique tree, in order to create a clique on the graph and then, be able to split it into two smaller graphs. Nevertheless, to have this method perform efficiently, we need to separate the graph so that the remaining edges to be added are evenly distributed between each graph. Several choice functions have been tested based on different properties of the augmented clique tree. The choices are based on two main ideas: find an edge of the augmented clique tree with small thickness in order to break the tree quickly, and break the augmented clique tree into two equivalent parts.

Thus, taking these two points into account when implementing the choice function, we can clearly improve our algorithm. However, it's unlikely that there is a best choice function for all graphs. Experimentation might be the key to fine-tune ad-hoc parameters to compute practically certain (class of) graphs. The impact of this choice function has been exhibited for cycles [5]. Using Algorithm 2, the number of nodes of the computation tree vary from $O(\rho^n)$, where ρ is the golden ratio to $O(n^2)$ when the edges are added in a clever way.

4.2   A Lower Bound of the Complexity

In this section, we provide a simple lower bound of the complexity of the previous algorithm directly connected to the structure of the triangulation set, viewed as a simple graph. In the following, we show that, within a same edge of the clique tree, the order in which we examine the edges in F is not significant in terms of number of nodes in the recursion tree. The proofs here are omitted but may be found in [5].

Lemma 3. *Let F be an edge set using labels in $\{1, \ldots, k\}$, and $n \geq k$. Let $\sum b_i^n \lambda^{(i)}$ be the chromatic polynomial of $G_n = K_n \setminus F$. Then, the value $s(n) = \sum b_i^n$ is independent of n $(= cc(F))$.*

Let F be an edge set using labels in $\{1, \ldots, k\}$. We define $cc(F)$ as the number of clique covers of F. Lemma 3 provides a way to compute this value using the chromatic polynomial. For example, it is easy to see that $cc(F) = 2$ is F is reduced to one edge, and 2^k if F is a matching of size k.

Theorem 3. *Let G be graph and F a triangulation of G. Let T be the augmented clique tree of G and F. Let F_1 be the label of an edge of T. Then, the number of nodes in the recursion tree induced by Algorithm 2 is greater than $2cc(F_1) - 1$.*

Thus, given a triangulation of a graph, we can evaluate a lower bound of the complexity, and thus of the general computation time of the chromatic polynomial. Note that $cc(F)$ is directly related to the chromatic polynomial of $K_m \setminus F$, where K_m is the smallest complete graph that contains F (e.g., two consecutive edges are included in K_3). Thus, the computation of the lower bound can be performed using our general technique, on smaller graphs as the input ones.

5   Experiments

In this section, we provide experimental results on several classes of graphs. The algorithms are coded in OCAML. The experiments have been carried out on a AMD Athlon XP 2800. In the following, we never consider triangulated graphs since Lemma 1 gives an efficient solution. All the computations of the thickness of triangulation parameter have been performed using the minimum degree heuristic. This only provides an upper bound of the parameter and a decomposition for this value.

Small Thickness of Triangulation
In this paragraph, we examine the behavior of this algorithm on graphs having a small thickness of triangulation. In this class, we only present cycles, cylinders and grids. Other experiments can be found in [5].

Cycle: C_n, n nodes, n edges, thickness of triangulation: 1. The clique tree is a chain, as we have seen in the previous sections. Even the closed form of the chromatic polynomial is known, this constitutes a good benchmark for generic algorithms.

n	100	200	300	400	500	600	700	800	900	1000	2000
Time (s)	0.23	1.36	2.78	5.43	8.92	13.22	19.81	26.01	33.84	43.05	241.20

Cylinders 4 times n: $Cyl(4, n)$, $4n$ nodes, $8n-4$ edges, thickness of triangulation: 4. The clique tree is a chain with extremities having two leaves.

n	5	10	15	20	25	30	40	50
$|V(G)|$	20	40	60	80	100	120	160	200
Time (s)	0.06	0.72	2.33	6.41	12.49	19.68	55.65	108

The thickness of triangulation increases a lot with the size of the cycle in the cylinder. The corresponding clique tree is more and more compact. All these tends to let more complex the chromatic polynomial.

Grids 3 times n: $M(3, n)$, $3n$ nodes, $5n - 3$ edges, thickness of triangulation: 2. The augmented clique tree in this case is the same as for the cylinders $Cyl(4, n)$.

n	10	20	30	40	50	60	70	80	90	100	150	200
Time (s)	0.03	0.17	0.42	0.87	1.40	2.05	2.96	4.09	5.04	6.73	16.87	32.15

Random Graphs
Many experiments on random graphs have been performed. However, we would like to emphasize several characteristics. The aim was to determine the link between the efficiency of computation, the size and the density of the graph. Each value is the mean over 100 test graphs.

From these experiments, we can see that the curves of the thickness of triangulation follows the time's. For the graphs having more than 22 edges, the line `Solved/10mn` presents the number of instances in the test that have been solved successfully without any time constraint and within 10 minutes (e.g., 73/53 means that 73 chromatic polynomials have been found and 53 within 10 mn of computation; -/100 means that all the instances have been solved within 10mn). The mean times are given relatively to the solved instances. It may not be significant when the number of solved instances is small.

20 vertices								
density	20	30	40	50	60	70	80	90
#edges Clique Tree	14.2	12.00	9.97	8.27	6.9	5.8	4.65	3.52
diameter Clique Tree	6.42	5.86	5.02	4.45	3.94	3.43	2.83	2.27
Size Max Clique	6.67	8.99	11.00	12.73	14.09	15.19	16.35	17.48
Thickness	8.7	15.86	22.26	25.61	25.51	23.17	17.8	10.26
Time (s)	0.25	3.02	16.30	39.17	36.62	24.29	6.68	0.49

21 vertices								
density	20	30	40	50	60	70	80	90
#edges Clique Tree	15.20	12.60	10.58	8.48	6.92	5.8	4.68	3.63
diameter Clique Tree	6.87	5.95	5.36	4.61	3.95	3.44	2.85	2.40
Size Max Clique	6.65	9.38	11.4	13.52	15.08	16.20	17.32	18.37
Thickness	9.1	18.01	24.35	30.05	29.94	26.79	20.49	11.47
Time (s)	0.44	9.73	38.43	169.82	167.69	76.83	17.99	1.03

22 vertices								
density	20	30	40	50	60	70	80	90
#edges Clique Tree	15.75	12.68	10.51	8.68	7.09	5.98	4.84	3.54
diameter Clique Tree	6.95	5.88	5.38	4.62	4.08	3.55	2.95	2.34
Size Max Clique	7.19	10.28	12.47	14.32	15.9	17.00	18.16	19.46
Thickness	10.38	21.75	29.69	34.63	33.97	30.3	22.43	12.78
Time (s)	1.31	24.59	188.06	556.29	652.60	256.30	42.73	2.07
Solved/10mn	-/100	-/100	99/95	98/69	100/69	100/89	-/100	-/100

23 vertices								
density	20	30	40	50	60	70	80	90
#edges Clique Tree	16.03	12.87	10.62	8.78	7.23	5.94	4.86	3.72
diameter Clique Tree	7.04	6.07	5.29	4.71	4.11	3.45	2.91	2.46
Size Max Clique	7.91	11.13	13.35	15.21	16.76	18.06	19.14	20.28
Thickness	13.98	26.82	35.49	39.74	39.23	33.45	25.17	13.96
Time (s)	2.98	112.95	465.65	1189.29	1540.93	867.54	121.00	4.29
Solved/10mn	-/100	99/97	85/63	74/24	78/21	100/54	100/98	-/100

24 vertices								
density	20	30	40	50	60	70	80	90
#edges Clique Tree	16.55	13.32	10.78	9.03	7.53	6.15	4.97	3.76
diameter Clique Tree	7.02	6.1	5.29	4.79	4.27	3.66	3.08	2.48
Size Max Clique	8.41	11.66	14.2	15.97	17.45	18.85	20.03	21.24
Thickness	16.01	30.75	40.47	45.22	43.82	38.23	28.41	15.79
Time (s)	8.16	304.10	1096.51	1070	1175	1316	351	9.2
Solved/10mn	-/100	96/82	56/21	16/5	15/2	60/12	100/84	-/100

25 vertices								
density	20	30	40	50	60	70	80	90
#edges Clique Tree	16.87	13.49	10.99	9.14	7.57	6.27	5.03	3.81
diameter Clique Tree	7.01	6.13	5.33	4.83	4.32	3.71	3.09	2.43
Size Max Clique	9.10	12.5	15.01	16.85	18.43	19.73	20.97	22.19
Thickness	19.84	35.21	46.2	49.83	48.11	42.63	31.32	18.23
Time (s)	48.20	474.87	1045	906	982	948	781.61	24.98
Solved/10mn	100/98	73/53	16/3	3/1	4/1	16/4	87/41	-/100

6   Conclusion

This work constitutes a first approach of the computation of the chromatic polynomial using the triangulation theory of the graphs. Doing this, we have enlightened some correlations between a new parameter of triangulation, namely the thickness of triangulation, and the computation time of our algorithm. Some non-trivial lower bounds, using the computation of chromatic polynomials on small graphs have been shown. In Section 5, we present some computational results using this technique. The main point is that the choice function has to be improved in order to compute the chromatic polynomial of random graphs of medium sized graphs (up to 30 or 40 nodes) for any density.

Many questions arise from this work. First of all, is this new parameter can be optimized independently from the other classical parameters of triangulation, as the minimum fill-in. In the example of Figure 4, we show that both parameters are not optimal for the same triangulations. In this example, only two minimal triangulations are possible (up to symmetry). The minimal degree heuristic finds the triangulation that minimizes the fill-in. Thus, it may be of interest studying this new parameter. Concerning the computation of the chromatic polynomial itself, this work constitutes a first step in the elaboration of a new heuristic. Studies on the choice function have to be performed and analysed. Another direction of experimentation could be to use the two opposite approaches for this computation. First, use the remove and contract edges paradigm (Equation 2)

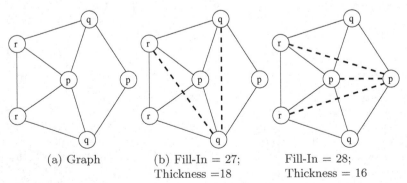

(a) Graph   (b) Fill-In = 27;   Fill-In = 28;
                        Thickness =18   Thickness = 16

Fig. 4. A graph for which the minimum fill-in and the thickness of triangulation does not exist for the same triangulations. Each circle denotes a complete graph and each edge represents a complete bipartite graph, and $p = 2$, $q = 3$ and $r = 6$.

in order to break large cliques, inducing large thickness, then use our approach when the triangulation induces a small enough thickness.

Acknowledgments

The authors thank Ioan Todinca for the counterexample of Figure 4, the anonymous referees for their helpful comments. They also thank Irénée Briquel for his deep improvement of the final implementation.

References

1. B. Aspvall and P. Heggernes. Finding minimum height elimination trees for interval graphs in polynomial time. *BIT*, 34(4):484–509, 1994.
2. C. Berge. *Graphs and Hypergraphs*. North Holland, Amsterdam, 1973.
3. P.A. Bernstein and N. Goodman. The power of natural semijoins. *SIAM Journal of Computing*, 10(4):751–771, February 1981.
4. A. Berry, P. Heggernes, and G. Simonet. The minimum degree heuristic and the minimal triangulation process. In H.L. Bodlaender, editor, *Graph Theoretical Concepts in Computer Science (WG) 2003*, volume 2880 of *Lecture Notes in Computer Science*, pages 58–70, October 2003.
5. P. Berthomé, S. Lebresne, and K. Nguyễn. Computation of chromatic polynomials using triangulation and clique trees. Technical report, LRI-1403, March 2005. Available at http://www.lri.fr/~berthome/biblio.html.
6. G.D. Birkhoff and D.C. Lewis. Chromatic polynomials. *Transactions of the American Mathematical Society*, 60:355–451, 1946.
7. N. Chandrasekharan, C.E.V. Madhavan, and R. Laskar. Chromatic polynomials of chordal graphs. *Congressus Numerantium*, 61:133–142, 1988.
8. S-C. Chang. Exact chromatic polynomials for toroidal chain of complete graphs. *Physica A*, 313:397–426, 2002.
9. F.M. Dong, K.L. Tep, K.M. Koh, and M.D. Hendy. Non-chordal graphs having integral-root chromatic polynomial II. *Discrete Mathematics*, 245:247–253, 2002.
10. G. Haggard and T.R. Mathies. Note on the computation of chromatic polynomials. *Discrete Mathematics*, 199:227–231, 1999.
11. B. Jackson. Zeros of chromatic and flow polynomials of graphs. *Journal of Geometry*, 76:95–109, 2003.
12. A. Natanzon, R. Shamir, and R. Sharan. A polynomial approximation algorithm for the minimum fill-in problem. In ACM, editor, *ACM Symposium On Theory of Computing*, pages 41–47, New York, NY, USA, 1998. ACM Press.
13. J. Oxley and D. Welsh. Chromatic, flow and reliability polynomials: the complexity of their coefficients. *Combinatorics, Probability and Computing*, 11:403–426, 2002.
14. D.R Shier and N. Chandrasekharan. Algorithms for computing the chromatic polynomial. *Journal of Combinatorial Mathematics and Combinatorial Computing*, 4:213–222, 1988.
15. M. Yannakakis. Computing the minimum fill-in is NP-complete. *SIAM Journal on Algebraic and Discrete Methods*, 2(1):77–79, March 1981.

Computing Branchwidth Via Efficient Triangulations and Blocks

Fedor Fomin[1], Frédéric Mazoit[2], and Ioan Todinca[3]

[1] Department of Informatics, University of Bergen, N-5020 Bergen, Norway
fomin@ii.uib.no
[2] LIF, Université de provence 13453 Marseille Cedex 13 France
Frederic.Mazoit@lif.univ-mrs.fr
[3] LIFO, Université d'Orléans 45067 Orléans Cedex 2 France
Ioan.Todinca@lifo.univ-orleans.fr

Abstract. Minimal triangulations and potential maximal cliques are the main ingredients for a number of polynomial time algorithms on different graph classes computing the treewidth of a graph. Potential maximal cliques are also the main engine of the fastest so far $\mathcal{O}(1.9601^n)$-time exact treewidth algorithm. Based on the recent results of Mazoit, we define the structures that can be regarded as minimal triangulations and potential maximal cliques for branchwidth: efficient triangulations and blocks. We show how blocks can be used to construct an algorithm computing the branchwidth of a graph on n vertices in time $(2 + \sqrt{3})^n \cdot n^{O(1)}$.

1 Introduction

Treewidth is one of the most basic parameters in graph algorithms and it plays an important role in structural graph theory. Treewidth serves as the main tools in Robertson and Seymour's Graph Minors project [19]. It is well known that many intractable problems can be solved in polynomial (and very often in linear time) when the input is restricted to graphs of bounded treewidth. See [3] for a comprehensive survey.

The branchwidth is strongly related to treewidth. It is known that for any graph G, $\mathrm{bw}(G) \leq \mathrm{tw}(G) + 1 \leq 1.5 \cdot \mathrm{bw}(G)$. Both bounds are tight and achievable on trees and complete graphs. Branchwidth was introduced by Robertson & Seymour and it appeared to be even more appropriate tool than treewidth for Graph Minor Theory. Since both parameters are so close, one can expect that the algorithmic behaviour of the problems is also quite similar. However, this is not true. For example, on planar graphs branchwidth is solvable in polynomial time [22] while computing the treewidth of a planar graph in polynomial time is a long standing open problem. Even more striking example was observed by Kloks et al. in [15]: it appeared that computing branchwidth is NP hard even on split graphs. Note that the treewidth of a split graph can be found in linear time.

The last decade has led to much research in fast exponential-time algorithms. Examples of recently developed exponential algorithms are algorithms for Maximum Independent Set [14,20], (Maximum) Satisfiability [8,13,18,21,24], Coloring [2,6,9], and many others (see the recent survey written by Woeginger [25] for an overview). There are several relatively simple algorithms based on dynamic programming computing the

D. Kratsch (Ed.): WG 2005, LNCS 3787, pp. 374–384, 2005.
© Springer-Verlag Berlin Heidelberg 2005

treewidth of a graph on n vertices in time $2^n \cdot n^{O(1)}$ which with more careful analysis can be sped-up to $O(1.9601^n)$ [10]. No such algorithm is known for branchwidth. The only nontrivial algorithm for branchwidth we were aware can be obtained by using deep min-max theorems of Robertson & Seymour [19] relating branchwidth and tangles. Then by playing with tangle axioms one can reduce the search space and perform dynamic programming to construct optimal tangles in time $4^n \cdot n^{O(1)}$. (We leave the details in this extended abstract.)

Thus treewidth seems to be more simple problem for design of exponential time algorithms than branchwidth. The explanation to that can be that all known exact algorithms for treewidth exploit the relations between treewidth, minimal triangulations, minimal separators and potential maximal cliques. Mazoit in [16] observed that the branchwidth also can be seen as a triangulation problem. However, while for treewidth one can work only with minimal triangulations the situation with branchwidth is more complicated. Luckily enough we still can use some specific triangulations, which we call efficient triangulations. The efficient triangulations were first used, under a different name, in [4]. In this paper we adopt the techniques of Mazoit to discover the analogue of potential maximal cliques for branchwidth, we call these structures by blocks. Potential maximal cliques are extremely useful tools in work with treewidth [5,10]. We believe that blocks can also be useful to work with branchwidth. To exemplify that we show how blocks can be used to compute branchwidth in time $(2 + \sqrt{3})^n \cdot n^{O(1)}$. Note that this is the fastest known exact algorithm for this problem.

2 Basic Definitions

We denote by $G = (V, E)$ a finite undirected and simple graph with $|V| = n$ vertices and $|E| = m$ edges. Throughout this paper we use a modified big-Oh notation that suppresses all polynomially bounded factors. For functions f and g we write $f(n) = O^*(g(n))$ if $f(n) = g(n) \cdot n^{O(1)}$.

For any non-empty subset $W \subseteq V$, the subgraph of G induced by W is denoted by $G[W]$. If S is a set of vertices, we denote by $G - S$ the graph $G[V \setminus S]$. The *neighborhood* of a vertex v is $N(v) = \{u \in V : \{u, v\} \in E\}$ and for a vertex set $S \subseteq V$ we put $N(S) = \bigcup_{v \in S} N(v) \setminus S$. A *clique* C of a graph G is a subset of V such that all the vertices of C are pairwise adjacent. Let $\omega(G)$ denote the maximum clique size of G.

A graph G is *chordal* if every cycle of G with at least four vertices has a chord, that is an edge between two non-consecutive vertices of the cycle. Consider an arbitrary graph $G = (V, E)$, and a supergraph $H = (V, F)$ of G (i.e. $E \subseteq F$). We say that H is a *triangulation* of G if H is chordal. Moreover, if no strict sub-graph of H is a triangulation of G, then H is called a *minimal triangulation*.

The notion of branchwidth is due to Robertson and Seymour [19]. A *branch decomposition* of a graph $G = (V, E)$ is a pair (T, τ) in which $T = (V_T, E_T)$ is a ternary tree (i.e. each node is of degree one or three) and τ is a function mapping each edge of G on a leaf of T. The vertices of T will be called *nodes* and its edges will be called *branches*. For any branch $e \in E_T$, let $T_1(e)$ and $T_2(e)$ be the subtrees obtained from T by removing e. Let $\mathrm{lab}(e)$ be the set of vertices of G both incident to edges mapped on $T_1(e)$ and $T_2(e)$. The maximum of $\{|\operatorname{lab}(e)|, e \in E_T\}$, is called the *width* of the branch decomposition. The *branchwidth* of a graph G $(\operatorname{bw}(G))$ is the minimum width over all branch decompositions of G. Note that the definitions of branch decomposition and branch-width also apply to hypergraphs. As pointed by Robertson and Seymour, the definition of branch decomposition can be relaxed. A *relaxed branch decomposition* of $G = (V, E)$ is a couple (T, τ) where T is an arbitrary tree and τ is an application mapping each edge of G to at least one leaf of T. The labels of the branches and the width of the decomposition are defined as before. From any relaxed branch decomposition we can construct a branch decomposition without increasing the width.

The branchwidth is strongly related to a well-known graph parameter introduced by Robertson and Seymour, namely the *treewidth*. One of the equivalent definitions for treewidth is $\operatorname{tw}(G) = \min\{\omega(H) - 1 \mid H \text{ is a triangulation of } G\}$. Robertson and Seymour show that the two parameters differ by at most a factor of 1.5. More precisely, for any graph G we have $\operatorname{bw}(G) \leq \operatorname{tw}(G) + 1 \leq 1.5 \operatorname{bw}(G)$. In particular, if G is a complete graph, its treewidth is $n - 1$, while its branchwidth is $\lceil 2n/3 \rceil$ (see [19]). Clearly, when computing the treewidth of a graph we can restrict to minimal triangulations. This observation and the study of minimal triangulations of graphs led to several results about treewidth computation, including an exact algorithm in $\mathcal{O}^*(1.961^n)$ time.

The branch decompositions of a graph can also be associated to triangulations. Indeed, given a branch decomposition (T, τ) of $G = (V, E)$, we can associate to each $x \in V$ the subtree of T covering all the leaves of T containing edges incident to x. It is well-known that the intersection graph of the sub-trees of a tree is chordal [11]. Thus the intersection graph of the trees T_x is a triangulation $H(T, \tau)$ of G. Note that for each branch $e \in E_T$, $\operatorname{lab}(e)$ is the set of vertices x such that e belongs to T_x. In particular, $\operatorname{lab}(e)$ induces a clique in $H(T, \tau)$, not necessarily maximal. (We shall point out later that, for each maximal clique Ω of $H(T, \tau)$, there exists a node u of T such that $u \in T_x$ for all $x \in \Omega$.)

The first big difference with treewidth is that there exist examples of graphs for which any optimal branch decomposition leads to non-minimal triangulations [16]. Therefore the many existing tools about minimal triangulations are not sufficient in our case. The second important difference is that the branchwidth problem remains NP-hard even for a restricted class of chordal graphs, the *split* graphs [15]. Nevertheless, our technique for computing the branchwidth relies on a structural result stating that, for any graph G, there is an optimal branch decomposition (T, τ) such that $H(T, \tau)$ is an *efficient* triangulation of G. The efficient decomposition, defined in the next section, behave somehow similarily to minimal decompositions. In order to obtain our exact algorithm for branchwidth, we will combine this observation with an exponential algorithm computing the branchwidth of hyper-cliques.

3   Branchwidth and Efficient Triangulations

Let a and b be two non adjacent vertices of a graph $G = (V, E)$. A set of vertices $S \subseteq V$ is an a, b-*separator* if in the graph $G - S$ a and b in are in different connected components. S is a *minimal* a, b-*separator* if no proper subset of S is an a, b-separator. We say that S is a *minimal separator* of G if there are two vertices a and b such that

S is a minimal a, b-separator. We denote by $\mathcal{C}(S)$ the set of connected components of $G - S$ and by Δ_G the set of all minimal separators of G.

Definition 1. *A triangulation H of G is* efficient *if*

1. *each minimal separator of H is also a minimal separator of G;*
2. *for each minimal separator S of H, the connected components of $H - S$ are exactly the connected components of $G - S$.*

The efficient triangulations were introduced in [4] (actually the authors used to call them "minimal triangulations"). In particular, all the minimal triangulations of G are efficient [17].

Theorem 1 ([16]). *There is an optimal branch decomposition (T, τ) of G such that the chordal graph $H(T, \tau)$ is an efficient triangulation of G. Moreover, each minimal separator of H is the label of some branch of T.*

Definition 2. *A set of vertices $B \subseteq V$ of G is called a* block *if, for each connected component C_i of $G - B$,*

- *its neighborhood $S_i = N(C_i)$ is a minimal separator;*
- *$B \setminus S_i$ is non empty and contained in a connected component of $G - S_i$.*

We say that the minimal separators S_i border the block B and we denote by $\mathcal{S}(B)$ the set of these separators.

Let $\mathcal{B}_G$ denote the set of blocks of G. Note that V is a block with $\mathcal{S}(V) = \emptyset$.

We prove that if H is an efficient triangulation of G, then any maximal clique K of H is a block of G.

Lemma 1 ([5]). *Let H be a chordal graph and Ω be a maximal clique of H. Then Ω is a block of H.*

Lemma 2. *Let H be an efficient triangulation of G and Ω be any maximal clique of H. Then Ω is a block of G. Conversely, for any block B of G, there is an efficient triangulation $H(B)$ of G such that B induces a maximal clique in H.*

Proof. If H is an efficient triangulation of G, by Lemma 1 every maximal clique Ω is a block of H. By definition of efficient triangulations, a block of H is also a block of G.

Conversely, if B is a block of G, let $C_1, \ldots, C_p$ be the connected components of $G - B$ and let $S_i = N(C_i)$, for all $1 \le i \le p$. Let $H(B)$ be the graph obtain from G by completing B and each set $S_i \cup C_i$ into a clique. The minimal separators of $H(B)$ are exactly $S_1, \ldots, S_p$. Moreover, for each S_i, the connected components of $H - S_i$ are exactly the components of $G - S_i$.

Note that the treewidth of a graph can be expressed by the following equation:

$$\mathrm{tw}(G) = \min_{H \text{ triangulation of } G} \max\{|\Omega| - 1 \mid \Omega \text{ maximal clique of } H\}. \quad (1)$$

The minimum can be taken over all minimal triangulations H of G. A similar formula can be obtained for branchwidth.

Definition 3 (block-branchwidth). *Let B be a block of G and $K(B)$ be the complete graph with vertex set B. A branch decomposition (T_B, τ_B) of $K(B)$ respects the block B if, for each minimal separator $S \in \mathcal{S}(B)$, there is a branch e of the decomposition such that $S \subseteq \mathrm{lab}(e)$. The block branchwidth $\mathrm{bbw}(B)$ of B is the minimum width over all the branch decompositions of $K(B)$ respecting B.*

Equivalently, $\mathrm{bbw}(B)$ is the branchwidth of the hypergraph obtained from the complete graph with vertex set B by adding a hyperedge S for each minimal separator S bordering B. The block-branchwidth allows us to express the branchwidth of G by a formula similar to Equation 1 (see Propositions 4.18 and 6.7 in [16]).

Theorem 2 ([16]).

$$\mathrm{bw}(G) = \min_{H \text{ efficient triangulation of } G} \max\{\mathrm{bbw}(\Omega) \mid \Omega \text{ maximal clique of } H\}.$$

(2)

Proof. Let (T, τ) be an optimal branch decomposition of G such that $H = H(T, \tau)$ is an efficient triangulation of G. Such a decomposition exists by Theorem 1. First, let us construct a branch decomposition (T', τ') of H having the same width as (T, τ). For each edge $\{x, y\}$ of $E(H) - E(G)$, the sub-trees T_x and T_y share a branch e. We divide the branch e by a node v, add a leaf w adjacent to v and map the edge $\{x, y\}$ on w. Clearly this will not increase the width of the decomposition. Consider any maximal clique Ω of G. By Lemma 2, Ω is a block of G and by Theorem 1 each minimal separator bordering Ω is contained in the label of some branch e_S of T'. For each S let (T_S, τ_S) be a arbitrary branch decomposition of the clique $K(S)$. We glue this decomposition to T' on the branch e_S. That is, we add a node on e_S and a node on some branch of T_S and make them adjacent. We call this new edge e'_S, in particular its label is exactly S. By this process we obtain a relaxed branch decomposition (T'', τ'') of H of same width as (T', τ'). By removing from T'' all the leaves that do not correspond to edges in the clique Ω, we obtained a relaxed clique decomposition of the complete graph $K(\Omega)$. For each minimal separator S bordering Ω, note that S is contained in the label of the edge e'_S, so the new decomposition respects Ω. Hence $\mathrm{bbw}(\Omega) \leq \mathrm{bw}(G)$ for each maximal clique Ω of H.

Conversely, let H be any efficient triangulation of G, let us show that $\mathrm{bw}(G) \leq \max\{\mathrm{bbw}(\Omega) \mid \Omega \text{ maximal clique of } H\}$. For each maximal clique Ω of G, we denote by (T_Ω, τ_Ω) an optimal branch decompoition of $K(\Omega)$, respecting the block Ω. We connect these decompositions into a relaxed branch decomposition of H. For this purpose we use a *clique tree* associated to the chordal graph graph H (see e.g. [12]). A clique tree is given by a tree $T = (V_T, E_T)$ and a one-to-one correspondence between the nodes of T and the maximal cliques of H such that, for each Ω, Ω' maximal cliques of H, their intersection is contained in all the cliques associated to nodes on the unique path from u_Ω to $u_{\Omega'}$ of T (u_Ω and $u_{\Omega'}$ denote the nodes associated to Ω and Ω' respectively). Moreover, for each branch $e = \{u_\Omega, u_{\Omega'}\}$ of T, $S = \Omega \cap \Omega'$ is a minimal separator bordering Ω and Ω' [12]. Let e_S (resp. e'_S) be a branch of T_Ω (resp. $T_{\Omega'}$) whose label contains S. We connect T_Ω and $T_{\Omega'}$ by adding a new branch between the middle of e_S and e'_S, for all branches $\{u_\Omega, u_{\Omega'}\}$ of T. Hence we obtain a relaxed branch decomposition of H. By the properties of the clique tree, the label of each newly created edge connecting T_Ω and $T_{\Omega'}$ is exactly $S = \Omega \cap \Omega'$. Consequently, the labels of the branches contained in some T_Ω do not change. Hence $\mathrm{bw}(H) \leq \max\{\mathrm{bbw}(\Omega) \mid \Omega \text{ maximal clique of } H\}$. G being a sub-graph of H, we have $\mathrm{bw}(G) \leq \mathrm{bw}(H)$ and the conclusion follows.

A potential maximal clique of a graph G is a set of vertices Ω such that there is a minimal triangulation H of G in which Ω introduces a maximal clique [5]. Using the Equation 1, Bouchitté and Todinca show that, given a graph and all its potential maximal cliques, the treewidth of the graph can be computed in polynomial time. The result is refined in [10], where the authors show the following:

Theorem 3. *There is an algorithm that, given a graph G and the set Π_G of its potential maximal cliques, computes the treewidth of G in $\mathcal{O}(nm|\Pi_G|)$ time.*

According to Lemma 2, a vertex subset Ω of G can be a maximal clique of an efficient triangulation H of G if and only if Ω is a block of G. Hence, in our case the blocks play the same role as the potential maximal cliques in Theorem 3.

Using Equation 2 instead of Equation 1 and blocks instead of potential maximal cliques, the algorithm cited in Theorem can be directly transformed into an algorithm taking G, the set $\mathcal{B}_G$ of all its blocks and the block-branchwidth of each block B, and computing the branchwidth of G in $\mathcal{O}(nm|\mathcal{B}_G|)$ time. In the rest of this section we give, without proofs, the new algorithm and the main tools for obtaining it.

Given a minimal separator S of G and a connected component C of $G - S$, let $R(S, C)$ denote the hypergraph obtained from $G[S \cup C]$ by adding the hyperedge S.

Lemma 3 (Similar to Corollary 4.5 in [5]). *For any graph G,*

$$\mathrm{bw}(G) = \min(\lceil 2n/3 \rceil, \min_{S \in \Delta_G} \max_{C \in \mathcal{C}(S)} \mathrm{bw}(R(S, C)))$$

Moreover, the minimum can be taken over the inclusion-minimal separators of G.

The case when $\mathrm{bw}(G) = \lceil 2n/3 \rceil$ corresponds to the fact that, for an optimal decomposition (T, τ) of G, the efficient triangulation $H(T, \tau)$ is the complete graph.

Lemma 4 (Similar to Corollary 4.8 in [5]). *Let S be a minimal separator of G and C be a component of $G - S$ such that $S = N(C)$. Then*

$$\mathrm{bw}(R(S, C)) = \min_{\text{blocks } \Omega \text{ s.t. } S \subset \Omega \subseteq S \cup C} \max(\mathrm{bbw}(\Omega), \mathrm{bw}(R(S_i, C_i)))$$

where C_i are the components of $G - \Omega$ contained in C and $S_i = N(C_i)$.

The algorithm for computing the branchwidth of G is a straightforward translation of Lemmas 3 and 4, and very similar to the one of [10].

```
Input: G, all its blocks and all its minimal separators
Output: bw(G)
begin
    compute all the pairs {S, C} where S is a minimal separator and C a component
```

of $G - S$ with $S = N(C)$; sort them by the size of $S \cup C$

```
for each {S, C} taken in increasing order
    bw(R(S, C)) := bbw(S ∪ C)
    for each block Ω with S ⊂ Ω ⊆ S ∪ C
        compute the components Ci of G − Ω contained in C and let Si = N(Ci)
        bw(R(S, C)) := min(bw(R(S, C)),
                                    max(bbw(Ω), bw(R(Si, Ci))))
                                     i
    end_for
end_for
```

let Δ_G^* be the set of inclusion-minimal separators of G

$\mathrm{bw}(G) := \min(\lceil 2n/3 \rceil, \min_{S \in \Delta_G^*} \max_{C \in \mathcal{C}(S)} \mathrm{bw}(R(S, C)))$

```
end
```

Theorem 4. *Given a graph G and the list $\mathcal{B}_G$ of all its blocks together with their block-branchwidth, the branchwidth of G can be computed in $\mathcal{O}(nm|\mathcal{B}_G|)$ time.*

Proof. The proof is very similar to the proof of [10], for treewidth and potential maximal cliques. We omit it here.

4 Computing the Block-Branchwidth

The main result of this section is that the block-branchwidth of a block B of G can be computed in $\mathcal{O}^*(\sqrt{3}^n)$ time. Computing the block-branchwidth is NP-hard, as it can be deduced directly from [15].

Let $n(B)$ denote the number of vertices of the block B of G and let $s(B)$ be the number of minimal separators bordering B. Note that $s(B)$ is at most the number of components of $G - B$, in particular $n(B) + s(B) \leq n$.

Lemma 5. $\mathrm{bbw}(B) \leq p$ *if and only if there is a partition of B into four parts A_1, A_2, A_3 and D such that*

1. $|B \setminus A_i| \leq p$, *for all $i \in \{1, 2, 3\}$;*
2. *for each minimal separator $S \in \mathcal{S}(B)$, S is contained in $B \setminus A_i$ for some $i \in \{1, 2, 3\}$.*

Proof. Suppose that $\mathrm{bbw}(B) \leq p$ and let (T, τ) be an optimal branch decomposition of B respecting the block. Recall that this branch decomposition corresponds to the complete graph $K(B)$ with vertex set B. For each $x \in B$ let T_x be the minimal sub-tree of T spanning all the leaves of T labeled with an edge incident to x. Let u represent B. Clearly u is a ternary node. Let e_1, e_2, e_3 be the branches of T incident to u. Let $T(i)$ be the sub-tree of T rooted in u, containing the branch e_i, for $i \in \{1, 2, 3\}$. Let $B_i = \{z \in B \mid z$ is incident to some edge of $K(B)$ mapped on a leaf of $T(i)\}$. Fix $D = B_1 \cap B_2 \cap B_3$, and $A_i = B_j \cap B_k \setminus D$ for all triples (i, j, k) with $i, j, k \in \{1, 2, 3\}$ and distinct. Observe that D, A_1, A_2, A_3 form a partition of B. The three sets are pairwise disjoint by construction. Since for all $x \in B, u \in T_x$, we have that $x \in B_i \cap B_j$ for distinct $i, j \in \{1, 2, 3\}$, so x is in one of the four sets A_1, A_2, A_3 or D. It remains to show that the partition satisfies the conditions of the theorem. Consider a separator $S \in \mathcal{S}(B)$ and a branch e in the decomposition with $S \subseteq \text{lab}(e)$. Suppose w.l.o.g. that $e \in T(i)$. Consequently $\text{lab}(e) \subseteq B_i$, and since $B_i = B \setminus A_i$ we have the second condition of the theorem. For proving the first condition, since A_1, A_2, A_3, D is a partition of B, note that $\text{lab}(e_i) = A_j \cup A_k \cup D = B \setminus A_i$. Therefore $|B \setminus A_i| \leq p$, for all $i \in \{1, 2, 3\}$.

Conversely, suppose that such a partition exists and let us construct a branch decomposition of $K(B)$ respecting the block B, of width at most p. Let $B_i = B \setminus A_i$, for each $i \in \{1, 2, 3\}$. For each i, construct an arbitrary branch decomposition (T_i, τ_i) of the complete graph with vertex set B_i. Let T be the tree obtained as follows : for each T_i, add a new node v_i on some branch of T_i, then glue the three trees by adding a new node u, adjacent to v_1, v_2, v_3. The tree T is a ternary tree and each edge of $K(B)$ is mapped on at least one leaf of T, so we obtained a relaxed tree decomposition (T, τ) of $K(B)$. Let e_i be the branch $\{u, v_i\}$. Note that $\text{lab}(e_i) = B_i \cap (B_j \cup B_k)$, where $\{i, j, k\} = \{1, 2, 3\}$. Hence $\text{lab}(e_i) = B_i$. Consequently, the relaxed branch decomposition respects the block B. Clearly for each branch e of T, $\text{lab}(e)$ is contained in some B_i, so $|\text{lab}(e)| \leq p$ and the conclusion follows.

Theorem 5. *The block-branchwidth of any block B can be computed in $\mathcal{O}^*(3^{s(B)})$ time.*

Proof. Let B be a block of G. Suppose that $\text{bbw}(B) \leq p$. By Lemma 5, there exists a partition of B in A_1, A_2, A_3 and D such that $|B \setminus A_i| \leq p$ and every $S \in \mathcal{S}(B)$ is a subset of $B \setminus A_i$. Denote by a_1, a_2, a_3 and d the sizes of A_1, A_2, A_3 and D. We can partition $\mathcal{S}(B)$ in three subsets $\mathcal{S}_i$ such that every $S \in \mathcal{S}_i$ is included in $B \setminus A_i$. Let S_i be the union of all the minimal separators of $\mathcal{S}_i$. The numbers a_1, a_2, a_3 and d satisfy the following inequalities:

1. $a_i \geq 0, d \geq 0, a_1 + a_2 + a_3 + d = |B|$;
2. $|S_1 \cap S_2 \cap S_3| \leq d, |(S_1 \cap S_2) \setminus S_3| \leq a_3, |(S_2 \cap S_3) \setminus S_1| \leq a_1, |(S_3 \cap S_1) \setminus S_2| \leq a_2$;
3. $a_1 + a_2 + d \leq p, a_2 + a_3 + d \leq p, a_3 + a_1 + d \leq p$.

The first inequalies express the fact that A_1, A_2, A_3 and D is a partition of B, the second express the fact that S_i is a subset of $B \setminus A_i$ and the last ones express the fact that $\text{bbw}(B) \leq p$.

Conversely, suppose there is a partition of $\mathcal{S}(B)$ in $\mathcal{S}_1$, $\mathcal{S}_2$ and $\mathcal{S}_3$ and four integers a_1, a_2, a_3, d satisfying the system above. Then there exist a partition of B into four sets A_1, A_2, A_3, D, of cardinalities a_1, a_2, a_3, d and such that D intersects $S_1 \cup S_2 \cup S_3$ exactly in $S_1 \cap S_2 \cap S_3$, and each A_i intersects $S_1 \cup S_2 \cup S_3$ exactly in $(S_j \cap S_k) \setminus S_i$, where $\{i, j, k\} = \{1, 2, 3\}$. Moreover $|B \setminus A_i| \leq p$ by the third series of inequalities, so by Lemma 5 we have $\text{bbw}(B) \leq p$.

Hence, there an efficient branch decomposition of $K(B)$ respecting B of branchwidth at most p if and only if there is a partition partition $\mathcal{S}_1, \mathcal{S}_2, \mathcal{S}_3$ of $\mathcal{S}(B)$ and four numbers a_1, a_2, a_3 and d satisfying the system. To decide whether $\text{bbw}(B) \leq p$ or not, we only have to try all the partitions of $\mathcal{S}(B)$ in $\mathcal{S}_1$, $\mathcal{S}_2$ and $\mathcal{S}_3$ and check all the n^4 possible values for the a_i's and d. This can be done in $\mathcal{O}^*(3^{|\mathcal{S}(B)|}) = \mathcal{O}^*(3^{s(B)})$ time as claimed.

Theorem 6. *The block-branchwidth of any block B can be computed in $\mathcal{O}^*(3^{n(B)})$ time.*

Proof. We show that for any number p, the existence of a partition like in Lemma 5 can be tested in $\mathcal{O}^*(3^{n(B)})$.

For this purpose, instead of partitioning B into four parts, we try all the partitions of B into three parts A_1, X, D, where X corresponds to $A_2 \cup A_3$. If $|B \setminus A_1| \leq p$, we check in polynomial time if X can be partitioned into A_2 and A_3 as required. Since there are at most $3^{n(B)}$ three-partitions of B, it only remains to solve this last point.

We say that two vertices $x, y \in X$ are equivalent if there exist $z \in A_1$ and a minimal separator S bordering B such that $x, y, z \in S$. In particular, $x \sim y$ implies that x and y must be both in A_2 or both in A_3. Let $X_1, \ldots, X_q$ be the equivalence classes of X. Then X can be partitioned into A_2 and A_3 as required if and only if $\{|X_1|, \ldots, |X_q|\}$ can be partitioned into two parts of sum at most $p - |A_1| - |D|$ vertices. Consider now the EXACT SUBSET-SUM problem, whose instance is a set of positive integers $I = \{i_1, \ldots, i_q\}$ and a number t, and the problem consists in finding a subset of I whose sum is exactly t. Though NP-hard in general, it becomes polynomial when t and the numbers i_j are polynomially bounded in n (see e.g. the chapter on approximation algorithms, the subset-sum problem in the book of Cormen, Leiserson, Rivest [7]). By taking $I = \{|X_1|, \ldots, |X_q|\}$ and trying all possible values of t between 1 and n^2, we can check in polynomial time if X can be partitioned as required.

Since at least one of $s(B)$ or $n(B)$ is smaller or equal to $n/2$, we deduce:

Theorem 7. *For any block B of G, the block-branchwidth of B can be computed in $\mathcal{O}^*(\sqrt{3}^n)$ time.*

Theorems 4 and 7 imply our main result.

Theorem 8. *The branchwidth of graphs can be computed in $\mathcal{O}^*((2 + \sqrt{3})^n)$ time and $\mathcal{O}^*(2^n)$ space.*

Proof. The algorithm enumerates every subset B of V and checks if B is a block. Clearly, we can verify if B is a block in polynomial time. If so, we compute the block branchwidth of B using Theorem 7. The number of blocks is at most 2^n and for each block we need $\mathcal{O}^*(\sqrt{3}^n)$ for computing its block branchwidth. Hence the running time of this phase is $\mathcal{O}^*((2 + \sqrt{3})^n)$, and the space is $\mathcal{O}^*(2^n)$.

Eventually, we use Theorem 4 for computing the branchwidth of G. The second phase takes $\mathcal{O}^*(2^n)$ time and space.

5   Open Problems

Our algorithm is based on the enumeration of the blocks of a graph (in $\mathcal{O}^*(2^n)$ time) and on the computation of the block-branchwidth of a block (in $\mathcal{O}^*(\sqrt{3}^n)$ time). It is natural to ask whether one of these steps can be improved.

Computing the block-branchwidth is the same problem as computing the branchwidth of a complete hypergraph with n' vertices and s' hyper-edges of cardinality at least three. Can we obtain an algorithm faster than our $O(\max(3^{n'}, 3^{s'}))$-time algorithm?

Note that there exist graphs with n vertices having $2^n/n^{O(1)}$ blocks. Indeed, consider the disjoint union of a clique K and an independent set I, both having $n/2$ vertices, and add a perfect matching between K and I. We obtain a graph G_n such that for any $I' \subseteq I$, $G_n - I'$ is a block. Thus G_n has at least $\binom{n}{n/2} \geq 2^n/n$ blocks. The interesting question here is if we can define a new class of triangulations, smaller than the efficient triangulations but also containing $H(T, \tau)$ for some optimal branch decompositions of the graph.

References

1. R. Beigel and D. Eppstein. 3-coloring in time $O(1.3446^n)$: a no-MIS algorithm. *Proceedings of the 36th IEEE Symposium on Foundations of Computer Science (FOCS 1995)*, pp. 444–452.
2. R. Beigel and D. Eppstein. 3-coloring in time $O(1.3289^n)$. *Journal of Algorithms*, 54:444–453, 2005.
3. H. L. Bodlaender, A partial k-arboretum of graphs with bounded treewidth, *Theoret. Comput. Sci.*, 209:1–45, 1998.
4. H.L. Bodlaender, T. Kloks, and D. Kratsch. Treewidth and pathwidth of permutation graphs. *SIAM J. on Discrete Math.*, 8:606–616, 1995.
5. V. Bouchitté and I. Todinca. Treewidth and minimum fill-in: grouping the minimal separators. *SIAM J. on Computing*, 31(1):212 – 232, 2001.
6. J. M. Byskov. Enumerating maximal independent sets with applications to graph colouring. *Operations Research Letters*, 32:547–556, 2004.
7. T. Cormen, C. Leiserson, and R. Rivest. *Introduction to algorithms*. The MIT press, 1990.
8. E. Dantsin, A. Goerdt, E. A. Hirsch, R. Kannan, J. Kleinberg, C. Papadimitriou, P. Raghavan, and U. Schöning. A deterministic $(2-2/(k+1))^n$ algorithm for k-SAT based on local search. *Theoretical Computer Science*, 289(1):69–83, 2002.
9. D. Eppstein. Improved algorithms for 3-coloring, 3-edge-coloring, and constraint satisfaction. *Proceedings of the 12th ACM-SIAM Symposium on Discrete Algorithms (SODA 2001)*, pp. 329–337.
10. F. Fomin, D. Kratsch, and I. Todinca. Exact (exponential) algorithms for treewidth and minimum fill-in. In *Proceedings 31st International Colloquium on Automatas, Languages and Programming (ICALP'04)*, volume 3142 of *Lecture Notes in Computer Science*, pages 568–580. Springer, 2004.
11. F. Gavril. The intersection graphs of a path in a tree are exactly the chordal graphs. *Journal of Combinatorial Theory*, 16:47–56, 1974.
12. M. C. Golumbic. *Algorithmic Graph Theory and Perfect Graphs*. Academic Press, New York, 1980.
13. K. Iwama and S. Tamaki. Improved upper bounds for 3-SAT. *Proceedings of the 15th ACM-SIAM Symposium on Discrete Algorithms (SODA 2004)*, p.328.
14. T. Jian. An $O(2^{0.304n})$ algorithm for solving maximum independent set problem. *IEEE Transactions on Computers*, 35(9):847–851, 1986.
15. T. Kloks, J. Kratochvíl, and H. Müller. New branchwidth territories. In *Proceedings 16th Annual Symposium on Theoretical Aspects of Computer Science (STACS '99)*, volume 1563 of *Lecture Notes in Computer Science*, pages 173–183. Springer, 1999.
16. F. Mazoit. *Décompositions algorithmiques des graphes*. PhD thesis, École normale supérieure de Lyon, 2004. In French.
17. A. Parra and P. Scheffler. Characterizations and algorithmic applications of chordal graph embeddings. *Discrete Appl. Math.*, 79(1-3):171–188, 1997.

18. R. Paturi, P. Pudlak, M. E. Saks, and F. Zane. An improved exponential-time algorithm for k-SAT. Proceedings of the *39th IEEE Symposium on Foundations of Computer Science (FOCS 1998)*, pp. 628–637.
19. N. Robertson and P. Seymour. Graph minors X. Obstructions to tree decompositions. *Journal of Combinatorial Theory Series B*, 52:153–190, 1991.
20. J. M. Robson. Algorithms for maximum independent sets. *Journal of Algorithms*, 7(3):425–440, 1986.
21. U. Schoning. A Probabilistic Algorithm for k-SAT and Constraint Satisfaction Problems. Proceedings of the *40th IEEE Symposium on Foundations of Computer Science (FOCS 1999)*, pp. 410-414.
22. P. D. Seymour and R. Thomas, Call routing and the ratcatcher, *Combinatorica*, 14:217–241, 1994.
23. R. Tarjan and A. Trojanowski. Finding a maximum independent set. *SIAM Journal on Computing*, 6(3):537–546, 1977.
24. R. Williams. A new algorithm for optimal constraint satisfaction and its implications. Proceedings of the *31st International Colloquium on Automata, Languages and Programming (ICALP 2004)*, Springer LNCS vol. 3142, 2004, pp. 1227–1237.
25. G. J. Woeginger. Exact algorithms for NP-hard problems: A survey. *Combinatorial Optimization – Eureka, You Shrink*, Springer LNCS vol. 2570, 2003, pp. 185–207.

Algorithms Based on the Treewidth of Sparse Graphs

Joachim Kneis, Daniel Mölle, Stefan Richter, and Peter Rossmanith

Computer Science Department, RWTH Aachen University, Fed. Rep. of Germany
{kneis, moelle, richter, rossmani}@cs.rwth-aachen.de

Abstract. We prove that given a graph, one can efficiently find a set of no more than $m/5.217 + 1$ nodes whose removal yields a partial two-tree. As an application, we immediately get simple algorithms for several problems, including MAX-CUT, MAX-2-SAT and MAX-2-XSAT. All of these take a record-breaking time of $O^*(2^{m/5.217})$, where m is the number of clauses or edges, while only using polynomial space. Moreover, the existence of the aforementioned node sets implies an upper bound of $m/5.217 + 3$ on the treewidth of a graph with m edges. Letting go of polynomial space restrictions, this can be improved to a bound of $m/5.769 + O(\log n)$ on the pathwidth, leading to algorithms for the above problems that take $O^*(2^{m/5.769})$ time.

1 Introduction

Recently, there has been a wave of effort in proving exponential-time worst-case upper bounds for NP-hard problems—in particular for the exact solution of MAX-SNP-hard problems [11]. One of the most intensely investigated problems in this area seems to be SAT, the problem of satisfiability of a propositional formula in conjunctive normal form (CNF). The maximum satisfiability problem—MAX-SAT—is an important generalization of SAT. Given a formula in CNF, it asks for the maximum number of simultaneously satisfiable clauses. The decision variant of this problem is complete for both NP and MAX-SNP, even if each clause contains at most two literals—this restriction is called MAX-2-SAT. Recently, numerous results regarding worst-case time bounds for the exact solution of MAX-SAT and MAX-2-SAT have been published. The best bound that has been achieved for MAX-SAT [3] with regards to m, the number of clauses, is $O^*(2^{m/2.46})$.[1] The best algorithm developed particularly for MAX-2-SAT [6] has time complexity $O^*(2^{m/5})$. With respect to the number of variables, the trivial $O^*(2^n)$ algorithm has not been improved until recently, when Williams came up with a new algorithm solving MAX-2-SAT in $O^*(2^{\omega n/3})$ steps [10] for $\omega \approx 2.379$.

In general, a MAX-SAT instance is represented by a multiset rather than a set of clauses, since a clause may occur more than once. In order to account for this, we let m denote the number of clause occurrences—the total *weight*.

---

[1] The O^*-notation was introduced by Woeginger and suppresses all polynomial factors; e.g., $2^k n^5 = O^*(2^k)$.

D. Kratsch (Ed.): WG 2005, LNCS 3787, pp. 385–396, 2005.

Furthermore, we declare t to stand for the number of *clause types*. A *clause type* will be understood as a maximum distinct set of variables occuring together in at least one clause, disregarding negations. It is easy to see that t can be much smaller than m, even in formulæ that do not have multiple identical clauses.

In this paper, we present a very simple algorithm for MAX-2-SAT that has time complexity $O^*(2^{t/5})$, and thus $O^*(2^{m/5})$ just like the involved one by Gramm et al. [6]. Moreover, we analyze a slightly more complicated version of the algorithm, lowering the bound to $O^*(2^{t/5.217})$. The latter improves upon the best known upper bounds for solving MAX-2-SAT [6], MAX-2-XSAT [7], and MAX-CUT [8]—while still using only polynomial space. Furthermise, we present the fastest polynomial space algorithm for DOMINATING SET on cubic graphs, which takes $O^*(2^{n/2})$ time. Whereas the runtime bound for this problem has recently been improved from $O^*(1.51^n)$ [5] to $O^*(1.21^n)$ [4], the latter algorithm is very involved and requires exponential space.

Impressive as these new record bounds may seem, they are just the tip of the iceberg. In fact, they represent little more than mere by-products of a much more general technique. It relies on our main graph theoretical result, which states that the removal of at most $|E|/5.217+1$ nodes from a graph $G = (V, E)$ yields a series-parallel graph. The method that stems from this observation enables a narrowing of the search space for many important NP-hard problems. In particular, a simple application yields the above-mentioned record-breaking bounds.

Moreover, it follows that the treewidth and pathwidth of G are at most $|E|/5.217 + 3$ and $|E|/5.217 + 3 + \log n$, respectively. Employing a recent result by Fomin and Høie, we can improve these bounds to $|E|/5.769 + O(\log n)$, yielding $O^*(2^{m/5.769})$ time algorithms for MAX-2-SAT, MAX-2-XSAT, and MAX-CUT based on dynamic programming [9]—at the price of exponential space.

2   Preliminaries

We assume the reader to be familiar with the notion of treewidth [1]. Throughout this paper, we adhere to the notation for boolean formulæ used by Gramm et al. [6].

Let F be a formula in 2-CNF whose set of variables will be called V. The corresponding *connectivity graph* is $G_F = (V, E)$ where

$$E = \{ \, \{x, y\} \mid \text{the distinct variables } x \text{ and } y \text{ occur together in a clause} \, \},$$

representing the way variables interact in a formula. Notice that it does not make a difference in how many clauses a pair of variables occurs, or whether a variable is negated or not. For instance, the two graphs $G_{F[x]}$ and $G_{F[\bar{x}]}$ are identical. As a consequence, the formula F cannot be reconstructed from G_F.

3   An Algorithm with Only One Reduction Rule

In what follows, we prove our foundational result: a graph with m edges has treewidth at most $m/5 + 2$, and we can quickly find a set of no more than $m/5$

nodes whose removal leaves a very simply structured graph, namely a special case of a partial 2-tree. As an application of this technique we can solve several optimization problems efficiently.

More precisely, our method becomes applicable when problems can be expressed in graph terms as follows: There is a graph $G = (V, E)$ for every instance, and given a node v in the graph, we can reduce the instance to a smaller one whose graph is $G[V \setminus \{v\}]$. Moreover, the problem must be easy to solve when the corresponding graphs are partial 2-trees. Finally, reduction steps on nodes of degree two or more may be *expensive*, whereas nodes of degree one have to be *easy* to deal with in the problem context. Then, the algorithm derived from our graph-theoretical result takes at most $m/5$ expensive operations to reduce any input instance with m egdes to one whose graph is a special case of a partial 2-tree.

Many problems have the aforementioned properties, where the expensive operations usually originate from case distinctions that lead to branching in the recursion tree. Consider MAX-CUT as an example: Vertices of degree one can be deleted, since they will increase the overall size of a maximum cut by one in any case, whereas nodes of higher degree require branching.

The algorithm presented in this section is rather simple and broadly applicable. An even simpler algorithm will emerge in the next section; however, it will have the additional requirement that nodes of degree two are easy to deal with as well. Hence, if this condition does not hold for a problem, we have to stick to the more general algorithm from this section; otherwise, the simpler algorithm from the next section is preferable.

The special rôle of degree-one nodes in the first algorithm is reflected in the following definition:

Definition 1. Let $G = (V, E)$ be a graph. Then $R(G)$ is the graph obtained by deleting vertices v with $\deg(v) = 1$ repeatedly until there are no such vertices left.

Observe that $R(G)$ is well-defined, since it does not make any difference in what order nodes are chosen for deletion. What is more, the following lemma shows that even when we delete arbitrary nodes between reductions, the order is irrelevant. This property greatly simplifies algorithmic application of the rule. From now on, we shorten $G[V \setminus \{v\}]$ to $G - v$ as well as $G[V \setminus D]$ to $G - D$.

Lemma 1. Let $G = (V, E)$ be a graph and $D = \{v_1, \ldots, v_k\}$ a set of vertices from V. Then, $R(G - D) = R(R(\ldots(R(R(G - v_1) - v_2) - v_3) \cdots - v_{k-1}) - v_k)$.

The previous as well as the following lemma are straightforward. In all interesting cases, R-reducing a graph does not affect its treewidth:

Lemma 2. Let G be a graph containing a cycle. Then $tw(G) = tw(R(G))$.

Having investigated the properties of our only reduction rule, we turn our attention to the simple family of hot dog graphs. Surprisingly, any graph can be turned into a hot dog graph by deleting a small set of nodes and applying the R-reduction.

Definition 2. A path of length at least one between two possibly identical nodes s and t in a graph G is called a *leg* if all its nodes other than s and t have degree two in G. A *hot dog graph* consists of nodes $v_1, \ldots, v_k$ such that v_i and v_{i+1} are connected by arbitrarily many legs. Additionally, v_k and v_1 may be connected in this fashion as well.

Definition 3. Let $G = (V, E)$ be a graph whose nodes have degree at least two. A *4-spider* is a subgraph that consists of a *head* $h \in V$ with degree four, three or four distinct *feet* $u_1, \ldots, u_l \in V \setminus \{h\}$ of degree at least three, and four disjoint legs connecting head and feet.

A *3-spider* is defined similarly for a head of degree three and exactly three distinct feet connected to it via three legs. In any case, the *body* of a spider consists of all its nodes except the feet.

The nice thing about spiders is that their bodies can be removed from a graph quite easily: First remove the head, which is a node with relatively high degree, and then remove the remainder of the body by consecutively removing nodes of degree one.

It is interesting to note that hot dog graphs cannot contain spiders. The following lemma shows that the converse is also true in a fairly general setting. This enables us to turn any graph into a hot dog graph using relatively cheap operations.

Lemma 3. *Let $G = (V, E)$ be a connected graph whose nodes have degree between two and four. G is a hot dog graph iff it does not contain a 3- or 4-spider.*

Proof. It is obvious that a hot dog graph cannot contain spiders. On the other hand, let G be a graph as postulated in the premise that does not contain a spider. Let H be the set of nodes that do not have exactly two neighbors. Observe that every $v \in H$ may be connected to at most two more nodes from H via legs, because otherwise v_i would be the head of a spider. Thus, we can arrange the nodes from H in a linear or cyclical fashion as in the definition of a hot dog graph. $\qquad\square$

Interestingly, if a node v has been the head of a spider in G, it keeps this rôle in the contracted graph. In what follows, we want to estimate the spider bodycount required to carve out a hot dog graph. In effect, we need to look for the number of edges that have to be removed. As it turns out, this feat is substantially eased if we analyze in terms of a potential function of nodes instead.

Definition 4. Let $G = (V, E)$ be a graph, $v \in V$, and

$$\deg_3(v) = \big|\{u \in V \mid \text{ there is a leg connecting } u \text{ and } v, \text{ and } \deg(u) \geq 3\}\big|.$$

We define the potential functions $\psi \colon V \to \mathbf{N}$ and $\Psi \colon \mathcal{G} \to \mathbf{N}$ as follows:

$$\psi(v) = \begin{cases} 0 & \text{if } \deg(v) \leq 2 \\ 0 & \text{if } \deg(v) = 3 \text{ and } \deg_3(v) = 1 \\ 5/4 & \text{if } \deg(v) = 3 \text{ and } \deg_3(v) > 1 \\ 2 & \text{if } \deg(v) \geq 4 \end{cases}$$

We extend the definition to graphs via $\Psi(G) = \sum_{v \in V} \psi(v)$.

Lemma 4. *Let $G = (V, E)$ be a graph. Then $\Psi(G) \leq |E|$.*

Proof.

$$|E| = \frac{1}{2} \sum_{v \in V} \deg(v) = \frac{1}{2} \sum_{i=1}^{|V|-1} \sum_{\substack{v \in V \\ \deg(v)=i}} i \geq \sum_{\substack{v \in V \\ \deg(v)=3}} \frac{5}{4} + \sum_{\substack{v \in V \\ \deg(v) \geq 4}} 2 \geq \Psi(G). \;\square$$

Lemma 5. *Let $G = (V, E)$ be a graph whose nodes have degree between two and four. If G contains a 4-spider with head h, then $\Psi(R(G - h)) \leq \Psi(G) - 5$.*

Proof. Let S be a 4-spider with head h. We have to distinguish several cases.

In the first case, S has four different feet $u_1, \ldots, u_4$ with $3 \leq \deg(u_i) \leq 4$. Removing h and all nodes of degree one consecutively has the following effect: Because h is erased, the potential decreases by 2. As a consequence, the degree of each foot is lowered by one. This means that the potential decreases by $2 - 5/4 = 3/4$ or $5/4 - 0 = 5/4$ per foot. The total loss of potential thus amounts to at least $2 + 4(3/4) = 5$.

In the second case, there are only three feet $u_1, \ldots, u_3$, and the situation is slightly more complicated. W.l.o.g. two paths are leading to u_1 and one path each to u_2 and u_3. If $\deg(u_1) = 4$, removing the body of S does the following: The potential of u_1 is lowered by 2, the potential of h decreased by 2 as well, and the potentials of u_2 and u_3 shrink by $2 - 5/4 = 3/4$ or $5/4 - 0 = 5/4$ each. Altogether, these values sum up to a loss of potential greater than 5.

Otherwise, if $\deg(u_1) = 3$, only one other leg starts from u_1. Let z denote the node this leg ends in. Note that z and h have to be different, since otherwise S would not be a spider at all: There would be three paths to u_1, but only two feet.

If z, u_2, u_3 are all different, the potential of h decreases by 2, the potential of u_1 by $5/4$, and the potentials of z, u_2, and u_3 by at least $3/4$ each, which is again more than 5 in total. If z, u_2, u_3 are not all different, say $z = u_3 \neq u_2$, then the potential of h is lowered by 2, the potential of u_1 by $5/4$, the potential of $z = u_3$ by at least $5/4$, and the potential of u_2 by at least $3/4$, which is more than 5 altogether. $\qquad\square$

Lemma 6. *Let $G = (V, E)$ be a connected graph whose nodes have degree between two and four. If G does not contain any 4-spider, but a 3-spider with head h, then $\Psi(R(G - h)) \leq \Psi(G) - 5$.*

Proof. Let h be the head of a 3-spider with feet u_1, u_2, u_3. Removing this spider causes the potential to decrease by at least 5, since $\psi(h) = 5/4$, and we lose at least $5/4$ on each foot, too. To see this, distinguish the following two cases: Either, $\deg(u_i) = 3$—this leads to a decrease in potential of exactly $5/4$—or, $\deg(u_i) = 4$ for some i. In the latter case, observe that there is exactly one leg between u_i and h, as u_i is a foot of the 3-spider with head h. Since u_i cannot be the head of a 4-spider, the three other legs starting in u_i end in the same node z. Then, however, we have that $\psi(u_i) = 0$ in $R(G - h)$ due to the definition of ψ and $\deg_3$. $\qquad\square$

Let us now begin putting the pieces together.

Theorem 1. *Let $G = (V, E)$ be a graph. There is a set $D \subseteq V$ such that $R(G - D)$ is a hot dog graph and $|D| \leq |E|/5$.*

Proof. We construct a set of nodes D such that $R(G - D)$ is a hot dog graph. As long as G contains a node v with degree at least five, remove v from G and set $D := D \cup \{v\}$. Now delete the bodies of all 4-spiders from G, and then do the same for 3-spiders. Add the heads of all these spiders to D. Note that removing a spider's body is the same as removing its head and applying the reduction rule R afterwards.

We obtain a set D such that $R(G - D)$ is a hot dog graph. Using Lemmata 4, 5, and 6, it is easy to see that $|D| \leq m/5$. $\square$

Theorem 2. *The treewidth of a graph $G = (V, E)$ is at most $|E|/5 + 2$.*

Proof. Let D be the set given by Theorem 1. By Lemma 2, R-reducing the graph $G - D$ leaves its treewidth intact, provided that $G - D$ contains a cycle. Hence, the treewidth of $G - D$ is at most that of a hot dog graph. It is easy to see that hot dog graphs constitute a special case of series-parallel graphs, which have treewidth at most two [2–p. 174]. Otherwise, $G - D$ is but a forest. Altogether, we have that $tw(G) \leq |D| + 2 = |E|/5 + 2$. $\square$

4   A Second Rule Simplifies the Algorithm

In this section, we develop a simpler algorithm which employs a second reduction rule in addition, that is, a rule that replaces a path (u, v, w) with $\deg(v) = 2$ by the path (u, w). We call this operation *contracting* v. Notice that this introduces another constraint on the set of possible applications: Degree-two nodes must be easy to handle in the problem translation. That is, the way they contribute to a solution should only depend on their two neighbors.

In short, we trade simplicity for applicability: As we will see in what follows, the refined method allows for a much simpler implementation, and thus eases the analysis. Moreover, in the place of hot dog graphs, it leaves a trivial graph without any edges.

On the other hand, there are problems that do not meet the above extra constraint, while the technique from the previous section can still be employed. Again, consider MAX-CUT: In the direct approach, it is not clear how to avoid branching on degree-two nodes. (Fortunately, MAX-CUT can be reduced to MAX-2-SAT by replacing an edge $\{x, y\}$ by two clauses $\{x, y\}$ and $\{\bar{x}, \bar{y}\}$. The number of edges becomes the number of clause types.)

Definition 5. *Let $G = (V, E)$ be a graph and $v \in V$. Let $R'(G)$ be the graph that we get from G by repeatedly removing degree one vertices and contracting degree two vertices until no such operation is possible. Whenever a contraction leads to a double edge, only a single edge is retained. We also define $R'_v(G) := R'(G - v)$.*

Algorithm A
Input: A graph $G = (V, E)$
Output: $C \subseteq V$, $|D| \leq |E|/5$, such that $R'(G - D)$ has no edges
$D \leftarrow \emptyset$;
while there is a node v with $\deg(v) \geq 3$ **do**
    choose a node v with maximum degree;
    $D \leftarrow D \cup \{v\}$; $G \leftarrow R'_v(G)$
od;
return D

Fig. 1. A simpler algorithm that uses R' rather than R

Lemma 7. *Let $G = (V, E)$ be a graph with minimum degree three and maximum degree four. If $v \in V$ and $\deg(v) = 4$, then $\Psi(R'_v(G)) \leq \Psi(G) - 5$.*

Proof. Let $u_1, \ldots, u_4$ be the neighbors of v. We have $\psi(v) = 2$, and removing v decreases the degree of each u_i by one. In total, the operation lowers the potential by at least $2 + 4(3/4) = 5$. Since neither the removal of a degree one node nor the contraction of a degree two node can increase the potential, this implies $\Psi(R'_v(G)) \leq \Psi(G) - 5$. $\qquad\square$

Lemma 8. *Let $G = (V, E)$ be a 3-regular graph. For every $v \in V$ we have that $\Psi(R'_v(G)) \leq \Psi(G) - 5$.*

Proof. Every node in a 3-regular graph has a potential of $\Psi(3) = 5/4$. Removing v hence lowers the potential by 5. $\qquad\square$

Theorem 3. *Algorithm A finds a set $D \subseteq V$ such that $|D| \leq m/5$ and the reduced graph $R'(G - D)$ has no edges.*

Proof. As long as there are nodes of degree at least five, the body of the **while**-loop increases the size of D by one while removing at least five edges. As soon as all nodes have degree at most four, $\Psi(G) \leq |E|$ by Lemma 4. From then on, the potential $\Psi(G)$ decreases by at least five in the body of the **while**-loop according to Lemmata 7 and 8. Since $R'(G)$ never contains nodes of degree one or two, the graph cannot have any edges when the algorithm terminates.

It is, however, not obvious that $R'(G - D)$ is the same graph. We only know that removing the nodes of D in the right order and applying reduction rules in between yields a graph without edges. However, analogously to Lemma 1, it is easy to see that indeed $R'_{x_k}(R'_{x_{k-1}}(\cdots R'_{x_1}(G) \cdots)) = R'(G - \{x_1, \ldots, x_k\})$. $\qquad\square$

In order to use Algorithm A for solving MAX-2-SAT, we must find reduction rules for formulæ that correspond to removing a node of degree one and contracting a node of degree two. A simple reduction rule can be used on a formula F to remove a node of degree one from G_F. But what do we need to do with F in order to contract a node of degree two in G_F? It is easy to see that we have to eliminate a variable x that occurs with *exactly two* other variables y and z in 2-clauses, introducing new clauses of the type $\{y, z\}$ in return.

Definition 6. Let F be a 2-SAT-formula. A variable x is a *double companion* if and only if the degree of x in G_F is two.

To ease the introduction of a double companion reduction rule, we now generalize the notion of a clause. We are used to defining a clause to be a pair (ω, C) where C is a set of (non-complementary) literals and ω a positive integer. In this section, we allow ω to be a negative integer as well. When we add up formulæ, the weights of identical clauses add up. Let F be a formula. Then $F[A]$ denotes the formula we get by setting all literals in A to true. This includes simplification and typically yields some satisfied clauses that take the form $(\omega, \mathbf{T})$. Moreover, let $OptVal(F)$ be the maximum (weighted) number of satisfiable clauses in F under any assignment [6]. For the following lemma, remember the definition of our new parameter t, the number of clause types.

Lemma 9 (The double companion reduction rule). *Let F be an arbitrary 2-SAT formula. If x is a double companion, then we can transform F into an equivalent formula F' which contains the same variables as F except x, and possibly clauses of negative weight, in polynomial time. The formula F' does not have more clause types than F. Moreover, $G_{F'}$ is the graph obtained from G_F by contracting x.*

Proof. Let x be a double companion that occurs together with y and z. Let $F = F' + F''$, where F' consists of all the clauses that contain x and F'' holds all the other clauses. We define $a = OptVal(F'[y, z])$, $b = OptVal(F'[y, \bar{z}])$, $c = OptVal(F'[\bar{y}, z])$, and $d = OptVal(F'[\bar{y}, \bar{z}])$. Let

$$G = \{(a + b + c + d, \mathbf{T}), (-d, \{y, z\}), (-c, \{y, \bar{z}\}), (-b, \{\bar{y}, z\}), (-a, \{\bar{y}, \bar{z}\})\}.$$

We easily see $a = OptVal(G[y, z])$, $b = OptVal(G[y, \bar{z}])$, $c = OptVal(G[\bar{y}, z])$, and $d = OptVal(G[\bar{y}, \bar{z}])$. Therefore, $OptVal(F' + F'') = OptVal(G + F'')$. Moreover, x does obviously not occur in $G + F''$. $\square$

We now have reduction rules for formulæ in 2-CNF that enable us to eliminate all nodes with degree up to two in the corresponding connectivity graph. A naïve approach uses this machinery on the connectivity graph of a 2-CNF formula to find the number of satisfiable clauses. The algorithm can be easily modified to return an optimal assignment, too. The running time is again $O^*(2^{t/5})$, where $t \leq m$ is the number of different clause types.

It turns out that we need not use the connectivity graph explicitly. Instead, we can employ a recursive procedure as described in Algorithm B. In this form it corresponds to classical satisfiability algorithms starting with the Davis–Putnam procedure: Apply reduction rules as long as possible and then choose a variable for branching. In the past, better and better algorithms included more and more complicated rules. This involves reduction rules as well as rules for choosing a variable (or a group of variables) to branch on, combined with clever pruning of cases that cannot lead to an optimal assignment. In contrast, Algorithm B is very simple: It is comprised of only two reduction rules and one rule to choose a variable for branching, none of which are complicated.

Algorithm B
Input: A MAX-2-SAT-formula F
Output: $OptVal(F)$
Reduce F by the (double) companion reduction rule while possible;
if $F = \{(k, \mathbf{T})\}$ then return k
else
      choose a variable x that occurs in a maximum number of clause types;
      return max{Algorithm B($F[x]$), Algorithm B($F[\bar{x}]$)}
fi

Fig. 2. A very simple algorithm for MAX-2-SAT that does not use the connectivity graph directly

5   Improving Beyond $t/5$

In this section, we apply a tiny modification to the algorithm discussed above. More precisely, we introduce the additional rule to avoid picking a node of degree four all of whose neighbors have degree four as well, whenever possible.

We begin by looking at a special case for graphs of low degree. This theorem is of independent interest, and its proof serves to introduce the methods we apply in Theorem 5.

Theorem 4. *Let* $G = (V, E)$ *be a graph with* m *edges and maximum degree four. Then there is a set* $D \subseteq V$, $|D| \leq \frac{3}{16} m + 1$, *such that* $R'(G - D)$ *has no edges.*

Proof. Given $G = (V, E)$, construct $D \subseteq V$ as follows. Pick a vertex of maximum degree, and while choosing vertices of degree four, only take a vertex all of whose neighbors have degree four if no other type of degree-four node remains. Note that the latter is only the case if the graph is 4-regular. Remove the chosen vertex, apply the two reduction rules, and repeat the procedure until the maximum degree in the remaining graph drops below three.

$$\psi(v) = \begin{cases} 0 & \text{if } \deg(v) \leq 2 \\ 4/3 & \text{if } \deg(v) = 3 \\ 2 & \text{if } \deg(v) \geq 4 \end{cases}$$

We redefine the potential function ψ as seen on the left. Let $\langle n_1, \ldots, n_d \rangle$ denote the case that we pick a node v of degree d whose neighbors have degree n_1 through n_d. The respective losses of potential caused by the removal of such nodes v can be computed easily: the potential of v drops to zero, whereas the degree of each of its neighbors decreases by one. For instance, the loss of potential in the case $\langle 4, 4, 4, 3 \rangle$ amounts to $2 + 3 \cdot (2 - 4/3) + 4/3$. The resulting values are listed in the following table.

case	$\langle 4, 4, 4, 4 \rangle$	$\langle 4, 4, 4, 3 \rangle$	$\langle 4, 4, 3, 3 \rangle$	$\langle 4, 3, 3, 3 \rangle$	$\langle 3, 3, 3, 3 \rangle$	$\langle 3, 3, 3 \rangle$
loss	$4\frac{2}{3}$	$5\frac{1}{3}$	6	$6\frac{2}{3}$	$7\frac{1}{3}$	$5\frac{1}{3}$

Observe that the special case $\langle 4, 4, 4, 4 \rangle$ can only occur in the first iteration, which causes at most one extra step, or if preceded by $\langle 3, 3, 3, 3 \rangle$: Clearly, it cannot be preceded by $\langle 3, 3, 3 \rangle$, because we always pick a vertex of maximum degree. Furthermore, a node of degree three is created in all the remaining cases, preventing the graph from becoming 4-regular and thus excluding the case $\langle 4, 4, 4, 4 \rangle$.

Except for the first step, the good case $\langle 3, 3, 3, 3 \rangle$ countervails against the bad case $\langle 4, 4, 4, 4 \rangle$. Since the average loss of potential in these two cases amounts to 6, we have that the potential decreases by an average of at least $5\frac{1}{3}$ per step. Hence, the overall potential will drop to zero after at most $\frac{3}{16}m$ additional iterations.   □

Note that, analogously to Lemma 4, it is easily checked that $\Psi(G) \leq |E|$ for continuations of the potential functions in both the previous and the upcoming proof.

Theorem 5. *Let $G = (V, E)$ be a graph with m edges. Then there is a set $D \subseteq V$, $|D| \leq \frac{23}{120}m + 1$, such that $R'(G - D)$ has no edges.*

Proof. We use both the algorithm and the notation described in the proof to the previous theorem. Again, we redefine the potential function ψ:

$$\psi(v) = \begin{cases} 0 & \text{if } \deg(v) \leq 2 \\ 30/23 & \text{if } \deg(v) = 3 \\ 45/23 & \text{if } \deg(v) = 4 \\ 5/2 & \text{if } \deg(v) \geq 5 \end{cases}$$

Obviously, we get rid of at least six edges per iteration as long as the algorithm removes nodes of degree at least six. It hence suffices to switch to an analysis via potential as soon as the maximum degree in the remaining graph has decreased to at most five. When a node of degree five is deleted, this lowers the potential by at least $5/2 + 5 \cdot (5/2 - 45/23) = 5\frac{5}{23}$. The other cases are listed below.

case	$\langle 4, 4, 4, 4 \rangle$	$\langle 4, 4, 4, 3 \rangle$	$\langle 4, 4, 3, 3 \rangle$	$\langle 4, 3, 3, 3 \rangle$	$\langle 3, 3, 3, 3 \rangle$	$\langle 3, 3, 3 \rangle$
loss	$4\frac{13}{23}$	$5\frac{5}{23}$	$5\frac{20}{23}$	$6\frac{12}{23}$	$7\frac{4}{23}$	$5\frac{5}{23}$

As detailed above, the good case $\langle 3, 3, 3, 3 \rangle$ countervails against the bad case $\langle 4, 4, 4, 4 \rangle$; their average loss of potential is $5\frac{20}{23}$. Hence, only nodes of degree at most two remain after at most $\frac{23}{120}m + 1$ iterations.   □

Modifiying Algorithm A according to the above result leads to the following improved running times.

Corollary 1. MAX-2-SAT *and* MAX-2-XSAT *can be solved in* $O^*(2^{t/5.217})$ *and thus in* $O^*(2^{m/5.217})$ *time.* MAX-CUT *can be solved in* $O^*(2^{m/5.217})$ *time. All algorithms require only polynomial space.*

In order to give an upper bound on the treewidth of a graph $G = (V, E)$ using the above results, it suffices to check that $tw(G - D) \leq 2$. This is because $tw(R'(G - D)) = 0$, and R' does not trivialize graphs of treewidth at least three [2–p. 174].

Corollary 2. *The treewidth of a graph $G = (V, E)$ is at most $|E|/5.217 + 3$.*

6 Dominating Set on Graphs of Maximum Degree Three

In the case of maximum degree three, Theorem 1 can be improved by redefining the potential function ψ as $\psi(v) = 1.5$ for degree three vertices v and $\psi(v) = 0$ otherwise. The refined theorem yields a set D of size at most $|E|/6$.

Consider a node v in such a graph. There are at most four ways to dominate v: by itself or not by itself, but by one of its neighbors. We try these four possibilities for each node from D, which accounts for at most $4^{m/6} = 2^{n/2}$ possibilities altogether. In each case, we mark the selected dominators as *dominating* and their neighbors as *dominated*. Then we remove the nodes in D from the graph, remembering the dominators in D as D'. An optimal dominating set D'' for the remaining annotated graph can be easily found in polynomial time. In any case, $D' \cup D''$ is a dominating set for the entire graph, and we obviously find a globally optimal dominating set by going through all the $2^{n/2}$ cases.

The overall algorithm is quite simple and has a running time of $O^*(2^{n/2}) = O^*(1.42^n)$, improving the result of $O^*(1.51^n)$ by Fomin, Kratsch, and Woeginger [5]. It is the fastest algorithm that uses polynomial space.

7 Trading Space for Time

The result on the size of D in the preceding section implicitly establishes an upper bound of $m/6 + O(1)$ on the treewidth of cubic graphs. In a more general setting, such an upper bound might be interesting in its own right. According to Fomin and Høie [4], the pathwidth of a cubic graph is at most $(1 + \epsilon)n/6 + O(\log n)$, where $\epsilon > 0$ is an arbitrarily small constant.

We can exploit this result in order to give an even tighter bound on the pathwidth of general graphs. Unfortunately, this method comes at the cost of losing several benefits of the approaches discussed so far. In particular, the resulting algorithms are no longer intuitive, and require exponential instead of polynomial space.

Theorem 6. *The pathwidth of a graph with m edges and n nodes is at most $m/5.769 + O(\log n)$. A corresponding path decomposition can be found in polynomial time. Thus MAX-CUT, MAX-2-SAT, and MAX-2-XSAT can be solved in $O^*(2^{m/5.769})$ steps.*

Proof. Consider the following algorithm, which employs the aforementioned reduction rules for nodes of degree at most two without further notice: While there are nodes with degree at least six, we delete them from the graph. When only nodes of degree at most five remain, we have $\Psi(G) \le m$ for the continuation of the potential function ψ, when we choose $\alpha = 25/26$, $\beta = 25/13$, and $\gamma = 5/2$ in

$$\psi(v) = \begin{cases} \alpha & \text{if } \deg(v) = 3 \\ \beta & \text{if } \deg(v) = 4 \\ \gamma & \text{if } \deg(v) = 5 \\ 0 & \text{if } \deg(v) < 3. \end{cases}$$

It is easy to verify that the potential decreases by at least $75/13$ whenever a node of degree four or five is removed. The sole exception is the removal of a degree-five node all of whose neighbors have degree five as well, which leads to a loss of only $145/26$. We can avoid this case unless the graph is five-regular. As in the proof of Theorem 4, this can only occur if the node deleted in the previous step was of a special kind: Here, this would be nodes of degree at least six, and degree-five nodes with only degree-three neighbors. In the latter case the potential decreases by $90/13$, implying a loss of $325/52 > 75/13$ on average.

When the algorithm terminates, a cubic graph remains. If p denotes its potential, it consists of exactly p/α nodes. According to the result of Fomin and Høie [4], its pathwidth is at most $(1 + \epsilon)p/6\alpha + O(\log n) = (1 + \epsilon)13p/75 + O(\log n)$. For the original graph and ϵ small enough, this implies a bound of $(1 + \epsilon)13m/75 + O(\log n) \leq m/5.769 + O(\log n)$.

Acknowledgements. We would like to thank Hans Bodlaender for valuable pointers on literature and several anonymous referees for helpful comments.

References

1. H. L. Bodlaender. A tourist guide through treewidth. *Acta Cybernetica*, 11:1–21, 1993.
2. A. Brandstädt, V. B. Le, and J. P. Spinrad. *Graph Classes: A Survey*. SIAM monographs on discrete mathematics and applications. SIAM, 1999.
3. J. Chen and I. A. Kanj. Improved exact algorithms for Max-Sat. *Discrete Applied Mathematics*, 142(1–3):17–27, 2004.
4. F. V. Fomin and K. Høie. Pathwidth of cubic graphs and exact algorithms. Technical Report 298, Department of Informatics, University of Bergen, May 2005.
5. F. V. Fomin, D. Kratsch, and G. J. Woeginger. Exact (exponential) algorithms for the dominating set problem. In *Proc. of 30th WG*, number 3353 in LNCS, pages 245–256. Springer, 2004.
6. J. Gramm, E. A. Hirsch, R. Niedermeier, and P. Rossmanith. New worst-case upper bounds for MAX-2-SAT with application to MAX-CUT. *Discrete Applied Mathematics*, 130(2):139–155, 2003.
7. B. A. Madsen and P. Rossmanith. Maximum exact satisfiability: NP-completeness proofs and exact algorithms. Technical Report RS-04-19, BRICS, Oct. 2004.
8. A. Scott and G. B. Sorkin. Faster algorithms for Max-CUT and Max-CSP, with polynomial expected time for sparse instances. In *Proc. of 7th RANDOM*, number 2764 in LNCS, pages 382–395. Springer, 2003.
9. J. A. Telle and A. Proskurowski. Algorithms for vertex partitioning problems on partial k-trees. *SIAM Journal on Discrete Mathematics*, 10(4):529–550, 1997.
10. R. Williams. A new algorithm for optimal constraint satisfaction and its implications. In *Proc. of 31st ICALP*, number 3142 in LNCS, pages 1227–1237. Springer, 2004.
11. G. Woeginger. Exact algorithms for NP-hard problems: A survey. In M. Junger, G. Reinelt, and G. Rinaldi, editors, *Combinatorial Optimization—Eureka, You Shrink!*, number 2570 in LNCS, pages 185–207. Springer, 2003.

Extending the Tractability Border
for Closest Leaf Powers*

Michael Dom, Jiong Guo, Falk Hüffner, and Rolf Niedermeier

Institut für Informatik, Friedrich-Schiller-Universität Jena,
Ernst-Abbe-Platz 2, D-07743 Jena, Germany
{dom, guo, hueffner, niedermr}@minet.uni-jena.de

Abstract. The NP-complete CLOSEST 4-LEAF POWER problem asks, given an undirected graph, whether it can be modified by at most ℓ edge insertions or deletions such that it becomes a 4-leaf power. Herein, a 4-leaf power is a graph that can be constructed by considering an unrooted tree—the 4-leaf root—with leaves one-to-one labeled by the graph vertices, where we connect two graph vertices by an edge iff their corresponding leaves are at distance at most 4 in the tree. Complementing and "completing" previous work on CLOSEST 2-LEAF POWER and CLOSEST 3-LEAF POWER, we show that CLOSEST 4-LEAF POWER is fixed-parameter tractable with respect to parameter ℓ.

1 Introduction

Graph powers form a classical concept in graph theory, and the rich literature dates back to the sixties of the previous century. The k-*power* of an undirected graph $G = (V, E)$ is the undirected graph $G^k = (V, E')$ with $(u, v) \in E'$ iff there is a path of length at most k between u and v in G. We say G is the k-*root* of G^k. While it is NP-complete to decide whether a given graph is a k-power [10], one can decide in $O(|V|^3)$ time whether a graph is a k-power of a tree for any fixed k [6], and it can be decided in linear time whether a graph is a square of a tree [9].

Here, we concentrate on certain practically motivated variants of tree powers. Whereas Kearney and Corneil [6] study the problem where every tree node one-to-one corresponds to a graph vertex, Nishimura, Ragde, and Thilikos [12] introduce the notion of *leaf powers* where exclusively the tree leaves stand in one-to-one correspondence with the graph vertices. In addition, Lin, Kearney, and Jiang [8] and Chen, Jiang, and Lin [2] examine the variant of leaf powers where all inner nodes of the root tree have degree at least three. Both problems find applications in computational evolutionary biology [12,8,2]. The corresponding recognition problems are called k-LEAF POWER [12] and k-PHYLOGENETIC

---

* Research supported by the Deutsche Forschungsgemeinschaft (DFG), Emmy Noether research group PIAF (fixed-parameter algorithms), NI 369/4.

D. Kratsch (Ed.): WG 2005, LNCS 3787, pp. 397–408, 2005.

ROOT [8], respectively.[1] For $k \leq 4$, both problems are solvable in polynomial time [12,8]. The complexities of both recognition problems for $k \geq 5$ are open.

Several groups of researchers [2,6,8] strongly advocate the consideration of a more relaxed or "approximate" version of the graph power recognition problem: Now, look for roots whose powers are *close* to the input graphs, thus turning the focus of study to the corresponding *graph modification* problems. In this "error correction setting" the question is whether a given graph can be modified by adding or deleting at most ℓ edges such that the resulting graph has a k-tree root. This problem turns out to be NP-complete for $k \geq 2$ [6,5]. One also obtains NP-completeness for the corresponding problems CLOSEST k-LEAF POWER [7,3] and CLOSEST k-PHYLOGENETIC ROOT [2].

All nontrivial ($k \geq 2$) "approximate recognition" problems in our context turn out to be NP-complete [2,3,5,6,7,14]. Hence, the pressing quest is to also show positive algorithmic tractability results such as polynomial-time approximation or non-trivial (exponential-time) exact algorithms. So far, only the most simple version of CLOSEST k-LEAF POWER, $k = 2$, has been algorithmically attacked with somewhat satisfactory success. In this context recently intricate polynomial-time constant-factor approximation algorithms have been developed [1].[2] Moreover, it is fairly easy to show that the problem is fixed-parameter tractable with respect to the parameter ℓ denoting the number of allowed edge modifications. At least with respect to this fixed-parameter tractability result, the success is surely due to the fact that there is a very simple characterization by a forbidden subgraph: a graph is a 2-leaf power iff it contains no induced 3-vertex subgraph forming a path. Observe that, in this way, also the recognition problem for 2-leaf powers is solvable in linear time by just checking whether the given graph is a disjoint union of cliques. By way of contrast, the recognition problem for 3-leaf and 4-leaf powers is much harder and only intricate cubic-time algorithms are known [12]. The key idea we put forward here and in a companion paper [3] is to again develop and employ forbidden subgraph characterizations of the respective graph classes. In [3], we describe a forbidden subgraph characterization for 3-leaf powers, consisting of five graphs of small size. Here, we employ a forbidden subgraph characterization for 4-leaf powers—it already requires numerous forbidden subgraphs.

Let us discuss the algorithmic use of these forbidden subgraph characterizations. First, both characterizations immediately imply polynomial-time recognition algorithms for 3- and 4-leaf powers which are conceptually simpler than those in [12]. However, they are of purely theoretical interest because the running times of these straightforward algorithms are much worse than that of the

---

[1] Both problems k-LEAF POWER and k-PHYLOGENETIC ROOT ask whether a given graph is a leaf power resp. a phylogenetic power. We find it more natural to use the term *power* instead of the term *root*, although we used the term *root* in the conference version of our previous considerations concerning the case $k = 3$ [3].

[2] Note that in the various papers (partially not referring to each other) CLOSEST 2-LEAF POWER appears under various names such as CLUSTER EDITING [14] and CORRELATION CLUSTERING [1].

known cubic-time algorithms from [12]. More important, the characterizations open up the way to the first tractability results for the harder problems CLOSEST k-LEAF POWER for $k = 3, 4$. Using the forbidden subgraphs for 3-leaf powers, in [3] we show that CLOSEST 3-LEAF POWER is fixed-parameter tractable with respect to the parameter "number ℓ of edge modifications." Due to the significantly increased combinatorial complexity of 4-leaf powers, analogous results for CLOSEST 4-LEAF POWER remained open in [3]. We close this gap here. We show that CLOSEST 4-LEAF POWER can be solved in polynomial time for $\ell = O(\log n / \log \log n)$; that is, it is fixed-parameter tractable with respect to parameter ℓ. Moreover, the variants of CLOSEST 4-LEAF POWER where only edge insertions or only edge deletions are allowed are fixed-parameter tractable as well.

Due to the lack of space, we omit all proofs.

2 Preliminaries

We consider only undirected graphs $G = (V, E)$ with $n := |V|$ and $m := |E|$. Edges are denoted as tuples (u, v), ignoring any ordering. For a graph $G = (V, E)$ and $u, v \in V$, let $d_G(u, v)$ denote the length of the shortest path between u and v in G. With $E(G)$, we denote the edge set E of a graph $G = (V, E)$. We call a graph $G' = (V', E')$ an *induced subgraph* of $G = (V, E)$ and denote G' with $G[V']$ if $V' \subseteq V$ and $E' = \{(u, v) \mid u, v \in V' \text{ and } (u, v) \in E\}$. For a collection of graphs $\mathcal{G}$, a graph is said to be $\mathcal{G}$-*free* if it does not contain any graph in $\mathcal{G}$ as induced subgraph. A cycle with n vertices is denoted as C_n. An edge between two vertices of a cycle that is not part of the cycle is called *chord*. An induced cycle of length at least four is called *hole*—note that a hole is chordless. A *chordal graph* then is a hole-free graph. Let a *minimum edge cut*, denoted $\text{MINCUT}(G, V_1, V_2)$, be a minimum weight set of edges in $G = (V, E)$ that disconnects all vertices in $V_1 \subseteq V$ from those in $V_2 \subseteq V$. We say a set is *maximal* with respect to some property if it is not a proper subset of another set with that property. For two sets A and B, $A \triangle B$ denotes the *symmetric difference* $(A \setminus B) \cup (B \setminus A)$.

Definition 1 ([12]). *Consider an unrooted tree T with leaves one-to-one labeled by the elements of a set V. The k-leaf power of T is a graph, denoted T^k, with $T^k := (V, E)$, where $E := \{(u, v) \mid u, v \in V \text{ and } d_T(u, v) \leq k\}$. We call T a k-leaf root of T^k.*

The k-LEAF POWER (LPk) problem then is to decide, given a graph G, whether there is a tree T such that $T^k = G$.

One may view the leaf power concept as a "Steiner extension" of the standard notion of tree powers [2,8]. The more general, *approximate version* of LPk we focus on in this work, called CLOSEST k-LEAF POWER (CLPk), then reads as follows. Consider a graph $G = (V, E)$ and a nonnegative integer ℓ, is there a tree T such that T^k and G differ by at most ℓ edges, that is, $|E(T^k) \triangle E(G)| \leq \ell$? CLP$k$ is NP-complete for $k \geq 2$ [7,3].

In this paper we also study two variations of CLPk referring to only one-sided errors: CLPk EDGE INSERTION only allows insertion of edges and CLPk EDGE DELETION only allows deletion of edges to obtain T^k. CLPk EDGE DELETION is NP-complete for $k \geq 2$ [11,3], and CLPk EDGE INSERTION is NP-complete for $k \geq 3$ but trivially polynomial-time solvable for $k = 2$.

A central technical tool within this work are critical cliques and critical clique graphs as Lin et al. [8] introduce them.

Definition 2. *A* critical clique *of a graph* G *is a clique* K *where the vertices of* K *all have the same set of neighbors in* $G\backslash K$, *and* K *is maximal under this property. Consider a graph* $G = (V, E)$. *Let* V_C *be the collection of its critical cliques. Then the* critical clique graph $CC(G)$ *is a graph* (V_C, E_C) *(we use the term* nodes *for its vertices) with* $(K_i, K_j) \in E_C \iff \forall u \in K_i, v \in K_j : (u, v) \in E$. *That is, the critical clique graph has the critical cliques as nodes, and two nodes are connected iff the corresponding critical cliques together form a larger clique.*

Definition 3. *Consider a graph* $G = (V, E)$ *and an arbitrary set of vertices* A *with* $A \cap V = \emptyset$. *An unrooted tree* $T = (A \cup V, E')$ *is called a* k-Steiner root *of* G *if* $E = \{(u, v) \mid u, v \in V \text{ and } d_T(u, v) \leq k\}$.

Note that if $A = \emptyset$, then a k-Steiner root simply is a k-tree root. Similarly, if A is the set of inner nodes of T, then a k-Steiner root is the same as a k-leaf root. This means that the set of graphs that have k-Steiner roots is a superset of the set of graphs that have k-tree roots or k-leaf roots. The following lemma is easy to show (a similar statement was already made by Lin et al. [8]).

Lemma 1. *A graph* G *has a* k-leaf root *iff* $CC(G)$ *has a* $(k-2)$-Steiner root.

We show that CLP4 and both its edge insertion and edge deletion variant are *fixed-parameter tractable* (FPT) with respect to parameter ℓ. That is, we show that CLP4 can be solved in $f(\ell) \cdot n^{O(1)}$ time, where f is a computable function only depending on ℓ, and n denotes the number of vertices of the input graph.

3   Forbidden Subgraph Characterization of 4-Leaf Powers

In this section we give a characterization of 4-leaf powers using a set of eight forbidden induced subgraphs for the critical clique graphs of 4-leaf powers. This set can be extended to a larger set of forbidden subgraphs for the 4-leaf powers themselves by a simple iterative algorithm. Independently and by different proof techniques, Rautenbach [13] achieves the same results. Our approach, however, is tailored towards the algorithmic treatment following in the next section. The eight forbidden subgraphs for critical clique graphs of 4-leaf powers are shown in Fig. 1. Let $\mathcal{F} := \{F_1, F_2, \ldots, F_8\}$ as given there.

Theorem 1. *For a graph* G, *the following are equivalent: (1)* G *is a 4-leaf power. (2)* G *is chordal and its critical clique graph* $CC(G)$ *is* $\mathcal{F}$-*free.*

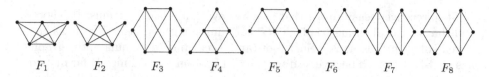

Fig. 1. The eight forbidden subgraphs for critical clique graphs of 4-leaf powers

---

$\mathrm{SRG}(\mathrm{CC}(G) = (V_C, E_C))$
Input: $(\mathcal{F} \cup \{C_4, C_5\})$-free critical clique graph $\mathrm{CC}(G) = (V_C, E_C)$
Output: Pseudo Steiner root graph of $\mathrm{CC}(G)$

1    $S \leftarrow (\{b_c \mid c \in V_C\}, \emptyset)$
2    $L \leftarrow$ list of all maximal cliques of $\mathrm{CC}(G)$
3   **while** there is a K in L which shares edges (c_1, c_2) and (c_1, c_3) with two
        other maximal cliques K' and K'' in $\mathrm{CC}(G)$:
4   Delete K from L
5   **for each** $c \in K, c \neq c_1$:
6   Insert an edge between b_{c_1} and b_c
7   **while** there is a K in L which shares only one edge (c_1, c_2) with only one
        other maximal clique K' in $\mathrm{CC}(G)$:
8   Delete K from L
9   **if** K' is in L:
10   **for each** $c \in K, c \neq c_1$:
11   Insert an edge between b_{c_1} and b_c
12   **else:**
13           $c' \leftarrow$ a node in $K' \setminus K$
14   **if** there is an edge $(b_{c_1}, b_{c'})$ in S:
15   **for each** $c \in K, c \neq c_2$:
16   Insert an edge between b_{c_2} and b_c
17   **else:**
18   **for each** $c \in K, c \neq c_1$:
19   Insert an edge between b_{c_1} and b_c
20   **while** there is a K in L:
21   Delete K from L
22   Add a new node s_K into S
23   **for each** $c \in K$:
24   Insert an edge between s_K and b_c
25   **while** there are at least two connected components S_1 and S_2 in S:
26   Add two new edge-connected Steiner nodes s_1 and s_2 to S and connect s_1
        by an edge to an arbitrary node in S_1 and s_2 to an arbitrary node in S_2
27   **return** S

Fig. 2. Algorithm to construct the pseudo Steiner root graph S of a critical clique graph $\mathrm{CC}(G)$

Corollary 1. *All graphs that are 4-leaf powers are chordal and can be characterized by a finite set of forbidden subgraphs.*

It is relatively easy to see that graphs having a 4-leaf root must be chordal, and that a critical clique graph $CC(G)$ containing a graph in $\mathcal{F}$ (Fig. 1) as an induced subgraph has no 2-Steiner root (and, according to Lemma 1, the graph G has no 4-leaf root). The reverse direction of Theorem 1 is technically far more difficult. We show constructively that every $\mathcal{F}$-free and chordal critical clique graph indeed has a 2-Steiner root by using Algorithm SRG (Fig. 2). This algorithm extends a method by Lin et al. [8] for constructing 2-Steiner roots: While their algorithm only computes an output graph if the input graph has a 2-Steiner root and says "no" otherwise, our Algorithm SRG (Fig. 2) also generates an output graph with some guaranteed properties for inputs that are $(\mathcal{F} \cup \{C_4, C_5\})$-free but nonchordal graphs. This will be of use for our fixed-parameter algorithms in Sect. 4.

For a given critical clique graph $CC(G) = (V_C, E_C)$, Algorithm SRG constructs a *pseudo Steiner root graph* $S = (V', E')$ with $V' := A \cup B$, where $B := \{b_c \mid c \in C\}$ and $A \cap B = \emptyset$. The nodes in A and B are called *Steiner* and *non-Steiner* nodes, respectively. Each non-Steiner node one-to-one corresponds to a node in $CC(G)$, whereas Steiner nodes do not correspond to nodes in $CC(G)$. If $CC(G)$ is $\mathcal{F}$-free and chordal, then S is a 2-Steiner root of $CC(G)$. (The term "pseudo Steiner root graph" expresses that if the input graph is $(\mathcal{F} \cup \{C_4, C_5\})$-free but nonchordal, then the output S has some, but not all properties of a 2-Steiner root.)

The idea of Algorithm SRG is to consider every maximal clique of the input graph $CC(G)$ and to connect the corresponding nodes in the output graph to form a star. More specifically, if a maximal clique K in $CC(G)$ has an edge e in common with another maximal clique K', then the node in the output graph corresponding to one of the endpoints of e is connected by edges with the other nodes corresponding to K and the node in the output graph corresponding to the other endpoint of e is connected by edges with the other nodes corresponding to K'. If otherwise K has no edge in common with another maximal clique, a Steiner node s_K is inserted into the output graph, and every node corresponding to a node of K is connected by an edge to s_K (see Fig. 3 for an example).

We can show that Algorithm SRG fulfills the following claims, which implies that the constructed pseudo Steiner root graph of an $\mathcal{F}$-free chordal critical clique graph $CC(G)$ actually is a 2-Steiner root of $CC(G)$, which together with

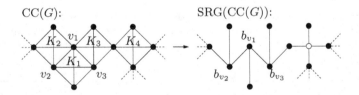

Fig. 3. Example of a subgraph of a critical clique graph $CC(G)$ and the pseudo Steiner root graph computed for this subgraph. Algorithm SRG first considers the maximal clique K_1 with $c_1 = v_1$ (see Fig. 2) and inserts edges between b_{v_1} and the other nodes corresponding to K_1. Thereafter, the cliques K_2 and K_3 are considered. When considering K_4, Algorithm SRG inserts a Steiner node (drawn white).

Lemma 1 proves the missing direction of Theorem 1. Note that the first two claims do not require chordality of the input graph. We will make use of this fact in Sect. 4 when we have to modify a critical clique graph to make it chordal. The claims are:

1. Every maximal clique K of an $(\mathcal{F} \cup \{C_4, C_5\})$-free critical clique graph $\mathrm{CC}(G)$ is considered at least once by Algorithm SRG, and for every node pair u, v in K, a path of length at most two is generated between the corresponding nodes of u and v in the output graph.
2. For an $(\mathcal{F} \cup \{C_4, C_5\})$-free critical clique graph $\mathrm{CC}(G) = (V_C, E_C)$ Algorithm SRG outputs a graph with the following property: If two nodes $u, v \in V_C$ are not adjacent in $\mathrm{CC}(G)$, then the distance between the nodes corresponding to u and v in the output graph is at least three.
3. For a chordal and $\mathcal{F}$-free critical clique graph $\mathrm{CC}(G)$ the output graph of Algorithm SRG is a tree.

Note that Algorithm SRG runs in polynomial time, as there are at most $2 \cdot |E_C|$ maximal cliques in an $(\mathcal{F} \cup \{C_4, C_5\})$-free critical clique graph $\mathrm{CC}(G) = (V_C, E_C)$.

4   Fixed-Parameter Tractability of CLP4

In this section we show the fixed-parameter tractability of CLP4 EDGE DELETION, CLP4 EDGE INSERTION, and CLP4 with respect to the parameter "number of edge editing operations" ℓ. The basic approach resembles our previous work for CLP3 [3]; however, for the case of CLP4 EDGE DELETION new, more intricate methods are necessary. Therefore, we focus on the CLP4 EDGE DELETION case in this section.

Note that graphs that have 3-leaf roots have a characterization similar to that of Theorem 1: they are graphs that are chordal and contain none of the induced subgraphs "bull," "dart," and "gem" [3]. Therefore, the basic idea for CLP3 EDGE DELETION as well as for CLP4 EDGE DELETION is to use the forbidden subgraph characterization in a depth-bounded search tree algorithm: find a forbidden subgraph, and recursively branch into several cases according to the possible edge deletions that destroy the forbidden subgraph. If we can upper-bound the number of branching cases by a function depending only on ℓ, since the depth can be bounded from above by ℓ, we obtain a run time that proves fixed-parameter tractability.

Since the forbidden subgraph characterization from Theorem 1 for the critical clique graph $\mathrm{CC}(G)$ is much simpler than the implied characterization for G (Corollary 1), we would like to apply modifications directly on $\mathrm{CC}(G)$. This is possible by the following lemma, which is a straightforward extension of Lemma 6 in [3].

Lemma 2. *For a graph G, there is always an optimal solution for CLP4 that is represented by edge editing operations on $\mathrm{CC}(G)$. That is, one can find an optimal solution that does not delete any edges within a critical clique; furthermore,* *in this optimal solution, between two critical cliques either all or no edges are inserted or deleted.*

Now, working with $CC(G) = (V_C, E_C)$ instead of G has two consequences: First, a deletion of an edge in $CC(G)$ can represent several edge deletions in G. Consider an edge e in $CC(G)$ between two nodes that represent critical cliques of sizes c_1 and c_2. Deleting e implies deleting all $c_1 \cdot c_2$ edges between the vertices of the critical cliques in G. Therefore, we give the edge e the weight $c_1 \cdot c_2$. Note that this means that an edge modification on $CC(G)$ can decrease the parameter ℓ in the depth-bounded search tree algorithm by more than one. Second, if two adjacent nodes in $CC(G)$ obtain an identical neighborhood after deleting edges in $CC(G)$, then $CC(G)$ needs to be updated, since each node in $CC(G)$ has to represent a critical clique in G. In this situation a *merge* operation is needed, which replaces these nodes in $CC(G)$ by a new node with the same neighborhood as the original nodes. Subsequently, assume that after each modification of $CC(G)$, all pairs of nodes in $CC(G)$ are checked as to whether a merge operation between them is required. This can be done in $O(|V_C| \cdot |E_C|)$ time.

The main obstacle in obtaining fixed-parameter tractability for both CLP3 EDGE DELETION and CLP4 EDGE DELETION is that the holes in $CC(G)$ can have arbitrary length, and, therefore, one cannot simply find some hole and branch for each edge of the hole that is to be deleted—the number of branching cases would not be a function only depending on ℓ. For CLP3 EDGE DELETION, the key observation is that the critical clique graph $CC(G)$ of a graph G containing neither a bull nor a dart nor a gem nor a C_4 contains no triangles. This allows to show that, after destroying the forbidden subgraphs bull, dart, gem, and C_4 in G, no hole in $CC(G)$ can be "accidentally" destroyed by merge operations between its nodes and, therefore, one has to delete at least one edge of every hole. Since moreover making a triangle-free graph chordal means to make it a forest, a minimum weight set of edges to be deleted to make $CC(G)$ chordal can be obtained in polynomial time by searching for a maximum weight spanning tree. Unfortunately, there *can* be triangles in an $\mathcal{F}$-free (Fig. 1) $CC(G)$ as we obtain it for CLP4 after deleting the forbidden subgraphs. Thus, the main technical contribution of this section is to show how to circumvent these difficulties.

The idea is to examine the output graph $SRG(CC(G))$ of Algorithm SRG (Fig. 2) for the critical clique graph $CC(G)$. If it is a tree, we are done. Otherwise, the output is a pseudo Steiner root graph S that contains a cycle which corresponds to a hole in $CC(G)$. By repeatedly deleting degree-1 nodes and contracting consecutive degree-2 nodes in S we get a graph S' in which there is no path that consists of three or more consecutive degree-2 nodes. By finding the shortest cycle in this reduced graph S', we can obtain an "FPT hole" in $CC(G)$, that is, a hole for which we can bound the possibilities to delete edges to get rid of the hole in an optimal way by a function only depending on ℓ (see Fig. 4).

For the pseudocode of this algorithm, which is presented in Fig. 5, we introduce some notation for the mapping between the nodes of a critical clique graph and the nodes of its pseudo Steiner root graph.

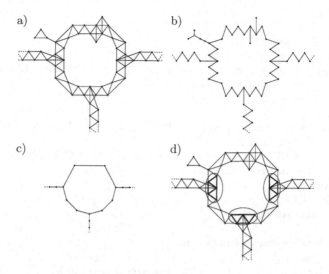

Fig. 4. Illustration of finding and destroying holes in an $(\mathcal{F} \cup \{C_4, C_5\})$-free critical clique graph: a) A nonchordal critical clique graph $CC(G)$. b) The pseudo Steiner root graph S constructed by Algorithm SRG for $CC(G)$. c) The reduced pseudo Steiner root graph S'. d) The sets marked with an ellipsis correspond to the degree-3 nodes in S'. Our algorithm for CLP4 EDGE DELETION either deletes one of the bold edges or it deletes a minimum weight set of edges between two of the node sets marked with an ellipsis (these sets are called "big node areas" in Def. 5).

Definition 4. *Consider a critical clique graph* $CC(G) = (V_C, E_C)$ *and a pseudo Steiner root graph* $S = (V_S, E_S)$ *constructed by Algorithm SRG for* $CC(G)$. *For* $v \in V_C$ *we use* $S(v)$ *to denote the node from* V_S *that corresponds to* v, *and for* $v_S \in V_S$, *we define* $S^{-1}(v_S)$ *as the node in* V_C *corresponding to* v_S *if* v_S *is a non-Steiner node, or* $\perp$ *if* v_S *is a Steiner node. We extend this notation to sets: for* $V_C' \subseteq V_C$, $S(V_C') := \{S(v) \mid v \in V_C'\}$, *and for* $V_S' \subseteq V_S$, $S^{-1}(V_S') := \{S^{-1}(v) \mid v \in V_S'\}$.

To define the branching set D in line 18 of Algorithm CLP4DEL-BRANCH, we need some notation.

Definition 5. *A* big node *is a node of a pseudo Steiner root graph* S *that is not deleted by the data reduction in lines 11–19 of Algorithm* CLP4DEL-BRANCH *(Fig. 5) and that has degree at least 3 in the constructed pseudo Steiner root graph* S' *(see Fig. 6).*

For a cycle Q *in a pseudo Steiner root graph* S *as constructed by Algorithm* CLP4DEL-BRANCH *in line 16, let* $v_0, \ldots, v_{q-1}$ *be the big nodes in* Q, *ordered by their appearance in* Q, *and for every node* v_i *with* $0 \leq i < q$ *let* P_i *be the path in* Q *between* v_i *and* $v_{(i+1) \bmod q}$.

With P_i^+ *we denote the path* P_i *plus its attached trees, that is, the maximal set of nodes in* S *such that* P_i^+ *contains the nodes of* P_i *and such that* P_i^+ *induces a connected component in* $S \setminus \{v_i, v_{(i+1) \bmod q}\}$.

CLP4DEL-BRANCH(G, ℓ)
Input: A graph $G = (V, E)$ and an integer ℓ
Output: A set of at most ℓ edges in G whose removal makes G a 4-leaf power, or **nil** if no such set exists

1   **if** $\ell < 0$: **return nil**
2   Compute $CC(G)$
3   **if** $CC(G)$ contains an induced forbidden subgraph $F \in \mathcal{F} \cup \{C_4, C_5\}$:
4   **for each** edge e in F:
5           $X \leftarrow$ CLP4DEL-BRANCH(CC-DEL$(G, \{e\})$, ℓ−CC-WEIGHT$(G, \{e\})$)
6   **if** $X \neq$ **nil**: **return** $X \cup \{e\}$
7   **return nil**
8   $S \leftarrow$ SRG$(CC(G))$
9   **if** S is a tree: **return** $\emptyset$
10 $S' \leftarrow S$
11 **while** there is a degree-1-node u in S':
12   delete u
13 **while** there is a path (u, v, w) of three degree-2-nodes in S':
14   delete v and insert an edge between u and w
15 $Q' \leftarrow$ shortest cycle in S'
16 $Q \leftarrow$ cycle in S corresponding to Q'
17 $H \leftarrow S^{-1}(Q) \setminus \{\bot\}$
18 Determine a set D (see Lemma 4) of edge sets in $CC(G)[H]$ such that at least one edge set $d \in D$ is a subset of an optimal solution
19 **for each** $d \in D$:
20      $X \leftarrow$ CLP4DEL-BRANCH(CC-DEL(G, d), ℓ−CC-WEIGHT(G, d))
21   **if** $X \neq$ **nil**: **return** $X \cup d$
22 **return nil**

Fig. 5. Algorithm for CLP4 EDGE DELETION. The subroutine CC-DEL(G, d) takes a graph G and a set d of edges in $CC(G)$ as input. For every edge $(K_1, K_2) \in d$, all edges from G that have one endpoint in the clique represented by K_1 and the other endpoint in the clique represented by K_2 are deleted by CC-DEL(G, d). The function CC-WEIGHT(G, d) returns the sum of the weights of the edges in d.

We further denote with A_i, $0 \leq i < q$, *the* big node areas *that are defined as*

$$A_i := S^{-1}(\{v \in Q \mid d_S(v_i, v) \leq 2\}) \setminus \{\bot\}.$$

The following lemma will help us to show that the cycle Q determined by Algorithm CLP4DEL-BRANCH (Fig. 5) in line 16 indeed induces at least one hole in $CC(G)$.

Lemma 3. *Consider a cycle Q in a pseudo Steiner root graph S as constructed by Algorithm CLP4DEL-BRANCH (Fig. 5) in line 16. Let $v_0, \ldots, v_{p-1}$ be the nodes of Q, ordered by their appearance in Q. Then there is no edge $(S^{-1}(v_i), S^{-1}(v_j))$ with $0 \leq i, j < p$ in $CC(G)$ such that v_i and v_j have a distance of more than 2 on Q.*

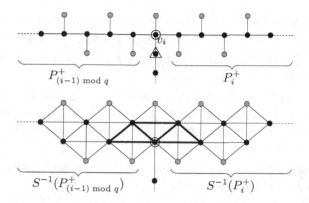

Fig. 6. Illustration for Definition 5. The upper picture shows a part of the pseudo Steiner root graph S. The encircled node v_i is a big node; black nodes are part of a cycle in S. The grey nodes are deleted by the data reduction in lines 11–14 of Algorithm CLP4DEL-BRANCH. The only Steiner node in this example is the node marked with a triangle. The lower picture shows the corresponding part of $CC(G)$. The bold edges are those between vertices of the big node area A_i.

The main observation that helps to bound the number of branching cases and, hence, leads to our fixed-parameter algorithm is that for a cycle Q in a pseudo Steiner root graph S the number of branching cases is independent of the lengths of the paths in Q between the big nodes: If we want to disconnect two big node areas, then it is always optimal to take an edge set with minimum weight whose removal disconnects the two big node areas. Such an edge set can be found in polynomial time by maximum flow techniques.

Lemma 4. *In* CLP4DEL-BRANCH, *the branching set D chosen as follows contains at least one subset of an optimal solution: Either delete an edge in a big node area, that is, an edge (u, v) with $u, v \in A_i$ for some $0 \le i < q$, or delete a set of edges*

$$\text{MINCUT}(CC(G)[S^{-1}(P_i^+) \setminus \{\bot\}], A_i, A_{(i+1) \bmod q}),$$

that is, delete a minimum weight set of edges such that all paths between two neighboring big node areas are destroyed.

It remains to show the complexity of CLP4DEL-BRANCH. All steps within a single invocation of CLP4DEL-BRANCH can be done in polynomial time. We therefore focus on the number of recursive calls. In line 4, there can be at most 10 recursive calls corresponding to at most 10 edges to delete in a forbidden subgraph (for example F_3 in Fig. 1); as we will see, this is dominated by the number of recursive calls in line 20 for destroying a long cycle.

A well-known result by Erdős and Pósa [4] states that any graph with minimum vertex degree at least 3 has a cycle of length at most $2 \log n + 1$, where n denotes the number of graph vertices. Using this result we can give an upper bound on the size of the shortest cycle in S' and show the following lemma:

Lemma 5. *When choosing D in line 18 of Algorithm* CLP4DEL-BRANCH *as described by Lemma 4, we can upper-bound its size by $|D| \leq 96 \cdot \log|V| + 24$.*

Theorem 2. CLP4 EDGE DELETION *with ℓ edge deletions allowed is fixed-parameter tractable with respect to ℓ.*

Proof. By Lemma 5 and the fact that the height of the search tree is bounded from above by ℓ, CLP4DEL-BRANCH runs in $(96 \cdot \log|V| + 24)^\ell \cdot |V|^{O(1)} \leq c^\ell \cdot (\ell \log \ell)^\ell \cdot n^{O(1)}$ time for a constant c (the inequality holds because $(\log n)^\ell \leq (3\ell \log \ell)^\ell + n$ for all values of n and ℓ).   □

With Theorem 2 and using the same techniques as applied for CLP3 EDGE INSERTION and CLP3 [3], we achieve the following result:

Theorem 3. *1. CLP4 EDGE INSERTION with ℓ edge insertions allowed is fixed-parameter tractable with respect to ℓ.*
2. CLP4 with ℓ edge insertions and deletions is fixed-parameter tractable with respect to ℓ.

References

1. M. Charikar, V. Guruswami, and A. Wirth. Clustering with qualitative information. In *Proc. 44th FOCS*, pages 524–533. IEEE Computer Society, 2003. To appear in *J. Comput. System Sci.*
2. Z.-Z. Chen, T. Jiang, and G. Lin. Computing phylogenetic roots with bounded degrees and errors. *SIAM J. Comput.*, 32(4):864–879, 2003.
3. M. Dom, J. Guo, F. Hüffner, and R. Niedermeier. Error compensation in leaf root problems. In *Proc. 15th ISAAC*, volume 3341 of *LNCS*, pages 389–401. Springer, 2004. Long version to appear under the title *Error compensation in leaf power problems* in *Algorithmica*.
4. P. Erdős and L. Pósa. On the maximal number of disjoint circuits of a graph. *Publ. Math. Debrecen*, 9:3–12, 1962.
5. T. Jiang, G. Lin, and J. Xu. On the closest tree kth root problem. Manuscript, Department of Computer Science, University of Waterloo, 2000.
6. P. E. Kearney and D. G. Corneil. Tree powers. *J. Algorithms*, 29(1):111–131, 1998.
7. M. Křivánek and J. Morávek. NP-hard problems in hierarchical-tree clustering. *Acta Inform.*, 23(3):311–323, 1986.
8. G. Lin, P. E. Kearney, and T. Jiang. Phylogenetic k-root and Steiner k-root. In *Proc. 11th ISAAC*, volume 1969 of *LNCS*, pages 539–551. Springer, 2000.
9. Y.-L. Lin and S. S. Skiena. Algorithms for square roots of graphs. *SIAM J. Discrete Math.*, 8(1):99–118, 1995.
10. R. Motwani and M. Sudan. Computing roots of graphs is hard. *Discrete Appl. Math.*, 54(1):81–88, 1994.
11. A. Natanzon. Complexity and approximation of some graph modification problems. Master's thesis, Department of Computer Science, Tel Aviv University, 1999.
12. N. Nishimura, P. Ragde, and D. M. Thilikos. On graph powers for leaf-labeled trees. *J. Algorithms*, 42(1):69–108, 2002.
13. D. Rautenbach. 4-leafroots. Manuscript, Forschungsinstitut für Diskrete Mathematik, Universität Bonn, June 2004.
14. R. Shamir, R. Sharan, and D. Tsur. Cluster graph modification problems. *Discrete Appl. Math.*, 144:173–182, 2004.

Bounding the Misclassification Error in Spectral Partitioning in the Planted Partition Model*

Joachim Giesen and Dieter Mitsche

Institute for Theoretical Computer Science, ETH Zürich, CH-8092 Zürich
{giesen, dmitsche}@inf.ethz.ch

Abstract. A partitioning of a set of n items is a grouping of these items into k disjoint, equally sized classes. Any partition can be modeled as a graph. The items become the vertices of the graph and two vertices are connected by an edge if and only if the associated items belong to the same class. In a planted partition model a graph that models a partition is given, which is obscured by random noise, i.e., edges within a class can get removed and edges between classes can get inserted. The task is to reconstruct the planted partition from this graph. We design a spectral partitioning algorithm and analyze how many items it misclassifies in the worst case. The number of classes k is one parameter in the model that allows to control the difficulty of the problem. Our analysis extends the range of k for which any non-trivial quality guarantees can be given.

1 Introduction

The partition reconstruction problem, which we study in this paper, is related to the k-partition problem. In the latter problem the task is to partition the vertices of a given graph into k equally sized classes such that the number of edges between the classes is minimized. This problem is already NP-hard for $k = 2$, i.e., in the graph bisection case [6]. Thus researchers, see for example [4,2] and the references therein, started to analyze the problem in specialized but from an application point of view (e.g., parallel scheduling or mesh partitioning) still meaningful, graph families - especially families of random graphs. The random graph families typically assume a given partition of the vertices of the graph (planted partition), which is obscured by random noise. The families are parameterized by a set of parameters, e.g., the number of vertices n and classes k. The goal now becomes to assess the ability of a partitioning algorithm to reconstruct the planted classes. Two measures to assess the quality of a partitioning algorithm in terms of the parameters of the random graph families are

(1) the probability that the algorithm can reconstruct the planted partition, and
(2) the number of items that the algorithm misclassifies (with a suited definition of misclassification).

---

* Partly supported by the Swiss National Science Foundation under the grant "Non-linear manifold learning".

D. Kratsch (Ed.): WG 2005, LNCS 3787, pp. 409–420, 2005.

The best studied random graph family for the partition reconstruction problem is the following: an edge in the graph appears with probability p if its two incident vertices belong to the same planted class and with probability $q < p$ otherwise, independently from all other edges. In general the probabilities p and q can depend on the number n of vertices in the graph and on the number k of classes of the planted partition. The known theoretical guarantees in this model state that certain algorithms can with high probability reconstruct the planted partition correctly for a certain range of the parameters of the random graph family. In these cases the number of misclassifications is concentrated at zero. Here we show that in certain situations where no guarantees on perfect reconstruction are known for any algorithm we can at least meaningfully bound the number of misclassifications for an algorithm that falls within the category of *spectral partitioning algorithms*. Spectral partitioning algorithms make use of the eigenvalues and eigenvectors of the similarity matrix in order to perform the partitioning.

Related Work. The partitioning problem in the planted partition model that we have described above gets more difficult if the difference $p - q$ gets small and/or k gets large. If we assume that p and q are fixed the only parameter left to control the difficulty of the problem is k. The algorithm of Shamir and Tsur [10] which builds on ideas of Condon and Karp [4] can with high probability reconstruct correctly up to $k = O(\sqrt{n/\log n})$ planted classes. The same guarantees can be given for an algorithm due to McSherry [8]. Both algorithms are polynomial in time and even allow the classes to differ in size (only a lower bound on the size of the classes is needed), i.e., they deal with the more general planted clustering problem. The algorithm of McSherry falls in the category of spectral clustering algorithms. The use of spectral methods for clustering has become increasingly popular in recent years. The vast majority of the literature points out the experimental success of spectral methods, see for example the review by Meila et al. [9] where also the measure for the number of misclassifications that we are going to use here was introduced. On the theoretical side much less is known about the reasons why spectral algorithms perform well. In 1987 Boppana [3] presented a spectral algorithm for recovering the optimal bisection of a graph. Much later Alon et al. [1] showed how the entries in the second eigenvector of the adjacency matrix of a graph can be used to find a hidden clique of size $\Omega(\sqrt{n})$ in a random graph. Spielman and Teng [11] showed how bounded degree planar graphs and d-dimensional meshes can be partitioned using the signs of the entries in the second eigenvector of the adjacency matrix of the graph or mesh, respectively.

Our Result. We design an efficient (polynomial in n) spectral algorithm and analyze the number of misclassifications it makes when k can be as large as $c\sqrt{n}$, where c is a constant that we specify. Note that so far nothing was known on how well or badly any partitioning algorithm performs for $k = \omega(\sqrt{n/\log n})$. We get that the relative number of misclassifications for $k = o(\sqrt{n})$ goes to zero with high probability when n goes to infinity.

2   Planted Partitions

In this section we introduce the planted partition reconstruction problem and define two quality measures that can be used to compare different partitioning algorithms. We first introduce the $A(\varphi, p, q)$ distribution, see also McSherry [8].

$A(\varphi, p, q)$ **Distribution.** Given a surjective function $\varphi : \{1, \ldots, n\} \to \{1, \ldots, k\}$ and probabilities $p, q \in (0,1)$ with $p > q$. The $A(\varphi, p, q)$ distribution is a distribution on the set of $n \times n$ symmetric, 0-1 matrices with zero trace. Let $\hat{A} = (\hat{a}_{ij})$ be a matrix drawn from this distribution. It is $\hat{a}_{ij} = 0$ if $i = j$ and for $i \neq j$,

$$
\begin{aligned}
P(\hat{a}_{ij} = 1) &= p & \text{if } \varphi(i) = \varphi(j) \\
P(\hat{a}_{ij} = 0) &= 1 - p & \text{if } \varphi(i) = \varphi(j) \\
P(\hat{a}_{ij} = 1) &= q & \text{if } \varphi(i) \neq \varphi(j) \\
P(\hat{a}_{ij} = 0) &= 1 - q & \text{if } \varphi(i) \neq \varphi(j),
\end{aligned}
$$

independently. The *matrix of expectations* $A = (a_{ij})$ corresponding to the $A(\varphi, p, q)$ distribution is given as

$$
\begin{aligned}
a_{ij} &= 0 & \text{if } i = j \\
a_{ij} &= p & \text{if } \varphi(i) = \varphi(j) \text{ and } i \neq j \\
a_{ij} &= q & \text{if } \varphi(i) \neq \varphi(j)
\end{aligned}
$$

Lemma 1 (Füredi and Komlós [5], van Vu [13], Krivelevich and van Vu [7]). *Let $\hat{A}$ be a matrix drawn from the $A(\varphi, p, q)$ distribution and A be the matrix of expectations corresponding to this distribution. Let $c = \min\{p(1 - p), q(1 - q)\}$ and assume that $c^2 \gg (\log n)^6 / n$. Then*

$$
|A - \hat{A}| \leq \sqrt{n}
$$

with probability at least $1 - 2e^{-c^2 n / 8}$. Here $|\cdot|$ denotes the L_2 matrix norm, i.e., $|B| = \max_{|x|=1} |Bx|$.

Planted Partition Reconstruction Problem. Given a matrix $\hat{A}$ drawn from the $A(\varphi, p, q)$ distribution. Assume that all classes $\varphi^{-1}(l), l \in \{1, \ldots, k\}$ have the same size n/k. Then the function φ is called a *partition function*. The planted partition reconstruction problem asks to reconstruct φ up to a permutation of $\{1, \ldots, k\}$ only from $\hat{A}$ (up to permutations of of $\{1, \ldots, k\}$).

Quality of a Reconstruction Algorithm. A planted partition reconstruction algorithm takes a matrix $\hat{A}$ drawn from the distribution $A(\varphi, p, q)$ as input and outputs a function $\psi : \{1, \ldots, n\} \to \{1, \ldots, k'\}$. There are two natural measures to assess the quality of the reconstruction algorithm.

(1) The probability of correct reconstruction, i.e.,

$$
P[\varphi = \psi \text{ up to a permutation of } \{1, \ldots, k\}].
$$

(2) The distribution of the number of elements in $\{1, \ldots, n\}$ misclassified by the algorithm. The definition for the number of misclassifications used here (see also Meila et al. [9]) is as the size of a maximum matching on the weighted, complete bipartite graph whose vertices are the classes $\varphi^{-1}(i), i \in \{1, \ldots, k\}$ and the classes $\psi^{-1}(j), j \in \{1, \ldots, k'\}$ produced by the algorithm. The weight of the edge $\{\varphi^{-1}(i), \psi^{-1}(j)\}$ is $|\varphi^{-1}(i) \cap \psi^{-1}(j)|$, i.e. the size of the intersection of the classes. The matching gives a pairing of the classes defined by φ and ψ. Assume without loss of generality that always $\varphi^{-1}(i)$ and $\psi^{-1}(i)$ are paired. Then the number of misclassifications is given as

$$n - \sum_{i=1}^{\min\{k,k'\}} |\varphi^{-1}(i) \cap \psi^{-1}(i)|.$$

3   Spectral Properties

Any real symmetric $n \times n$ matrix has n real eigenvalues and $\mathbb{R}^n$ has a corresponding eigenbasis. Here we are concerned with two types of real symmetric matrices. First, any matrix $\hat{A}$ drawn from an $A(\varphi, p, q)$ distribution. Second, the matrix A of expectations corresponding to the distribution $A(\varphi, p, q)$.

We want to denote the eigenvalues of $\hat{A}$ by $\hat{\lambda}_1 \geq \hat{\lambda}_2 \geq \ldots \geq \hat{\lambda}_n$ and the vectors of a corresponding orthonormal eigenbasis of $\mathbb{R}^n$ by $v_1, \ldots, v_n$, i.e., it is $\hat{A}v_i = \hat{\lambda}_i v_i$, $v_i^T v_j = 0$ if $i \neq j$ and $v_i^T v_i = 1$, and the $v_1, \ldots, v_n$ span the whole $\mathbb{R}^n$.

For the sake of analysis we want to assume here without loss of generality that the matrix A of expectations has a block diagonal structure, i.e., the elements in the i-th class have indices from $\frac{n}{k}(i-1)+1$ to $\frac{n}{k}i$ in $\{1, \ldots, n\}$. It is easy to verify that the eigenvalues $\lambda_1 \geq \ldots \geq \lambda_n$ of A are $(\frac{n}{k}-1)p+(n-\frac{n}{k})q$, $\frac{n}{k}(p-q)-p$ and $-p$ with corresponding multiplicities 1, $k-1$ and $n-k$, respectively. A possible orthonormal basis of the eigenspace corresponding to the k largest eigenvalues of A is u_i, $i = 1, \ldots, k$, whose j-th coordinates are given as follows,

$$u_{ij} = \begin{cases} \sqrt{\frac{k}{n}}, & j \in \{\frac{n}{k}(i-1)+1, \ldots, \frac{n}{k}i\} \\ 0, & \text{else.} \end{cases}$$

Theorem 1 (Weyl).

$$\max\{|\lambda_i - \hat{\lambda}_i| \mid i \in \{1, \ldots, n\}\} \leq |A - \hat{A}|.$$

Spectral Separation. The *spectral separation* $\delta_k(A)$ of the eigenspace of the matrix A of expectations corresponding to its k largest eigenvalues from its complement is defined as the difference of the k-th and the $(k+1)$-th eigenvalue, i.e., it is $\delta_k(A) = \frac{n}{k}(p-q)$.

Projection Matrix. The matrix $\hat{P}$ that projects any vector in $\mathbb{R}^n$ to the eigenspace corresponding to the k largest eigenvalues of a matrix $\hat{A}$ drawn from the distribution $A(\varphi, p, q)$, i.e., the projection onto the space spanned by the vectors $v_1, \ldots, v_k$, is given as

$$\hat{P} = \sum_{i=1}^{k} v_i v_i^T.$$

The matrix P that projects any vector in $\mathbb{R}^n$ to the eigenspace corresponding to the k largest eigenvalues of the matrix A of expectations can be characterized even more explicitly. Its entries are given as

$$p_{ij} = \begin{cases} \frac{k}{n}, & \varphi(i) = \varphi(j) \\ 0, & \varphi(i) \neq \varphi(j) \end{cases}$$

Lemma 2. *All the k largest eigenvalues of $\hat{A}$ are larger than $\sqrt{n}$ and all the $n - k$ smallest eigenvalues of $\hat{A}$ are smaller than $\sqrt{n}$ with probability at least $1 - 2e^{-c^2 n/8}$ provided that n is sufficiently large and $k < \frac{p-q}{4}\sqrt{n}$.*

Proof. Plugging in our assumption that $k < \frac{p-q}{4}\sqrt{n}$ gives that the k largest eigenvalues of A are larger than $4\sqrt{n} - p > 2\sqrt{n}$. By the lemma of Füredi and Komlós it is $|A - \hat{A}| \leq \sqrt{n}$ with probability at least $1 - 2e^{-c^2 n/8}$. Now it follows from Weyl's theorem that the k largest eigenvalues of $\hat{A}$ are larger than $\sqrt{n}$ with probability at least $1 - 2e^{-c^2 n/8}$. Since the $n - k$ smallest eigenvalues of A are $-p$ it also follows that the $n - k$ smallest eigenvalues of $\hat{A}$ are smaller than $\sqrt{n}$ with probability at least $1 - 2e^{-c^2 n/8}$. □

Lemma 3. *With probability at least $1 - 2e^{-c^2 n/8}$ it holds*

$$\frac{n}{k}(p - q) - p - \sqrt{n} \leq \hat{\lambda}_2 \quad and \quad \hat{\lambda}_2 \frac{k}{n} - \frac{k}{\sqrt{n}} \leq p - q,$$

provided n is sufficiently large.

Proof. It holds $\lambda_2 = \frac{n}{k}(p - q) - p$. By combining Weyl's theorem and the lemma of Füredi and Komlós we get that with probability at least $1 - 2e^{-c^2 n/8}$ it holds

$$\hat{\lambda}_2 \in \left[\frac{n}{k}(p - q) - p - \sqrt{n}, \frac{n}{k}(p - q) - p + \sqrt{n} \right].$$

Hence with the same probability

$$\hat{\lambda}_2 \frac{k}{n} - \frac{k}{\sqrt{n}} \leq p - q \leq \hat{\lambda}_2 \frac{k}{n} + \frac{k}{\sqrt{n}} + \frac{k}{n}.,$$

where we used $p \leq 1$ for the upper bound and $p \geq 0$ for the lower bound. □

Theorem 2 (Stewart [12]). *Let $\hat{P}$ and P be the projection matrices as defined above. It holds*

$$|P - \hat{P}| \leq \frac{2|A - \hat{A}|}{\delta_k(A) - 2|A - \hat{A}|}$$

if $\delta_k(A) > 4|A - \hat{A}|$ where $|\cdot|$ is the L_2 matrix norm.

4   A Spectral Algorithm

Now we have all prerequisites at hand that we need to describe our spectral algorithm to solve the planted partition reconstruction problem.

SPECTRALRECONSTRUCT($\hat{A}$)
1   $k' :=$ number of eigenvalues of $\hat{A}$ that are larger than $\sqrt{n}$.
2   $\hat{P} :=$ projection matrix computed from the k' largest eigenvectors
        $v_1, \ldots, v_{k'}$ of $\hat{A}$.
3   **for** $i = 1$ **to** n **do**
4       $R_i :=$ set of row indices which are among the $\frac{n}{k'}$ largest entries of
            the i-th column of $\hat{P}$.
5   **for** $j = 1$ **to** n **do**
6           $c_{ij} := \begin{cases} 1, & j \in R_i \\ 0, & \text{else} \end{cases}$
7   **end for**
8       $c_i := (c_{i1}, \ldots, c_{in})^T$
9   **end for**
10 $I := \{1, \ldots, n\}; l := 1$
11 **while** exists an unmarked index $i \in I$ **do**
12      $C_l := \emptyset$
13   **for each** $j \in I$ **do**
14   **if** $c_i^T c_j > \frac{4n}{5k'}$ **do**
15              $C_l := C_l \cup \{j\}$
16   **end if**
17   **end for**
18   **if** $|C_l| \geq \left(1 - \sqrt{\frac{160\sqrt{n}}{\lambda_2 - 3\sqrt{n}}}\right) \frac{n}{k'}$ **do**
19          $I := I \setminus C_l; l := l + 1$
20   **else**
21   mark index i.
22   **end if**
23 **end while**
24 $C_l := I$
25 **return** $C_1, \ldots, C_l$

In line 1 the number of planted classes k' is estimated. The estimate is motivated by Lemma 2. In line 2 the projection matrix $\hat{P}$ that belongs to $\hat{A}$ is computed. From line 3 to line 9 for each column i of $\hat{A}$ a vector $c_i \in \{0,1\}^n$ with exactly $\frac{n}{k'}$ entries that are one is computed. In lines 10 to 24 the actual partitioning takes place. Roughly speaking, two indices i, j are put into the same class if the Hamming distance of the corresponding vectors c_i and c_j is small (test in line 14). A class as created in lines 12 to 17 is not allowed to be too small (test in line 18), otherwise its elements get distributed into other classes that are going to be constructed in future executions of the body of the while-loop.

Notice that the algorithm runs in time polynomial in n and only makes use of quantities that can be deduced from $\hat{A}$, i.e., it does not need to know the values of p, q and k.

5   Bounding the Number of Misclassifications

Safe Vector. A vector $c_i \in \{0,1\}^n$ as produced by the algorithm SPECTRAL-RECONSTRUCT is called *safe* with respect to φ if more than $\frac{9}{10}$ of the indices in $\{1, \ldots, n\}$ that correspond to the one entries in c_i are mapped by φ to $\varphi(i)$, i.e., all these elements belong to the same class. A vector c_i is called *unsafe* if it is not safe.

Lemma 4. *In the algorithm* SPECTRALRECONSTRUCT *with probability at least* $1 - 2e^{-c^2 n/8}$ *at most*

$$\frac{160n}{\frac{\sqrt{n}}{k}(p-q) - 2}$$

vectors $c_i \in \{0,1\}^n$ *are constructed that are unsafe if* $k < \frac{p-q}{4}\sqrt{n}$.

Proof. Let x be the number of unsafe vectors c_i that are computed within the algorithm SPECTRALRECONSTRUCT from the projection matrix $\hat{P}$. If c_i is an unsafe vector then at least $\frac{n}{10k}$ of the $\frac{n}{k}$ largest entries in the i-th column of $\hat{P}$ correspond to row indices $j \in \{1, \ldots, n\}$ such that the entries p_{ij} in P are zero, i.e., these entries are not among the $\frac{n}{k}$ largest entries in the i-th column of P. That is, at least $x\frac{n}{10k}$ of the large entries in P become small entries in $\hat{P}$, i.e., they do no longer belong to the $\frac{n}{k}$ largest entries in their column. We denote the number of such entries by y and can bound it by using the Frobenius norm of the matrix $P - \hat{P}$. The Frobenius norm of a real $n \times n$ matrix B is defined as

$$|B|_F = \sqrt{\sum_{i,j=1}^{n} b_{ij}^2}.$$

The Frobenius norm and the L_2 norm are related by $|B|_F^2 \leq r|B|^2$, where r is the rank of B. Thus in order to bound the Frobenius norm of $P - \hat{P}$ we first bound its L_2 norm,

$$|P - \hat{P}| \leq \frac{2|A - \hat{A}|}{\delta_k(A) - 2|A - \hat{A}|}$$

$$= \frac{2|A - \hat{A}|}{\frac{n}{k}(p-q) - 2|A - \hat{A}|}$$

$$\leq \frac{2\sqrt{n}}{\frac{n}{k}(p-q) - 2\sqrt{n}} = \frac{2}{\frac{\sqrt{n}}{k}(p-q) - 2} < 1,$$

where we use the theorem of Stewart in the first inequality, the definition of the spectral gap in the first equality, the lemma of Füredi and Komlós in the second inequality and our assumption on k in the last inequality. Note, that the second inequality only holds with probability at least $1 - 2e^{-c^2 n/8}$. From Lemma 2 it follows that the k' chosen in line 1 of the algorithm SPECTRALRECONSTRUCT is exactly k with probability at least $1 - 2e^{-c^2 n/8}$. Hence the rank of $P - \hat{P}$ is at most $2k$ with probability at least $1 - 2e^{-c^2 n/8}$. That gives

$$|P - \hat{P}|_F^2 \leq 2k|P - \hat{P}|^2 < 2k|P - \hat{P}| \leq \frac{4k}{\frac{\sqrt{n}}{k}(p - q) - 2},$$

where the first inequality only holds with probability at least $1 - 2e^{-c^2 n/8}$.

In order for a large entry in P to become a small entry in $\hat{P}$ this entry must become at least as small in $\hat{P}$ as some other entry in the same column which is zero in P. The number of large/small pairs in a column of P that become small/large pairs in $\hat{P}$ can be maximized for a given bound on the Frobenius norm $|P - \hat{P}|_F^2$ if the large entry, which is $\frac{k}{n}$ in P, and the small entry, which is zero in P, both become $\frac{k}{2n}$ in $\hat{P}$. By this argument the number y of such pairs can be bounded from above by

$$(\frac{k}{2n})^2 y \leq \frac{4k}{\frac{\sqrt{n}}{k}(p - q) - 2}, \quad \text{that is} \quad y \leq \frac{16\frac{n^2}{k}}{\frac{\sqrt{n}}{k}(p - q) - 2}.$$

Putting everything together we get

$$x\frac{n}{10k} \leq y \leq \frac{16\frac{n^2}{k}}{\frac{\sqrt{n}}{k}(p - q) - 2}, \quad \text{that is} \quad x \leq \frac{160n}{\frac{\sqrt{n}}{k}(p - q) - 2}.$$

This inequality holds with the same probability that the bound on the Frobenius norm $|P - \hat{P}|_F^2$ holds. The latter probability is at least $1 - 2e^{-c^2 n/8}$.   □

Notation. To shorten our exposition we set in the following

$$\alpha = \frac{160}{\frac{\sqrt{n}}{k}(p - q) - 2}.$$

Lemma 5. *With probability at least* $1 - 2e^{-c^2 n/8}$ *in at least* $(1 - \sqrt{\alpha})k$ *classes we have at least* $(1 - \sqrt{\alpha})\frac{n}{k}$ *associated safe vectors if* $k < \frac{p-q}{4}\sqrt{n}$.

Proof. The proof of the lemma is equivalent to showing that the complementary event that there are less than $(1 - \sqrt{\alpha})k$ classes with at least $(1 - \sqrt{\alpha})\frac{n}{k}$ associated safe vectors occurs with probability at most $2e^{-c^2 n/8}$. In case of the complementary event there are more than $\sqrt{\alpha}k$ classes, which contain more than $\sqrt{\alpha}\frac{n}{k}$ unsafe vectors each. Thus we get in total more than $\sqrt{\alpha}k\sqrt{\alpha}\frac{n}{k} = \alpha n$ unsafe vectors. By Lemma 4, however, this happens only with probability at most $2e^{-c^2 n/8}$.   □

Covered Vectors and Split Classes. A vector c_i *covers* a vector c_j and vice versa if $c_i^T c_j > \frac{4n}{5k}$. A class $C = \varphi^{-1}(l), l \in \{1, \ldots, k\}$ is split by an unsafe vector c_j if there exists a safe vector c_i with $i \in C$ that is covered by c_j. An unsafe vector c_h *almost splits* C if it does not split C, but there exists an unsafe vector c_j which splits C and covers c_h.

Lemma 6. *Every unsafe vector can split or almost split at most one class.*

Proof. Let c_j be an unsafe vector and let c_i be a safe vector covered by c_j, i.e., $c_i^T c_j > \frac{4n}{5k}$. By the definition of safe vectors more than $\frac{9}{10}\frac{n}{k}$ of the indices corresponding to the one entries in c_i are mapped by φ to $\varphi(i)$. That is, at least

$$\left(\frac{4}{5} - \frac{1}{10}\right)\frac{n}{k} = \frac{7}{10}\frac{n}{k} > \frac{n}{2k}$$

of the indices corresponding to the one entries of c_j are mapped by φ to $\varphi(i)$. That shows there cannot be another safe vector c_h with $\varphi(i) \neq \varphi(h)$ covered by c_j. Thus c_j can split at most one class.

It remains to show that an unsafe vector can almost split at most one class. Let c_j be an unsafe vector and let c_i be an unsafe vector that splits a class and is covered by c_j. That is, $c_i^T c_j > \frac{4n}{5k}$ and there exists a safe vector c_h such that at least $\frac{7}{10}\frac{n}{k}$ of the indices corresponding to the one entries of c_i are mapped by φ to $\varphi(h)$. That is, more than

$$\left(\frac{7}{10} - \frac{1}{5}\right)\frac{n}{k} = \frac{n}{2k}$$

of the indices corresponding to the one entries of c_j are mapped by φ to $\varphi(h)$. That shows that c_j can split or almost split at most one class. □

Lemma 7. *With probability at least $1 - 2e^{-c^2 n/8}$ at most $\sqrt{\alpha}k$ classes are split or almost split by more than $\sqrt{\alpha}\frac{n}{k}$ unsafe vectors, provided $k < \frac{p-q}{4}\sqrt{n}$.*

Proof. The proof of the lemma is equivalent to showing that the complementary event that more than $\sqrt{\alpha}k$ classes are split or almost split by more than $\sqrt{\alpha}\frac{n}{k}$ unsafe vectors occurs with probability at most $2e^{-c^2 n/8}$. If this is the case, since by Lemma 6 every unsafe vector splits or almost splits at most one class, the total number of unsafe vectors that split or almost split a class is more than

$$\sqrt{\alpha}k\sqrt{\alpha}\frac{n}{k} = \alpha n.$$

That is, the number of unsafe vectors is more than αn. But this happens according to Lemma 4 only with probability at most $2e^{-c^2 n/8}$. □

Safe Class. A class $\varphi^{-1}(m), m \in \{1, \ldots, k\}$ is called safe if it contains more than $(1 - \sqrt{\alpha})\frac{n}{k}$ indices of safe vectors and if it is split or almost split by at most $\sqrt{\alpha}\frac{n}{k}$ unsafe vectors.

Lemma 8. *For any safe class* $C = \varphi^{-1}(m), m \in \{1, \ldots, k\}$ *the algorithm* SPEC-TRALRECONSTRUCT *with probability at least* $1 - 2e^{-c^2 n/8}$ *outputs a class* C_l *that contains at least* $\left(1 - \sqrt{\alpha} - \sqrt{2\alpha}\right) \frac{n}{k}$ *indices in* C *corresponding to safe vectors, provided* $k < \frac{p-q}{8} \sqrt{n}$.

Proof. Let c_i be vector which is used in line 14 of the algorithm SPECTRALRE-CONSTRUCT to create a class C_l. A safe vector c_j with $j \in C$ can only be put into the class C_l if either c_i is another safe vector with $i \in C$ or if c_i is an unsafe vector that splits C. We discuss the two cases now.

Assume c_i is a safe vector with $i \in C$. Then all safe vectors c_h with $h \in C$ will also be drawn into C_l since we have

$$c_i^T c_h > \left(1 - 2\frac{1}{10}\right) \frac{n}{k} = \frac{4n}{5k}.$$

That is, C_l will contain all safe vectors whose index is in C. It remains to show that C_l will pass the test in line 18 of the algorithm. But this follows from our definition of safe class and

$$1 - \sqrt{\alpha} = 1 - \sqrt{\frac{160}{\frac{\sqrt{n}}{k}(p-q) - 2}} \geq 1 - \sqrt{\frac{160}{\frac{\sqrt{n}}{k}\left(\hat{\lambda}_2 \frac{k}{n} - \frac{k}{\sqrt{n}}\right) - 2}} = 1 - \sqrt{\frac{160\sqrt{n}}{\hat{\lambda}_2 - 3\sqrt{n}}},$$

where we used the lower bound on $p - q$ from Lemma 3.

Now assume that c_i is an unsafe vector that splits C. Then c_i can draw some of the safe vectors whose index is in C into C_l and it can draw some unsafe vectors that either split or almost split C. Assume that C_l passes the test in line 18. Since by the definition of a safe class it can draw at most $\sqrt{\alpha} \frac{n}{k}$ unsafe vectors, it has to draw at least

$$\left(1 - \sqrt{\frac{160\sqrt{n}}{\hat{\lambda}_2 - 3\sqrt{n}}}\right) \frac{n}{k} - \sqrt{\alpha} \frac{n}{k}$$

safe vectors with an index in C. Using that with probability at least $1 - 2e^{-c^2 n/8}$ it holds

$$\frac{n}{k}(p-q) - p - \sqrt{n} \leq \hat{\lambda}_2,$$

see Lemma 3, we find that with the same probability

$$1 - \sqrt{\frac{160\sqrt{n}}{\hat{\lambda}_2 - 3\sqrt{n}}} \geq 1 - \sqrt{\frac{160}{\frac{\sqrt{n}}{k}(p-q) - \frac{p}{\sqrt{n}} - 4}} \geq 1 - \sqrt{\frac{160}{\frac{1}{2}\left(\frac{\sqrt{n}}{k}(p-q) - 2\right)}} = 1 - \sqrt{2\alpha},$$

here we used

$$\frac{1}{2}\frac{\sqrt{n}}{k}(p-q) \geq \frac{p}{\sqrt{n}} + 3,$$

which follows from $k < \frac{p-q}{8}\sqrt{n}$. Combining everything we get that C_l with probability at least $1 - 2e^{-c^2 n/8}$ contains at least

$$\left(1 - \sqrt{\alpha} - \sqrt{2\alpha}\right)\frac{n}{k}$$

indices in C corresponding to safe vectors. □

Theorem 3. *With probability at least* $1 - 2e^{-c^2 n/8}$ *at most*

$$\left(3\sqrt{\alpha} + \sqrt{2\alpha}\right)n$$

indices are misclassified by the algorithm SPECTRALRECONSTRUCT *provided* $k < \frac{p-q}{8}\sqrt{n}$.

Proof. By combining Lemmas 5 and 7 we get with probability at least $1 - 2e^{-c^2 n/8}$ at least

$$\left(1 - \sqrt{\alpha} - \sqrt{\alpha}\right)k = \left(1 - 2\sqrt{\alpha}\right)k$$

safe classes. From each safe class at least

$$\left(1 - \sqrt{\alpha} - \sqrt{2\alpha}\right)\frac{n}{k}$$

indices of safe vectors are grouped together by the algorithm with probability at least $1 - 2e^{-c^2 n/8}$, see Lemma 8. Thus in total with probability at least $1 - 2e^{-c^2 n/8}$ at least

$$\left(1 - \sqrt{\alpha} - \sqrt{2\alpha}\right)\frac{n}{k}\left(1 - 2\sqrt{\alpha}\right)k > \left(1 - 3\sqrt{\alpha} - \sqrt{2\alpha}\right)n$$

indices of safe vectors are grouped together correctly. Hence, by the definition of the number of misclassifications via a maximum weight matching, with probability at least $1 - 2e^{-c^2 n/8}$ at most

$$\left(3\sqrt{\alpha} + \sqrt{2\alpha}\right)n$$

elements are misclassified. □

Discussion. Note that the theorem is non-trivial only if

$$k < \frac{p-q}{1762 + 960\sqrt{2}}\sqrt{n}.$$

The theorem implies that if $k = o(\sqrt{n})$ then the relative number of misclassifications goes to zero with high probability as n goes to infinity. That is the first non-trivial result for $k = \omega(\sqrt{\frac{n}{\log n}})$. But also in the case $k = c\sqrt{n}$ for a small constant c the theorem provides useful information. It basically says that on average the percentage of elements per class that get misclassified by the algorithm becomes arbitrarily small if c is small enough.

6   Concluding Remarks

We presented and analyzed a spectral partitioning algorithm. The analysis provided non-trivial guarantees for a range of parameters where such guarantees were not known before. As we have presented it, the algorithm and its analysis are restricted to the case that all classes have exactly the same size. It is an interesting question whether classes of different size can be handled in the same way.

References

1. N. Alon, M. Krivelevich, and B. Sudakov. Finding a large hidden clique in a random graph. *Random Structures and Algorithms*, 13:457–466, 1998.
2. B. Bollobas and A.D. Scott. Max cut for random graphs with a planted partition. *Combinatorics, Probability and Computing*, 13:451–474, 2004.
3. R. B. Boppana. Eigenvalues and graph bisection: An average-case analysis. *Proceedings of 28th IEEE Symposium on Foundations on Computer Science*, pages 280–285, 1987.
4. A. Condon and R. Karp. Algorithms for graph partitioning on the planted partition model. *Random Structures and Algorithms 8*, 2:116–140, 1999.
5. Z. Füredi and J. Komlós. The eigenvalues of random symmetric matrices. *Combinatorica I*, 3:233–241, 1981.
6. M. R. Garey, D. S. Johnson, and L. Stockmeyer. Some simplified NP-complete graph problems. *Theoretical Computer Science*, 1:237–267, 1976.
7. M. Krivelevich and V. H. Vu. On the concentration of eigenvalues of random symmetric matrices. *Microsoft Technical Report*, 60, 2000.
8. F. McSherry. Spectral partitioning of random graphs. *Proceedings of 42nd IEEE Symosium on Foundations of Computer Science*, pages 529–537, 2001.
9. M. Meila and D. Verma. A comparison of spectral clustering algorithms. *UW CSE Technical report 03-05-01*.
10. R. Shamir and D. Tsur. Improved algorithms for the random cluster graph model. *Proceedings 7th Scandinavian Workshop on Algorithm Theory*, pages 230–259, 2002.
11. D. Spielman and S.-H. Teng. Spectral partitioning works: Planar graphs and finite element meshes. *Proceedings of 37th IEEE Symposium on Foundations on Computer Science*, pages 96–105, 1996.
12. G. Stewart and J. Sun. *Matrix perturbation theory*. Academic Press, Boston, 1990.
13. V. H. Vu. Spectral norm of random matrices. In *STOC '05: Proceedings of the thirty-seventh annual ACM symposium on Theory of computing*, pages 423–430, New York, NY, USA, 2005. ACM Press.

Algebraic Operations on PQ Trees
and Modular Decomposition Trees

Ross M. McConnell[1] and Fabien de Montgolfier[2]

[1] Colorado State University
rmm@cs.colostate.edu
[2] Université Paris 7
fm@liafa.jussieu.fr

Abstract. Partitive set families are families of sets that can be quite large, but have a compact, recursive representation in the form of a tree. This tree is a common generalization of PQ trees, the modular decomposition of graphs, certain decompositions of boolean functions, and decompositions that arise on a variety of other combinatorial structures. We describe natural operators on partitive set families, give algebraic identities for manipulating them, and describe efficient algorithms for evaluating them. We use these results to obtain new time bounds for finding the common intervals of a set of permutations, finding the modular decomposition of an edge-colored graph (also known as a two-structure), finding the PQ tree of a matrix when a consecutive-ones arrangement is given, and finding the modular decomposition of a permutation graph when its permutation realizer is given.

1 Introduction

A 0-1 matrix has the *consecutive-ones property* if there exists a permutation of the set of columns such that the 1's in each row occupy a consecutive block. Such a permutation is called a *consecutive-ones ordering*.

In general, the number of consecutive-ones orderings need not be polynomial; there may be $|V|!$ of them. However, the *PQ tree* of a family that has the consecutive-ones property gives a way to represent all of its consecutive-ones orderings using $O(|V|)$ space. The PQ tree is a rooted, ordered tree whose leaves are the elements of V, and whose internal nodes are each labeled either P or Q. The left-to-right leaf order gives a consecutive-ones ordering, and any new leaf order that can be obtained by permuting arbitrarily the children of a P node or reversing the order of children of a Q node is also a consecutive-ones ordering. There are no other consecutive-ones orderings.

One of the most significant applications of PQ trees is in finding planar embeddings of planar graphs. Booth and Lueker used PQ trees to develop an algorithm for determining whether a family of sets has the consecutive-ones property [2]. The algorithm runs in $O(|V| + l(\mathcal{F}))$ time, where $l(\mathcal{F})$ is the sum of cardinalities of members of $\mathcal{F}$, or *length* of $\mathcal{F}$.

D. Kratsch (Ed.): WG 2005, LNCS 3787, pp. 421–432, 2005.
© Springer-Verlag Berlin Heidelberg 2005

A set family $\mathcal{F}$ with the consecutive-ones property gives rise to an *interval graph*, which has one vertex for each member of $\mathcal{F}$, and an adjacency between two vertices if and only if the corresponding members of $\mathcal{F}$ intersect. Booth and Lueker's result gave a linear-time algorithm for determining whether a given graph is an interval graph, and, if so, finding such a set family $\mathcal{F}$ for it. This problem played a key role during the 1950's in establishing that DNA has a linear topology [1], though linear-time algorithms were unavailable at that time. Variations on this problem come up in the physical mapping of a genome, using laboratory data that can be modeled with a graph [19,25].

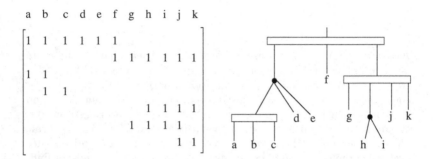

Fig. 1. A consecutive-ones ordering of a matrix, and the corresponding PQ tree. The zeros in the matrix are omitted. The ordering of the columns is a consecutive-ones ordering because the 1's in each row are consecutive. The left-to-right leaf order of the PQ tree gives this ordering. Reversing the left-to-right order of children of a Q node (rectangles) or permuting arbitrarily the left-to-right order of children of a P node (points) induces a new leaf order, which is also a consecutive-ones ordering. For instance, permuting the order of children of the left child of the root and reversing the order of children of the right child gives $(d, a, b, c, e, f, k, j, h, i, g)$ as a consecutive-ones ordering. An ordering of columns of the matrix is a consecutive-ones ordering iff it is the leaf order of the PQ tree induced by reversing the children of some set of Q nodes and permuting the children of some set of P nodes.

A *module* of an undirected graph $G = (V, E)$ is a set X of vertices such that each vertex $y \in V - X$ is either adjacent to all members of X or adjacent to none of them. The number of modules can be exponential in the size of G. However, there exists a compact $O(|V|)$ representation of all the modules, called the *modular decomposition*. The modular decomposition, first described by Gallai [7], is a tree that has the members of V as its leaves, and where the internal nodes are all labeled *prime* or *degenerate*. Details of the representation are given below. The modular decomposition can be computed in $O(|V| + |E|)$ time [15].

A close relationship between the modular decomposition and a variety of combinatorial problems on graphs have been described. Gallai [7] showed a close relationship to the *transitive orientation problem*, which is the problem of orienting the edges of an undirected edge so that the resulting digraph is transitive (*i.e.* a poset relation). Using the modular decomposition, a transitive orientation, if it exists, can be found in $O(|V| + |E|)$ time [15]. This result has led to linear time bounds for maximum clique and minimum coloring on transitively orientable graphs (*i.e.* comparability graphs), and recognizing permutation graphs and co-interval graphs. Surveys on applications can be found in [20,21,22]. The modular decomposition has a straightforward extension to directed graphs, and linear time bounds have recently been given for finding it [17].

The modules of a graph are an example of a *partitive set family* [3,20]. All partitive set families have a compact representation by means of a tree; the modular decomposition is just an example of it when the set family is the modules of a graph. The PQ tree is another example of this phenomenon. In [14], it is shown that the PQ tree is this representation of a certain partitive family defined by the 0-1 matrix, and, more generally that, like the modular decomposition, the PQ tree is an example of a *substitution decomposition* [21], a combinatorial abstraction that has partitive families as a central ingredient.

Other partitive families have played a role in linear time bounds for recognizing circular-arc graphs [13,16], $O(n + m \log n)$ bounds for recognizing probe interval graphs [18], and arise in decompositions of boolean expressions [21].

In this paper, we describe natural algebraic operators on decomposition trees of partitive families, give identities for manipulating them, and develop algorithms for evaluating them. We use these results to obtain new time bounds for combinatorial problems that involve partitive families, such as finding the *common intervals* of a set of permutations [24,10], finding the modular decomposition of edge-colored graphs, or *two-structures* [6], finding the PQ tree of a matrix when a consecutive-ones arrangement is given, and finding the modular decomposition of a permutation graph when its realizer in the form of a permutation is given.

2   Preliminaries

Two sets X and Y *overlap* if they intersect, but neither is a subset of the other. That is, they overlap if $X - Y$, $Y - X$, and $X \cap Y$ are all nonempty.

Let $\mathcal{F}$ is a family of subsets of a set V. Then let $|\mathcal{F}|$ denote the number of sets in $\mathcal{F}$; this contrasts with $l(\mathcal{F})$, which is the sum of cardinalities of the sets in $\mathcal{F}$. In general, it takes $\Omega(l(\mathcal{F}))$ space to represent $\mathcal{F}$ in the computer. However, suppose that $\mathcal{F}$ satisfies the following: $V \in \mathcal{F}$, $\{x\} \in \mathcal{F}$ for all $x \in V$, and no two members of $\mathcal{F}$ overlap. In this case, it is easy to see that $l(\mathcal{F})$ can be $\Omega(|V|^2)$, but $\mathcal{F}$ can be represented in $O(|V|)$ space. The Hasse diagram of the subset relation on members of $\mathcal{F}$ is a tree whose root is V and whose leaves are its one-element subsets. Labeling only the leaves of this tree with the corresponding set gives a representation of $\mathcal{F}$. Given a node of the tree, the set X that it represents can be returned in $O(|X|)$ time by traversing its subtree and assembling the disjoint union of its leaf descendants. This is as efficient as any representation of X, but takes $O(1)$ space to represent X.

Let us call such a set family a *tree-like* family, and its tree representation its *inclusion tree*. Partitive families are a generalization of tree-like families, called *partitive families*, that may have a number of members that is exponential in the size of V, yet still has an $O(|V|)$ representation.

Definition 1. *[7,3,21,6] A set family $\mathcal{F}$ on domain V is partitive iff it has the following properties:*

- $V \in \mathcal{F}$, $\emptyset \notin \mathcal{F}$, *and for all $v \in V$, $\{v\} \in \mathcal{F}$*
- *For all $X, Y \in \mathcal{F}$, if X and Y overlap, then $X \cap Y \in \mathcal{F}$, $X \cup Y \in \mathcal{F}$, $X - Y \in \mathcal{F}$, and $Y - X \in \mathcal{F}$.*

Let the *strong members* of a partitive family be those that overlap with no other member of $\mathcal{F}$, and let the *weak members* be the remaining members.

Theorem 1. *[3,21] The strong members of a partitive family $\mathcal{F}$ are a tree-like family where the Hasse diagram T of the subset relation has the following properties:*

1. *Every weak member of $\mathcal{F}$ is a union of siblings in T;*
2. *Each internal node X can be classified as one of the following types:*
   - *(a) Degenerate: Every union of more than one child is a member of $\mathcal{F}$;*
   - *(b) Prime: Other than X itself, no union of more than one child is a member of $\mathcal{F}$;*
   - *(c) Linear: There exists a linear order on the children such that a union of more than one child is a member of $\mathcal{F}$ if and only if the children are consecutive in the linear order.*

Conversely, a set family that has such a representative is partitive. Let us call the tree representation of $\mathcal{F}$ given by the theorem the *decomposition tree* of $\mathcal{F}$.

Example 1. A nonempty set X of vertices of a directed graph $G = (V, E)$ is a *module* iff it satisfies the following conditions for to every $y \in V - X$:

1. Either every element of X or no element of X is a neighbor of y;
2. y is either a neighbor of every element of X or of no element of X.

It is not hard to show that the modules of a graph satisfy the requirements of Definition 1. It follows that the modules of a graph can be represented in $O(|V|)$ space with a tree [7,21].

It takes $O(|V| + |E|)$ time to compute the modular decomposition of an arbitrary directed graph [17]; linear time bounds for the special case of undirected graphs were given in [15].

If $\mathcal{F}$ is a partitive set family, let $T(\mathcal{F})$ denote its decomposition tree, and if T is a partitive decomposition tree, let $\mathcal{F}(T)$ denote the set family that it represents. There is no way to distinguish whether a node with two children is prime, degenerate or linear, but the classification is unique for nodes with three or more children. Henceforth, we will consider a node to be classified as prime, degenerate, or linear only if it has at least three children.

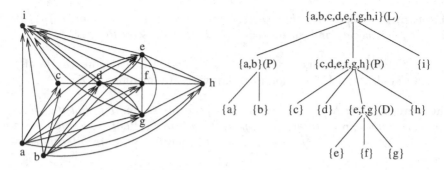

Fig. 2. A graph and its modular decomposition tree. Each strong module of the graph corresponds to a node of the tree, labeled *linear* (L), *prime* (P), or *degenerate*(D). The other modules are unions of children of $\{e, f, g\}$, namely, $\{e, f\}$, $\{e, g\}$, and $\{f, g\}$, and unions of consecutive children of $\{a, b, c, d, e, f, g, h, i\}$, namely, $\{a, b, c, d, e, f, g, h\}$ and $\{c, d, e, f, g, h, i\}$.

Definition 2. *A partitive set family is* symmetric *if, whenever X and Y are overlapping members of $\mathcal{F}$, the symmetric difference $X \Delta Y = (X - Y) \cup (Y - X)$ is a member of $\mathcal{F}$. It is* antisymmetric *if $X \Delta Y$ is never a member when X and Y are overlapping members.*

It is not hard to see that if a graph is symmetric (undirected), its modules are a symmetric partitive family, and that if it is antisymmetric, its modules are an antisymmetric partitive family, unless it has modules that induce disconnected subgraphs. In particular, the modules of a tournament are an antisymmetric partitive family. A partitive set family is symmetric if and only if its decomposition tree has no linear nodes, and it is antisymmetric if and only if its decomposition tree has no degenerate nodes.

If $\mathcal{F}$ is an arbitrary set family, let $\mathcal{C}(\mathcal{F})$ denote the *partitive closure* of $\mathcal{F}$, namely, the smallest partitive family $\mathcal{F}'$ such that $\mathcal{F} \subseteq \mathcal{F}'$. Let $\mathcal{S}(\mathcal{F})$ denote the *symmetric partitive closure* of $\mathcal{F}$, namely, the smallest symmetric partitive family $\mathcal{F}''$ such that $\mathcal{F} \subseteq \mathcal{F}''$. It is shown in [14] that each of these closures is unique.

If M is a zero-one matrix, then let V denote its columns, and let $\mathcal{F}(M)$ denote the set family on V that has one set for each row of M, namely, the one obtained by interpreting the row as the bit-vector representation of a set. That is, the set represented by a row is the set of columns where the row has a 1. Conversely, if $\mathcal{F}$ is a family of subsets of a domain V, we may obtain a representation of $\mathcal{F}$ with M, such that $\mathcal{F}(M) = \mathcal{F}$. M has the consecutive-ones property if and only if there exists an ordering of V such that every member of $\mathcal{F}$ is consecutive, and, in this case, we may refer to the PQ tree of M as the the PQ tree of $\mathcal{F}$. The following gives a generalization of the PQ tree to arbitrary set families or zero-one matrices:

Definition 3. *[14] Let $\mathcal{F}$ be an arbitrary set family. Let the* PQR tree *of $\mathcal{F}$ be the decomposition tree of $\mathcal{C}(\mathcal{F})$, where the prime nodes are labeled P, the linear nodes are labeled Q, and the degenerate nodes are labeled R.*

Theorem 2. *[14] $\mathcal{F}$ has the consecutive-ones property if and only if its PQR tree has no R nodes, and, in this case, its PQR tree is its PQ tree.*

Let $\mathcal{F}$ be an arbitrary set family on V, and $\mathcal{N}(\mathcal{F})$ denote the family of nonempty subsets of V that don't overlap with any member of $\mathcal{F}$.

Theorem 3. *[11] $\mathcal{N}(\mathcal{F})$ is a symmetric partitive set family, and if $\mathcal{F}$ has the consecutive-ones property, its decomposition tree is the PQ tree, where the prime nodes are interpreted as Q nodes and the degenerate nodes are interpreted as P nodes.*

The proof of the following is elementary:

Theorem 4. *[17] If $\mathcal{F}_1$ and $\mathcal{F}_2$ are two partitive families, then so is $\mathcal{F}_1 \cap \mathcal{F}_2$. If they are both symmetric partitive families, then so is $\mathcal{F}_1 \cap \mathcal{F}_2$, and if they are both antisymmetric partitive families, then so if $\mathcal{F}_1 \cap \mathcal{F}_2$.*

Definition 4. *If T_1 and T_2 are partitive decomposition trees, then let $T_1 \cap T_2$ be the decomposition tree of $\mathcal{F}(T_1) \cap \mathcal{F}(T_2)$, which exists by Theorem 4.*

Theorem 5. *[17] Given decomposition trees T_1 and T_2 of symmetric partitive families, it takes time proportional to the sum of cardinalities of their nodes to find $T_1 \cap T_2$.*

3 New Results

3.1 Intersection of Arbitrary Partitive Families

Theorem 5 applies only to symmetric partitive families. The case where they are not symmetric is more difficult.

The additional difficulties posed by linear nodes are illustrated by the simple case of two trees T_1 and T_2 that each have $V = \{1, 2, .., 8\}$ as their only internal node. If T_1 and T_2 are decomposition trees of symmetric partitive families, then V is prime or degenerate in each. In $T_1 \cap T_2$, V is the only internal node, and it is degenerate if it is degenerate in both trees and prime otherwise. The intersection is trivial to compute in this case.

On the other hand, suppose V is linear in each of T_1 and T_2, and $(\{1\}, \{2\}, ..., \{8\})$ is the order of its children in T_1 and $(\{6\}, \{5\}, \{8\}, \{7\}, \{2\}, \{1\}, \{4\}, \{3\})$ is the order of children in T_2. Then $\{1, 2\}$, $\{3, 4\}$, $\{5, 6\}$, $\{7, 8\}$, $\{1, 2, 3, 4\}$, and $\{5, 6, 7, 8\}$ are internal nodes. In general, two linear nodes in two partitive trees can give rise to a complicated subtree in the intersection.

We improve the bound of Theorem 4 and generalize it to arbitrary partitive families, not just symmetric ones:

Theorem 6. *Given arbitrary partitive decomposition trees T_1 and T_2 on domain V, it takes $O(|V|)$ time to find $T_1 \cap T_2$.*

Let the *intervals* of a permutation $(v_1, v_2, ..., v_n)$ be a nonempty set of the form $\{v_i, v_{i+1}, ..., v_j\}$. The *common intervals* of a set of permutations of the same set are the intervals that are common to all of them.

If π is a linear ordering of V, then its intervals are an antisymmetric partitive family: their decomposition tree $T(\pi)$ is the tree with one internal linear node, and leaf set V ordered in the order given by π. It follows that the common intervals of a set $\{\pi_1, \pi_2, ..., \pi_k\}$ is a partitive set family whose decomposition tree is given by $T(\pi_1) \cap T(\pi_2) \cap ... \cap T(\pi_k)$.

Application 1. The following is immediate from this example and Theorem 6:

Theorem 7. *It takes $O(kn)$ time to find the common intervals of a set $\{\pi_1, \pi_2, ...\pi_k\}$ of permutations of a set V.*

The previous time bound for this problem was $O(nk + K)$, where K is the number of common intervals [10]. (Note that K can be quadratic in n).

Application 2. The conceptual complexity of many linear-time algorithms for computing the PQ tree is well-known. However, Theorem 6 gives a simple $O(nm)$ approach to finding the PQ tree of an $n \times m$ matrix. Let M be a 0-1 matrix with m columns and n rows, let M_1 be the submatrix given by the top $\lfloor m/2 \rfloor$ rows and let M_2 be the submatrix given by the remaining $\lceil m/2 \rceil$ rows. To find the PQ tree of M, we may find the PQ trees T_1 and T_2 of M_1 and M_2 by recursion, and then return $T_1 \cap T_2$ as the PQ tree of M. The correctness follows from Theorem 3.

Application 3. [12] If T is a PQ tree, let $\Pi(T)$ denote the set of permutations represented by T. If T and T' are PQ trees on domain V, then let $T \preceq T'$ denote that $\Pi(T) \subseteq \Pi(T')$. Given the PQ trees T_1 and T_2 of two set families on the same domain V, let the the *join* of T_1 and T_2 be the minimal PQ tree T_3 (with respect to $\preceq$) such that $T_1 \preceq T_3$ and $T_2 \preceq T_3$. The join of two PQ trees was first described by Landau, Parida, and Weimann [12], who have used it in an application to genomics, and who obtained an $O(|V|^3)$ algorithm to compute it. When we communicated Theorem 7, they used it to improve the bound to $O(|V|)$.

A *two-structure* is a directed graph whose edges are colored. A module of a two-structure is a module X of the underlying graph that satisfies the following additional requirement: whenever $y \in V - X$, all edges from members of X to y are the same color, and all edges from y to members of X are the same color. The modules of a two-structure are a partitive family [6].

The following give a relationship between modular decomposition of two-structures and problems in other areas that have not been observed before.

Example 2. A distance function d on a set V is an *ultrametric* if, for all $x, y, z \in V$, either $d(x, y)$, $d(y, z)$ and $d(x, z)$ are all equal, or two are equal and the third is smaller. Ultrametrics arise in many clustering applications, such as the problem of inferring phylogenetic trees. An example is the distance metric in a graph with edge weights, where the *height* of a path is the maximum weight of an edge on the path, and where the *distance* between two vertices is the height of the minimum-height path between them. An ultrametric can be modeled as a two-structure, where for $x, y \in V$, the "color" of edge xy is $d(x, y)$. In this case, the modular decomposition of the two structure is the tree returned by the well-known UPGMA clustering algorithm [23].

Example 3. Given a string $a_1 a_2, ..., a_n$, let us define a two-structure with vertices $\{1, 2, ..., n\}$, and for vertices i and j, let the label ("color") of edge ij be the longest common prefix of $a_i a_{i+1} ... a_n$ and $a_j a_{j+1} ... a_n$. It is not hard to show that the modular decomposition of this two-structure is the well-known *suffix tree* of the string, which is used in efficient solutions to a variety of combinatorial problems on strings [4,9].

Though these last two examples yield an interesting structural relationship, they do not yield more efficient algorithms. However, in [17], we give a linear-time algorithm for finding the modular decomposition of a symmetric (undirected) two-structure. This is a key step in the linear time bounds we show there for finding the modular decomposition of a directed graph.

Because of the added difficulties posed by linear nodes in the decomposition, the best bound until now for finding the modular decomposition of arbitrary two-structures has been $O(|V|^2)$ [5]. However, given Theorem 6, we can now improve this quite easily:

Proposition 1. *It takes $O(k|V| + |E|)$ time to find the modular decomposition of a two-structure that has vertex set V, edge set E, and k edge colors.*

Proof. Let G_i denote the graph on V given by edges of color i. Find the modular decomposition T_i of each G_i for each i from 1 to k using the linear-time modular decomposition algorithm for directed graphs given in [17]. Since the edge sets are disjoint, this takes a total of $O(k|V| + |E|)$ time. The modular decomposition of the two-structure is given by $T_1 \cap T_2 \cap ... \cap T_{k-1}$, which takes $O(k|V|)$ time to find, by Theorem 6. $\square$

By an only slightly more involved proof, we can get a linear time bound, as follows. Let the *essential subtree* of a partitive decomposition tree T be the tree T' obtained by deleting leaf children of the root if the root is degenerate, and let its *size* be the number of leaves in the tree.

Lemma 1. *Let $T_1, T_2, ..., T_k$ be partitive trees on domain V, and let $T_1', T_2', ..., T_k'$ be their essential subtrees. It takes time proportional to the sum of sizes of $T_1', T_2', ..., T_k'$ to find $T_1 \cap T_2 \cap ... \cap T_k$.*

Theorem 8. *It takes $O(|V| + |E|)$ time to find the modular decomposition of an arbitrary two-structure.*

Proof. Let $G_1, G_2, ..., G_k$ be as in the proof of Proposition 1. It is easy to see that the essential subtree of the modular decomposition of G_i can be obtained from the modular decomposition of the subgraph induced by non-isolated vertices, which has $O(|E_i|)$ vertices. Therefore, given E_i, the modular decomposition of G_i can be found in $O(|E_i|)$ time using the linear-time modular decomposition algorithm of [17]. Using this observation, and replacing Theorem 7 with Lemma 1 in the proof of Proposition 1, yields the linear time bound.

3.2 New Algebraic Operators on Symmetric Partitive Families

By Theorem 3, $\mathcal{N}(\mathcal{F})$ has a decomposition tree even when $\mathcal{F}$ does not have the consecutive-ones property. Therefore, $\mathcal{N}(\mathcal{F})$ is defined even when $\mathcal{F}$ is itself a symmetric partitive family.

Theorem 9. *If $\mathcal{F}$ is a symmetric partitive family, then $T(\mathcal{N}(\mathcal{F}))$ is obtained from $T(\mathcal{F})$ be relabeling each degenerate node as prime and each prime node as degenerate.*

Definition 5. *If T is the decomposition tree of a symmetric partitive family, let its complement $\overline{T}$ denote $T(\mathcal{N}(\mathcal{F}(T)))$. That is, $\overline{T}$ is the result of exchanging the roles of prime and degenerate nodes.*

Theorem 10. *If $\mathcal{F}$ is an arbitrary set family, then $T(\mathcal{S}(\mathcal{F})) = \overline{T(\mathcal{N}(\mathcal{F}))}$.*

Definition 6. *Let $\mathcal{F}_1$ and $\mathcal{F}_2$ be symmetric partitive families on V, and let T_1 and T_2 be their decomposition trees. The union $\mathcal{F}_1 \cup \mathcal{F}_2$ is not necessarily partitive, so let $T_1 \cup T_2$ denote the smallest symmetric partitive family that has $\mathcal{F}_1 \cup \mathcal{F}_2$ as a subfamily, that is, let $T_1 \cup T_2 = T(\mathcal{S}(\mathcal{F}_1 \cup \mathcal{F}_2))$.*

These definitions of intersection and union therefore define a lattice on the set of all symmetric partitive trees on domain V. The minimal element of the lattice is the tree with V as its only internal node, with V labeled as prime, and the maximal element is this same tree, but with V labeled degenerate.

The following shows that the definitions satisfy familiar properties expected of these operators; the proof will appear in the journal version.

Theorem 11. *Let T_1 and T_2 be decomposition trees of symmetric partitive families. Then: $\overline{\overline{T_1}} = T_1$, $T_1 \cap T_2 = \overline{\overline{T_1} \cup \overline{T_2}}$, and $T_1 \cup T_2 = \overline{\overline{T_1} \cap \overline{T_2}}$*

Corollary 1. *It takes $O(|V|)$ time to find the union of two symmetric partitive trees.*

Clearly, the intersection operator is commutative and associative, as is the union operator. However, together, they are not distributive. That is, it is not true in general that $T_1 \cap (T_2 \cup T_3) = (T_1 \cap T_2) \cup (T_1 \cap T_3)$, as the following example illustrates. Let $V = \{1, 2, 3\}$, let T_1, T_2, T_3 be decomposition trees on V where $\{1, 2\}$ is the only non-root internal node of T_1, $\{1, 3\}$ is the only non-root internal node of T_2, and $\{2, 3\}$ is the only non-root internal node of T_3. Then $T_2 \cup T_3$ is the maximal element of the lattice, hence $T_1 \cap (T_2 \cup T_3) = T_1$.

However, $T_1 \cap T_2 = T_1 \cap T_3$ is the minimal element of the lattice, hence so is $(T_1 \cap T_2) \cup (T_1 \cap T_3)$.

Let a symmetric decomposition tree on domain V be *elementary* if it has at most one non-root internal node.

Theorem 12. *If T is the decomposition tree of a symmetric partitive family on domain V and T has $k \geq 1$ non-root internal nodes, then T can be written as $T = T_1 \, op_1 \, T_2 \, op_2 \, , ..., \, op_{k-1} T_k$, where each op_i is either $\cup$ or $\cap$, the operators are evaluated left-to-right, and each T_i is elementary.*

The proof is by induction on the number of non-root internal nodes. Theorem 12 is a key element in our time bound for intersecting partitive trees.

Modules and the quotients they induce in a graph are examples of a *substitution decomposition* on the domain of graphs [21]. We can define a substitution decomposition on the domain of decomposition trees of symmetric partitive families where where the roots carry a bit that it as *strong* or *weak*. Let an *autonomous set* denote a node of T or a union of siblings in T. *Note: It is not necessary for the parent of C to be degenerate.* If X is autonomous, then if it is a node of T, the *factor* $T[X]$ is the subtree rooted at X, and if it is a union of a set C of siblings, the factor $T[X]$ is the tree where X is the root, and its subtrees are the subtrees of T rooted at members of C; in this case, if X has at least three children, then it has the same prime/degenerate label as the parent of C. Let the *quotient* T/X denote the operation of nodes that are subsets of X and replacing them with a single leaf. The quotient is *strong* if X is a node of T, and *weak* if it is a union of siblings that is not a node of T.

Clearly, these operations are invertible: T can be uniquely reconstructed from a quotient T/W and factor $T[W]$ if the leaf w of T/W that corresponds to W is indicated, and a bit at the root of $T[W]$ identifies whether the quotient was strong or weak.

Several algebraic properties have been described previously for substitution decompositions, but the introduction of union and intersection operators on partitive trees yields the following new identities, which we use in obtaining the new time bounds in this paper:

Theorem 13. *If T_a and T_b are decomposition trees of symmetric partitive families on domain V and A is autonomous in both T_a and T_b, then:*

- $(T_a \cap T_b)/A = T_a/A \cap T_b/A$ - $(T_a \cap T_b)[A] = T_a[A] \cap T_b[A]$
- $(T_a \cup T_b)/A = T_a/A \cup T_b/A$ - $(T_a \cup T_b)[A] = T_a[A] \cup T_b[A]$

If X and Y are disjoint autonomous sets, $(T/X)/Y = (T/Y)/X$. Therefore, we can write this as $T/\{X, Y\}$, and, more generally, if $\{A_1, A_2, ..., A_k\}$ are disjoint autonomous sets, the quotient $T/\{A_1, A_2, ..., A_k\}$ is uniquely defined.

3.3   Algorithmic Uses of Compact Representations

Any algorithm can be made to run in time linear in the size of its input simply by selecting a suitably space-inefficient representation for the input. For instance, many algorithms for NP-complete problems can be made to run in "linear" time by choosing a unary representation for integer inputs. Linearity of an algorithm does not imply an optimal time bound unless the representation of the input is also asymptotically optimal.

When Booth and Lueker's algorithm [2] for finding the PQ tree is applied to a set family that is not known to have the consecutive-ones property, the algorithm either returns the PQ tree, or else rejects the family as not having the consecutive-ones property. The running time of $O(|V| + l(\mathcal{F}))$ is an optimum time bound, since it uses a space-efficient representation of arbitrary set families.

However, when it is applied to a set family that is already known to have the consecutive-ones property, the proof of optimality of the time bound is no longer valid because it assumes an input of size $\Theta(|V| + l(\mathcal{F}))$. Families with the consecutive-ones property have a representation that is more compact than the standard listing of elements of each member of the family. A consecutive-ones family $\mathcal{F}$ can be represented in $O(|V| + |\mathcal{F}|)$ space by giving a consecutive-ones ordering, and representing each member X of $\mathcal{F}$ in $O(1)$ space by giving the first and last member of the interval occupied by X in this ordering.

Theorem 14. *It takes $O(|V| + |\mathcal{F}|)$ time to find the PQ tree of a consecutive-ones family $\mathcal{F}$, given a consecutive-ones ordering and, for each $X \in \mathcal{F}$, the first and last element of X in the ordering.*

It is worth noting that Theorem 14 is the key starting point in the proofs of all of the remaining results of this paper. It also implies that, given the interval representation of an interval graph, the graph's PQ tree can be obtained in $O(|V|)$ time if the endpoints of the intervals are integers from 1 to $O(1)$, and in $O(|V| \log |V|)$ time if they are given as real numbers.

A similar type of result can be obtained for modular decomposition of *permutation graphs*. A permutation graph is obtained from two permutations of V, by letting the members of V be the vertices and letting two vertices x and y be adjacent if x is before y in one of the permutations and after it in the other [8]. Recognizing permutations and deriving their modular decomposition takes linear time [15]. However, it turns out that this bound for finding the modular decomposition is not optimal if the input graph is known to be a permutation graph:

Theorem 15. *Given an $O(|V|)$ representation of a permutation graph using two permutations of V, it takes $O(|V|)$ time to find its modular decomposition.*

References

1. S. Benzer. On the topology of the genetic fine structure. *Proc. Nat. Acad. Sci. U.S.A.*, 45:1607–1620, 1959.
2. S. Booth and S. Lueker. Testing for the consecutive ones property, interval graphs, and graph planarity using PQ-tree algorithms. *J. Comput. Syst. Sci.*, 13:335–379, 1976.
3. M. Chein, M. Habib, and M. C. Maurer. Partitive hypergraphs. *Discrete Mathematics*, 37:35–50, 1981.

4. M. Crochemore and W. Rytter. *Text Algorithms*. Oxford University Press, 1994.
5. A. Ehrenfeucht, H. N. Gabow, R. M. McConnell, and S. J. Sullivan. An $O(n^2)$ divide-and-conquer algorithm for the prime tree decomposition of two-structures and modular decomposition of graphs. *Journal of Algorithms*, 16:283–294, 1994.
6. A. Ehrenfeucht and G. Rozenberg. Theory of 2-structures, part 2: Representations through labeled tree families. *Theoretical Computer Science*, 70:305–342, 1990.
7. T. Gallai. Transitiv orientierbare Graphen. *Acta Math. Acad. Sci. Hungar.*, 18:25–66, 1967.
8. M. C. Golumbic. *Algorithmic Graph Theory and Perfect Graphs*. Academic Press, New York, 1980.
9. Dan Gusfield. *Algorithms on Strings, Trees, and Sequences*. Cambridge University Press, Cambridge, 1997.
10. Steffen Heber and Jens Stoye. Finding all common intervals of k permutations. In *CPM*, pages 207–218, 2001.
11. W.L. Hsu and R.M. McConnell. PC trees and circular-ones arrangements. *Theoretical Computer Science*, 296:59–74, 2003.
12. Gad M. Landau, Laxmi Parida, and Oren Weimann. Using pq trees for comparative genomics. In *CPM'05*, pages 128–143, 2005.
13. R. M. McConnell. Linear-time recognition of circular-arc graphs. *Proceedings of the 42nd Annual IEEE Symposium on Foundations of Computer Science (FOCS01)*, 42:386–394, 2001.
14. R. M. McConnell. A certifying algorithm for the consecutive-ones property. *Proceedings of the 15th Annual ACM-SIAM Symposium on Discrete Algorithms (SODA04)*, 15:to appear, 2004.
15. R. M. McConnell and J. P. Spinrad. Modular decomposition and transitive orientation. *Discrete Mathematics*, 201(1-3):189–241, 1999.
16. R.M. McConnell. Linear-time recognition of circular-arc graphs. *Algorithmica*, 37:93–147, 2003.
17. R.M. McConnell and F de Montgolfier. Linear-time modular decomposition of directed graphs. *Discrete Applied Mathematics*, 2005.
18. R.M. McConnell and J.P. Spinrad. Construction of probe interval models. *Proceedings of the Thirteenth Annual ACM-SIAM Symposium on Discrete Algorithms*, pages 866–875, 2002.
19. F.R. McMorris, C. Wang, and P. Zhang. On probe interval graphs. *Discrete Applied Mathematics*, 88:315–324, 1998.
20. R. H. Möhring. Algorithmic aspects of comparability graphs and interval graphs. In I. Rival, editor, *Graphs and Order*, pages 41–101. D. Reidel, Boston, 1985.
21. R. H. Möhring. Algorithmic aspects of the substitution decomposition in optimization over relations, set systems and boolean functions. *Annals of Operations Research*, 4:195–225, 1985.
22. R. H. Möhring and F. J. Radermacher. Substitution decomposition for discrete structures and connections with combinatorial optimization. *Annals of Discrete Mathematics*, 19:257–356, 1984.
23. R.R. Sokal and C.D. Michener. A statistical method for evaluating systematic relationships. *The University of Kansas Scientific Bulletin*, 38:1409–1438, 1958.
24. T. Uno and M. Yagiura. Fast algorithms to enumerate all common intervals of two permutations. *Algorithmica*, 26(2):290–309, 2000.
25. P. Zhang. United states patent: Method of mapping DNA fragments. Technical report, Available at www.cc.columbia.edu/cu/cie/techlists/patents/5667970.htm, July 2000.

Linear-Time Counting Algorithms for Independent Sets in Chordal Graphs

Yoshio Okamoto[1], Takeaki Uno[2], and Ryuhei Uehara[3]

[1] Department of Information and Computer Sciences, Toyohashi University of Technology,
Hibarigaoka 1-1, Tempaku, Toyohashi, Aichi 441-8580, Japan
okamotoy@ics.tut.ac.jp
[2] National Institute of Informatics, Hitotsubashi 2-1-2,
Chiyoda-ku, Tokyo 101-8430, Japan
uno@nii.jp
[3] School of Information Science, JAIST, Asahidai 1-1,
Nomi, Ishikawa 923-1292, Japan
uehara@jaist.ac.jp

Abstract. We study some counting and enumeration problems for chordal graphs, especially concerning independent sets. We first provide the following efficient algorithms for a chordal graph: (1) a linear-time algorithm for counting the number of independent sets; (2) a linear-time algorithm for counting the number of maximum independent sets; (3) a polynomial-time algorithm for counting the number of independent sets of a fixed size. With similar ideas, we show that enumeration (namely, listing) of the independent sets, the maximum independent sets, and the independent sets of a fixed size in a chordal graph can be done in constant amortized time per output. On the other hand, we prove that the following problems for a chordal graph are #P-complete: (1) counting the number of maximal independent sets; (2) counting the number of minimum maximal independent sets. With similar ideas, we also show that finding a minimum weighted maximal independent set in a chordal graph is NP-hard, and even hard to approximate.

Keywords: Chordal graph, counting, enumeration, independent set, NP-completeness, #P-completeness, polynomial time algorithm.

1 Introduction

How can we cope with computationally hard graph problems? There are several possible answers, and one of them is to utilize the special graph structures arising from a particular context. This has been motivating the study of special graph classes in algorithmic graph theory [3,12]. This paper deals with counting and enumeration problems from this perspective. Recently, counting and enumeration of some specified sets in a graph have been widely investigated, e.g., in the data mining area. In general, however, from the graph-theoretic point of view, those problems are hard even if input graphs are quite restricted. For example, counting the number of independent sets in a planar bipartite graph of maximum degree 4 is #P-complete [17]. Therefore, we wonder what kind of graph structures makes counting and enumeration problems tractable.

D. Kratsch (Ed.): WG 2005, LNCS 3787, pp. 433–444, 2005.

Table 1. Summary of the results. We denote the number of vertices and edges by n and m respectively. The running times for enumeration algorithms refer to amortized time per output.

Chordal graphs	Counting	[ref.]	Enumeration	[ref.]
independent sets	$O(n+m)$	[this paper]	$O(1)$	[this paper]
maximum independent sets	$O(n+m)$	[this paper]	$O(1)$	[this paper]
independent sets of size k	$O(k^2(n+m))$	[this paper]	$O(1)$	[this paper]
maximal independent sets	#P-complete	[this paper]	$O(n+m)$	[7,14]
minimum maximal independent sets	#P-complete	[this paper]		

In this paper, we consider chordal graphs. A *chordal graph* is a graph in which every cycle of length at least four has a chord. From the practical point of view, chordal graphs have numerous applications in, for example, sparse matrix computation (e.g., see Blair & Peyton [2]), relational databases [1], and computational biology [4]. Chordal graphs have been widely investigated, and they are sometimes called triangulated graphs, or rigid circuit graphs (see, e.g., Golumbic's book [12–Epilogue 2004]). A chordal graph has various characterizations; for example, a chordal graph is an intersection graph of subtrees of a tree, and a graph is chordal if and only if it admits a special vertex ordering, called perfect elimination ordering [3]. Also, the class of chordal graphs forms a wide subclass of perfect graphs [12].

It is known that many graph optimization problems can be solved in polynomial time for chordal graphs; to list a few of them, the maximum weighted clique problem, the maximum weighted independent set problem, the minimum coloring problem [11], the minimum maximal independent set problem [8]. There are also parallel algorithms to solve some of these problems efficiently [13]. However, relatively fewer problems have been studied for enumeration and counting in chordal graphs; the only algorithms we are aware of are the enumeration algorithms for all maximal cliques [10], all maximal independent sets [7,14], all minimum separators and minimal separators [5], and all perfect elimination orderings [6].

In this paper, we investigate the problems concerning the number of independent sets in a chordal graph. Table 1 lists the results of the paper. We first give the following efficient algorithms for a chordal graph; (1) a linear-time algorithm to count the number of independent sets, (2) a linear-time algorithm to count the number of maximum independent sets, and (3) a polynomial-time algorithm to count the number of independent sets of a given size. The running time of the third algorithm is linear when the size is constant. Note that in general counting the number of independent sets and the number of maximum independent sets in a graph is #P-complete [15], and counting the number of independent sets of size k in a graph is #W[1]-complete [9] (namely, intractable in a parameterized sense). Let us also note that the time complexity here refers to the arithmetic operations, not to the bit operations.

The basic idea of these efficient algorithms is to invoke a clique tree associated with a chordal graph and perform a bottom-up computation via dynamic programming on the clique tree. A clique tree is based on the characterization of a chordal graph as an intersection graph of subtrees of a tree. Since a clique tree can be constructed in linear time and the structure of clique tree is simple, this approach leads to simple and efficient algorithms for the problems above. However, a careful analysis is necessary to obtain the linear-time complexity.

Along the same idea, we can also enumerate all independent sets, all maximum independent sets, and all independent sets of constant size in a chordal graph in $O(1)$ amortized time per output.

On the other hand, we show that the following counting problems are #P-complete: (1) counting the number of maximal independent sets in a chordal graph, and (2) counting the number of minimum maximal independent sets in a chordal graph. Using a modified reduction, we furthermore show that the problem to find a minimum weighted maximal independent set is NP-hard. We also show that the problem is even hard to approximate. More precisely speaking, there is no randomized polynomial-time approximation algorithm to find such a set within a factor of $c \ln |V|$, for some constant c, unless NP $\subseteq$ ZTIME$(n^{O(\log \log n)})$. This is in contrast with a linear-time algorithm by Farber that finds a minimum weighted maximal independent set in a chordal graph when the weights are 0 or 1 [8].

Due to space limitation, some proofs are omitted.

2   Preliminaries

In this article, we assume that the reader has a moderate familiarity with graph theory. This section aims at fixing the notation and introducing a chordal graph and concepts around that. Let $G = (V, E)$ be a graph, which we always assume to be simple and finite, and also we assume that graphs are connected without loss of generality. The *neighborhood* of a vertex v in a graph $G = (V, E)$ is the set $N_G(v) = \{u \in V \mid \{u, v\} \in E\}$. For a vertex subset U of V, we denote by $N_G(U)$ the set $\{v \in V \mid v \in N(u)$ for some $u \in U\}$. If no confusion can arise we will omit the subscript G. We denote the closed neighborhood $N(v) \cup \{v\}$ by $N[v]$. A vertex set I is an *independent set* of G if any pair of vertices in I is not an edge of G, and a vertex set C is a *clique* if every pair of vertices in C is an edge of G. An independent set is *maximum* if it has the largest size among all independent sets. An independent set is *maximal* if none of its proper supersets is an independent set. An independent set is *minimum maximal* if it is maximal and has the smallest size among all maximal independent sets. A maximum clique, a maximal clique and a minimum maximal clique are defined analogously. An edge which joins two vertices of a cycle but is not itself an edge of the cycle is a *chord* of the cycle. A graph is *chordal* if each cycle of length at least 4 has a chord.

To a chordal graph $G = (V, E)$, we associate a tree T, called a *clique tree* of G, satisfying the following two properties. (A) The nodes of T are the maximal cliques of G. (B) For every vertex v of G, the subgraph T_v of T induced by the maximal cliques containing v is a tree. (In the literature, the condition (A) is sometimes weakened as each node is a vertex subset of G.) It is well known that a graph is chordal if and only if it has a clique tree, and in such a case a clique tree can be constructed in linear time. Some details are explained in books [3,16].

3   Linear-Time Algorithm to Count the Independent Sets

In this section, we describe an algorithm for counting the number of independent sets in a chordal graph. The basic idea of our algorithm is to divide the input graph into subgraphs induced by subtrees of the clique tree. Any two of these subtrees share a vertex of a clique if they are disjoint in the clique tree. This property is very powerful for counting the number of independent sets since any independent set can include at most one vertex of a clique. We compute the number of independent sets including each vertex of the clique, or no vertex of the clique by using the recursions.

First, we introduce some notations and state some lemmas. Given a chordal graph $G = (V, E)$, we construct a clique tree T of G. We now pick up any node in the clique tree T, regard the node as the root of T, and denote it by K_r. This is what we call a *rooted clique tree*. For a maximal clique K in a chordal graph G and a rooted clique tree T of G, a maximal clique K' in G is a *descendant* of K (with respect to T) if K' is a descendant of K in T. For convenience, we consider K itself a descendant of K as well, and when no confusion arises we omit saying "with respect to T." Let PRT(K) be the parent of K in T. For convenience, we define PRT(K_r) by $\emptyset$. We denote by $T(K)$ the subtree of T rooted at the node corresponding to the maximal clique K. Let $G(K)$ denote the subgraph of G induced by the vertices included in at least one node in $T(K)$. Observe that $G(K)$ is a chordal graph of which $T(K)$ is a clique tree.

For a graph G, let $\mathcal{IS}(G)$ be the family of independent sets in G. For a vertex v, let $\mathcal{IS}(G, v)$ be the family of independent sets in G including v, i.e., $\mathcal{IS}(G, v) := \{S \mid S \in \mathcal{IS}(G), v \in S\}$. For a vertex set U, let $\overline{\mathcal{IS}}(G, U)$ be the family of independent sets in G including no vertex of U, i.e., $\overline{\mathcal{IS}}(G, U) := \{S \mid S \in \mathcal{IS}(G), S \cap U = \emptyset\}$.

Lemma 1. *Let G be a chordal graph and T be a rooted clique tree of G. Choose a maximal clique K of G, and let $K_1, \ldots, K_\ell$ be the children of K in T. (If K is a leaf of the clique tree, we set $\ell := 0$.) Furthermore let $v \in K$ and $S \subseteq V(G(K))$. Then, $S \in \mathcal{IS}(G(K), v)$ if and only if S is represented by the union of $\{v\}$ and $S_1, \ldots, S_\ell$ such that $S_i \in \mathcal{IS}(G(K_i), v)$ if v belongs to K_i, and $S_i \in \overline{\mathcal{IS}}(G(K_i), K \cap K_i)$ otherwise. Furthermore, such a representation is unique.*

By a close inspection of the proof, we can observe that for every $i, j \in \{1, \ldots, \ell\}$, $i \neq j$, it holds that $V(G(K_i)) \setminus K$ is disjoint from $V(G(K_j)) \setminus K$. This property gives a nice decomposition of the problem into several independent parts, and enables us to perform the dynamic programming on a clique tree.

By similar discussion, we obtain the following lemma.

Lemma 2. *Let G be a chordal graph and T be a rooted clique tree of G. Choose a maximal clique K of G, and let $K_1, \ldots, K_\ell$ be the children of K in T. (If K is a leaf of the clique tree, we set $\ell := 0$.)*

1. We have $S \in \overline{\mathcal{IS}}(G(K), K)$ if and only if S is the union of $S_1, \ldots, S_l$ such that $S_i \in \overline{\mathcal{IS}}(G(K_i), K \cap K_i)$. Furthermore, such a representation is unique.
2. For each $i \in \{1, \ldots, \ell\}$, we have $S_i \in \overline{\mathcal{IS}}(G(K_i), K \cap K_i)$ if and only if S_i belongs either to $\mathcal{IS}(G(K_i), v)$ for some $v \in K_i \setminus K$ or to $\overline{\mathcal{IS}}(G(K_i), K_i)$. Furthermore, S_i belongs to exactly one of them.

From these lemmas, we have the following recursive equations for $\mathcal{IS}$.

Equations 1. *Let G be a chordal graph and T be a rooted clique tree of G. For a maximal clique K of G which is not a leaf of the clique tree, let $K_1, \ldots, K_\ell$ be the children of K in T. Furthermore, let $v \in K$. Then, the following identities hold. (We remind that $\dot{\cup}$ means "disjoint union.")*

Algorithm 1: #IndSets

    Input : A chordal graph $G = (V, E)$;
    Output: The number of independent sets in G;
1 construct a rooted clique tree T of G with root K_r;
2 call #IndSetsIter(K_r);
3 **return** $\left|\overline{IS}(G, K_r)\right| + \sum_{v \in K_r} |IS(G(K_r), v)|$.

Procedure #IndSetsIter (K)

    Input : A maximal clique K of the chordal graph G;
4 **if** K *is a leaf of* T **then**
5 | set $\left|\overline{IS}(G(K), K)\right| := 0$ and $|IS(K, v)| := 1$ for each $v \in K$;
6 **else**
7 | **foreach** *child* K' *of* K **do** call #IndSetsIter(K');
8 | **foreach** *child* K' *of* K **do** compute $\left|\overline{IS}(G(K'), K \cap K')\right|$ by
          $\left|\overline{IS}(G(K'), K')\right| + \sum_{u \in K' \setminus K} |IS(G(K'), u)|$;
9 | compute $\left|\overline{IS}(G(K), K)\right|$ by $\prod_{K' \in \mathrm{CHD}(K)} \left|\overline{IS}(G(K'), K \cap K')\right|$;
10 | **foreach** $v \in K$ **do** compute $|IS(G(K), v)|$ by
          $\prod_{K' \in \mathrm{CHD}(K), v \in K'} |IS(G(K'), v)| \times \prod_{K' \in \mathrm{CHD}(K), v \notin K'} \left|\overline{IS}(G(K'), K \cap K')\right|$.

Fig. 1. Algorithm to count the number of independent sets in a chordal graph

$$IS(G(K)) = \overline{IS}(G(K), K) \,\dot{\cup}\, \bigcup_{v \in K} IS(G(K), v);$$

$$IS(G(K), v) = \{S \cup \{v\} \mid S = \bigcup_{i=1}^{\ell} S_i, S_i \in \left\{ \begin{array}{ll} IS(G(K_i), v) & if\ v \in K_i \\ \overline{IS}(G(K_i), K \cap K_i) & otherwise \end{array} \right\} \};$$

$$\overline{IS}(G(K), K) = \{S \mid S = \bigcup_{i=1}^{\ell} S_i, S_i \in \overline{IS}(G(K_i), K \cap K_i)\};$$

$$\overline{IS}(G(K_i), K \cap K_i) = \overline{IS}(G(K_i), K_i) \,\dot{\cup}\, \bigcup_{u \in K_i \setminus K} IS(G(K_i), u) \quad for\ each\ i \in \{1, \dots, \ell\}.$$

These equations lead us to the algorithm in Fig. 1 to count the number of independent sets in a chordal graph. For a maximal clique K of a chordal graph G, we denote the set of children of K in a rooted clique tree of G by $\mathrm{CHD}(K)$.

Theorem 1. *The algorithm* #IndSets *outputs the number of independent sets in a chordal graph* $G = (V, E)$ *in* $O(|V| + |E|)$ *time.*

4 Linear-Time Algorithm to Count the Maximum Independent Sets

In this section, we modify Algorithm #IndSets to count the number of maximum independent sets in a chordal graph. For a set family S, we denote by $\max(S)$ the cardinality of a largest set in S, and argmax(S) denotes the family of largest sets in S. For a graph G, let $MIS(G)$ be the family of maximum independent sets in G. For a vertex v, let $MIS(G, v)$ be the family of maximum independent sets in G including v, i.e., $MIS(G, v) := \{S \in MIS(G) \mid v \in S\}$. For a vertex set U, let $\overline{MIS}(G, U)$ be the family of maximum independent sets in G including no vertex of U, i.e., $\overline{MIS}(G, U) := \{S \in MIS(G) \mid S \cap U = \emptyset\}$.

From lemmas stated in the previous section and Equations 1, we immediately have the following equations.

Equations 2. *With the same set-up as Equations 1, the following identities hold.*

$$MIS(G(K)) = \text{argmax}(\overline{MIS}(G(K), K) \dot\cup \bigcup_{v \in K} MIS(G(K), v));$$

$$MIS(G(K), v) = \text{argmax}(\{S \mid S = \bigcup_{i=1}^{\ell} S_i, S_i \in \left\{ \begin{matrix} MIS(G(K_i), v) & \text{if } v \in K_i \\ \overline{MIS}(G(K_i), K \cap K_i) & \text{otherwise} \end{matrix} \right\}\});$$

$$\overline{MIS}(G(K), K) = \text{argmax}(\{S \mid S = \bigcup_{i=1}^{\ell} S_i, S_i \in \overline{MIS}(G(K_i), K \cap K_i)\});$$

$$\overline{MIS}(G(K_i), K \cap K_i) = \text{argmax}(\overline{MIS}(G(K_i), K_i) \dot\cup \bigcup_{u \in K_i \setminus K} MIS(G(K_i), u)).$$

Since the sets of each family on the left hand side have the same size in each equation, the cardinality of the set can be computed in the same order as Algorithm #IndSets. For example, $MIS(G(K))$ can be computed as follows.

1. Set $N := 0$ and $M := \max(\overline{MIS}(G(K), K) \cup \bigcup_{v \in K} MIS(G(K), v))$;
2. if the size of a member of $\overline{MIS}(G(K), K)$ is equal to M, then $N := N + |\overline{MIS}(G(K), K)|$;
3. for each $v \in K$, if the size of a member of $MIS(G(K), v))$ is equal to M, then $N := N + |MIS(G(K), v))|$;
4. output N.

In this way we have the following theorem.

Theorem 2. *The number of maximum independent sets in a chordal graph $G = (V, E)$ can be computed in $O(|V| + |E|)$ time.*

5   Efficient Algorithm to Count the Independent Sets of Size k

In this section, we modify Algorithm #IndSets to count the number of independent sets of size k. For a graph G and a number k, let $IS(G; k)$ be the family of independent sets in G of size k. For a vertex v, let $IS(G, v; k)$ be the family of independent sets in G of size k including v, i.e., $IS(G, v; k) := \{S \in IS(G; k) \mid v \in S\}$. For a vertex set U, let $\overline{IS}(G, U; k)$ be the family of independent sets in G of size k including no vertex of U, i.e., $\overline{IS}(G, U; k) = \{S \in IS(G; k) \mid S \cap U = \emptyset\}$.

From lemmas stated in Section 3 and Equations 1, we immediately obtain the following equations.

Equations 3.

$$IS(G(K); k) = \overline{IS}(G(K), K; k) \,\dot{\cup}\, \bigcup_{v \in K} IS(G(K), v; k);$$

$$IS(G(K), v; k) = \left\{ S \mid S = \bigcup_{i=1}^{\ell} S_i, |S| = k, S_i \in \left\{ \begin{array}{ll} IS(G(K_i), v) & \text{if } v \in K_i \\ \overline{IS}(G(K_i), K \cap K_i) & \text{otherwise} \end{array} \right\} \right\};$$

$$\overline{IS}(G(K), K; k) = \{ S \mid S = \bigcup_{i=1}^{\ell} S_i, |S| = k, S_i \in \overline{IS}(G(K_i), K \cap K_i) \};$$

$$\overline{IS}(G(K_i), K \cap K_i; k) = \overline{IS}(G(K_i), K_i; k) \,\dot{\cup}\, \bigcup_{u \in K_i \setminus K} IS(G(K_i), u; k).$$

In contrast to Equations 1, the second and third equations of Equations 3 do not give a straightforward way to compute $|IS(G(K), v; k)|$ and $\left|\overline{IS}(G(K), K; k)\right|$, respectively, since we have to count the number of combinations of $S_1, \ldots, S_\ell$ which generate an independent set of size k. To compute them, we use a more detailed algorithm.

Here we only explain a method to compute $|IS(G(K), v; k)|$ since $\left|\overline{IS}(G(K), K; k)\right|$ can be computed in a similar way. Fix an arbitrary vertex $v \in K$. Then, according to v, we give indices to the children of K such that $K_1, \ldots, K_p$ include v and $K_{p+1}, \ldots, K_\ell$ do not. For $k' \leq k$ and $\ell' \leq p$, let $\mathrm{NUM}(\ell'; k') := \{ S \mid S = \bigcup_{i=1}^{\ell'} S_i, S_i \in IS(K_i, v), |S| = k' \}$. For $k' \leq k$ and $\ell' \geq p+1$, let $\overline{\mathrm{NUM}}(\ell'; k') := \{ S \mid S = \bigcup_{i=\ell'}^{\ell} S_i, S_i \in IS(K_i, K_i \setminus K), |S| = k' \}$. Then, it holds that $|IS(G(K), v; k)| = \sum_{h=0}^{k} (|\mathrm{NUM}(p; h)| \times |\overline{\mathrm{NUM}}(p+1; k-h)|)$.

For each ℓ' and k', $|\mathrm{NUM}(\ell'; k')|$ can be computed in $O(k \times p)$ time based on the following recursive equation:

$$\left|\mathrm{NUM}(\ell'; k')\right| = \left\{ \begin{array}{ll} \sum_{h=0}^{k'} |\mathrm{NUM}(\ell'-1; h)| \times |IS(G(K_{\ell'}), v; k'-h)| & \text{if } \ell' > 1, \\ |IS(G(K_1), v; k')| & \text{otherwise.} \end{array} \right.$$

Similarly, $\left|\overline{\mathrm{NUM}}(\ell'; k')\right|$ can be computed in $O(k')$ time. The computation of $|\mathrm{NUM}(\ell'; k')|$ and $\left|\overline{\mathrm{NUM}}(\ell'; k')\right|$ for all combinations of ℓ' and k' can be done in $O(k^2 |\mathrm{CHD}(K)|)$ time, thus we can count the number of independent sets of size k in a chordal graph in $O(k^2 |V|^2)$ time. In the following, we reduce the computation time by the same technique used in the previous sections.

Observe that $\left|\overline{IS}(G(K), K; k')\right| = \sum_{h=0}^{k'} \left|\overline{\mathrm{NUM}}(p; h)\right| \times \left|\overline{\mathrm{NUM}}(p+1; k'-h)\right|$, which gives $\left|\overline{\mathrm{NUM}}(p+1; k')\right| \times \left|\overline{\mathrm{NUM}}(p; 0)\right| = \left|\overline{IS}(G(K), K; k')\right| - \sum_{h=1}^{k'} \left|\overline{\mathrm{NUM}}(p; h)\right| \times \left|\overline{\mathrm{NUM}}(p+1; k'-h)\right|$. This implies that we can compute $\left|\overline{\mathrm{NUM}}(k'; p+1)\right|$ from $\left|\overline{IS}(G(K), K; h)\right|$ and $\left|\overline{\mathrm{NUM}}(p; h)\right|$ in the increasing order of k'. The computation time for this task is $O(k \times p)$.

In summary, we can compute $|IS(G(K), v; k')|$ for all $v \in K$ and $k' \in \{0, \ldots, k\}$ in $O(k^2 \sum_{v \in K} |\{K' \in \mathrm{CHD}(K) \mid v \in K'\}|)$ time. Therefore, the total computation time over all iterations can be bounded in the same way as the above section, and we obtain the following theorem.

Theorem 3. *1. The number of independent sets of size k in a chordal graph $G = (V, E)$ can be computed in $O(k^2(|V| + |E|))$ time.*

2. *The numbers of independent sets of all sizes from* 0 *to* $|V|$ *in a chordal graph* $G =$ (V, E) *can be simultaneously computed in* $O(|V|^2(|V| + |E|))$ *time.*

6   Enumeration

Equations 1 in Section 3 directly give the following algorithm for enumerating the independent sets of a given chordal graph, in which each procedure corresponds to an equation of Equations 1.

Algorithm 3: EnumIS(G)

   Input : a chordal graph $G = (V, E)$;
   Output: all independent sets in G;
1 construct a clique tree T of G with root K;
2 **foreach** $u \in K$ **do** enumerate all independent sets in $IS(G, u)$ by EnumIS2(K, u);
3 enumerate all independent sets in $\overline{IS}(G, K)$ by EnumIS3(K).

Procedure EnumIS2 (K, u)

   Input : A maximal clique K of G, a vertex $u \in K$;
4 **if** K *has no child* **then**
5 | **output** $\{u\}$; //output an independent set if the bottom level is reached
6 **else**
7 | **foreach** *child* K_i *of* K *such that* $u \in K_i$ **do** enumerate all independent sets in
       $IS(G(K_i), u)$ by EnumIS2(K_i, u);
8 | **foreach** *child* K_i *of* K *such that* $u \notin K_i$ **do** enumerate all independent sets in
       $\overline{IS}(G(K_i), K \cap K_i)$ by EnumIS4(K_i);
9 | **output** all independent sets in $IS(G(K), u)$ by combining the independent sets in
       $IS(G(K_i), u)$ and in $\overline{IS}(G(K_j), K \cap K_j)$ for all i, j;

Procedure EnumIS3 (K)

   Input : A maximal clique K of G;
10 **if** K *has no child* **then**
11 | **output** $\emptyset$; //output an independent set if the bottom level is reached
12 **else**
13 | **foreach** *child* K_i *of* K **do** enumerate all independent sets in $\overline{IS}(G(K_i), K \cap K_i)$ by
       EnumIS4(K_i);
14 | **output** all independent sets in $\overline{IS}(G(K), K)$ by combining the independent sets in
       $\overline{IS}(G(K_i), K \cap K_i)$;

Procedure EnumIS4 (K)

   Input : A maximal clique K of G;
15 call EnumIS3(K);
16 **foreach** $u \in K \setminus \text{PRT}(K)$ **do** enumerate all independent sets in $IS(G(K), u)$ by
   EnumIS2($G(K), u$);
17 **output** all independent sets in $\overline{IS}(G(K), K \cap \text{PRT}(K))$ by combining the independent sets
   in $IS(G(K), u)$;

From the lemmas and theorems in the previous sections, EnumIS(G) surely enumerates all independent sets in G. However, we cannot bound its time complexity by constant for each output. In the following, we present a slight modification to obtain a constant-time enumeration algorithm.

Let us consider the computation tree of this algorithm. A *computation tree* is a rooted-tree representation of a recursive structure, in which the vertices are recursive calls, and the edges connect two vertices if and only if one vertex recursively calls the other. We define an *iteration* of the algorithm by the operations done in a vertex of the computation tree. In other words, an iteration is the computation in some procedure P recursively called by another procedure, in which the computation in the recursive calls generated by P is excluded.

We first reduce the number of iterations by the following two modifications. (1) If an iteration I generated by an iteration I_p recursively calls just one iteration I_c, we modify the algorithm so that I_p recursively calls I_c directly. (2) If an iteration I outputs just one independent set, merge I and the iteration which recursively calls I into one.

For a given chordal graph $G = (V, E)$ and a rooted clique tree of G, the number of possible inputs for each procedure is at most $O(|E|)$, as in our counting algorithms. Thus, we can enumerate all of these cases in $O(|E|)$ time, and keep the results of modifications (1) and (2) in the memory. It can be done as a preprocessing within $O(|E|)$ time.

By these modifications, we can see that any iteration which is a leaf of the computation tree outputs at least two independent sets, thus the number of iterations is not greater than the number of independent sets in G. We can also see that if an iteration outputs just one independent set, then, the input clique must be a leaf of the clique tree. Hence, the size of the output independent set is at most one.

We next consider how to compute all combinations of independent sets in, for example, Step 9 of the algorithm. In the procedures, the independent sets for K are generated by combining the independent recursive calls for several maximal cliques, say K_1 and K_2. This step can be implemented as follows. First, we compute an indenendent set I_1 for K_1, and for this I_1, we compute all independent sets I_2 for K_2, and output $I_1 \cup I_2$. Next we compute another independent set I_1' for K_1, and compute all independent sets I_2 for K_2, and output $I_1 \cup I_2$, then compute yet another independent set for K_1, and so on. Then the computation time in one iteration is proportional to (the number of recursive calls generated) times (the maximum number of vertices added to the current independent set). Because of modification (2), any iteration adds at most one vertex to the current independent set. Therefore, the total time complexity of the algorithm is linear in the number of independent sets.

Theorem 4. *All independent sets in a chordal graph can be enumerated in constant time for each on average with additional $O(|V| + |E|)$ time for preprocessing.*

Similar algorithms can be developed to enumerate the maximum independent sets and the independent sets of size k. However, some iterations may add to the current independent set several vertices not bounded by a constant. Since there are at most $|E|$ kinds of inputs for each procedure, we can enumerate all such sets of vertices that will be added in an iteration, and put an identical name to each set of vertices in short time. By adding the name instead of adding vertices in a vertex set, we can execute the addition in constant time. Thus, the maximum independent sets and the independent sets of size k can be enumerated in constant time for each on average with additional $O((|V| + |E|)|V|^2)$ time for preprocessing.

7   Hardness of Counting the Maximal Independent Sets

In this section, we show the hardness results for counting the number of maximal independent sets in a chordal graph. Although finding a maximal independent set is easy even in a general graph, we show that the counting version of the problem is actually hard.

Theorem 5. *Counting the number of maximal independent sets in a chordal graph is #P-complete.*

The proof is based on a reduction from the counting problem of the number of set covers. Let X be a finite set, and $S \subseteq 2^X$ be a family of subsets of X. A *set cover* of X is a subfamily $\mathcal{F} \subseteq S$ such that $\bigcup \mathcal{F} = X$. Counting the number of set covers is #P-complete [15].

Proof of Theorem 5 (Sketch). The membership in #P is immediate. To show the #P-hardness, we use a polynomial-time reduction of the problem for counting the number of set covers to our problem.

Let X be a finite set and $S \subseteq 2^X$ be a family of subsets of X, and consider them as an instance of the set cover problem. Let us put $S := \{S_1, \ldots, S_t\}$. From X and S, we construct a chordal graph $G = (V, E)$ in the following way.

We set $V := X \cup S \cup S'$, where $S' := \{S'_1, \ldots, S'_t\}$. Namely, S' is a copy of S. Now, we draw edges. There are three kinds of edges. (1) We connect every pair of vertices in X by an edge. (2) For every $S \in S$, we connect $x \in X$ and S by an edge if and only if $x \in S$. (3) For every $S \in S$, we connect S and S' (a copy of S) by an edge. Formally speaking, we define $E := \{\{x, y\} \mid x, y \in X\} \cup \{\{x, S\} \mid x \in X, S \in S, x \in S\} \cup \{\{S, S'\} \mid S \in S\}$. This completes our construction, which can be done in polynomial time. The constructed graph G is indeed chordal.

Now, we look at the relation between the set covers of X and the maximal independent sets of G. Let U be a maximal independent set of G. We distinguish two cases.

Case 1. Consider the case in which U contains a vertex $x \in X$. Let $G_x := G \setminus N_G[x]$. By the construction, we have that $V(G_x) = \{S \in S \mid x \notin S\} \cup S'$ and $E(G_x) = \{\{S, S'\} \mid S \in S, x \notin S\}$. Then the number of maximal independent sets containing x is exactly $2^{|\{S \in S \mid x \notin S\}|}$.

Case 2. Consider the case in which U contains no vertex of X. Then, the number of maximal independent sets containing no vertex of X is equal to the number of set covers of X.

To summarize, we obtained that the number of maximal independent sets of G is equal to the number of set covers of X plus $\sum_{x \in X} 2^{|\{S \in S \mid x \notin S\}|}$. Since the last sum can be computed in polynomial time, this concludes the reduction.   □

As a variation, let us consider the problem for counting the minimum maximal independent sets in a chordal graph. Note that a minimum maximal independent set in a chordal graph can be found in polynomial time [8]. In contrast to that, the counting version is hard.

Theorem 6. *Counting the minimum maximal independent sets in a chordal graph is #P-complete.*

8   Hardness of Finding a Minimum Weighted Maximal Independent Set

In this section, we consider an optimization problem to find a minimum weighted maximal independent set in a chordal graph. Namely, given a chordal graph G and a weight for each vertex, we are asked to find a maximal independent set of G with minimum weight. Here, the weight of a vertex subset is the sum of the weights of its vertices.

Notice that there is a linear-time algorithm for this problem when the weight of each vertex is zero or one [8]. On the contrary, we show that the problem is actually hard when the weight is arbitrary.

Theorem 7. *Finding a minimum weighted maximal independent set in a chordal graph is NP-hard.*

The proof is similar to what we saw in the previous section. We use the optimization version of the set cover problem, namely the minimum set cover problem. It is known that the minimum set cover problem is NP-hard.

Proof of Theorem 7. For a given instance of the minimum set cover problem, we use the same construction of a graph G as in the proof of Theorem 5. We define a weight function w as follows: $w(x) := 2|S| + 1$ for every $x \in X$; $w(S) := 2$ for every $S \in \mathcal{S}$; $w(S') := 1$ for every $S' \in \mathcal{S}'$. This completes the construction.

Now, observe that $\mathcal{S}$ is a maximal independent set of the constructed graph G, and the weight of $\mathcal{S}$ is $2|S|$. Therefore, no element of X takes part in any minimum weighted maximal independent set of G. Then, from the discussion in the proof of Theorem 5, if M is a maximal independent set of G satisfying $M \cap X = \emptyset$, then $M \cap \mathcal{S}$ is a set cover of X. The weight of M is $|M \cap \mathcal{S}| + |\mathcal{S}|$. Therefore, if M is a minimum weighted independent set of G, then M minimizes $|M \cap \mathcal{S}|$, which is the size of a set cover. Hence, $M \cap \mathcal{S}$ is a minimum set cover. This concludes the reduction.   □

We can further show the hardness to get an approximation algorithm running in polynomial time. The precise statement is as follows ($\mathsf{ZTIME}(t)$ is the class of languages which have a randomized algorithm running in expected time t with zero error).

Theorem 8. *There is no randomized polynomial-time algorithm for the minimum weight maximal independent set problem in a chordal graph with approximation ratio $c \ln |V|$, for some fixed constant c, unless $\mathsf{NP} \subseteq \mathsf{ZTIME}(n^{O(\log \log n)})$.*

Acknowledgement. The authors are grateful to L. Shankar Ram for pointing out a paper [5].

References

1. C. Beeri, R. Fagin, D. Maier, and M. Yannakakis. On the Desirability of Acyclic Database Schemes. *Journal of the ACM*, 30:479–513, 1983.
2. J.R.S. Blair and B. Peyton. An Introduction to Chordal Graphs and Clique Trees. In *Graph Theory and Sparse Matrix Computation*, volume 56 of *IMA*, pages 1–29. (Ed. A. George and J.R. Gilbert and J.W.H. Liu), Springer, 1993.
3. A. Brandstädt, V.B. Le, and J.P. Spinrad. *Graph Classes: A Survey*. SIAM, 1999.
4. P. Buneman. A Characterization of Rigid Circuit Graphs. *Discrete Mathematics*, 9:205–212, 1974.
5. L.S. Chandran. A Linear Time Algorithm for Enumerating All the Minimum and Minimal Separators of a Chordal Graph. In *COCOON 2001*, pages 308–317. Lecture Notes in Computer Science Vol. 2108, Springer-Verlag, 2001.
6. L.S. Chandran, L. Ibarra, F. Ruskey, and J. Sawada. Generating and Characterizing the Perfect Elimination Orderings of a Chordal Graph. *Theoretical Computer Science*, 307:303–317, 2003.
7. D. Eppstein. All Maximal Independent Sets and Dynamic Dominance for Sparse Graphs. In *Proc. 16th Ann. ACM-SIAM Symp. on Discrete Algorithms*, pages 451–459. ACM, 2005.
8. M. Farber. Independent Domination in Chordal Graphs. *Operations Research Letters*, 1(4):134–138, 1982.
9. J. Flum and M. Grohe. The Parameterized Complexity of Counting Problems. *SIAM Journal on Computing*, 33(4):892–922, 2004.
10. D.R. Fulkerson and O.A. Gross. Incidence Matrices and Interval Graphs. *Pacific J. Math.*, 15:835–855, 1965.
11. F. Gavril. Algorithms for Minimum Coloring, Maximum Clique, Minimum Covering by Cliques, and Maximum Independent Set of a Chordal Graph. *SIAM Journal on Computing*, 1(2):180–187, 1972.
12. M.C. Golumbic. *Algorithmic Graph Theory and Perfect Graphs*. Annals of Discrete Mathematics 57. Elsevier, 2nd edition, 2004.
13. P.N. Klein. Efficient Parallel Algorithms for Chordal Graphs. *SIAM Journal on Computing*, 25(4):797–827, 1996.
14. J.Y.-T. Leung. Fast Algorithms for Generating All Maximal Independent Sets of Interval, Circular-Arc and Chordal Graphs. *Journal of Algorithms*, 5:22–35, 1984.
15. J.S. Provan and M.O. Ball. The Complexity of Counting Cuts and of Computing the Probability that a Graph is Connected. *SIAM Journal on Computing*, 12:777–788, 1983.
16. J.P. Spinrad. *Efficient Graph Representations*. American Mathematical Society, 2003.
17. S.P. Vadhan. The Complexity of Counting in Sparse, Regular, and Planar Graphs. *SIAM Journal on Computing*, 31(2):398–427, 2001.

Faster Dynamic Algorithms for Chordal Graphs, and an Application to Phylogeny

Anne Berry[1], Alain Sigayret[1], and Jeremy Spinrad[2]

[1] LIMOS UMR, bat. ISIMA, 63173 Aubière cedex, France
{berry, sigayret}@isima.fr
[2] EECS dept, Vanderbilt University, Nashville, TN 37235, USA
spin@vuse.vanderbilt.edu

Abstract. We improve the current complexities for maintaining a chordal graph by starting with an empty graph and repeatedly adding or deleting edges.

1 Introduction

Our motivation for this paper stems from the biology-based problem of improving the matrix representing an evolutionary tree (phylogeny) which contains errors. To solve this, Berry, Sigayret and Sinoquet in [2] needed to start with an independent set and repeatedly add an edge of minimum weight, while maintaining a chordal graph. However, the most efficient algorithms for dynamically maintaining a chordal graph (see Ibarra [8]) were insufficient to ensure a complexity which could be used in practice.

In this paper, we improve the time complexity for dynamic algorithms for chordal graphs. Ibarra studied the problem of maintaining a chordal graph as edges are inserted or deleted. Operations considered were insert, delete, insert query and delete query; the last two operations ask whether deletion/addition of a given edge xy preserves chordality. He gave three implementations of the algorithm. In the first implementation, all operations take $O(n)$ time. In the second, deletions take $O(n \log n)$ time, deletion queries and insertion run in $O(n)$ time, while insertion queries run in $O(\log^2 n)$ time. The third variant was designed for sparse chordal graphs and will not be addressed in this paper.

Our new running times are $O(n)$ for insertion and deletion, $O(1)$ for insertion queries, and $O(n)$ for deletion queries. All data structures used are simple. We do make one extra assumption, which is not made in the Ibarra paper. We assume that we start with an empty graph; if this assumption is not made, there is an initial start-up cost of $O(m\Delta)$, where Δ is the maximum degree of a vertex, or $O(n^\alpha)$, where n^α is the cost of doing matrix multiplication. We use this result to improve the time bound for the original phylogeny problem, from $O(n^4)$ to $O(n^3)$.

2 Preliminaries

In this paper, $G = (V, E)$ will be a graph with n vertices and m edges. We will use non-formal notations such as $G - S$ instead of $G(V - S)$. An xy-separator in

D. Kratsch (Ed.): WG 2005, LNCS 3787, pp. 445–455, 2005.

a connected graph G is a non-empty set S of vertices such that there is no path from x to y in $G-S$. S is a minimal xy-separator if S does not properly contain any xy-separator. Whenever there exists a pair $\{x, y\}$ of vertices such that S is a minimal xy-separator, S is called a minimal separator.

A graph is *chordal* if every cycle of length greater than three has a chord. Chordal graphs have a long history of study; see, for example [1,6]. A *clique tree* of a chordal graph G is a tree T such that nodes have a 1-1 correspondence with maximal cliques of G, edges correspond to non empty intersections of pairs of maximal cliques, and for all vertices v in G, the set of maximal cliques which contain v induces a subtree of T. A graph is chordal iff it has a clique tree ([5], [3], [12]). Each edge $S = K_1 \cap K_2$ of T corresponds to one minimal separator S of G; conversely, each minimal separator of G is represented by at least one edge of T. A chordal graph may thus have several clique trees. A clique tree has O(n) nodes. There are known algorithms (see for example [10]) which find a clique tree of a chordal graph in O($m + n$) time.

The discussions in this paper will use the following theorems which are easy to prove using well-known results on chordal graphs and a characterization from [2]:

Characterization 1. *([2]) Let $G = (V, E)$ be a chordal graph, $xy \notin E$. Then $G + xy$ is chordal iff $\{x, y\}$ is a 2-pair of G (i.e. all chordless paths between x and y are of length 2).*

Theorem 2. *Edge xy can be deleted from a chordal graph G without causing a chordless cycle iff x and y are not together in any minimal separator of G.*

Theorem 3. *Edge xy can be added to a connected chordal graph G without causing a chordless cycle iff x and y are both adjacent to every vertex in some minimal xy-separator S. Furthermore $S = N(x) \cap N(y)$.*

Theorem 4. *The number of maximal cliques containing a vertex x in a chordal graph is at most $|N(x)|$. The total number of vertices in all maximal cliques of a clique tree is O(m).*

3   Data Structures

We maintain a clique tree of the current chordal graph G with the following modifications. For each maximal clique K and minimal separator S in the clique tree, and each vertex x, we keep variables $neighnum(x, K)$ and $neighnum(x, S)$ denoting the number of neighbors of x in K and S respectively.

We also maintain an array *Insertable*; $Insertable(x, y) = 1$ iff xy can be inserted while maintaining chordality.

We will discuss how to calculate initial values if we are given a start graph G after Theorem 7. If we start with an edgeless graph, all values are initially 0.

4   Algorithms

We now discuss how to implement the various operations. Some of these operations are identical to those in [8], but are repeated here so that the reader can have easy access to the full algorithm.

The simplest primitive, given our data structures, is **Insert Query**. We look up in our array $Insertable$ whether $Insertable(x, y) = 1$. The other simple primitive to describe is **Delete Query**; to achieve an $O(n)$ bound, step through the tree and test whether more than one maximal clique contains both x and y. If it is desired, this can be reduced to $O(\min\{degree(x), degree(y)\})$ by using Theorem 4 and letting each vertex x keep a list pointing to each clique which contains x.

Operations **Insert** and **Delete** are very similar to each other. In each case, we modify the clique tree and then update the array $Insertable$ to decide which edges may now be added to the graph while preserving chordality.

We first deal with the insertion of xy.

The first part of the modification, finding a clique tree for $G + xy$, may work exactly as in Ibarra's paper, though we must make sure to update our new data structures as well. For completeness, we will give the full process here.

Let xy be the edge to be inserted. There will be exactly one new maximal clique, which is $y + x + (N(x) \cap N(y))$. At most 2 maximal cliques of G are deleted, which are $x + (N(x) \cap N(y))$ and $y + (N(x) \cap N(y))$ if these cliques are currently maximal in G. For simplicity, we will treat the cases of 2, 1 or 0 of $x + (N(x) \cap N(y))$, $y + (N(x) \cap N(y))$ existing in the clique tree separately. To determine whether one or both of these cliques are in the current tree, let K_1 be any maximal clique containing x and K_2 be any maximal clique containing y. We find the path from K_1 to K_2 in the clique tree. Let K_x be the last clique on this path which contains x, and let K_y be the first clique on the path which contains y. It is not hard to see that if $x + (N(x) \cap N(y))$ is in the tree, then it must be K_x and if $y + (N(x) \cap N(y))$ is in the tree, it must be K_y. Thus, we can determine which cliques appear and disappear with the addition of edge xy in $O(n)$ time.

- Case 1. both $x + N(x) \cap N(y)$ and $y + N(x) \cap N(y)$ are maximal cliques of G. There is a separator S_{xy} on the path from $K_x = x + N(x) \cap N(y)$ to $K_y = y + N(x) \cap N(y)$ which is exactly equal to $N(x) \cap N(y)$, else xy could not have been inserted while maintaining chordality. We delete this edge and unify the two nodes representing K_x and K_y into node $K_{xy} = x + y + (N(x) \cap N(y))$ as in Figure 1.
- Case 2. we now consider the case in which one of the two possible maximal cliques ceases to be maximal due to the addition of xy; w.l.o.g., let us assume that $K_x = x + (N(x) \cap N(y))$ is a maximal clique of G.
  Let K_y be the first clique containing y on the path from K_1 to K_2; as noted earlier, $K_x = x + (N(x) \cap N(y))$ will be the last clique containing x on this path. We add an edge from the K_x to K_y, and delete edge $N(x) \cap N(y)$ on the path from x to y. We then add y to K_x as in Figure 2.

Fig. 1. Insert — case 1

Fig. 2. Insert — case 2

Fig. 3. Insert — case 3

- Case 3. in this case, we add a new node to the clique tree corresponding to $x + y + (N(x) \cap N(y))$, and add edges from this node to K_x and K_y, and delete S_{xy} as in Figure 3.

We need to update the variables $neighnum(v, K)$ and $neighnum(v, S)$ for each maximal clique K and minimal separator S of the tree to reflect the changes caused by insertion of xy. For each clique K and separator S containing y, add 1 to $neighnum(x, K)$ and $neighnum(x, S)$; similarly, increment the values of $neighnum(y, K)$ and $neighnum(y, S)$ for each clique or separator containing y. We have at most one new maximal clique K_{xy} in the tree: $x+y+(N(x) \cap N(y))$; for each vertex v, we let $neighnum(v, K_{xy}) = neighnum(v, S_{xy})$ plus the number of neighbors of v in $\{x, y\}$.

We now have to update the values for separators which were changed by the insertion of xy. Note that in case 1, no separators change, in case 2 a single separator (which goes between K_x and K_y) has a vertex added to it, and in case 3 two separators are added corresponding to the edges around the new node of the clique tree.

The only separators changed are those adjacent to the new maximal clique $x+y+(N(x)\cap N(y))$, and correspond to $N(x)\cap N(y)$ plus possibly a vertex from $\{x,y\}$. Since we know the number of neighbors of each vertex w.r.t. $N(x)\cap N(y)$, it is easy to calculate the number of neighbors of each vertex w.r.t. to the new separators in constant time.

The array Insertable must be updated after addition of edge xy. We will defer discussion of this step until after we discuss deletion of an edge, since both steps make use of a routine which takes an input vertex x and a clique tree, and finds all z such that xz is insertable in $O(n)$ time.

We will now examine deletion.

The deletion of an edge xy causes the maximal clique $x+y+(N(x)\cap N(y))$ to disappear from the clique tree. At most two new maximal cliques may appear: $x+(N(x)\cap N(y))$ and $y+(N(x)\cap N(y))$. As in the case of insertion, we discuss the cases of 0, 1 or 2 maximal cliques appearing separately. As observed by Ibarra [8], it is easy to determine whether each of these cliques appears in $O(n)$ time. We find the single maximal clique containing x and y, and look for edges from this node which correspond to separators containing $1+|N(x)\cap N(y)|$ vertices; we then test whether x or y is the vertex missing in such a separator.

- Case 1. Two maximal cliques appear: $x+(N(x)\cap N(y))$ and $y+(N(x)\cap N(y))$. In this case, the tree node corresponding to $x+y+(N(x)\cap N(y))$ is split into 2 adjacent nodes $x+(N(x)\cap N(y))$ and $y+(N(x)\cap N(y))$. The neighboring cliques Y_* containing y are made neighbors of $y+(N(x)\cap N(y))$, other cliques which were neighbors of $x+y+(N(x)\cap N(y))$ become neighbors of $x+(N(x)\cap N(y))$ in the new tree, as in Figure 4.
- Case 2. If one maximal clique, which we will w.l.o.g. assume is $x+(N(x)\cap N(y))$ appears, the following changes are made.

  Let K_y be the neighbor of $x+y+(N(x)\cap N(y))$ which is separated by an edge separator with $1+|N(x)\cap N(y)|$ vertices. We remove y from the node $x+y+(N(x)\cap N(y))$ and for every neighbor Y_* of $x+y+(N(x)\cap N(y))$ containing y in the clique tree except for K_y, we remove the connection from Y_* to $x+y+(N(x)\cap N(y))$ and add an edge from Y_* to K_y, as in Figure 5. Note that since none of these Y_* correspond to cliques containing x, the separator between Y_* and K_y is the same as the old separator between K_y and $x+y+(N(x)\cap N(y))$.
- Case 3. In the remaining case, no new maximal clique appears; we remove $x+y+(N(x)\cap N(y))$ from the clique tree, then we find K_y and an analogous K_x as in the previous case; these are cliques which contain $y+N(x)\cap N(y)$ and $x+N(x)\cap N(y)$ respectively. We add an edge between K_x and K_y. All former neighbors of $x+y+(N(x)\cap N(y))$ containing y are given edges to K_y, while other neighbors are given edges to K_x, as in Figure 6. Since no remaining clique contains both x and y, the separators remain the same in the new clique tree, except for the edge separating K_x and K_y.

We now describe how to modify the variables maintained after the modification of the clique tree.

Fig. 4. Delete — case 1

Fig. 5. Delete — case 2

Fig. 6. Delete — case 3

The only new maximal cliques to appear are either $x + (N(x) \cap N(y))$ or $y + (N(x) \cap N(y))$. Since we know the number of neighbors in each vertex in $x + y + (N(x) \cap N(y))$, it is easy to compute the number of neighbors in the new clique in constant time per vertex. The only new separator which could have been created was $N(x) \cap N(y)$. Again, it is easy to compute the number of neighbors of each vertex w.r.t. the new separator $N(x) \cap N(y)$, since we know the number of its neighbors in $x + y + N(x) \cap N(y)$.

We now come to the updating of the array *Insertable* after deletion or addition of the edge xy:

Theorem 5. *Let G be a connected chordal graph. Let x, y, v and w be vertices such that neither v nor w is equal to x or y.*

1. If $G+xy$ is chordal: vw can be inserted into $G+xy$ while preserving chordality iff vw could be added to G while preserving chordality.
2. If $G-xy$ is chordal: vw can be inserted into $G-xy$ while preserving chordality iff vw could be added to G while preserving chordality.

Proof.

1. xy is not an edge of G. G and $G + xy$ are chordal.
    $\Longrightarrow (G+xy)+vw$ is chordal. Suppose $G + vw$ is not chordal: there exists a chordless cycle $C = v \sim w \sim v$. In $G + xy + vw$, which is chordal, C

does not remain chordless and this must be due to chord xy. Then x and y are in C which is w.l.o.g. $v-s-w{\sim}x{\sim}y{\sim}v$. As a consequence, cycle $v-s-w{\sim}x-y \sim v$ is chordless in $G + xy + vw$ — a contradiction.

$\Longleftarrow$ $G+vw$ is chordal. By Theorem 3 there exists a minimal vw-separator $S = N(v) \cap N(w)$. The addition of edge xy does not change the common neighborhood of v and w; then there exists in $G + xy$ a minimal vw-separator $S' \supseteq S$. Suppose $S' \neq S$; then there exists a new path between v and w; this path must use edge xy, and is w.l.o.g. $v{\sim}x-y{\sim}w$. As there is also a path $v-s-w$ for some $s \in S$, chordal graph $G + xy$ contains a chordless cycle of length ≥ 4 — a contradiction. Then $S' = S = N(v) \cap N(w)$ remains a minimal vw-separator in $G + xy$ and, by Theorem 3, vw is insertable in $G + xy$.

2. Apply part 1 of this theorem with $G' = G + xy$ and thus $G = G' - xy$. $\square$

Given the above theorem, we only need to find which vertex pairs $\{x, z\}$ and $\{y, z\}$ can be inserted to preserve chordality, and update these values in the array *Insertable*.

We give an algorithm which takes an arbitrary single vertex v and finds all non-neighbors w of v such that vw can be inserted and preserve chordality in $O(n)$ time given the information maintained on the clique tree. By Theorem 5, we can simply run this for x and y when xy is inserted or deleted, and we will have updated our *Insertable* list correctly.

Theorem 6. *Given a vertex v, we can find all vertices w such that vw can be inserted while preserving chordality in $O(n)$ time.*

Proof. The following algorithm takes a vertex v and finds all w such that adding vw to G will preserve chordality.

For each non-neighbor w of v, place w on the clique tree at any clique which contains w. We select any clique containing v, and traverse the clique tree in a depth-first fashion. When a non-neighbor w is reached on the clique tree, we will decide whether adding vw to the tree would preserve chordality.

We keep one extra data structure during our traversal. Recall that vw can be added iff w and v are both completely adjacent to some minimal vw-separator. The extra structure is an array *posssep* of size n, which holds pointers to possible separators S such that S might be a minimal vw-separator meeting the criteria of Theorem 2 for some w we may encounter on the path. Initially, *posssep* is empty.

Suppose that we are at clique node K in our traversal of T, and our DFS traversal of T leaves K by an edge corresponding to separator S to a new node K' of T.

We test using $neighnum(v, S)$ whether v is adjacent to all vertices of S. If $neighnum(v, S) = |S|$, then S could be a possible vw-separator meeting the conditions of Theorem 2. We look at position $|S|$ of *posseps*. If this is already non-empty, then there is already a separator S' with the same vertices as S encountered earlier than S on the path from x. In this case, if S meets the conditions of Theorem 2, S' also meets the conditions of Theorem 2 for any w

encountered on the path, and S is not stored in *posseps*. If position $|S|$ is empty, we mark this edge e as a candidate separator, and put a pointer in $posseps(|S|)$ to edge e. We add 1 to a count of the number of separators in *posseps*, and if this number of separators in *posseps* becomes 1, we call this separator *startsep*.

Suppose that we encounter a vertex w which was placed on the clique tree. We want to test whether v and w are separated by any vw-separator S such that both v and w are completely adjacent to S.

It is not hard to see that all minimal vw-separators are on the path from v to the current node, though (since w may be in many cliques on this path) not all edges on the path correspond to minimal vw-separators. We test how many neighbors of w there are in separator *startsep*; call this number *vwneighbors*. Clearly, any separator with fewer than *vwneighbors* vertices cannot separate v from w. In addition, any separator with more than *vwneighbors* vertices cannot be completely adjacent to w, since neighbors of v can only disappear as we traverse the path from v to w. Thus, we only need to check if the edge pointed to by $posseps(vwneighbors)$ is a minimal vw-separator satisfying the conditions of Theorem 2. If the array points to edge $K_i - S - K_j$, with K_i closer to v, we check that w is not in K_i (or this would not be a vw-separator), and that $neighnum(w, S) = |S|$.

We make $Insertable(v, w) = 1$ if these conditions hold, and 0 otherwise.

As we back up across an edge $e = K_i - S - K_j$ in the DFS, if e is marked as a separator, we delete the pointer in $posseps(|S|)$, and decrement the number of current possible separators.   □

Combining the theorems above, we get the desired result.

Theorem 7. *The algorithm maintains a chordal graph under the operations insert, delete, delete query, and insert query, taking $O(n)$ time for the first three operations and $O(1)$ time for insert query.*

If a connected graph G is given as input, it is necessary to construct a clique tree, and compute $neighnum(v, S)$ and $neighnumN(v, K)$ for each vertex v, each minimal separator S, and each clique K of the clique tree.

As a clique tree of a chordal graph can be found in $O(m + n)$ time and has an $O(m)$ overall number of vertices in the nodes, we can step through all maximal cliques K, and for each v in K add 1 to $neighnum(w, K)$ for all neighbors w of v, thus finding all these variables in $O(m\,n)$ time.

Although there are also $O(m)$ vertices over all minimal separators of the clique tree (so the argument above could also be used to count all variables in $O(mn)$ time), the algorithms for finding a clique tree of a chordal graph usually do not explicitly label the separators. Therefore, we describe briefly how these separators could be labeled in $O(m + n)$ time. Choose an arbitrary root for the clique tree. For each node K of the clique tree, perform the following operation. Mark all positions of an array which are vertices of K. For each child K' of K, step through all vertices of K', putting them on the separator between K and K' if these are marked in the array. Since vertices of each maximal clique K are traversed at most twice (once when K is a parent, and once when K is a child), the total time spent constructing separators is $O(m + n)$.

Thus, all initial variables can be computed in $O(mn)$ time. Alternatively, we can use matrix multiplication to get the variables. Construct a graph G' with a vertex for each minimal separator S, a vertex for each maximal clique K, and two copies v_1 and v_2 of each vertex v of the graph. Add an edge from v_1 to w_2 if and only if v and w are adjacent in G, and an edge from v_2 to K or S if and only if v is in this maximal clique or minimal separator. The number of neighbors of v in K (or S) in G is the number of paths of length two from v_1 to K (or S) in G'. It is well known that the number of paths of length two from i to j is equal to $M^2[i, j]$, where M is the adjacency matrix of the graph. Since G' has $O(n)$ vertices, all variables can be constructed in $O(n^\alpha)$ time, where α is the coefficient of n in a matrix multiplication algorithm. The best known bound for α is 2.376 [4], and the variables can be computed in this time bound if one is willing to allow the (complex) algorithms used for matrix multiplication in that paper.

Recall that our algorithm assumes that our initial graph is edgeless. We thus need to deal with a chordal graph which is not connected. We begin with a forest of elementary clique trees. While the graph is not connected, the insertability of xy is determined by first testing whether x and y are in the same connected component of G. If not, xy is insertable, and the corresponding cliques trees will be merged (process similar to Case 1 of insertion, but without initial edge $N(x) \cap N(y)$); clearly, the only changes we have then to perform on array *Insertable* are w.l.o.g. $Insertable(x, z) = 1$ for each neighbor z of y. For operation Delete, when the graph is disconnected by an edge deletion, its clique tree is split (by deleting edge $K_x - K_y$ in Case 1 or 3 of deletion) and the child trees are managed separately.

5 Phylogeny

We now discuss the problem of efficient construction of a chordal graph by repeatedly adding a minimum weight edge xy to a current chordal graph G such that $G + xy$ is chordal. The fastest known algorithm for this problem runs in $O(n^4)$ time ([2]); we will reduce the time complexity to $O(n^3)$. The previous algorithm relied on Characterization 1 and used the algorithm ([11]) for maintaining all 2-pairs in a general graph as edges are added in $O(n^4)$ overall time. In this paper, we show that new 2-pairs can be found more efficiently if the graph is known to be chordal.

The problem discussed in this section arose in the context of computational biology. The problem, called phylogeny, involves reconstructing an evolutionary tree, given genetic information of modern species. A correspondence between the phylogeny problem and chordal graphs was first noted in [7].

To summarize very briefly the work most relevant to this paper, one can construct a matrix computing phylogenetic dissimilarity between different pairs of species. If we assume that this matrix is an 'additive tree distance' (which corresponds to the notion that if species A branches off from B and C, and then later B branches off from C, the phylogenetic difference between A and C is equal to the difference between A and B plus the difference between B and C), then the graph G_i formed by including all edges with difference less than each threshold i will be a chordal graph ([7]). In practice, [7] found that the data tends to give graphs which are not chordal, but are "almost" chordal. We want to take this data, and modify it as little as possible to get a chordal graph. [2] proposes an edge addition scheme, starting from an edgeless graph, and repeatedly adding the smallest weight edge which preserves chordality, as an effective way of processing the phylogenetic data.

In that paper, an $O(n^4)$ algorithm was given to solve the problem. We use the results of the previous section to reduce the time complexity to $O(n^3)$. Recall from Theorem 5 that when an edge xy which maintains chordality is added, the only pairs which can change status as far as eligibility for addition are pairs containing x or y.

Therefore, we can maintain a list of edges eligible for addition to the structure, and there will be at most n^3 modifications of the list throughout the running of the algorithm. If the list of eligible edges is stored in increasing order of weight, we will simply choose the first eligible edge for addition at any step, and use the algorithm of the previous section to determine which changes to make in the list in $O(n)$ time.

If the list is stored as a balanced tree, additions and deletions can be made in $O(\log n)$ time, leading to an $O(n^3 \log n)$ algorithm for finding the order in which edges will be added. We will show how to accomplish the same task in $O(n^3)$ time.

Theorem 8. *We can find the complete sequence in which we will add minimum weight edges which preserve chordality in $O(n^3)$ time.*

Proof. As a first step, we sort all possible pairs by weight, and label each pair xy with the position of xy in the sorted list. At each step, we want to find the eligible pair with smallest label to add to our graph.

Instead of keeping the entire list of eligible pairs sorted, we group the eligible pairs as follows. We maintain unordered lists L_i of eligible pairs with thresholds $in+1$ through $(i+1)n$ for each i from 0 to $n-1$; i.e., we keep a list of eligible edges with thresholds in the ranges $[1..n]$, $[n+1..2n]$, ... $[n(n-1)+1..n^2]$. Each eligible pair xy has a pointer to the position of xy in the appropriate list.

To choose the next eligible edge for addition, step through the lists until we find some non-empty list. Since there are $O(n)$ lists, we can find the next eligible pair in $O(n)$ time. Once the appropriate list is found, examine all pairs in the list to find the smallest eligible pair. Since each list contains at most n elements, the total time to find the next eligible pair is $O(n)$.

Using the algorithm of the previous section, we can find all pairs which must be added and deleted from the list of eligible edges in $O(n)$ time. Since each modification clearly takes constant time using our data structures, the total time for adding an edge is $O(n)$.

Since there are $O(n^2)$ edges added, the total time taken to find the sequence of edge additions is $O(n^3)$. $\square$

6 Conclusion

This paper shows that if we start with an edgeless graph, we can maintain chordality as edges are added and deleted using $O(n)$ time for insertions, deletions, and delete queries, and constant time for insert queries. As an application, we show that a triangulation problem arising out of phylogeny can be solved in $O(n^3)$ time.

Acknowledgement

The authors thank Jens Gustedt for drawing their attention to this possible complexity improvement.

References

1. Brandstädt A., Le V. B., Spinrad J. P.: *Graph Classes: A Survey.* Society for Industrial and Applied Mathematics, Philadelphia (USA), (1999).
2. Berry A., Sigayret A., Sinoquet C.: Maximal Sub-Triangulation as Improving Phylogenetic Data. *Proceedings of JIM'03. Soft Computing – Recent Advances in Knowledge Discovery*, G. Govaert, R. Haenle and M. Nadif (eds), **1900**:01 (2005).
3. Buneman P.: A characterization of rigid circuit graphs. *Discrete Mathematics*, **9** (1974) 205–212.
4. Coppersmith D., Winograd S.: On the Asymptotic Complexity of Matrix Multiplication. *SIAM J. Comput.*, **11**:3 (1982) 472–492.
5. Gàvril F.: The intersection graphs of subtrees of trees are exactly the chordal graphs. *Journal of Combinatorial Theory B*, **16** (1974) 47–56.
6. Golumbic M. C.: *Algorithmic Graph Theory and Perfect Graphs.* Academic Press, New York, (1980).
7. Huson D., Nettles S., T. Warnow T.: Obtaining highly accurate topology estimates of evolutionary trees from very short sequences. *Proc. RECOMB'99, Lyon (France)*, (1999) 198–207.
8. Ibarra L.: Fully Dynamic Algorithms for Chordal and Split Graphs. *Proc. 10th Annual ACM-SIAM Synposium on Discrete Algorithms (SODA'99)*, (1999) 923–924.
9. Kearney P., Hayward R., Meijer H.: Inferring evolutionary trees from ordinal data. *Proc. 8th Annual ACM-SIAM Symposium on Discrete Algorithms (SODA'97)*, (1997) 418–426.
10. Spinrad J. P.: *Efficient Graph Representation.* Fields Institute Monographs 19. American Mathematics Society, Providence (RI, USA), (2003) 324p.
11. Spinrad J., Sritharan R.: Algorithms for Weakly Triangulated Graphs. *Discrete Applied Mathematics*, **59** (1995) 181-191.
12. Walter J. R.: *Representations of Rigid Circuit Graphs.* PhD. Dissertation, Wayne State University, Detroit (USA), (1972).

Recognizing HHDS-Free Graphs

Stavros D. Nikolopoulos and Leonidas Palios

Department of Computer Science, University of Ioannina,
GR-45110 Ioannina, Greece
{stavros, palios}@cs.uoi.gr

Abstract. In this paper, we consider the recognition problem on the HHDS-free graphs, a class of homogeneously orderable graphs, and we show that it has polynomial time complexity. In particular, we describe a simple $O(n^2 m)$-time algorithm which determines whether a graph G on n vertices and m edges is HHDS-free. To the best of our knowledge, this is the first polynomial-time algorithm for recognizing this class of graphs.

Keywords: HHD-free graphs, HHDS-free graphs, sun, homogeneously orderable graphs, perfectly orderable graphs, recognition.

1 Introduction

In the late 1990s, Brandstädt, Dragan, and Nicolai [2] defined the *homogeneously orderable graphs* as those graphs admitting a homogeneous elimination order (a vertex ordering $v_1, v_2, \ldots, v_n$ is a *homogeneous elimination ordering* if for every i, v_i is h-extremal in the subgraph induced by $v_i, v_{i+1}, \ldots, v_n$; a vertex v is h-extremal in a graph G if the set $D_2(v)$ of vertices at distance at most 2 from v in G contains a proper homogeneous dominating set, i.e., there exists a set $H \subset D_2(v)$ such that H is a homogeneous set in G and $D_2(v) \subseteq N[H]$). They showed that the class of homogeneously orderable graphs contains the class of *homogeneous graphs* introduced by D'Atri, Moscarini, and Sassano [7]. The larger class of homogeneously orderable graphs seems to be more interesting for several reasons; among these are algorithmic reasons, e.g., the (cardinality) Steiner tree problem is solvable in polynomial time on homogeneously orderable graphs [7].

In this paper, we consider a subclass of homogeneously orderable graphs, namely, the HHDS-free graphs. A graph is *HHDS-free* if it contains no induced hole (i.e., a chordless cycle on ≥ 5 vertices), house, domino (see Figure 1), or sun. In [2], Brandstädt, Dragan, and Nicolai proved that a graph G is HHDS-free if and only if G is hereditary homogeneously orderable, i.e., every induced subgraph of G is homogeneously orderable.

The definition of the class of homogeneously orderable graphs implies that this class is a generalization of both the class of dually chordal and the class of distance-hereditary graphs [2,3]. Bandelt and Mulder [1] showed that a graph G is distance-hereditary if and only if it contains no induced house, hole, domino, or gem; then, since every sun contains a gem [2,3], distance-hereditary graphs

D. Kratsch (Ed.): WG 2005, LNCS 3787, pp. 456–467, 2005.

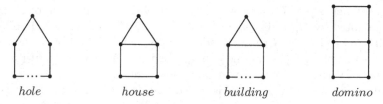

| hole | house | building | domino |

Fig. 1. Some useful graphs

are HHDS-free. Additionally, the HHD-free graphs properly generalize the class of chordal (or triangulated) graphs [9]; a graph is {house,hole,domino}-free or HHD-free if it contains no induced house, hole, or domino. In [11], Hoàng and Khouzam proved that the HHD-free graphs admit a *perfect order*, and thus are *perfectly orderable* [4,13,16]; as a result, the HHDS-free graphs are perfectly orderable as well. A superclass of the HHD-free graphs, which also properly generalizes the class of chordal graphs, is the class of {house,hole}-free or HH-free graphs; Chvátal conjectured [5] and later Hayward [10] proved that the complement $\overline{G}$ of an HH-free graph G is perfectly orderable.

In [3], it is mentioned that the recognition complexity of HHDS-free graphs is open. Yet, several recognition algorithms have been proposed for graph classes that are defined or characterized by forbidden induced holes, houses, or dominos (see [3,9]). Indeed, Hoàng and Khouzam [11], while studying the class of brittle graphs (a well known class of perfectly orderable graphs which contains the HHD-free graphs), showed that the HHD-free graphs can be recognized in $O(n^4)$ time, where n denotes the number of vertices of the input graph. An improved result was obtained by Hoàng and Sritharan [12] who presented an $O(n^3)$-time algorithm for recognizing HH-free graphs and showed that HHD-free graphs can be recognized in $O(n^3)$ time as well; one of the key ingredients in their algorithms is the reduction of a subproblem to the recognition of chordal graphs. Based on the result in [12], recently, Nikolopoulos and Palios [14] presented an $O(\min\{nm\,\alpha(n), nm + n^2\log n\})$-time and $O(n+m)$-space algorithm for recognizing HHD-free graphs, where m is the number of edges of the input graph and $\alpha(n)$ is the very slowly growing inverse of the Ackerman's function.

The main result of this paper is that an HHD-free graph G is also HHDS-free if and only if there is no vertex v of G such that v is the top of a house or a "building" in an auxiliary graph which is a modification of G; a building, which is a generalization of a house, is a cycle on at least 5 vertices with a single chord (i.e., an edge joining two nonconsecutive vertices of the cycle) connecting two vertices of the cycle which are at distance 2 (see Figure 1). This result enables us to describe an $O(n^2m)$-time algorithm for recognizing whether an input graph on n vertices and m edges is HHDS-free. The space required by the algorithm is $O(n^2)$.

2   Theoretical Framework

We consider finite undirected graphs with no loops or multiple edges. Let G be such a graph; then, $V(G)$ and $E(G)$ denote the set of vertices and of edges of G, respectively. Let $S \subseteq V(G)$ be a set of vertices of G; the subgraph of G induced by S is denoted by $G[S]$. The *neighborhood* $N(x)$ of a vertex $x \in V(G)$ is the set of all the vertices of G that are adjacent to x. We use $M(x)$ to denote the set $V(G) - (N(x) \cup \{x\})$ of non-neighbors of x in G. An *independent* (or *stable*) *set* is a set of vertices no two of which are adjacent.

A path $v_0 v_1 \ldots v_k$ of a graph G is called *simple* if none of its vertices occurs more than once; it is called a *cycle* (*simple cycle*) if $v_0 v_k \in E(G)$. A simple path (cycle) is *chordless* if $v_i v_j \notin E(G)$ for any two non-consecutive vertices v_i, v_j in the path (cycle). A chordless path (chordless cycle, respectively) on n vertices is commonly denoted by P_n (C_n, respectively).

A graph is *chordal* (or *triangulated*) if and only if every cycle of length strictly greater than 3 possesses a chord (i.e., an edge joining two nonconsecutive vertices of the cycle) [3,9,17]. The following definition is taken from [3].

Definition 1. [6,8] *A* sun (*or* trampoline) *is a chordal graph G on $2n$ vertices for some $n \geq 3$ whose vertex set can be partitioned into two sets, $U = \{u_0, u_1, \ldots, u_{n-1}\}$ and $W = \{w_0, w_1, \ldots, w_{n-1}\}$, such that W is an independent set and for each i and j, w_j is adjacent to u_i if and only if $i = j$ or $i \equiv j + 1$ mod n.*

A sun on $2k$ vertices is often called a *k-sun*. A sun such that the set U induces a complete graph is called a *complete sun*. It has been shown that every sun contains a complete sun [6,8]; yet, determining whether a graph contains a complete sun does not seem easier than determining whether it contains a sun. We prove the following lemma.

Lemma 1. *Let H be a graph whose vertices can be partitioned into two sets $U = \{u_0, u_1, \ldots, u_{k-1}\}$ and $W = \{w_0, w_1, \ldots, w_{k-1}\}$ of $k \geq 3$ vertices each, such that W is an independent set and for each i and j, w_j is adjacent to u_i if and only if $i = j$ or $i \equiv j + 1$ mod k. Then, H is a sun with partition sets U and W if and only if the subgraph $H[U]$ is chordal and the vertices $u_0, u_1, \ldots, u_{k-1}$ form a cycle $u_0 u_1 \cdots u_{k-1}$.*

Proof. ($\Longrightarrow$) Since H is a sun, then H is chordal and thus the subgraph $H[U]$ is chordal as well. Moreover, for all $i = 0, 1, \ldots, k-1$, the vertices u_i and $u_{i+1 \bmod k}$ are adjacent in H since a chordless path from $u_{i+1 \bmod k}$ to u_i in the (connected) graph induced by $\{u_{i+1 \bmod k}, w_{i+1 \bmod k}, \ldots, u_{i-1}, w_{i-1}, u_i\}$ in H has to be of length 1; otherwise, the vertices of the path along with vertex w_i would induce a chordless cycle on 4 or more vertices, a contradiction to the chordality of H. ($\Longleftarrow$) Since $H[U]$ is chordal, the lemma follows easily from the fact that no w_i ($0 \leq i < k$) participates in a chordless cycle on 4 or more vertices since w_i's only neighbors, u_i and $u_{i+1 \bmod k}$, are adjacent in H. ∎

Let G be a graph and let v be an arbitrary vertex of G. Let us define the following set of non-edges of G

$$E_v = \{ xz \mid x, z \in M(v) \text{ and } \exists y \in M(v) \text{ such that } xyz \text{ is a } P_3 \text{ of } G \}$$

which we call P_3-*edges*. Then, we construct the graph $\widehat{G}_v$ from G as follows:

$$V(\widehat{G}_v) \;=\; V(G) \qquad\text{and}\qquad E(\widehat{G}_v) \;=\; E(G) \cup E_v.$$

Note that the definition of P_3-edges implies that $E(G) \cap E_v = \emptyset$. If the graph G has n vertices and m edges, then the graph $\widehat{G}_v$ has n vertices and $O(n^2)$ edges.

Definition 2.

▷ *We collectively call a house or a building a* generalized house *or* g-house *for short.*

▷ *If vertex v is the top of a house or a building, then v is the* top *of the g-house. If v at the top is adjacent to vertices u, w in the g-house, we say that the* roof *of the g-house is $(v; u, w)$. The vertices of the g-house that do not belong to its roof form a chordless path which we call the* base *of the g-house.*

▷ *A g-house is* shorter *than another g-house if it involves fewer vertices.*

Our HHDS-free graph recognition algorithm relies on the following theorem.

Theorem 1. *Let G be an HHD-free graph. The graph G contains a sun if and only if there exists a vertex v such that the graph $\widehat{G}_v$ defined above with respect to v contains a house or a building with v at its top.*

Proof. ($\Longrightarrow$) Suppose that the graph G contains a sun induced by the sets of vertices $U = \{u_0, u_1, \ldots, u_{k-1}\}$ and $W = \{w_0, w_1, \ldots, w_{k-1}\}$, where $k \geq 3$ (see Definition 1). Then, in the graph $\widehat{G}_{w_0}$, the vertices $w_0, u_0, u_1, w_1, w_2, \ldots, w_{k-1}$ induce a house or a building with vertex w_0 at its top (see Figure 2 for an example where $k = 5$; dashed edges indicate P_3-edges); note that $u_0 u_1 \in E(G)$ (see Lemma 1), that the vertices u_0 and u_1 are not adjacent to any of the vertices $w_1, w_2, \ldots, w_{k-2}$ and $w_2, w_3, \ldots, w_{k-1}$, respectively, and that, for all $i = 1, 2, \ldots, k - 2$, the vertices w_i and w_{i+1} induce a P_3-edge.

($\Longleftarrow$) Suppose that there exists a vertex v which is the top of a house or a building in $\widehat{G}_v$, i.e., v is the top of a g-house. Then, the following holds:

Fact 1. If the vertex v is the top of a g-house in the graph $\widehat{G}_v$, with roof $(v; u, w)$, then every edge in the base of a *shortest* g-house with roof $(v; u, w)$ is a P_3-edge.

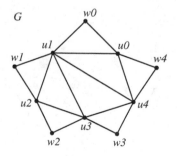
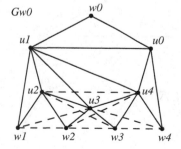

Fig. 2

Fact 1 is established in Lemma 2. Thus, if a shortest g-house with roof $(v; u, w)$ has base $p_1 p_2 \cdots p_k$, then each $p_i p_{i+1}$ $(1 \leq i \leq k-1)$ is a P_3-edge; let us replace each such edge with a corresponding P_3 $p_i q_i p_{i+1}$ in G. Then, from the fact that we are considering a shortest g-house, we conclude that for $i = 1, 2, \ldots, k-1$, the vertex q_i is not adjacent to any of the vertices in $\{p_1, p_2, \ldots, p_{i-1}, p_{i+2}, \ldots, p_k\}$ (as in the proof of Lemma 2), which implies that the q_is are all distinct (note that the q_is may be arbitrarily adjacent to one other); the situation is depicted in Figure 3 where dashed lines indicate potential edges.

Additionally, vertex u is adjacent to at least one of the vertices $q_1, q_2,$ $\ldots, q_{k-1}$. If u were not adjacent to any of them, then if x is the leftmost neighbor of w among $q_1, q_2, \ldots, q_{k-1}, p_k$ and if ρ is a chordless path from p_1 to x in the (connected) graph induced by the vertices $\{p_1, q_1, p_2, q_2, \ldots, x\}$ in G, the vertices v, u, w, and the vertices of the path ρ induce a house or a building in G (with v at its top), which contradicts the fact that the graph G is HHD-free. Thus, u is adjacent to at least one q_i. In fact, we can show the following:

Fact 2. There exists an integer r, where $1 \leq r \leq k - 1$, such that the vertex u is adjacent to precisely $q_1, q_2, \ldots, q_r$ among the q_is, otherwise the graph G contains a sun.

Fact 2 is established in Lemma 6 (case (b)) with the aid of Lemma 4: since u is adjacent to both p_1 and a vertex q_i, then Lemma 4 implies that it is also adjacent to q_1; then, for $r = \max\{\, j \mid u q_j \in E(G)\,\}$, Lemma 6 (case (b)) implies that if there exists a vertex q_i $(2 \leq i \leq r - 1)$ which is not adjacent to u, then the graph G contains a sun, as desired.

So, let us consider the case where the vertex u is adjacent to each of the vertices $q_1, q_2, \ldots, q_r$, where $1 \leq r \leq k-1$. Similarly, we assume that there exists an integer ℓ, where $1 \leq \ell \leq k-1$, such that the vertex w is adjacent to each of the vertices $q_\ell, q_{\ell+1}, \ldots, q_{k-1}$. Then, it has to be that $r \geq \ell$; if $r < \ell$, then the vertices v, u, w, and the vertices of a chordless path from q_r to q_ℓ in the (connected) graph induced by $\{q_r, p_{r+1}, q_{r+1}, \ldots, p_\ell, q_\ell\}$ induce a house or a building in G, a contradiction. In fact, $r = k - 1$ and $\ell = 1$, i.e., the vertices u, w are adjacent to each of the vertices $q_1, q_2, \ldots, q_{k-1}$. Suppose for contradiction that $r \leq k - 2$ which implies that $k \geq 3$ since $r \geq 1$; then, because $r \geq \ell$, the vertex w is

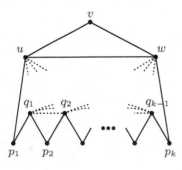

Fig. 3

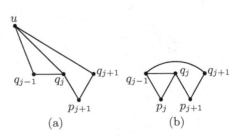

Fig. 4

adjacent to both q_{k-2} and q_{k-1}. Moreover, $q_{k-2}q_{k-1} \in E(G)$ (for otherwise the vertices $w, q_{k-2}, p_{k-1}, q_{k-1}, p_k$ would induce a house in G with vertex p_k at its top, a contradiction); then, the vertices $p_{k-2}, q_{k-2}, q_{k-1} \in M(v)$ induce a P_3 in G, that is, $p_{k-2}q_{k-1}$ would be a P_3-edge in $\widehat{G}_v$, which implies that the vertices $v, u, p_1, p_2, \ldots, p_{k-2}, q_{k-1}, w$ induce a g-house in $\widehat{G}_v$ with roof (v, u, w); note that q_{k-1} is not adjacent to $p_1, p_2, \ldots, p_{k-3}$ nor to u. This, however, contradicts the minimality of the g-house induced by $v, u, p_1, p_2, \ldots, p_k, w$. Thus, the assumption that $r \leq k-2$ led us to a contradiction. Hence, $r = k-1$ (i.e., vertex u is adjacent to each of the vertices $q_1, q_2, \ldots, q_{k-1}$); similarly, vertex w is adjacent to each of these vertices as well.

If there exists a vertex q_i that is adjacent to a vertex q_j but is not adjacent to a vertex $q_{j'}$, where $1 \leq i < j' < j \leq k-1$, then clearly $k \geq 4$ and Lemma 6 along with Lemma 4 imply that the graph G contains a sun: since q_i is adjacent to both p_{i+1} and q_j, then Lemma 4 implies that it is also adjacent to q_{i+1} (note that the graph G is HHD-free and contains the path $p_{i+1}q_{i+1}p_{i+2}q_{i+2}\cdots p_jq_j$, with chords only between q_is, and the vertex q_i is not adjacent to any of $p_{i+2}, p_{i+3}, \ldots, p_j$); then, Lemma 6 (case (b)) implies that since vertex q_i is not adjacent to vertex $q_{j'}$, where $i + 2 \leq j' \leq j - 1$, the graph G contains a sun.

Suppose now that no vertex q_i as in the previous paragraph exists; that is, for all $i = 1, 2, \ldots, k-2$, if q_i is adjacent to a vertex q_j, where $1 \leq i < j \leq k-1$, then q_i is adjacent to each of $q_{i+1}, q_{i+2}, \ldots, q_j$. Then Lemma 5 implies that the subgraph of G induced by the vertices $w, u, q_1, q_2, \ldots, q_{k-1}$ is chordal; recall that $uw \in E(G)$ and both u and w are adjacent to each of the vertices $q_1, q_2, \ldots, q_{k-1}$. Additionally, we take advantage of the fact that u is adjacent to each of the vertices $q_1, q_2, \ldots, q_{k-1}$ in order to show by induction on i that $q_iq_{i+1} \in E(G)$ for all $i = 1, 2, \ldots, k-2$. For the basis step, we observe that if $q_1q_2 \notin E(G)$ then the vertices u, p_1, q_1, p_2, q_2 induce a house in G (with vertex p_1 at its top), a contradiction. For the inductive step, we assume that $q_{j-1}q_j \in E(G)$ where $j \geq 2$, and suppose for contradiction that $q_jq_{j+1} \notin E(G)$; if $q_{j-1}q_{j+1} \notin E(G)$, then the vertices $u, q_{j-1}, q_j, p_{j+1}, q_{j+1}$ induce a house in G with vertex q_{j-1} at its top (Figure 4(a)), which leads to a contradiction, whereas if $q_{j-1}q_{j+1} \in E(G)$, then the vertices $q_{j-1}, p_j, q_j, p_{j+1}, q_{j+1}$ induce a house in G with vertex p_j at its top (Figure 4(b)), a contradiction again. Therefore, $q_jq_{j+1} \in E(G)$, and from the induction, $q_iq_{i+1} \in E(G)$ for all $i = 1, 2, \ldots, k-2$. This result, the chordality of the subgraph $G[\{w, u, q_1, q_2, \ldots, q_{k-1}\}]$, the fact that $uw \in E(G)$, $uq_1 \in E(G)$, and $wq_{k-1} \in E(G)$, and Lemma 1 imply that the subgraph of G induced by the vertices $v, u, p_1, q_1, p_2, q_2, \ldots, p_{k-1}, q_{k-1}, p_k, w$ is a sun with partition sets $U = \{u, q_1, q_2, \ldots, q_{k-1}, w\}$ and $W = \{v, p_1, p_2, \ldots, p_k\}$. ∎

Lemma 2. *Let G be an HHD-free graph, v a vertex of G, and $\widehat{G}_v$ be the auxiliary graph defined above with respect to v. If the vertex v is the top of a g-house in the graph $\widehat{G}_v$ and if u and w are the neighbors of v in the g-house, then every edge in the base of a shortest g-house with roof $(v; u, w)$ is a P_3-edge.*

Proof. Let a shortest g-house with roof $(v; u, w)$ have base $p_1p_2 \cdots p_k$, where $k \geq 2$ (Figure 5(a)). Since G does not contain a house or a hole, the path $p_1 \cdots p_k$

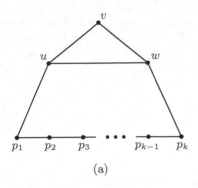
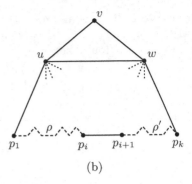

(a)   (b)

Fig. 5

contains P_3-edges; let us replace each P_3-edge p_ip_{i+1} $(1 \le i < k)$ by a corresponding P_3 $p_iq_ip_{i+1}$ of G. Then, each such vertex q_i is not adjacent to any vertex in $\{p_1, \ldots, p_{i-1}, p_{i+2}, \ldots, p_k\}$: if q_i were adjacent to p_j, for some $j \in \{1, 2, \ldots, i-1\}$ then the vertices p_j, q_i, p_{i+1} would induce a P_3 in G, and thus p_jp_{i+1} would be a P_3-edge, which would imply that the vertices $v, u, p_1, \ldots, p_j, p_{i+1}, \ldots, p_k, w$ would induce a g-house with roof $(v; u, w)$ in $\widehat{G}_v$, in contradiction to the minimality of the g-house induced by $v, u, p_1, p_2, \ldots, p_k, w$; a similar argument leads to a contradiction if q_i were adjacent to p_j, for some $j \in \{i+2, i+3, \ldots, k\}$. The fact that q_i is not adjacent to any vertex in $\{p_1, \ldots, p_{i-1}, p_{i+2}, \ldots, p_k\}$ also implies that the vertices q_i are all different.

We will show next that every edge p_ip_{i+1} is a P_3-edge. Suppose for contradiction that p_ip_{i+1} is not a P_3-edge; hence, it is an edge of G instead. Consider a chordless path ρ in G from p_1 to p_i in the (connected) graph induced by $\{p_1, q_1, p_2, \ldots, q_{i-1}, p_i\}$ and a chordless path ρ' from p_{i+1} to p_k in the (connected) graph induced by $\{p_{i+1}, q_{i+1}, p_{i+2}, \ldots, q_{k-1}, p_k\}$. We show that the concatenation of the path ρ, the edge p_ip_{i+1}, and the path ρ' forms a chordless path in G (see Figure 5(b)). If there were a chord, this would have been an edge $q_\ell q_r$, where $\ell < i$ and $r \ge i+1$. Let us consider an edge $q_\ell q_r$ that minimizes the difference $r - \ell$; then, the vertices of the path ρ from q_ℓ to p_i, and the vertices of the path ρ' from p_{i+1} to q_r induce a cycle in G. In fact, they induce a chordless cycle due to the minimality of $q_\ell q_r$, the chordlessness of ρ and ρ', and the fact that p_i sees none of the vertices of ρ' except for p_{i+1}, and that p_{i+1} sees none of the vertices of ρ except for p_i. Additionally, because G contains no hole, it must be the case that $\ell = i-1$ and $r = i+1$, i.e., the vertices $q_\ell, p_i, p_{i+1}, q_r$ form a C_4. Then, the vertices q_ℓ, q_r, p_{r+1} induce a P_3 in G and thus the edge $q_\ell p_{r+1}$ is a P_3-edge in $\widehat{G}_v$. If neither u nor w see q_ℓ then the vertices $v, u, p_1, p_2, \ldots, p_\ell, q_\ell, p_{r+1}, p_{r+2}, \ldots, p_k, w$ would form a g-house in $\widehat{G}_v$ with roof $(v; u, w)$ which is shorter than the g-house induced by $v, u, p_1, \ldots, p_k, w$, in contradiction to the minimality of the latter g-house; hence, at least one of u, w sees q_ℓ, and similarly at least one of u, w sees q_r. On the other hand, neither u nor w see both q_ℓ and q_r, since G does not contain a house. Therefore, either u sees q_ℓ and w sees q_r or u sees q_r and w sees q_ℓ; in either case, the vertices

v, u, q_ℓ, q_r, w induce a house (recall that $uw \in E(G)$); a contradiction. Thus, no chord exists, and the concatenation of the path ρ, the edge $p_i p_{i+1}$, and the path ρ' forms a chordless path π in G (Figure 5(b)).

The vertex u is not adjacent to any vertex in the path ρ'. If it were, let t' be the leftmost such vertex; clearly, $t' \neq p_{i+1}$. Moreover, let t be the rightmost vertex of ρ which is adjacent to u; t is well defined since $up_1 \in E(G)$ and $t \neq p_i$. But then, the vertex u and the vertices in the part of the path π from t to t' induce a hole in G, which leads to a contradiction; thus, u is not adjacent to any vertex in ρ'. Similarly, w is not adjacent to any vertex in ρ. But then G contains a hole: it is induced by the vertices u, w, and the vertices of the path π from the rightmost neighbor of u in ρ (which is to the left of p_i) to the leftmost neighbor of w in ρ' (which is to the right of p_{i+1}). This however contradicts the fact that G is HHD-free, and therefore we conclude that the base of the g-house induced by $u, v, w, p_1, p_2, \ldots, p_k$ consists entirely of P_3-edges. ∎

Lemma 3. *Let G be a graph which contains a C_4 $abcd$ and a path ρ from c to d (different from the path cd) whose vertices other than its endpoints c and d are adjacent neither to a nor to b. Then, the graph G contains a hole, a house, or a domino.*

Lemma 4. *Let G be an HHD-free graph that contains a path $p_s q_s p_{s+1} q_{s+1} \cdots p_t q_t$, where $t \geq s + 1$, with chords only between $q_i s$, and let x be a vertex of G that is adjacent to p_s and is not adjacent to any of $p_{s+1}, p_{s+2}, \ldots, p_t$. If the vertex x is adjacent to q_t, then it is also adjacent to q_s.*

Proof. Suppose for contradiction that $xq_s \notin E(G)$. Let $t' = \min\{i \mid s + 1 \leq i \leq t$ and $xq_i \in E(G)\}$; the vertex $q_{t'}$ is well defined since x is adjacent to q_t. Then, $q_s q_{t'} \in E(G)$, otherwise the length of a chordless path from q_s to $q_{t'}$ in the (connected) graph induced by $\{q_s, p_{s+1}, q_{s+1}, \ldots, p_{t'}, q_{t'}\}$ in G would be of length at least 2 and the vertices of the path along with x and p_s would induce a hole in G, a contradiction. But then, the vertices $x, p_s, q_s, q_{t'}$ induce a C_4 in G and G contains the path $q_s p_{s+1} q_{s+1} \cdots p_{t'} q_{t'}$ whose vertices other than its endpoints are adjacent neither to x nor to p_s. Thus, Lemma 3 applies, implying that the graph G contains a hole, a house, or a domino, in contradiction to the fact that G is HHD-free. Therefore, the vertex x is adjacent to q_s. ∎

Lemma 5. *Let H be a graph that does not contain holes, and $v_1, v_2, \ldots, v_k$ ($k \geq 3$) be an ordering of a subset of vertices of H such that, for all $i = 1, 2, \ldots, k-1$, if v_i is adjacent to v_j, where $i < j \leq k$, then v_i is adjacent to each of the vertices $v_{i+1}, v_{i+2}, \ldots, v_j$. Then, the subgraph of H induced by the vertices $v_1, v_2, \ldots, v_k$ is chordal.*

Proof. Since the graph H does not contain holes, we only need to show that the subgraph induced by the vertices $v_1, v_2, \ldots, v_k$ does not contain a C_4. Suppose for contradiction that it contained a C_4, say, $v_a v_b v_c v_d$, and suppose without loss of generality that $a = \min\{a, b, c, d\}$. Then, we distinguish the following cases:

(i) $b = \max\{a, b, c, d\}$: then, v_a is adjacent to v_b but is not adjacent to v_c and yet $c < b$ (see Figure 6(a)), a contradiction;

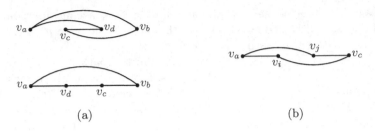

Fig. 6. Different cases for the C_4 $v_a v_b v_c v_d$

(ii) $c = \max\{a, b, c, d\}$: then, if $i = \min\{b, d\}$ and $j = \max\{b, d\}$, v_i is adjacent to v_c but is not adjacent to v_j and yet $i < j < c$ (see Figure 6(b)), a contradiction;

(iii) $d = \max\{a, b, c, d\}$: the case is similar to case (i) and leads to a contradiction.

In all cases, we reached a contradiction, which implies that the subgraph of H induced by the vertices $v_1, v_2, \ldots, v_k$ is chordal. ∎

Lemma 6. *Let G be an HHD-free graph that contains a path $q_s p_{s+1} q_{s+1} \cdots p_t q_t$, where $t \geq s + 2$, with chords only between q_is, and let x be a vertex of G that is adjacent to q_s and q_t, and is not adjacent to any of $p_{s+1}, p_{s+2}, \ldots, p_t$.*

(a) *Suppose that the vertex x is not adjacent to the vertices $q_{s+1}, q_{s+2}, \ldots, q_{t-1}$, and that for $i = s, s+1, \ldots, t-1$, if the vertex q_i is adjacent to q_j (where $i < j \leq t$) then it is adjacent to each of the vertices $q_{i+1}, q_{i+2}, \ldots, q_j$. Then, the vertices $x, q_s, p_{s+1}, q_{s+1}, \ldots, p_t, q_t$ induce a sun in G.*

(b) *If there exists a vertex q_i ($s + 1 \leq i \leq t - 1$) that is not adjacent to x, then the graph G contains a sun.*

Proof. (a) First, the set $\{q_s, q_{s+1}, \ldots, q_t\}$ contains at least 3 vertices. Next, due to the property of the q_is, Lemma 5 implies that the subgraph of G induced by the vertices $q_s, q_{s+1}, \ldots, q_t$ is chordal. In light of Lemma 1 and of the fact that the vertex x is adjacent to q_s and q_t only, and each vertex p_i ($s + 1 \leq i \leq t$) is adjacent to q_{i-1} and q_i only, we need only prove that the vertices $q_s, q_{s+1}, \ldots, q_t$ induce a cycle $q_s q_{s+1} \cdots q_t$ in G.

We begin by showing that the vertex q_s is adjacent to at least one vertex in $\{q_s, q_{s+1}, \ldots, q_t\}$; if it were not, then the vertices x, q_s, p_{s+1}, and the vertices of a chordless path from q_{s+1} to q_t in the (connected) graph induced by $\{q_{s+1}, p_{s+2}, q_{s+2}, \ldots, p_t, q_t\}$ would induce a hole in G, a contradiction. If q_ℓ is that vertex, i.e., $q_s q_\ell \in E(G)$, then $q_s q_t \in E(G)$: this is trivially true if $q_\ell = q_t$; if $q_\ell \neq q_t$, then because the graph G contains the path $x q_t p_t q_{t-1} \cdots p_{\ell+1} q_\ell$, where $\ell \leq t - 1$, with chords only between q_is, and the vertex q_s is adjacent to x and q_ℓ but is not adjacent to any of $p_t, p_{t-1}, \ldots, p_{\ell+1}$, Lemma 4 applies, implying that q_s is adjacent to q_t in G. From this fact and from the property of the vertices q_i ($s \leq i < t$) that if q_i is adjacent to q_j, where $i < j \leq t$, then q_i is adjacent to each of the vertices $q_{i+1}, q_{i+2}, \ldots, q_j$, we conclude that q_s is adjacent to each of the vertices $q_{s+1}, q_{s+2}, \ldots, q_t$; this in turn enables us to additionally show (by induction on i) that $q_i q_{i+1} \in E(G)$ for all $i = s+1, s+2, \ldots, t-1$. For the basis step, we note that if $q_{s+1} q_{s+2} \notin E(G)$, then the vertices $q_s, p_{s+1}, q_{s+1}, p_{s+2}, q_{s+2}$ induce a house in G with vertex p_{s+1} at its top, a contradiction. For the inductive step, assume that $q_{j-1} q_j \in E(G)$ where $j \geq s+2$. We show that $q_j q_{j+1} \in E(G)$; if not, then the vertices $q_s, q_{j-1}, q_j, p_{j+1}, q_{j+1}$ induce a house in G with vertex q_{j-1} at its top, a contradiction. Our inductive proof is complete implying that $q_i q_{i+1} \in E(G)$ for all $i = s+1, s+2, \ldots, t-1$; then, because $q_s q_{s+1} \in E(G)$ and $q_s q_t \in E(G)$, we have that the vertices $q_s, q_{s+1}, \ldots, q_t$ indeed induce a cycle $q_s q_{s+1} \cdots q_t$ in G.

(b) Since the vertex x is adjacent to q_s and q_t, and is not adjacent to a vertex in $\{q_{s+1}, q_{s+1}, \ldots, q_{t-1}\}$, we can find vertices q_ℓ, q_r, where $s \leq \ell < r \leq t$, such that x is adjacent to q_ℓ and q_r but is not adjacent to any of $q_{\ell+1}, q_{\ell+2}, \ldots, q_{r-1}$. Then, if for each vertex q_i ($\ell \leq i \leq r-1$), the fact that q_i is adjacent to a vertex q_j, where $i < j \leq r$, implies that q_i is adjacent to each of the vertices $q_{i+1}, q_{i+2}, \ldots, q_j$, Lemma 6 (case (a)) applies, implying that the vertices $x, q_\ell, p_{\ell+1}, q_{\ell+1}, \ldots, p_r, q_r$ induce a sun in G. Suppose now that there exists a vertex q_i ($\ell \leq i \leq r-1$) that is adjacent to a vertex q_j and is not adjacent to a vertex $q_{j'}$, where $i < j' < j \leq r$. Let us collect all such vertices in a (non-empty) set S.

For each vertex q_i in S (which is adjacent, say, to q_{j_i} where $i + 1 < j_i$), Lemma 4 implies that q_i is adjacent to q_{i+1}; note that G is HHD-free and contains the path $p_{i+1} q_{i+1} \cdots p_{j_i} q_{j_i}$, and q_i is adjacent to p_{i+1} and q_{j_i}. Then, for each vertex $q_i \in S$, we can find indices ℓ_i and r_i where $i < \ell_i < r_i \leq r$, such that q_i is adjacent to q_{ℓ_i} and q_{r_i} but is not adjacent to any of the vertices $q_{\ell_i+1}, q_{\ell_i+2}, \ldots, q_{r_i-1}$, and the difference $r_i - \ell_i$ is minimized. Let $q_{\hat{i}}$ be a vertex in S such that $r_{\hat{i}} - \ell_{\hat{i}} = \min_{q_i \in S}\{r_i - \ell_i\}$; the minimality of $q_{\hat{i}}$ implies that for $i = \ell_{\hat{i}}, \ell_{\hat{i}} + 1, \ldots, r_{\hat{i}} - 1$, if the vertex q_i is adjacent to q_j (where $i < j \leq r_{\hat{i}}$) then it is adjacent to each of the vertices $q_{i+1}, q_{i+2}, \ldots, q_j$. This, the fact that the graph G contains the path $q_{\ell_{\hat{i}}} p_{\ell_{\hat{i}}+1} q_{\ell_{\hat{i}}+1} \cdots p_{r_{\hat{i}}} q_{r_{\hat{i}}}$, where $r_{\hat{i}} \geq \ell_{\hat{i}} + 2$, with chords only between q_is, and the fact that vertex $q_{\hat{i}}$ is adjacent to $q_{\ell_{\hat{i}}}$ and $q_{r_{\hat{i}}}$ but is not adjacent to any of $q_{\ell_{\hat{i}}+1}, q_{\ell_{\hat{i}}+2}, \ldots, q_{r_{\hat{i}}-1}$ imply that Lemma 6 (case (a)) applies, and therefore, the vertices $q_{\hat{i}}, q_{\ell_{\hat{i}}}, p_{\ell_{\hat{i}}+1}, q_{\ell_{\hat{i}}+1}, \ldots, p_{r_{\hat{i}}}, q_{r_{\hat{i}}}$ induce a sun in G. ∎

3   The Algorithm

The recognition algorithm takes advantage of Theorem 1. We start by checking whether the input graph G is HHD-free. If it is not, then clearly G is not HHDS-free. Otherwise, for each vertex v of G, we construct the auxiliary graph $\widehat{G}_v$ and check whether v is the top of a house or a building in $\widehat{G}_v$; if this is so for any vertex v, then G is not HHDS-free. We note that in order to check whether v is the top of a house or a building in $\widehat{G}_v$, we can use the algorithms in [12] (Algorithm High) and [14] (Algorithm Not-in-HHB) which for a graph H and a vertex x return true if and only if the vertex x belongs to a hole or is the top of a house or a building in H; Lemma 7 establishes that v does not belong to a hole in $\widehat{G}_v$ if G is HHD-free.

Lemma 7. *Let G be an HHD-free graph, v a vertex of G, and $\widehat{G}_v$ be the auxiliary graph defined in Section 2 with respect to v. Then, the vertex v does not belong to a hole in the graph $\widehat{G}_v$.*

Formally, the recognition algorithm works as follows:

Algorithm Rec-HHDS-free

1. **if** G is not HHD-free
   then return "G is not HHDS-free";
2. **for** each vertex v of G **do**
   2.1   construct the auxiliary graph $\widehat{G}_v$;
   2.2   **if** v is the top of a house or a building in $\widehat{G}_v$
           then return "G is not HHDS-free"; $\{G$ contains a sun$\}$
3. **return** "G is HHDS-free".

The correctness of the algorithm follows from Theorem 1.

Time and Space Complexity.   Let n and m be the number of vertices and edges of the input graph G. Step 1 can be executed in $O(\min\{nm\alpha(n), nm + n^2 \log n\})$ time and $O(n + m)$ space [14]. In Step 2, the construction of the auxiliary graph $\widehat{G}_v$ can be done in $O(nm)$ time and requires $O(n^2)$ space. Then, we check whether vertex v is the top of a house or a building by means of the Algorithm Not-in-HHB [14], which for a graph on N vertices and M edges takes $O(N + \min\{M\alpha(N), M + N \log N\})$ time and $O(N + M)$ space; since $\widehat{G}_v$ has n vertices and $O(n^2)$ edges, Substep 2.2 takes $O(n^2)$ time and space. Thus, the entire execution of Step 2 for all the vertices of G takes $O(n^2 m)$ time and $O(n^2)$ space. Step 3 takes constant time and space.

Therefore, we obtain the following theorem.

Theorem 2. *Let G be an undirected graph on n vertices and m edges. Then, there exists an algorithm for determining whether G is an HHDS-free graph in $O(n^2 m)$ time and $O(n^2)$ space.*

4   Concluding Remarks

We have presented a recognition algorithm for the class of HHDS-free graphs running in $O(n^2 m)$ time with $O(n^2)$ space. To the best of our knowledge, it is the first polynomial-time algorithm for recognizing the class of HHDS-free graphs. The proposed recognition algorithm can be augmented to provide a certificate (an induced house, hole, domino, or sun) in linear additional time and space whenever it decides that the input graph is not HHDS-free: for a house, hole, or domino, see [15]; for a sun, we take advantage of the proof of Theorem 1, which is constructive. Finally, the use of P_3-edges enables us to recognize {house, hole, domino, 3-sun}-free graphs in $O(n^2 m)$ time and $O(n)$ space.

Acknowledgment. The authors would like to thank Andreas Brandstädt for bringing this problem to their attention and for useful discussions. They would also like to thank the anonymous referees for their constructive comments.

References

1. H.-J. Bandelt and H.M. Mulder, Distance-hereditary graphs, *J. Combin. Theory B* **41**, 182–208, 1986.
2. A. Brandstädt, F.F. Dragan, and F. Nicolai, Homogeneously orderable graphs, *Theoret. Comput. Sci.* **172**, 209–232, 1997.
3. A. Brandstädt, V.B. Le, and J.P. Spinrad, *Graph Classes: A Survey*, SIAM Monographs on Discrete Mathematics and Applications, 1999.
4. V. Chvátal, Perfectly ordered graphs, *Annals of Discrete Math.* **21**, 63–65, 1984.
5. V. Chvátal, A class of perfectly orderable graphs, Report 89573-OR, Forschungsinstitut für Diskrete Mathematik, Bonn, 1989.
6. G.J. Chang, *k-Domination and Graph Covering Problems*, Ph.D Thesis, School of OR and IE, Cornell University, Ithaca, NY, 1982.
7. A. D'Atri, M. Moscarini, and A. Sassano, The Steiner tree problem and homogeneous sets, *Lecture Notes in Comput. Sci.* **324**, 249–261, 1988.
8. M. Farber, Characterizations of strongly chordal graphs, *Discrete Math.* **43**, 173–189, 1983.
9. M.C. Golumbic, *Algorithmic Graph Theory and Perfect Graphs*, Academic Press, Inc., 1980.
10. R. Hayward, Meyniel weakly triangulated graphs I: co-perfect orderability, *Discrete Appl. Math.* **73**, 199–210, 1997.
11. C.T. Hoàng and N. Khouzam, On brittle graphs, *J. Graph Theory* **12**, 391–404, 1988.
12. C.T. Hoàng and R. Sritharan, Finding houses and holes in graphs, *Theoret. Comput. Sci.* **259**, 233–244, 2001.
13. M. Middendorf and F. Pfeiffer, On the complexity of recognizing perfectly orderable graphs, *Discrete Math.* **80**, 327–333, 1990.
14. S.D. Nikolopoulos and L. Palios, Recognizing HHD-free and Welsh-Powell opposition graphs, *Proc. 30th Workshop on Graph Theoretic Concepts in Computer Science (WG'04)*, LNCS 3353, 105–116, 2004.
15. S.D. Nikolopoulos and L. Palios, Recognizing HHD-free and Welsh-Powell opposition graphs, Technical Report TR-16-04, Dept. of Computer Science, University of Ioannina, 2004.
16. S. Olariu, All variations on perfectly orderable graphs, *J. Combin. Theory* Ser. B **45**, 150–159, 1988.
17. D.J. Rose, R.E. Tarjan, and G.S. Lueker, Algorithmic aspects of vertex elimination on graphs, *SIAM J. Comput.* **5**, 266–283, 1976.

Author Index

Lecture Notes in Computer Science

For information about Vols. 1–3748

please contact your bookseller or Springer

Vol. 3795: H. Zhuge, G.C. Fox (Eds.), Grid and Cooperative Computing - GCC 2005. XXI, 1203 pages. 2005.

Vol. 3794: X. Jia, J. Wu, Y. He (Eds.), Mobile Ad-hoc and Sensor Networks. XX, 1136 pages. 2005.

Vol. 3793: T. Conte, N. Navarro, W.-m.W. Hwu, M. Valero, T. Ungerer (Eds.), High Performance Embedded Architectures and Compilers. XIII, 317 pages. 2005.

Vol. 3792: I. Richardson, P. Abrahamsson, R. Messnarz (Eds.), Software Process Improvement. VIII, 215 pages. 2005.

Vol. 3791: A. Adi, S. Stoutenburg, S. Tabet (Eds.), Rules and Rule Markup Languages for the Semantic Web. X, 225 pages. 2005.

Vol. 3790: G. Alonso (Ed.), Middleware 2005. XIII, 443 pages. 2005.

Vol. 3789: A. Gelbukh, Á. de Albornoz, H. Terashima-Marín (Eds.), MICAI 2005: Advances in Artificial Intelligence. XXVI, 1198 pages. 2005. (Sublibrary LNAI).

Vol. 3788: B. Roy (Ed.), Advances in Cryptology - ASIACRYPT 2005. XIV, 703 pages. 2005.

Vol. 3787: D. Kratsch (Ed.), Graph-Theoretic Concepts in Computer Science. XIV, 470 pages. 2005.

Vol. 3785: K.-K. Lau, R. Banach (Eds.), Formal Methods and Software Engineering. XIV, 496 pages. 2005.

Vol. 3784: J. Tao, T. Tan, R.W. Picard (Eds.), Affective Computing and Intelligent Interaction. XIX, 1008 pages. 2005.

Vol. 3783: S. Qing, W. Mao, J. Lopez, G. Wang (Eds.), Information and Communications Security. XIV, 492 pages. 2005.

Vol. 3781: S.Z. Li, Z. Sun, T. Tan, S. Pankanti, G. Chollet, D. Zhang (Eds.), Advances in Biometric Person Authentication. XI, 250 pages. 2005.

Vol. 3780: K. Yi (Ed.), Programming Languages and Systems. XI, 435 pages. 2005.

Vol. 3779: H. Jin, D. Reed, W. Jiang (Eds.), Network and Parallel Computing. XV, 513 pages. 2005.

Vol. 3778: C. Atkinson, C. Bunse, H.-G. Gross, C. Peper (Eds.), Component-Based Software Development for Embedded Systems. VIII, 345 pages. 2005.

Vol. 3777: O.B. Lupanov, O.M. Kasim-Zade, A.V. Chaskin, K. Steinhöfel (Eds.), Stochastic Algorithms: Foundations and Applications. VIII, 239 pages. 2005.

Vol. 3776: S.K. Pal, S. Bandyopadhyay, S. Biswas (Eds.), Pattern Recognition and Machine Intelligence. XXIV, 808 pages. 2005.

Vol. 3775: J. Schönwälder, J. Serrat (Eds.), Ambient Networks. XIII, 281 pages. 2005.

Vol. 3774: G. Bierman, C. Koch (Eds.), Database Programming Languages. X, 295 pages. 2005.

Vol. 3773: A. Sanfeliu, M.L. Cortés (Eds.), Progress in Pattern Recognition, Image Analysis and Applications. XX, 1094 pages. 2005.

Vol. 3772: M. Consens, G. Navarro (Eds.), String Processing and Information Retrieval. XIV, 406 pages. 2005.

Vol. 3771: J.M.T. Romijn, G.P. Smith, J. van de Pol (Eds.), Integrated Formal Methods. XI, 407 pages. 2005.

Vol. 3770: J. Akoka, S.W. Liddle, I.-Y. Song, M. Bertolotto, I. Comyn-Wattiau, W.-J. van den Heuvel, M. Kolp, J. Trujillo, C. Kop, H.C. Mayr (Eds.), Perspectives in Conceptual Modeling. XXII, 476 pages. 2005.

Vol. 3769: D.A. Bader, M. Parashar, V. Sridhar, V.K. Prasanna (Eds.), High Performance Computing – HiPC 2005. XXVIII, 550 pages. 2005.

Vol. 3768: Y.-S. Ho, H.J. Kim (Eds.), Advances in Multimedia Information Processing - PCM 2005, Part II. XXVIII, 1088 pages. 2005.

Vol. 3767: Y.-S. Ho, H.J. Kim (Eds.), Advances in Multimedia Information Processing - PCM 2005, Part I. XXVIII, 1022 pages. 2005.

Vol. 3766: N. Sebe, M.S. Lew, T.S. Huang (Eds.), Computer Vision in Human-Computer Interaction. X, 231 pages. 2005.

Vol. 3765: Y. Liu, T. Jiang, C. Zhang (Eds.), Computer Vision for Biomedical Image Applications. X, 563 pages. 2005.

Vol. 3764: S. Tixeuil, T. Herman (Eds.), Self-Stabilizing Systems. VIII, 229 pages. 2005.

Vol. 3762: R. Meersman, Z. Tari, P. Herrero (Eds.), On the Move to Meaningful Internet Systems 2005: OTM 2005 Workshops. XXXI, 1228 pages. 2005.

Vol. 3761: R. Meersman, Z. Tari (Eds.), On the Move to Meaningful Internet Systems 2005: CoopIS, DOA, and ODBASE, Part II. XXVII, 653 pages. 2005.

Vol. 3760: R. Meersman, Z. Tari (Eds.), On the Move to Meaningful Internet Systems 2005: CoopIS, DOA, and ODBASE, Part I. XXVII, 921 pages. 2005.

Vol. 3759: G. Chen, Y. Pan, M. Guo, J. Lu (Eds.), Parallel and Distributed Processing and Applications - ISPA 2005 Workshops. XIII, 669 pages. 2005.

Vol. 3758: Y. Pan, D.-x. Chen, M. Guo, J. Cao, J.J. Dongarra (Eds.), Parallel and Distributed Processing and Applications. XXIII, 1162 pages. 2005.

Vol. 3757: A. Rangarajan, B. Vemuri, A.L. Yuille (Eds.), Energy Minimization Methods in Computer Vision and Pattern Recognition. XII, 666 pages. 2005.

Vol. 3756: J. Cao, W. Nejdl, M. Xu (Eds.), Advanced Parallel Processing Technologies. XIV, 526 pages. 2005.

Vol. 3754: J. Dalmau Royo, G. Hasegawa (Eds.), Management of Multimedia Networks and Services. XII, 384 pages. 2005.

Vol. 3753: O.F. Olsen, L.M.J. Florack, A. Kuijper (Eds.), Deep Structure, Singularities, and Computer Vision. X, 259 pages. 2005.

Vol. 3752: N. Paragios, O. Faugeras, T. Chan, C. Schnörr (Eds.), Variational, Geometric, and Level Set Methods in Computer Vision. XI, 369 pages. 2005.

Vol. 3751: T. Magedanz, E.R.M. Madeira, P. Dini (Eds.), Operations and Management in IP-Based Networks. X, 213 pages. 2005.

Vol. 3750: J.S. Duncan, G. Gerig (Eds.), Medical Image Computing and Computer-Assisted Intervention – MICCAI 2005, Part II. XL, 1018 pages. 2005.

Vol. 3749: J.S. Duncan, G. Gerig (Eds.), Medical Image Computing and Computer-Assisted Intervention – MICCAI 2005, Part I. XXXIX, 942 pages. 2005.